ORDINARY
AND PARTIAL
DIFFERENTIAL
EQUATIONS

Contents

Preface   xiii

Acknowledgments   xiv

PART I   **Ordinary Differential Equations, Boundary Value Problems, Fourier Series, and the Introduction to Integral Equations**

CHAPTER 1 ▪ First-Order Differential Equations   3

1.1   GENERAL CONSIDERATIONS   3

1.2   FIRST-ORDER EQUATIONS: EXISTENCE AND UNIQUENESS OF SOLUTION   5

1.3   INTEGRAL CURVES, ISOCLINES   7

1.4   SEPARABLE EQUATIONS   11

1.5   LINEAR EQUATIONS AND EQUATIONS REDUCIBLE TO LINEAR FORM   20

1.6   EXACT EQUATIONS   27

1.7   EQUATIONS UNRESOLVED FOR DERIVATIVE (INTRODUCTORY LOOK)   30

1.8   EXAMPLES OF PROBLEMS LEADING TO DIFFERENTIAL EQUATIONS   33

1.9   SOLUTION OF FIRST-ORDER DIFFERENTIAL EQUATIONS WITH ACCOMPANYING SOFTWARE   37

CHAPTER 2 ▪ Second-Order Differential Equations   43

2.1   GENERAL CONSIDERATION: NTH -ORDER DIFFERENTIAL EQUATIONS   43

2.2   SECOND-ORDER DIFFERENTIAL EQUATIONS   44

2.3   REDUCTION OF ORDER   46

2.4   LINEAR SECOND-ORDER DIFFERENTIAL EQUATIONS   49

     2.4.1   Homogeneous Equations   50

     2.4.2   Inhomogeneous Equations   52

2.5   LINEAR NTH-ORDER DIFFERENTIAL EQUATIONS   54

2.6   LINEAR SECOND-ORDER EQUATIONS WITH CONSTANT COEFFICIENTS   58

     2.6.1   Homogeneous Equations   58

     2.6.2   Inhomogeneous Equations: Method of Undetermined Coefficients   62

         2.6.2.1   *Function f(x) Is a Polynomial*   63

         2.6.2.2   *Function f(x) Is a Product of a Polynomial and Exponent*   64

         2.6.2.3   *Function f(x) Contains Sine and (or) Cosine Functions*   65

         2.6.2.4   *Function f(x) Is a Product of Polynomial, Exponent, Sine and (or) Cosine Functions*   67

2.7   LINEAR *N*TH-ORDER EQUATIONS WITH CONSTANT COEFFICIENTS   68

2.8   SOLUTION OF IVP FOR SECOND-ORDER EQUATIONS WITH ACCOMPANYING SOFTWARE   72

CHAPTER 3 ▪ Systems of Differential Equations   77

3.1   INTRODUCTION   77

3.2   SYSTEMS OF FIRST-ORDER DIFFERENTIAL EQUATIONS   79

3.3   SYSTEMS OF FIRST-ORDER LINEAR DIFFERENTIAL EQUATIONS   86

3.4   SYSTEMS OF FIRST-ORDER LINEAR HOMOGENEOUS DIFFERENTIAL EQUATIONS WITH CONSTANT COEFFICIENTS   88

     3.4.1   Roots of the Characteristic Equation Are Real and Distinct   90

     3.4.2   Roots of the Characteristic Equation Are Complex and Distinct   92

     3.4.3   Roots of the Characteristic Equation Are Repeated   95

3.5   SYSTEMS OF FIRST-ORDER LINEAR INHOMOGENEOUS DIFFERENTIAL EQUATIONS WITH CONSTANT COEFFICIENTS   108

CHAPTER 4 ▪ Boundary Value Problems for Second-Order ODE and Sturm-Liouville Theory   113

4.1   INTRODUCTION TO BVP   113

4.2   THE STURM-LIOUVILLE PROBLEM   118

4.3   EXAMPLES OF STURM-LIOUVILLE PROBLEMS   124

CHAPTER 5 ▪ Qualitative Methods and Stability of ODE Solutions   135

5.1   QUALITATIVE APPROACH FOR AUTONOMOUS FIRST-ORDER EQUATIONS: EQUILIBRIUM SOLUTIONS AND PHASE LINES   135

5.2   QUALITATIVE APPROACH FOR AUTONOMOUS SYSTEMS OF FIRST-ORDER EQUATIONS: EQUILIBRIUM SOLUTIONS AND PHASE PLANES   140

     5.2.1   Phase Planes of Linear Autonomous Equations   144

5.2.2	Real and Distinct Eigenvalues	146
5.2.3	Complex Eigenvalues	148
5.2.4	Repeated Real Eigenvalues	152
5.3	STABILITY OF SOLUTIONS	156
5.3.1	Lyapunov and Orbital Stability	156
5.3.2	Sensitivity to Small Perturbations	159
5.3.3	Stability Diagram in the Trace-Determinant Plane	159
5.3.4	Lyapunov Stability of a General System	160
5.3.5	Stability Analysis of Equilibria in Nonlinear Autonomous Systems	160

CHAPTER 6 ▪ Method of Laplace Transforms for ODE   165

6.1	INTRODUCTION	165
6.2	PROPERTIES OF THE LAPLACE TRANSFORM	166
6.3	APPLICATIONS OF LAPLACE TRANSFORM FOR ODE	169

CHAPTER 7 ▪ Integral Equations   179

7.1	INTRODUCTION	179
7.2	INTRODUCTION TO FREDHOLM EQUATIONS	182
7.3	ITERATIVE METHOD FOR THE SOLUTION OF FREDHOLM EQUATIONS OF THE SECOND KIND	191
7.4	VOLTERRA EQUATION	194
7.5	SOLUTION OF VOLTERRA EQUATIONS WITH THE DIFFERENCE KERNEL USING THE LAPLACE TRANSFORM	196

CHAPTER 8 ▪ Series Solutions of ODE and Bessel and Legendre Equations   201

8.1	SERIES SOLUTIONS OF DIFFERENTIAL EQUATIONS: INTRODUCTION	201
8.2	BESSEL EQUATION	204
8.3	PROPERTIES OF BESSEL FUNCTIONS	210
8.4	BOUNDARY VALUE PROBLEMS AND FOURIER-BESSEL SERIES	213
8.5	SPHERICAL BESSEL FUNCTIONS	219
8.6	THE GAMMA FUNCTION	221
8.7	LEGENDRE EQUATION AND LEGENDRE POLYNOMIALS	226
8.8	FOURIER-LEGENDRE SERIES IN LEGENDRE POLYNOMIALS	231
8.9	ASSOCIATE LEGENDRE FUNCTIONS $P_n^m(x)$	234
8.10	FOURIER-LEGENDRE SERIES IN ASSOCIATED LEGENDRE FUNCTIONS	239

CHAPTER 9 ■ Fourier Series 243

9.1 PERIODIC PROCESSES AND PERIODIC FUNCTIONS 243

9.2 FOURIER FORMULAS 245

9.3 CONVERGENCE OF FOURIER SERIES 248

9.4 FOURIER SERIES FOR NONPERIODIC FUNCTIONS 250

9.5 FOURIER EXPANSIONS ON INTERVALS OF ARBITRARY LENGTH 251

9.6 FOURIER SERIES IN COSINE OR IN SINE FUNCTIONS 253

9.7 EXAMPLES 257

9.8 THE COMPLEX FORM OF THE TRIGONOMETRIC SERIES 262

9.9 FOURIER SERIES FOR FUNCTIONS OF SEVERAL VARIABLES 266

9.10 GENERALIZED FOURIER SERIES 267

9.11 SOLUTION OF DIFFERENTIAL EQUATIONS USING FOURIER SERIES 269

9.12 FOURIER TRANSFORMS 273

PART II **Partial Differential Equations**

CHAPTER 10 ■ One-Dimensional Hyperbolic Equations 287

10.1 WAVE EQUATION 287

10.2 BOUNDARY AND INITIAL CONDITIONS 290

10.3 LONGITUDINAL VIBRATIONS OF A ROD AND ELECTRICAL OSCILLATIONS 294

10.3.1 Rod Oscillations: Equations and Boundary Conditions 294

10.3.2 Electrical Oscillations in a Circuit 296

10.4 TRAVELING WAVES: D'ALEMBERT'S METHOD 298

10.5 FINITE INTERVALS: THE FOURIER METHOD FOR HOMOGENEOUS EQUATIONS 303

10.6 THE FOURIER METHOD FOR NONHOMOGENEOUS EQUATIONS 316

10.7 EQUATIONS WITH NONHOMOGENEOUS BOUNDARY CONDITIONS 323

10.8 THE CONSISTENCY CONDITIONS AND GENERALIZED SOLUTIONS 330

10.9 ENERGY IN THE HARMONICS 331

CHAPTER 11 ■ Two-Dimensional Hyperbolic Equations 343

11.1 DERIVATION OF THE EQUATIONS OF MOTION 344

11.1.1 Boundary and Initial Conditions 346

11.2 OSCILLATIONS OF A RECTANGULAR MEMBRANE 348

11.2.1 The Fourier Method for Homogeneous Equations with Homogeneous Boundary Conditions 349

11.2.2 The Fourier Method for Nonhomogeneous Equations 358

11.2.3 The Fourier Method for Equations with Nonhomogeneous Boundary Conditions 366

11.3 THE FOURIER METHOD APPLIED TO SMALL TRANSVERSE OSCILLATIONS OF A CIRCULAR MEMBRANE 372

11.3.1 The Fourier Method for Homogeneous Equations with Homogeneous Boundary Conditions 373

11.3.2 Radial Oscillations of a Membrane 381

11.3.3 The Fourier Method for Nonhomogeneous Equations 388

11.3.4 Forced Radial Oscillations of a Circular Membrane 392

11.3.5 The Fourier Method for Equations with Nonhomogeneous Boundary Conditions 393

CHAPTER 12 ▪ One-Dimensional Parabolic Equations 405

12.1 HEAT CONDUCTION AND DIFFUSION: BOUNDARY VALUE PROBLEMS 405

12.1.1 Heat Conduction 405

12.1.2 Diffusion Equation 406

12.1.3 Boundary Conditions 408

12.2 THE FOURIER METHOD FOR HOMOGENEOUS EQUATIONS 411

12.3 THE FOURIER METHOD FOR NONHOMOGENEOUS EQUATIONS 419

12.4 THE FOURIER METHOD FOR NONHOMOGENEOUS EQUATIONS WITH NONHOMOGENEOUS BOUNDARY CONDITIONS 422

12.5 BOUNDARY PROBLEMS WITHOUT INITIAL CONDITIONS 430

CHAPTER 13 ▪ Two-Dimensional Parabolic Equations 445

13.1 HEAT CONDUCTION WITHIN A FINITE RECTANGULAR DOMAIN 445

13.1.1 The Fourier Method for the Homogeneous Heat Conduction Equation (Free Heat Exchange) 449

13.1.2 The Fourier Method for the Nonhomogeneous Heat Conduction Equation 453

13.1.3 The Fourier Method for the Nonhomogeneous Heat Conduction Equation with Nonhomogeneous Boundary Conditions 458

13.2 HEAT CONDUCTION WITHIN A CIRCULAR DOMAIN   470

    13.2.1   The Fourier Method for the Homogeneous Heat Conduction Equation   472

    13.2.2   The Fourier Method for the Nonhomogeneous Heat Conduction Equation   477

    13.2.3   The Fourier Method for the Nonhomogeneous Heat Conduction Equation with Nonhomogeneous Boundary Conditions   483

CHAPTER 14 ■ Elliptic Equations   493

14.1   ELLIPTIC DIFFERENTIAL EQUATIONS AND RELATED PHYSICAL PROBLEMS   493

14.2   BOUNDARY CONDITIONS   495

14.3   THE BVP FOR LAPLACE'S EQUATION IN A RECTANGULAR DOMAIN   498

14.4   THE POISSON EQUATION WITH HOMOGENEOUS BOUNDARY CONDITIONS   502

14.5   THE LAPLACE AND POISSON EQUATIONS WITH NONHOMOGENEOUS BOUNDARY CONDITIONS   504

    14.5.1   Consistent Boundary Conditions   505

    14.5.2   Inconsistent Boundary Conditions   507

14.6   LAPLACE'S EQUATION IN POLAR COORDINATES   518

14.7   LAPLACE'S EQUATION AND INTERIOR BVP FOR CIRCULAR DOMAIN   521

14.8   LAPLACE'S EQUATION AND EXTERIOR BVP FOR CIRCULAR DOMAIN   525

14.9   POISSON'S EQUATION: GENERAL NOTES AND A SIMPLE CASE   526

14.10   POISSON'S INTEGRAL   529

14.11   APPLICATION OF BESSEL FUNCTIONS FOR THE SOLUTION OF LAPLACE'S AND POISSON'S EQUATIONS IN A CIRCLE   532

14.12   THREE-DIMENSIONAL LAPLACE EQUATION FOR A CYLINDER   539

APPENDIX 1 ■ Eigenvalues and Eigenfunctions of One-Dimensional Sturm-Liouville Boundary Value Problem for Different Types of Boundary Conditions   549

APPENDIX 2 ■ Auxiliary Functions, $w(x,t)$, for Different Types of Boundary Conditions   553

APPENDIX 3 ■ Eigenfunctions of Sturm-Liouville Boundary Value Problem for the Laplace Equation in a Rectangular Domain for Different Types of Boundary Conditions   557

APPENDIX 4 ■ A Primer on the Matrix Eigenvalue Problems and the Solution of the Selected Examples in Section 5.25   563

APPENDIX 5 ■ How to Use the Software Associated with the Book   569

BIBLIOGRAPHY   623

INDEX   625

Preface

This book presents ordinary differential equations (ODE) and partial differential equations (PDE) under one cover. All topics that form the core of a modern undergraduate and the beginner's graduate course in differential equations are presented at full length. We especially strived for clarity of presenting concepts and the simplicity and transparency of the text. At the same time, we tried to keep all rigor of mathematics (but without the emphasis on proofs—some simpler theorems are proved, but for more sophisticated ones we discuss only the key steps of the proofs). In our best judgment, a balanced presentation has been achieved, which is as informative as possible at this level, and introduces and practices all necessary problem-solving skills, yet is concise and friendly to a reader. A part of the philosophy of the book is "teaching-by-examples" and thus we provide numerous carefully chosen examples that guide step-by-step learning of concepts and techniques. The level of presentation and the book structure allows its use in engineering, physics, and mathematics departments.

The primary motivation for writing this textbook is that, to our knowledge, there has not been published a comprehensive textbook that covers both ODE and PDE. A professor who teaches ODE using this book can use the PDE sections to complement the main ODE course. Professors teaching PDE very often face the situation when students, despite having an ODE prerequisite, do not remember the techniques for solving ODE and thus can't do well in the PDE course. A professor can choose the key ODE sections, quickly review them in the course of, say, three or four lectures, and then seamlessly turn to the main subject, i.e., PDE.

The ODE part of the book contains topics that can be omitted (fully or partially) from the basic undergraduate course, such as the integral equations, the Laplace transforms, and the boundary value problems. For the undergraduate PDE course the most technical sections (for instance, where the nonhomogeneous boundary conditions are discussed) can be omitted from lectures and instead studied with the accompanying software. At least Sections 1 through 7 from Chapter 8, Fourier Series, should be covered prior to teaching PDEs. For class time savings, a few of these sections can be studied using the software.

The software is a very special component of the textbook. Our software covers both fields. The ODE part of the software is fairly straightforward—i.e., the software allows readers to compare their analytical solution and the results of a computation. For PDE the software also demonstrates the sequence of all the steps needed to solve the problem. Thus it leads a user in the process of solving the problem, rather than informs of the result of solving the problem. This feature is completely or partially absent from all software that we have seen and tested. After the software solution of the problem, a deeper investigation is offered, such as the study of the dependence of the solution on the parameters, the accuracy of the solution, the speed of a series convergence, and related questions. Thus the software is a platform for learning and investigating all textbook topics, an inherent part of the learning experience rather than an interesting auxiliary. The software enables lectures, recitations, and homeworks to be research-like, i.e., to be naturally investigative, thus hopefully increasing the student rate of success. It allows students to study a *limitless* number of problems (a known drawback of a typical PDE course is that, due to time constraints, students are limited to practicing solutions of a small number of simple problems using the "pen and paper" method).

The software is very intuitive and simple to use, and it does not require students to learn a (quasi)programming language as do the computer algebra systems (CAS), such as Mathematica and Maple. Most CAS require a significant time investment in learning commands, conventions, and other features, and often the undergraduate students are very reluctant, especially if they have reasons to think that they will not use CAS in the future; furthermore, the instructors are often not willing to spend valuable classroom time teaching the basics of using CAS. Besides, where using CAS to solve an ODE is a matter of typing in one command (*dsolve* in Maple or *DSolve* in Mathematica), which students usually can learn how to do with a minor effort, solving a PDE in CAS is more complicated and puts the burden on the instructor to create a worksheet with several commands, often as many as ten or fifteen, where students just change the arguments to enable the solution of an assigned problem. Creation of a collection of such worksheets that covers all sections of the textbook is only possible when the instructor teaches the course multiple times.

The software and tutorials contain a few topics, such as the classical orthogonal polynomials, generalized Legendre functions, and others, which are not included in the book to avoid its overload with content that is presently rarely taught in PDE courses (at least in the U.S. academic system). These topics with the help of the software can be assigned for an independent study, essay, etc.

The software tutorials for different chapters are placed in the appendices.

Finally, we would like to suggest the book *Mathematical Methods in Physics* [1] as a more complete and advanced PDE textbook. That book is written in the same style and uses the previous version of the software.

ACKNOWLEDGMENTS

We wish to thank Prof. Harley Flanders (University of Michigan) for advice and support.

We are also very grateful to Prof. Kyle Forinash (Indiana University Southeast) for permission to use in the current book several large fragments of our jointly written book *Mathematical Methods in Physics* [1].

I

Ordinary Differential Equations, Boundary Value Problems, Fourier Series, and the Introduction to Integral Equations

First-Order Differential Equations

1.1 GENERAL CONSIDERATIONS

A *differential equation (DE)* is an equation that contains derivatives of an unknown function of one or more variables. When an unknown function depends on a single variable, an equation involves ordinary derivatives and is called an *ordinary differential equation (ODE)*. When an unknown function depends on two or more variables, partial derivatives of an unknown function emerge and an equation is called a *partial differential equation (PDE)*. In Part I of the book we study ODEs.

The *order* of a DE is the order of a highest derivative in the equation.

A *solution* of a differential equation is the function that when substituted for the unknown function in this equation makes it the identity. Usually a solution of a differential equation is sought in some domain D of an independent variable and an unknown function. The process of the solution ultimately boils down to integration, and therefore the solution is often called an *integral* of the DE.

For example, $y'(x) = e^x + x^2$ is the first-order ODE. To find the unknown function $y(x)$ we use *Leibnitz notation* $y'(x) = dy/dx$, write the equation as $dy = (e^x + x^2)dx$, and then integrate both sides. This gives $y(x) = e^x + x^3/3 + C$, where C is an arbitrary integration constant.

A solution that contains an arbitrary integration constant is called the *general solution* (or the *general integral*) of a DE. Such solution can be denoted as $y(x,C)$. For different C values the function $y = y(x,C)$ gives different curves in the (x,y)-plane. These curves are called the *integral curves*. All these curves are described by the same equation. In order to single out some particular curve (one that corresponds to a certain C value), a value of $y(x)$ can be specified at some $x = x_0$ —which gives a point $P(x_0, y(x_0))$. For instance, let in the above example $y(0) = 1$. Setting $x_0 = 0$, $y_0 = 1$ in the solution $y(x) = e^x + x^3/3 + C$, gives $1 = 1 + 0 + C$. Thus, $C = 0$, and the solution is $y(x) = e^x + x^3/3$. Solutions obtained in such a way are called the *particular* solutions. We also say that the condition in the form $y(x_0) = y_0$ is the *initial condition*. For our example, the initial condition used to obtain the particular solution $y(x) = e^x + x^3/3$ is $y(0) = 1$.

This example illustrates *an initial value problem (IVP)* for an ODE, also called the *Cauchy problem*. It shows that in order to obtain some concrete solution we need to specify initial conditions for the DE. The terminology initial conditions comes from mechanics, where the independent variable x represents time and is customarily symbolized as t.

Because the DE is first-order in our example, only one initial condition is necessary and sufficient in order to determine an arbitrary constant C. General solutions of higher-order ODEs contain more than one arbitrary constants and thus more than one initial conditions are needed. We will see later that the number of arbitrary constants in the general solution of an ODE equals the order n of the equation and the initial value problem for an nth-order ODE requires n initial conditions.

In Chapter 4 we will discuss problems of different type: the boundary value problems (BVP) for ordinary differential equations.

We can now generalize the example above and present the solution process for the first-order equation of a simple type:

$$y' = f(x).$$

After writing this equation as $dy = f(x)dx$, the general solution $y(x,C)$ is obtained by integration:

$$y(x) = \int f(x)dx + C.$$

(Note that the solution is pronounced successful even if $f(x)$ cannot be integrated down to the elementary functions.) If the initial condition is assigned, constant C is determined by plugging $y(x_0) = y_0$ in the general solution. Finally, the particular solution is obtained by replacing C in the general solution by its numerical value. This solution can be also written without an integration constant:

$$y = y_0 + \int_{x_0}^{x} f(x)dx.$$

This form of the particular solution explicitly includes the initial condition $y(x_0) = y_0$. For instance, the particular solution of the example above can be written as

$$y = 1 + \int_{0}^{x} (e^x + x^2)\, dx.$$

After this elementary discussion, consider a general nth-order ODE:

$$F(x, y, y' \ldots y^{(n)}) = 0, \tag{1.1}$$

where $F(x, y, y' \ldots y^{(n)})$ is some function.

Often nth-order ODE is given in the form resolved for the highest-order derivative:

$$y^{(n)} = f(x, y, y' \ldots y^{(n-1)}). \tag{1.2}$$

In either case (1.1) or (1.2), the general solution

$$y = y(x, C_1, \ldots, C_n) \tag{1.3}$$

contains n arbitrary constants.

1.2 FIRST-ORDER EQUATIONS: EXISTENCE AND UNIQUENESS OF SOLUTION

First-order equations have either the form

$$F(x, y, y') = 0, \tag{1.4}$$

or the form

$$y' = f(x, y). \tag{1.5}$$

The general solution contains one arbitrary constant: $y = y(x, C)$.

Often the general solution is not *explicit*, $y = y(x, C)$, but an *implicit* one:

$$\Phi(x, y, C) = 0.$$

A particular solution emerges when C takes on a certain numerical value.

Now consider the inverse problem: the determination of a differential equation that has a given solution $\Phi(x, y, C) = 0$. Here one has to differentiate the function Φ, considering y as a function of x (thus using the chain rule), and then eliminate C with the help of equation $\Phi(x, y, C) = 0$.

For example, let the general solution of some equation be $y = Cx^3$ and we wish to find the equation. Differentiation of the general solution gives $y' = 3Cx^2$. Substitution into this expression $C = y/x^3$ from the general solution, results in the differential equation $y' = 3y/x$.

In order to find the particular solution, a single initial condition

$$y(x_0) = y_0 \tag{1.6}$$

is needed. The problem (1.4), (1.6) (or (1.5), (1.6)) is the IVP (or Cauchy problem) for the first-order ODE. To be specific, we will assume in the forthcoming the IVP (1.5), (1.6).

The question that we now need to ask is this: Under what conditions on a function $f(x,y)$ does there exist a unique solution of the IVP? The following theorem provides the answer.

Picard's Theorem (Existence and Uniqueness of a Solution): Let a function f be continuous in a rectangle D: $x_0 - \alpha \le x \le x_0 + \alpha$, $y_0 - \beta \le y \le y_0 + \beta$ that contains the point (x_0, y_0). Also let the partial derivative $\partial f/\partial y$ exists and be bounded in D. Then the solution of IVP (1.5), (1.6) exists and is unique on $x_0 - \delta \le x \le x_0 + \delta$, where $0 < \delta \le \alpha$.

Proof of Picard's Theorem can be found in many books.

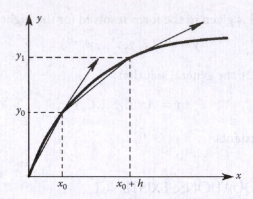

FIGURE 1.1 Illustration of Euler's method.

Conditions of Picard's Theorem can be illustrated by the simple method of the numerical integration of the ODE $y' = f(x, y)$ on the interval $[x_0, b]$. Let $[x_0, b]$ be divided into n subintervals $[x_0, x_1]$, $[x_1, x_2]$, ..., $[x_{n-1}, b]$ of equal length h, where $h = (b - x_0)/n$. The quantity h is the step size of the computation (Figure 1.1). The idea of the method is to approximate the integral curve by the set of the straight-line segments on the intervals $[x_i, x_{i+1}]$, such that each segment is tangent to the solution curve at one of the endpoints of the subintervals.

First, consider the interval $[x_0, x_0 + h]$ and find the value y_1 from equation $y' = f(x, y)$, as follows. Suppose h is small, then the derivative $y' \approx \frac{\Delta y}{\Delta x} = \frac{y_1 - y_0}{h}$, and thus $y_1 = y_0 + y'h$.

Also, if h is small and the function $f(x,y)$ is continuous and changes slowly, then y' can be replaced by $f(x_0, y_0)$ on this interval. Then,

$$y_1 = y_0 + f(x_0, y_0)h.$$

Next, let repeat this construction on the interval $[x_0 + h, x_0 + 2h]$ (which is $[x_1, x_2]$). Taking

$$y' \approx \frac{y_2 - y_1}{h} \approx f(x_1, y_1),$$

we find $y_2 = y_1 + hf(x_1, y_1)$. Then, repeating the construction for other subintervals, we finally arrive to the *Euler's formula*:

$$y_{i+1} = y_i + hf(x_i, y_i), \quad i = 0, 1, \ldots, n-1. \tag{1.7}$$

(Euler's formula can be applied also when $b < x_0$. In this case the step size $h < 0$.) As $|h|$ decreases, the line segments become shorter and they trace the integral curve better – thus the accuracy of the approximate solution increases as $h \to 0$. The integral curve through the point x_0, y_0 represents the particular solution.

Another important question arises: What will happen with the solution if the initial conditions slightly change? Will the solution change also slightly? This question has not only theoretical significance but also a big practical meaning: how an error in initial conditions (which often are obtained from the experimental data, or from calculations performed with some limited precision) can affect the solution of the Cauchy problem. The answer to this question is this: It can be shown that if the conditions of the Picard's theorem are satisfied, the solution continuously depends on the initial conditions. It also can be shown that if the equation contains some parameter λ,

$$y' = f(x, y, \lambda), \tag{1.8}$$

the solution continuously depends on λ if function f is a continuous function of λ. In Chapter 5 we will study the problem of *stability of solution*, which we only touched here.

In the following problems construct differential equations describing given families of integral curves.

Problems	Answers
1. $y = (x - C)^2$	$y' = 2\sqrt{y}$
2. $x^3 + Cy^2 = 5y$	$y' = \dfrac{3x^2 y}{2x^3 - 5y}$
3. $y^3 + Cx = x^2$	$y' = \dfrac{x^2 + y^3}{3xy^2}$
4. $y = e^{Cx+5}$	$y' = \dfrac{y(\ln y - 5)}{x}$
5. $y = Ce^x - 3x + 2$	$y' = y + 3x - 5$
6. $\ln y = Cx + y$	$y' = \dfrac{y(\ln y - y)}{x(1 - y)}$
7. $y = C \cos x + \sin x$	$y' \cos x + y \sin x = 1$
8. $e^{-y} = x + Cx^2$	$y' = e^y - 2/x$
9. $y^2 + Cx = x^3$	$y' = \dfrac{2x^3 + y^2}{2xy}$
10. $x = Cy^2 + y^3$	$y' = \dfrac{y}{2x + y^3}$

1.3 INTEGRAL CURVES, ISOCLINES

A differential equation $\frac{dy}{dx} = f(x, y)$ can be seen as a formula that provides a connection between the Cartesian coordinates of a point, (x,y), and the slope of the integral curve, $\frac{dy}{dx}$, at this point. To visualize this slope, we can actually draw at a point (x,y) a short line segment (called the *slope mark*) that has the slope $f(x,y)$. Repeating this for some other points in the Cartesian plane gives the *direction field*. (As a rule, tens or even hundreds of points are required for the construction of a quality direction field.) Thus, from a geometrical perspective, to solve a differential equation $y' = f(x, y)$ means to find curves that are tangent to the direction field at each point. As we pointed out in Section 1.2, these solution curves are called the integral curves.

When an analytical solution is unavailable, the construction of the direction field often can be made easier by first drawing lines of a constant slope, $y' = k$. Such lines are called *isoclines*. Since $y' = f(x, y)$, the equation of an isocline is $f(x,y) = k$. This means that an isocline is a *level curve* of the function $f(x,y)$.

Example 1.1

Consider equation $\frac{dy}{dx} = \frac{y}{x}$. We notice that at any point (x,y) the slope of the integral curve, y/x, is also the slope of straight lines leaving the origin $(0,0)$ (for the right half-plane), and entering the origin (for the left half-plane). The direction field is shown in Figure 1.2 with short arrows. (Since the function $f(x,y) = y/x$ is not defined for $x = 0$, the equation does not have a solution on the y-axis, but the direction field can be plotted everywhere in the plane. Note that along the y-axis the slope is infinite.) The isoclines are given by equation $y/x = k$. Thus, in this example the isoclines are the straight lines coinciding with the direction field. The rectangle is a chosen domain for the graphical representation of the solution.

Example 1.2

Consider equation $\frac{dy}{dx} = -\frac{x}{y}$. Isoclines equation $-x/y = k$ gives straight lines shown in Figure 1.3. Along each isocline the value of y' does not change. This means that all slope marks (black arrows) along a particular isocline are parallel. Connecting slope marks on neighboring isoclines we can plot integral curves, which are obviously the circles centered at the origin: $x^2 + y^2 = C^2$. For the particular solution, starting from the point (x_0, y_0), we obtain the value of C (the circle radius): $C = \sqrt{x_0^2 + y_0^2}$. Function $f(x,y) = -x/y$ is not defined at $y = 0$, therefore the direction field can be plotted everywhere in the plane (x,y) except the x-axis. Thus we have to consider the solution in the form $y = \sqrt{C - x^2}$ in the upper half-plane and $y = -\sqrt{C - x^2}$ in the bottom half-plane.

In these figures, as well as in Figure 1.4, the isoclines and direction fields are plotted by the program **ODE 1st order**. The description of the accompanying software is in Section 1.9.

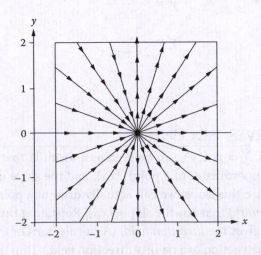

FIGURE 1.2   Direction field and isoclines for equation $y' = y/x$ (Example 1.1).

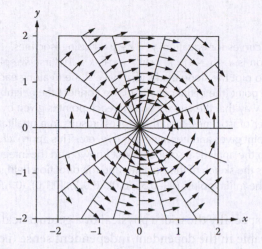

FIGURE 1.3   Direction field and isoclines for equation $y' = -x/y$ (Example 1.2).

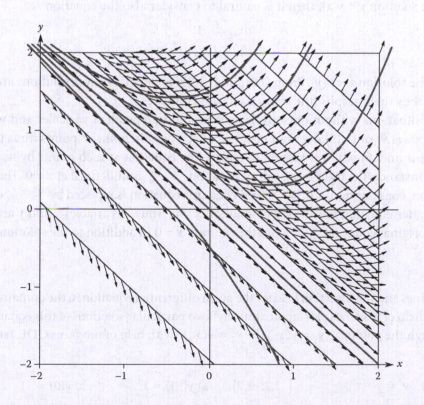

FIGURE 1.4   Direction field, isoclines, and five integral curves for equation $y' = x + y - 1$ (Example 1.3).

Example 1.3

Sketch several solution curves for equation $y' = x + y - 1$ using isoclines.

The isoclines equation is $x + y - 1 = k$, or $y = -x + (k + 1)$. First, we plot several isoclines for few values of k, and also plot the slope marks (that have slope k) along each isocline. Then, starting at a particular initial point (the initial condition), we connect the neighboring slope marks by a smooth curve in such a way that the curve's slope at each point is given by these marks. If we use reasonably large number of isoclines, then this process results in a qualitative plot of the integral curve. Another initial point gives another integral curve, etc. This approach can be useful in some situations, such as when the analytical solution is not known and the integral curves are relatively simple. Figure 1.4 shows the sketch of the isoclines and the direction field, as well as the five integral curves plotted for the initial points $(0, -0.5)$, $(0,0)$, $(0,0.5)$, $(0,1.0)$, $(0,1.5)$.

In many problems, especially those of a physical or a geometrical nature, the variables x and y are indistinguishable in the dependent/independent sense (i.e., y may be considered independent and x dependent, or vice versa). Thus if such problem is described by a differential equation

$$\frac{dy}{dx} = f(x, y),\qquad(1.9)$$

that has a solution $y = y(x)$, then it is natural to consider also the equation

$$\frac{dx}{dy} = \frac{1}{f(x, y)},\qquad(1.10)$$

that has the solution $x = x(y)$. From (1.9) and (1.10) it is clear that these solutions are equivalent, and thus their graphs (the integral curves) coincide.

In the situations without obvious dependent and independent variables and when one of equations (1.9) or (1.10) does not make sense at a certain point or points (thus the right-hand side is undefined), it is natural to replace the equation at such points by its counterpart. For instance, the right-hand side of equation $\frac{dy}{dx} = \frac{y}{x}$ is undefined at $x = 0$. The solution of equation, considering $x \neq 0$, is $y = Cx$. When this equation is replaced by $\frac{dx}{dy} = \frac{x}{y}$, one finds that the latter equation has the trivial solution $x = 0$. Thus, if variables x and y are equivalent, the original equation $\frac{dy}{dx} = \frac{y}{x}$ has the solution $x = 0$ in addition to the solution $y = Cx$.

Problems

Plot isoclines and the direction field for the given differential equation in the domain $[a,b;c,d]$. With the help of isoclines plot (approximately) two particular solutions of this equation passing through the points $a)$ $y_1 = y(x_1)$; $b)$ $y_2 = y(x_2)$ (with help of program **ODE 1st order**).

1. $y' = y - x^2$, $[-2,2;-2,2]$,   a) $y(-1) = 1$;   b) $y(0) = -1$

2. $y' = 2x - e^y$, $[-2,2;-2,2]$,   a) $y(-1) = 1$;   b) $y(0) = 0.5$

3. $y' = \dfrac{5x}{x^2 + y^2}$, $[1,5;1,5]$,   a) $y(1) = 2$;   b) $y(4) = 3$

4. $y' = x + 1 - 2y$, $[-1,1;-1,1]$,   a) $y(-0.5) = -1$;   b) $y(0) = 0.5$

5. $y' = \dfrac{y - 3x}{x + 3y}$, $[1,3;1,3]$,   a) $y(1.5) = 2.5$;   b) $y(2) = 1.5$

6. $y' = 2x^2 - y$, $[-2,2;-2,2]$,   a) $y(-1) = 1$;   b) $y(0) = -1.5$

7. $y' = \sin x \sin y$, $[-3,3;-3,3]$,   a) $y(-2) = -2$;   b) $y(2) = 1$

8. $y' = x^2 + \sin y$, $[-2,2;-2,2]$,   a) $y(-1) = -1$;   b) $y(1) = 1$

9. $y' = x(x - y)^2$, $[-1,1;-1,1]$,   a) $y(-1) = -0.5$;   b) $y(-0.5) = 0$

10. $y' = 3\cos^2 x + y$, $[-2,2;-2,2]$,   a) $y(-2) = 0$;   b) $y(2) = 1$

There is no general solution method applicable to all differential equations, and it is possible to solve analytically only the first-order equations of certain types. In the next sections we introduce these types and the solution methods.

1.4 SEPARABLE EQUATIONS

As we already pointed out, the first-order equation

$$\frac{dy}{dx} = f(x, y) \tag{1.11}$$

can be written in the equivalent form

$$dy = f(x, y)dx. \tag{1.12}$$

If function $f(x,y)$ can be represented as a fraction, $f_2(x)/f_1(y)$, then equation (1.12) takes the form

$$f_1(y)dy = f_2(x)dx. \tag{1.13}$$

In equation (1.13) variables x and y are separated to the right and to the left of the equality sign. Equations that allow transformation from (1.11) to (1.13) are called *separable,* and the process of transformation is called *separation of variables.* Next, by integrating both sides of (1.13), one obtains

$$\int f_1(y)dy = \int f_2(x)dx + C. \tag{1.14}$$

Expression (1.14) defines the general solution of the differential equation either explicitly, as a function $y(x) + C$, or implicitly, as a function $\Phi(x, y, C) = 0$. As we have noticed, even if the integrals in (1.14) cannot be evaluated in elementary functions, we pronounce that the solution of a separable equation has been determined.

The particular solution can be obtained from (1.14) by substituting in it the initial condition $y(x_0) = y_0$, which allows to determine the value of C. The particular solution can be written also as

$$\int_{y_0}^{y} f_1(y)\,dy = \int_{x_0}^{x} f_2(x)\,dx, \tag{1.15}$$

where no arbitrary constant is present. Obviously, formula (1.15) gives the equality $0 = 0$ at the point (x_0, y_0); thus this particular solution includes the initial condition $y(x_0) = y_0$.

Often, in order to transform (1.12) into (1.13), it is necessary to divide both sides of (1.12) by a certain function of x, y. Zero(s) of this function may be the solution(s) of (1.11), which are lost when the division is performed. Thus we must take a note of these solutions before division.

Example 1.4

Consider again the differential equation from Example 1.2, $\frac{dy}{dx} = -\frac{x}{y}$. Multiplication of the equation by $y\,dx$ gives $y\,dy = -x\,dx$. In the latter equation variables are separated. Next, integrate: $\int y\,dy = -\int x\,dx + C$. Evaluation of the integrals gives the general solution $x^2 + y^2 = C^2$. Let the initial condition be $y(3) = 4$. Substituting the initial condition in the general solution gives $C = 5$, thus the particular solution is the circle $x^2 + y^2 = 25$.

It is instructive to solve this exercise using the program **ODE 1st order**. The program offers a simple and convenient interface to the numerical solution procedures of first-order IVPs given in the form $y' = f(x, y)$, $y(x_0) = y_0$. (The description of **ODE 1st order** is in Section 1.9.) Using same interface one can plot a graph of a particular solution corresponding to the initial condition $y(x_0) = y_0$, that has been determined by the reader analytically. Numerical solution of IVP may encounter difficulties, if function $f(x, y)$ does not satisfy conditions of Picard's Theorem at some points. Thus in **ODE 1st order**, the region D for the integration of the differential equation must be chosen so that it does not contain such points. For instance, in Example 1.4 the domain D must include the point (3,4) (the initial condition), but it should not include any point with the zero y-coordinate. Since the particular solution is the circle of radius 5 centered at the origin, the interval of x-values for the numerical solution has to be $-5 < x < 5$. For the initial condition $y(3) = 4$ the numerical solution gives the upper semi-circle $y = +\sqrt{25 - x^2}$, and for the initial condition $y(3) = -4$ the numerical solution gives the bottom semi-circle $y = -\sqrt{25 - x^2}$.

Example 1.5

Next, we consider differential equation from Example 1.3: $y' = \frac{y}{x}$. Dividing both sides of equation by y and multiplying by dx, the equation becomes $\frac{dy}{y} = \frac{dx}{x}$. Dividing by y, we may have lost the solution $y = 0$. Indeed, plugging $y = 0$ into both sides of the original equation $y' = \frac{y}{x}$, we see that equation becomes the equality $0 = 0$. Thus $y = 0$ indeed is the solution. To find other solutions, we integrate $\frac{dy}{y} = \frac{dx}{x}$, and obtain: $\ln|y| = \ln|x| + C$. Exponentiation gives $|y(x)| = e^{\ln|x| + C} = e^{\ln|x|} e^{C} = C|x|$, where we used C again to denote arbitrary *positive* constant e^C. Now obviously, $y(x) = \pm Cx$. The trivial solution $y = 0$ can be included in the solution $y(x) = Cx$ by allowing $C = 0$. Thus finally, the general solution is $y = Cx$, where C is *arbitrary constant* (can be either positive, or negative, or zero).

To find particular solutions, any initial condition $y(x_0) = y_0$ is allowed, except x_0 cannot be zero. This is because the conditions of Picard's theorem are violated when $x_0 = 0$: the function $f(x,y) = y/x$ is unbounded. When $x_0 \neq 0$, the particular solution is $y = y_0 x / x_0$.

Numerical solutions are always conducted in a concrete domain D. In this example the line $x = 0$ (the y-axis) is the line of discontinuity. Therefore, for the initial condition taken at $x_0 > 0$, we can solve the equation in the domain $0 < x \leq a$ (with some positive a), if $x_0 < 0$—in the domain $a \leq x < 0$ (with some negative a).

Example 1.6

Consider differential equation $\frac{dy}{dx} = -\frac{y}{x}$. (Note that this differs from Example 1.5 only in that the right-hand side is multiplied by −1). Functions $f(x,y) = -y/x$ and $f_y(x,y) = -1/x$ are continuous at $x \neq 0$; thus equation satisfies the conditions of Picard's Theorem in the entire (x,y) plane, except again at the y-axis. Separating the variables we have $dy/y = -dx/x$ and after the integration obtain $\ln|y| = \ln C - \ln|x|$ (here we choose the arbitrary constant in the form $\ln C$, $C > 0$), which gives $|y| = C/|x|$. From there $y = \pm C/x$ and the general solution can be finally written as $y = C/x$, where C is an arbitrary constant ($C = 0$ is also allowed because $y = 0$ is a trivial solution of the given equation). This solution holds in the domains $x < 0$ and $x > 0$. The integral curves are the hyperbolas shown in Figure 1.5. An initial condition $y(x_0) = y_0$ taken at the point (x_0, y_0) in each of these

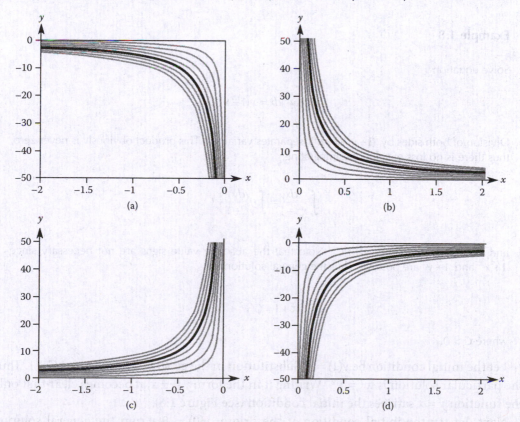

FIGURE 1.5 Integral curves for the equation $\frac{dy}{dx} = -\frac{y}{x}$ (Example 1.6). General solution is $y = C/x$. In (a) and (b) the curves for $C = 1, 2, \ldots, 9$ are shown; the particular solution $y = 5/x$ is a thick line. In (c) and (d) the curves for $C = -9, -8, \ldots, -1$ are shown; the particular solution $y = -5/x$ is a thick line.

four quadrants determines only one integral curve corresponding to the particular solution of the differential equation, as shown in Figure 1.5.

Example 1.7

Solve equation

$$y' = ky,$$

where k is a constant. For $k<0$ ($k>0$) this equation describes unlimited decay (growth) (see Section 1.8). Division by y and multiplication by dx gives $\frac{dy}{y} = kdx$. Integrating, $\ln|y| = kx + C$. Next, exponentiation gives $|y| = Ce^{kx}$, $C>0$, or equivalently $y = Ce^{kx}$, where C is arbitrary and non-zero. By allowing $C = 0$, this formula again includes the trivial solution $y = 0$ which has been lost when the equation was divided by y. Thus $y(x) = Ce^{kx}$ is the general solution, where C is arbitrary constant.

Since the right-hand side of the equation $y' = ky$ satisfies the conditions of Picard's theorem everywhere in the plane, there is no limitations on the initial condition (any x_0 and y_0 can be used). Substituting the initial condition $y(x_0) = y_0$ in the general solution, we obtain $y_0 = Ce^{kx_0}$, thus $C = y_0 e^{-kx_0}$. Substituting this C in the general solution, the particular solution (the solution of the IVP) takes the form $y = y_0 e^{k(x-x_0)}$.

Example 1.8

Solve equation

$$x(1+y^2)dx = y(1+x^2)dy.$$

Division of both sides by $(1+y^2)(1+x^2)$ separates variables. This product obviously is never zero, thus there is no loss of solutions. Integrating,

$$\int \frac{xdx}{x^2+1} = \int \frac{ydy}{y^2+1} + C,$$

and then $\ln(1+x^2) = \ln(1+y^2) + C$. Note that the absolute value signs are not necessary since $1+x^2$ and $1+y^2$ are positive. Thus the general solution is

$$x^2 + 1 = C(y^2 + 1),$$

where $C > 0$.

Let the initial condition be $y(1) = 1$. Substitution in the general solution gives $C = 1$. Thus the particular solution is $x^2 = y^2$. Writing it in the form $y = \pm x$, it becomes clear that only the function $y = x$ satisfies the initial condition (see Figure 1.6).

Next, let try the initial condition at the origin: $y(0) = 0$. From the general solution we still get $C = 1$. Now both solutions $y = \pm x$ match the initial condition. This seems to

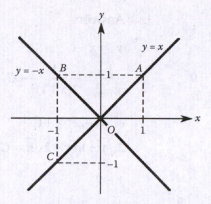

FIGURE 1.6 Example 1.8: the particular solution corresponding to the initial condition $y(1) = 1$.

contradict the uniqueness property stated in Picard's theorem. But from the equation in the form $y' = f(x, y)$ with

$$f(x, y) = \frac{x(1 + y^2)}{y(1 + x^2)}$$

it is clear that at $y = 0$ the function $f(x,y)$ has infinite discontinuity—thus the condition of Picard's theorem is violated and there is no guarantee of the uniqueness of the particular solution near the origin.

In practice this means that in the case of a nonunique solution, when an integral curve is traced, while passing the line of discontinuity one can slip from this curve to an adjacent one. For instance, let solve our example equation on the interval $[-1,1]$ with the initial condition $y(1) = 1$, i.e., start at point A in Figure 1.6. Moving along the solution $y = x$ toward the origin, at the destination point $(0,0)$ we have two possibilities for the following motion: continue along the line $y = x$ towards the final point C($-1,-1$), or along the line $y = -x$ towards the final point B($-1,1$). This demonstrates that when solution is not unique and one of the solution curves is traced, we may slip to an adjacent curve. Problems of this kind are especially serious for numerical solutions of differential equations, when there is a danger to miss a discontinuity point and not find several branches of solution. The choice of the proper integral curve in the case of a nonunique solution can be based on the real situation described by the differential equation. If, for instance, it is known that for $x < 0$ the values of a function y, describing a particular process, are positive, then on the interval $[-1,1]$ the line AOB should be taken for the solution; if it is known that for $x < 0$ the values of a function y are negative, then the line AOC should be taken. Often the choice can be made based on the knowledge of the asymptotes at $x \to \infty$ or $x \to -\infty$ of the function described by a differential equation. The example discussed above demonstrates the importance of the analysis of the properties of a differential equation before starting the solution.

Problems	Answers
1. $2y'y = 1$	$y^2 = x + C$
2. $(x+1)^3 dy - (y-2)^2 dx = 0$	$\dfrac{1}{y-2} = \dfrac{1}{2(x+1)^2} + C, \quad y = 2$
3. $e^{x+y} - 3e^{-x-2y}\, y' = 0$	$e^{2x} = -2e^{-3y} + C$
4. $y'(\sqrt{x} + \sqrt{xy}) = y$	$2\sqrt{y} + \ln y - 2\sqrt{x} = C$
	$y = 0$
5. $y'x^2 + y^2 = 0$	$\dfrac{1}{y} = -\dfrac{1}{x} + C$
6. $y'y + x = 1$	$(x-1)^2 + y^2 = C$
7. $y'x^2 y^2 - y + 1 = 0$	$\dfrac{y^2}{2} + y + \ln\lvert y - 1 \rvert = -\dfrac{1}{x} + C, \; y = 1$
8. $(x^2+1)y^3 dx = (y^2-1)x^3 dy$	$\ln\lvert x/y \rvert = (x^{-2} + y^{-2})/2 + C$
9. $xy\,dx + (x+1)dy = 0$	$y = C(x+1)e^{-x}$
10. $x\dfrac{dx}{dt} + t = 1$	$x^2 = -t^2 + 2t + C$
11. $\sqrt{1+y^2}\,dx = xy\,dy$	$x = Ce^{\sqrt{1+y^2}}, \; C > 0$
12. $(x^2-1)y' + 2xy^2 = 0$	$y(\ln\lvert x^2 - 1 \rvert + C) = 1; \; y = 0$
13. $y' = -\dfrac{xy}{x+2}$	$y = Ce^{-x}(x+2)^2; \; y = 0$
14. $\sqrt{y^2+5}\,dx = 2xy\,dy$	$\dfrac{1}{2}\ln\lvert x \rvert = C + \sqrt{y^2 + 5}; \; x = 0$
15. $2x^2 yy' + y^2 = 5$	$y^2 = 5 + Ce^{1/x}$
16. $y' - 2xy^2 = 3xy$	$y = \dfrac{3}{Ce^{-3x^2/2} - 2}; \; y = 0$
17. $xy' + y = y^2$	$y = \dfrac{1}{1 - Cx}, \; y = 0$
18. $2x\sqrt{1+y^2}\,dx + y\,dy = 0$	$y^2 = (x^2 + C)^2 - 1$
19. $y' = \dfrac{\sqrt{y}}{x}$	$2\sqrt{y} = \ln\lvert x \rvert + C; \; y = 0$
20. $\sqrt{1-y^2}\,dx + \sqrt{1-x^2}\,dy = 0$	$\arcsin x + \arcsin y = C$

In all these problems you can choose the initial condition and find the particular solution. Next, run the program **ODE 1st order** and compare the particular solution to the output. Using the program you can obtain the numerical solution of the IVP $y' = f(x, y)$, $y(x_0) = y_0$, and also plot the integral curve for the analytical solution. Keep in mind that numerical solution may run into a difficulty if the function $f(x,y)$ is discontinuous. Thus in the program, the domain of integration of the differential equation, D, must be chosen such that it does not contain points of discontinuity of $f(x, y)$. This guarantees solution uniqueness and trouble-free numerical integration.

To summarize, separable equations can be always solved analytically. It is important that some types of equations can be transformed into separable equations using the change of a variable. Two such types of equations are the equations $y' = f(ax + by)$ and $y' = f(y/x)$.

1. Consider equation

$$y' = f(ax + by), \tag{1.16}$$

where a, b are constants.

Let

$$z = ax + by \tag{1.17}$$

be the new variable. Differentiating z with respect to x, we obtain $\frac{dz}{dx} = a + b\frac{dy}{dx}$. Notice that $\frac{dy}{dx} = f(z)$ (eq. (1.16)), thus

$$\frac{dz}{dx} = a + bf(z). \tag{1.18}$$

This is a separable equation:

$$\frac{dz}{a + bf(z)} = dx.$$

Integration gives:

$$x = \int \frac{dz}{a + bf(z)} + C.$$

After the integration at the right-hand side is completed, replacing the variable z by $ax + by$ gives the general solution of (1.16).

Example 1.9 Solve $\frac{dy}{dx} = 2x - 3y$.

Using new variable $z = 2x - 3y$, we find $z' = 2 - 3y' = 2 - 3z$. Separating variables, $\frac{dz}{2-3z} = dx$. Integration gives

$$x + C = -\frac{1}{3}\ln|2 - 3z|, \quad \text{or} \quad \ln|2 - 3z| = -3x + C.$$

Then,

$$|2-3z| = e^{-3x+C} = Ce^{-3x}, \quad \text{where } C > 0.$$

This can be written as $2-3z = Ce^{-3x}$ with C either positive or negative. Substituting $z = 2x-3y$ gives $2-6x+9y = Ce^{-3x}$. Thus the general solution is

$$y = Ce^{-3x} + 2x/3 - 2/9.$$

It is easy to check that for $C = 0$ the function $y = 2x/3 - 2/9$ also satisfies the given equation; thus C in the general solution is an arbitrary constant.

Example 1.10   Solve $y' = \frac{2}{x-y} - 3$.

Let $z = x-y$ be the new variable. Differentiation gives $z' = 1-y'$. Since $y' = \frac{2}{z} - 3$, the equation for the function z is $1- z' = \frac{2}{z} - 3$.
    Separation of variables gives $dx = \frac{z\,dz}{4z-2}$. The right-hand side can be algebraically transformed: $\frac{z\,dz}{4z-2} = \frac{1}{4}\frac{4z-2+2}{4z-2}dz = \frac{1}{4}\left(1+\frac{2}{4z-2}\right)dz$. Integration of the latter equation gives $x+C = \frac{1}{4}z + \frac{1}{8}\ln|4z-2|$. Multiplying by 8, and letting $8C = C$ since C is arbitrary, finally gives the general solution in the implicit form:

$$8x + C = 2(x-y) + \ln|4x-4y-2|.$$

Problems

1. $(2x + y + 1)dx-(4x + 2y-3)dy = 0$

2. $x - y - 1 - (y - x + 2)y' = 0$

3. $y' = (y+x+3)(y+x+1)$

4. $y' = \sqrt{2x-y+3} + 2$

5. $y' = (y-4x+3)^2$

6. $y' - y = 3x - 5$

7. $y' = \dfrac{1}{x+2y}$

8. $y' = \sqrt{4x+2y-1}$

9. $y' = \dfrac{x+3y}{2x+6y-5}$

Answers

1. $2x + y - 1 = Ce^{2y-x}$

2. $-2x + 2y + 3 = Ce^{-2(x+y)}$

3. $-\dfrac{1}{x+y+2} = x+C$

4. $2\sqrt{2x-y+3} + x + C = 0$

5. $\dfrac{1}{4}\ln\left|\dfrac{y-4x+1}{y-4x+5}\right| = x+C$

6. $3x + y - 2 = Ce^x$

7. $x + 2y + 2 = Ce^y$

8. $\sqrt{4x+2y-1} - 2\ln(\sqrt{4x+2y-1}+2) = x+C$

9. $6y - 3x - 3\ln|x+3y-1| = C$

2. Consider equation

$$y' = f\left(\frac{y}{x}\right) \tag{1.19}$$

Here, function $f(x,y)$ contains variables x and y in the combination y/x.
  Let

$$z = \frac{y}{x} \tag{1.20}$$

be the new variable. Then, $y = zx$. Differentiating this in x, one obtains $y' = z + xz'$, or

$$\frac{dy}{dx} = z + x\frac{dz}{dx}.$$

Thus

$$f(z) = z + x\frac{dz}{dx}. \tag{1.21}$$

Separating variables $\frac{dx}{x} = \frac{dz}{f(z)-z}$ and integrating gives

$$\ln|x| + \ln C = \int \frac{dz}{f(z)-z},$$

or

$$x = C\exp\left[\int \frac{dz}{f(z)-z}\right].$$

Finally, replacement of z by y/x in the solution results in the general solution of (1.19).

Example 1.11

$$y' = \frac{x^2 + y^2}{2xy}$$

Division by $2xy$ gives $\frac{dy}{dx} = \frac{1}{2}\left(\frac{x}{y} + \frac{y}{x}\right)$. With the new variable $z = \frac{y}{x}$, we have $y' = z + xz'$. Equating this to $f(z) = \frac{1}{2}\left(\frac{1}{z} + z\right) = \frac{1+z^2}{2z}$ gives $z + xz' = \frac{1+z^2}{2z}$. Next, separate variables, $\frac{dx}{x} = \frac{2z\,dz}{1-z^2}$, and integrate. This gives $|x| = \frac{C}{|z^2-1|}$, $C > 0$.
  Thus, $x = \frac{C}{1-z^2}$, where C is arbitrary constant. Substitution $z = \frac{y}{x}$ gives $x = \frac{Cx^2}{x^2-y^2}$. Finally, $y^2 = x^2 - Cx$ is the general solution.

Example 1.12

$$y' = y/x + \tan(y/x)$$

First, introduce new variable z: $y = xz$, $y' = z + xz'$. Then the equation becomes $x\frac{dz}{dx} + z = z + \tan z$. Next, separate variables: $\frac{dx}{x} = \frac{\cos z}{\sin z}dz$, and integrate. This gives $\ln|\sin z| = \ln|x| + \ln C$, $C > 0$. Thus, $\sin z = Cx$, or $\sin(y/x) = Cx$, where C is arbitrary constant.

Problems	Answers
1. $(x + 2y)dx - xdy = 0$	$y = Cx^2 - x$
2. $xy' = y + \sqrt{x^2 + y^2}$	$y - Cx^2 + \sqrt{x^2 + y^2} = 0$
3. $xy' = y - xe^{y/x}$	$y = -x\ln\ln|Cx|$
4. $xy' - y = (x + y)\ln\dfrac{x+y}{x}$	$\ln\left(1 + \dfrac{y}{x}\right) = Cx$
5. $xdy = (\sqrt{xy} + y)dx$	$\sqrt{y} = \dfrac{1}{2}\sqrt{x}\ln Cx$
6. $ydx + (2\sqrt{xy} - x)dy = 0$	$\sqrt{x} + \sqrt{y}\ln Cy = 0$
7. $2x^3 dy - y(2x^2 - y^2)dx = 0$	$(x/y)^2 = \ln Cx, \quad y = 0$
8. $(y^2 - 2xy)dx + x^2 dy = 0$	$y(x-y) = Cx^3, \quad y = 0$
9. $y^2 + x^2 y' = xyy'$	$y = Ce^{y/x}$

1.5 LINEAR EQUATIONS AND EQUATIONS REDUCIBLE TO LINEAR FORM

An equation that is linear in y and y',

$$y' + p(x)y = f(x), \tag{1.22}$$

is called *linear first-order differential equation*. We assume that $p(x)$ and $f(x)$ are continuous in a domain where the solution is sought. The coefficient of y' in (1.22) is taken equal to one simply for convenience—if it is not equal to one, then the equation can be divided by this coefficient prior to the solution, which casts the equation in the form (1.22). Linearity of the expression $y' + p(x)y$ in y и y' means that when $y(x)$ is replaced by $C_1 y_1(x) + C_2 y_2(x)$, the left-hand side of equation (1.22) transforms into $C_1(y_1' + py_1) + C_2(y_2' + py_2)$. For instance, equations $y'y + 2y = x$ and $y' - \sin y = 1$ are nonlinear: the first equation due to the term $y'y$, the second one – due to the term $\sin y$. An example of a linear equation may be $x^2 y' + e^x y = 5\sin x$.

If a function $f(x)$ is not identically zero, the first-order equation (1.22) is called *linear inhomogeneous*.

In the opposite case when $f(x) \equiv 0$, equation

$$y' + p(x)y = 0 \tag{1.23}$$

is called *linear homogeneous*. In such equation variables can be separated:

$$\frac{dy}{y} = -p(x)dx.$$

Integration gives

$$\ln|y| = -\int p(x)dx + \ln C,$$

where we chose to denote an arbitrary constant as $\ln C$ (with $C > 0$). Removing the absolute value sign and taking into account that $y(x) = 0$ is also the solution of (5.2), we obtain the *general solution of the linear homogeneous equation*:

$$y(x) = Ce^{-\int p(x)dx}, \tag{1.24}$$

where C is arbitrary constant.

Recall that our goal is to determine the *general* solution of the *inhomogeneous* equation (1.22). Now that the homogeneous equation has been solved, this can be accomplished by using the *method of variation of the parameter*. The idea of the method is that the solution of (1.22) is sought in the form (1.24), *where constant C is replaced by a function C(x)*:

$$y(x) = C(x)e^{-\int p(x)dx}. \tag{1.25}$$

To find $C(x)$, we substitute (1.25) in (1.22). First, find the derivative y':

$$y' = C'(x)e^{-\int p(x)dx} - C(x)p(x)e^{-\int p(x)dx},$$

and substitute y' and y (see (1.25)) in (1.22):

$$y' + p(x)y = C'(x)e^{-\int p(x)dx} - C(x)p(x)e^{-\int p(x)dx} + C(x)p(x)e^{-\int p(x)dx} = f(x).$$

Terms with $C(x)$ cancel, and we obtain

$$C'(x) = f(x)e^{\int p(x)dx}.$$

Integrating, we find

$$C(x) = \int f(x)e^{\int p(x)dx}\, dx + C_1.$$

Substituting this $C(x)$ in (1.25), we finally obtain:

$$y(x) = C(x)e^{-\int p(x)dx} = C_1 e^{-\int p(x)dx} + e^{-\int p(x)dx}\int f(x)e^{\int p(x)dx}\, dx. \tag{1.26}$$

Expression (1.26) *is the general solution of* (1.22). Notice that the first term at the right-hand side is the general solution (1.24) of the homogeneous equation (of course, notations C_1 and C stand for the same arbitrary constant). The second term in (1.26) is the *particular solution* of the inhomogeneous equation (1.22) (this can be checked by plugging the second term in (1.26) for y in (1.22), and its derivative for y'; also note that particular solution does not involve the arbitrary constant).

In summary, the general solution of the inhomogeneous equation is the sum of the general solution of the homogeneous equation and the particular solution of the inhomogeneous equation. In this book we denote these solutions by $Y(x)$ and $\bar{y}(x)$, respectively. Thus, the general solution of (1.22) is written as

$$y(x) = Y(x) + \bar{y}(x). \tag{1.27}$$

When solving problems, the complicated formula (1.26) usually is not used. Instead, all steps of the solution scheme are applied as described above.

Example 1.13 Solve $y' + 5y = e^x$.

First we solve the homogeneous equation

$$y' + 5y = 0.$$

Separating variables, $dy/y = -5dx$, and integrating gives: $|y| = e^{-5x+C} = e^C e^{-5x}$, or $y = Ce^{-5x}$ with arbitrary C. This is the general solution of the homogeneous equation.

Next we proceed to find the particular solution of the inhomogeneous equation through the variation of the parameter:

$$y = C(x)e^{-5x}.$$

The derivative is $y' = C'(x)e^{-5x} - 5C(x)e^{-5x}$. Substitution of y and y' in $y' + 5y = e^x$ gives

$$C'(x)e^{-5x} + 5C(x)e^{-5x} - 5C(x)e^{-5x} = e^x,$$

or $C'(x) = e^{6x}$. Integration results in $C(x) = e^{6x}/6$. (We do not add arbitrary constant because we search for a particular solution.) The particular solution of the inhomogeneous equation is $y = C(x)\,e^{-5x} = e^x/6$. Easy to check that $e^x/6$ satisfies the given inhomogeneous equation. The general solution of the inhomogeneous equation is $y(x) = Y(x) + \bar{y}(x) = Ce^{-5x} + e^x/6$.

Example 1.14   Solve $y' - \frac{y}{x} = x$.

First solve the homogeneous equation

$$y' - y/x = 0.$$

Separating variables, we obtain $\frac{dy}{y} = \frac{dx}{x}$. Integration gives $\ln|y| = \ln|x| + \ln C$, $C > 0$. Then $y = Cx$, where C is arbitrary. This is the general solution of the homogeneous equation.

A particular solution of the original inhomogeneous equation is sought in the form:

$$y = C(x)x.$$

Substitution of y and its derivative $y' = C'(x)x + C(x)$ into $y' - y/x = x$ gives $C'(x)x = x$, and then $C(x) = x$, thus $\bar{y}(x) = x^2$. Finally, the general solution of the problem is $y = Cx + x^2$.

Example 1.15   Solve $(2e^y - x)y' = 1$.

This equation, which is not linear for function $y(x)$, becomes linear if we consider it as the equation for function $x(y)$. To do this, let us write it in the form $2e^y - x = \frac{dx}{dy}$, which is a linear inhomogeneous equation for $x(y)$.

The solution of homogeneous equation $dx/dy + x = 0$ is $x = Ce^{-y}$. Then seek a particular solution of inhomogeneous equation in the form $x = C(y)e^{-y}$. Substitution in $2e^y - x = x'$ gives $C'(y) = 2e^{2y}$, and integration gives $C(y) = e^{2y}$. Thus the particular solution of inhomogeneous equation is $x = C(y)e^{-y} = e^y$. Finally, the general solution of the problem is $x = Ce^{-y} + e^y$.

Example 1.16   Solve $(\sin^2 y + x\cot y)y' = 1$.

This equation becomes linear if we consider it as the equation for function $x(y)$:

$$\frac{dx}{dy} - x\cot y = \sin^2 y.$$

This is a linear inhomogeneous equation for $x(y)$. Integration of homogeneous equation $dx/dy - \cot y\, dy = 0$ gives $\ln|x| = \ln|\sin y| + \ln C$ from which $x = C\sin y$. Then seek a particular solution of inhomogeneous equation in the form $x = C(y)\sin y$. Substitution in inhomogeneous equation gives $C'(y) = \sin y$, thus $C(y) = -\cos y$. Thus the particular solution of inhomogeneous equation is $x = C(y)\sin y = -\sin y\cos y$. The general solution of the problem is the sum of solutions of homogeneous and inhomogeneous equations, thus $x = (C - \cos y)\sin y$.

Another widely used method for the solution of (1.22) is based on using an *integrating factor*. The integrating factor for a linear differential equation is a function

$$\mu(x) = e^{\int p(x)dx}. \tag{1.28}$$

When (1.22) is multiplied by $\mu(x)$, the equation becomes

$$\frac{d}{dx}[\mu(x)y] = \mu(x)f(x). \tag{1.29}$$

(You can *check* that the left-hand side in (1.29) equals $\mu y' + \mu p y$.)

Multiplication of (1.29) by dx, integration and division by $\mu(x)$ gives the solution:

$$y(x) = \frac{1}{\mu(x)}\left[\int \mu(x)f(x)dx + C\right]. \tag{1.30}$$

Substitution of $\mu(x)$ from (1.28) shows that (1.30) coincides with (1.26), which was obtained by the method of variation of the parameter.

Example 1.17   Solve equation $y' + 5y = e^x$ from Example 1.13 using the integrating factor.

Here $\mu(x) = e^{\int 5dx} = e^{5x}$, and solution (1.30) is

$$y(x) = \frac{1}{e^{5x}}\left[\int e^{5x}e^x \, dx + C\right] = \frac{1}{e^{5x}}\left[\frac{1}{6}e^{6x} + C\right] = Ce^{-5x} + \frac{e^x}{6}.$$

Next, consider the *Bernoulli equation*

$$y' + p(x)y = f(x)y^n, \tag{1.31}$$

where n is arbitrary real number (for $n = 0,1$ this equation is already linear). Let show how the Bernoulli equation can be transformed into a linear equation.

First, we divide (1.31) by y^n:

$$y'y^{-n} + p(x)y^{1-n} = f(x). \tag{1.32}$$

Let

$$z(x) = y^{1-n}, \tag{1.33}$$

and find the derivative $z'(x)$ by the chain rule:

$$z'(x) = (1-n)y^{-n}y'.$$

Substitution of y' and y^{1-n} in (1.32) gives the linear inhomogeneous equation for $z(x)$:

$$\frac{1}{1-n}\frac{dz}{dx} + p(x)z = f(x).$$

The general solution of the latter equation can be found by the methods just described above, and then $y(x)$ is determined from (1.33).

Example 1.18   Solve $y' - y = \frac{3}{y}$.

This is Bernoulli equation with $n = -1$. Division by y^{-1} (i.e., multiplication by y) gives $yy' - y^2 = 3$. Let $z = y^2$, then $z' = 2yy'$ and the equation becomes $\frac{1}{2}z' - z = 3$. First we solve the homogeneous equation $\frac{1}{2}z' - z = 0$. Its general solution is $z = Ce^{2x}$. Using variation of the parameter, we seek the particular solution of the inhomogeneous equation in the form $z = C(x)e^{2x}$, which gives $C'(x) = 6e^{-2x}$. Integration gives $C(x) = -3e^{-2x}$, thus the particular solution is $z = -3$ and the general solution of the inhomogeneous equation is $z = Ce^{2x} - 3$. Returning to the variable $y(x)$, the solution of the original equation is obtained: $y^2 = Ce^{2x} - 3$.

Finally, consider the *Riccati equation*

$$y' + p(x)y + q(x)y^2 = f(x). \tag{1.34}$$

This equation cannot be solved analytically in general case, but it can be transformed into the Bernoulli equation if one particular solution, $y_1(x)$, of equation (1.34) is known. In this case let $y = y_1 + z$ and substitute it in equation (1.34):

$$y_1' + z' + p(x)(y_1 + z) + q(x)(y_1 + z)^2 = f(x).$$

Then, because $y_1' + p(x)y_1 + q(x)y_1^2 = f(x)$, for function $z(x)$ we obtain the Bernoulli equation

$$z' + [p(x) + 2q(x)y_1]z + q(x)z^2 = 0.$$

Its general solution plus function $y_1(x)$ gives a general solution of equation (1.34).

Example 1.19 Solve $\frac{dy}{dx} = y^2 - \frac{2}{x^2}$.

Solution. A particular solution of this equation is easy to guess: $y_1 = \frac{1}{x}$. Then $y = z + \frac{1}{x}$ and $y' = z' - \frac{1}{x^2}$. Substituting these y and y' into equation, we obtain $z' - \frac{1}{x^2} = (z + \frac{1}{x})^2 - \frac{2}{x^2}$, or $z' = z^2 + 2\frac{z}{x}$. To solve this Bernoulli equation make the substitution $u = z^{-1}$, which brings to linear equation $\frac{du}{dx} = -\frac{2u}{x} - 1$. A general solution of the corresponding homogeneous equation is

$$\ln|u| = -2\ln|x| + \ln C, \quad \text{thus} \quad u = \frac{C}{x^2}.$$

Then seek a particular solution of inhomogeneous equation in the form $u = c(x)/x$. It leads to $C'(x)/x^2 = -1$, then $C(x) = -x^3/3$ and the particular solution is $u = -x/3$. A general solution is $u = \frac{C}{x^2} - \frac{x}{3}$. It gives $\frac{1}{z} = \frac{C}{x^2} - \frac{x}{3}$, then $\frac{1}{y - \frac{1}{x}} = \frac{C}{x^2} - \frac{x}{3}$ and a general solution of the given equation is

$$y = \frac{1}{x} + \frac{3x^2}{C - x^3}.$$

Problems	Answers
1. $xy' + x^2 + xy - y = 0$	$y = x(Ce^{-x} - 1)$
2. $(2x+1)y' = 4x + 2y$	$y = (2x+1)(\ln|2x+1| + C) + 1$
3. $x^2 y' + xy + 1 = 0$	$xy = C - \ln|x|$
4. $y'\cos x + y\sin x - 1 = 0$	$y = C\cos x + \sin x$
5. $y' = \dfrac{2y + \ln x}{x \ln x}$	$y = C\ln^2 x - \ln x$

6. $(x + y^2)dy = ydx$ $\qquad x = y^2 + Cy; \; y = 0$

7. $y'(y^3 + 2x) = y$ $\qquad x = Cy^2 + y^3$

8. $y' + 2y = y^2 e^x$ $\qquad y = \dfrac{1}{e^x + Ce^{2x}}, \; y = 0$

9. $xy' + 2y + x^5 y^3 e^x = 0$ $\qquad y^{-2} = x^4(2e^x + C), \; y = 0$

10. $xy^2 y' = x^2 + y^3$ $\qquad y^3 = Cx^3 - 3x^2$

11. $xy' + 2x^2\sqrt{y} = 4y$ $\qquad y = x^4(C - \ln x)^2, \; y = 0$

12. $xydy = (y^2 + x)dx$ $\qquad y^2 = Cx^2 - 2x$

13. $y' = \dfrac{y}{-2x + y - 4\ln y}$ $\qquad x = y/3 - 2\ln y + C/y^2 + 1$

14. $(3x - y^2)dy = ydx$ $\qquad x = Cy^3 + y^2; \; y = 0$

15. $(1 - 2xy)y' = y(y - 1)$ $\qquad (y-1)^2 x = y - \ln Cy; \; y = 0; \; y = 1$

16. $2y' - \dfrac{x}{y} = \dfrac{xy}{x^2 - 1}$ $\qquad y^2 = x^2 - 1 + C\sqrt{|x^2 - 1|}$

17. $y'(2x^2 y \ln y - x) = y$ $\qquad x = \dfrac{1}{y(C - \ln^2 y)}$

18. $y' = \dfrac{x}{x^2 - 2y + 1}$ $\qquad x^2 = Ce^{2y} + 2(y - 1)$

19. $x^2 y' + xy + x^2 y^2 = 4$ $\qquad y = \dfrac{2}{x} + \dfrac{4}{Cx^3 - x}$

20. $3y' + y^2 + \dfrac{2}{x^2} = 0$ $\qquad y = \dfrac{1}{x} + \dfrac{1}{Cx^{2/3} + x}$

21. $xy' - (2x+1)y + y^2 = -x^2$ $\qquad y = x + \dfrac{1}{Cx + 1}$

22. $y' - 2xy + y^2 = 5 - x^2$ $\qquad y = x + 2 + \dfrac{4}{4Ce^{4x} - 1}$

23. $y' + 2ye^x - y^2 = e^{2x} + e^x$ $\qquad y = \dfrac{1}{C - x} + e^x$

24. $y' = \dfrac{1}{2}y^2 + \dfrac{1}{2x^2}$ $\qquad y = -\dfrac{1}{x} + \dfrac{2}{x(C - \ln x)}$

1.6 EXACT EQUATIONS

First-order equation $y' = f(x, y)$ always can be written as

$$P(x, y)dx + Q(x, y)dy = 0, \tag{1.35}$$

where $P(x,y)$ and $Q(x,y)$ are some functions. If the left-hand side of (1.35) is a total differential of some function $F(x,y)$,

$$dF(x, y) = P(x,y)dx + Q(x,y)dy, \tag{1.36}$$

then equation (1.35) is written as

$$dF(x, y) = 0. \tag{1.37}$$

Such equations are called *exact*. Solution of (1.37) is

$$F(x,y) = C. \tag{1.38}$$

Recall from calculus that in order for $P(x,y)dx + Q(x,y)dy$ be a total differential, it is necessary and sufficient that

$$\frac{\partial P}{\partial y} = \frac{\partial Q}{\partial x}. \tag{1.39}$$

Now write $dF(x,y)$ as

$$dF(x, y) = \frac{\partial F}{\partial x}dx + \frac{\partial F}{\partial y}dy. \tag{1.40}$$

Comparison of (1.40) and (1.36) gives (since dx and dy are arbitrary)

$$\frac{\partial F(x, y)}{\partial x} = P(x, y), \tag{1.41}$$

$$\frac{\partial F(x, y)}{\partial y} = Q(x, y). \tag{1.42}$$

Next, integrate (1.41) in x to obtain

$$F(x, y) = \int P(x, y)dx + C(y), \tag{1.43}$$

where $C(y)$ is an arbitrary function of y. (Note that y is considered constant as far as integration in x is a concern.)

Substitution of (1.43) in (1.42) gives

$$\frac{\partial}{\partial y}\left(\int P(x,y)dx\right)+C'(y)=Q(x,y),$$

and thus $C'(y)=Q(x,y)-\frac{\partial}{\partial y}\int P(x,y)dx$. From the last equation $C(y)$ is determined by integration:

$$C(y)=\int\left[Q(x,y)-\frac{\partial}{\partial x}\left(\int P(x,y)dx\right)\right]dy+C_1.$$

Substitution of this $C(y)$ in (1.43) gives $F(x,y)$. Next, equating this $F(x,y)$ to a constant (see (1.38)) gives the final solution of (1.35). Note that the sum of C and C_1 should be combined in a single constant which can be again denoted as C.

Just as we did before for the method of variation of the parameters, in practice it is better not to use the derived formula for the solution. Instead, the solution scheme as described should be applied for each equation to be solved.

Example 1.20

Equation $(3x^2y^2+7)dx+2x^3ydy=0$ is exact because $\frac{d}{dy}(3x^2y^2+7)=\frac{d}{dx}(2x^3y)=6x^2y$. This equation is equivalent to the following equation: $dF(x,y)=0$, which has the solution $F(x,y)=C$.

From $dF(x,y)=\frac{\partial F}{\partial x}dx+\frac{\partial F}{\partial x}dy$ it follows that $\frac{\partial F(x,y)}{\partial x}=3x^2y^2+7$, thus $F(x,y)=\int(3x^2y^2+7)dx+C(y)$. Since y is considered constant in the integrand, $F(x,y)=x^3y^2+7x+C(y)$.

Substitution of $F(x,y)$ in $\frac{\partial F(x,y)}{\partial y}=2x^3y$ gives $2x^3y+C'(y)=2x^3y$, from which $C'(y)=0$. Thus $C'(y)=C_1$, and the function $F(x,y)$ is $F(x,y)=x^3y^2+7x+C_1$. Equating $F(x,y)$ to C gives the (implicit) solution of the problem:

$$x^3y^2+7x=C.$$

When the left-hand side of (1.35) is not a total differential (and thus the equation is not exact), in some cases it is possible to find the auxiliary function $\mu(x,y)$, called *the integrating factor*. When (1.35) is multiplied by such function, its left-hand side becomes a total differential. That is,

$$dF(x,y)=\mu(x,y)P(x,y)dx+\mu(x,y)Q(x,y)dy.$$

Unfortunately, there is no universally applicable method for the determination of the integrating factor.

As *Reading Exercise*, show that the linear inhomogeneous equation

$$y'+p(x)y=f(x),$$

or $[p(x)y-f(x)]dx+dy=0$, is transformed into an exact equation when it is multiplied by $\mu=e^{\int p(x)dx}$. It must be taken into consideration that the multiplication by $\mu(x,y)$ can lead to extraneous solutions that must be excluded from the final solution. When such extraneous solutions exist, they are simultaneously the solutions of equation $\mu(x,y) = 0$.

Example 1.21

Equation $[x + x^2(x^2 + y^2)]dx + ydy = 0$ is not exact, but after it is multiplied by $\mu = 1/(x^2 + y^2)$ it takes the form

$$\frac{xdx + ydy}{x^2 + y^2} + x^2dx = 0,$$

where the left-hand side is a total differential:

$$dF(x,y) = d\left[\frac{1}{2}\ln(x^2 + y^2) + \frac{x^3}{3}\right].$$

Thus now we are dealing with the equation $dF(x,y) = 0$, which has the solution $F(x,y) = \ln C$, where for convenience we adopt the integration constant $\ln C$ instead of C. Multiplying $F(x,y) = \ln C$ by 2 and choosing to write the arbitrary constant $2\ln C$ as $\ln C$, we obtain

$$\ln(x^2 + y^2) + \frac{2x^3}{3} = \ln C,$$

and then

$$(x^2 + y^2)e^{2x^3/3} = C \quad (C>0).$$

Note that $\mu = 1/(x^2 + y^2) \neq 0$; thus there are no extraneous solutions.

Remark: Equation (1.42):

$$\frac{\partial F(x,y)}{\partial y} = Q(x,y),$$

where $Q(x,y)$ is given and $F(x,y)$ is the unknown function to be determined, is the example of the first-order *partial differential equation* (PDE). Solution of this equation,

$$F(x,y) = \int P(x,y)dx + C(y),$$

contains $C(y)$—an arbitrary function of y. The general solution of a second-order PDE would contain two arbitrary functions, and so on—i.e., the number of arbitrary functions in the general solution is n, where n is the order of the equation.

Example 1.22

Solve second-order PDE

$$\frac{\partial^2 z(x,y)}{\partial x \partial y} = 0.$$

Integration in x gives $\partial z(x,y)/\partial y = \varphi(y)$, where $\varphi(y)$ is an arbitrary function of y. Next, integrate in y to obtain

$$z(x,y) = \int \varphi(y)dy + f_1(x),$$

where $f_1(x)$ is an arbitrary function of x. Let denote $\int \varphi(y)dy = f_2(y)$—this function is also an arbitrary function of y (since $\varphi(y)$ is arbitrary). At last,

$$z(x,y) = f_1(x) + f_2(y).$$

Problems	Answers
1. $2xydx + (x^2 - y^2)dy = 0$	$3yx^2 - y^3 = C$
2. $(2 - 9xy^2)xdx + (4y^2 - 6x^3)ydy = 0$	$x^2 - 3x^3y^2 + y^4 = C$
3. $3x^2(1 + \ln y)dx = (2y - x^3/y)dy$	$x^3(1 + \ln y) - y^2 = C$
4. $2x(1 + \sqrt{x^2 - y})dx - \sqrt{x^2 - y}dy = 0$	$x^2 + \dfrac{2}{3}(x^2 - y)^{3/2} = C$
5. $e^{-y}dx - (2y + xe^{-y})dy = 0$	$xe^{-y} - y^2 = C$
6. $(y\cos x - x^2)dx + (\sin x + y)dy = 0$	$y\sin x - x^3/3 + y^2/2 = C$
7. $(e^y + 2xy)dx + (e^y + x)xdy = 0$	$e^yx + x^2y = C$
8. $(1 + y^2\sin 2x)dx - 2y\cos^2 xdy = 0$	$x - y^2\cos^2 x = C$
9. $\dfrac{y}{x}dx + (y^3 + \ln x)dy = 0$	$4y\ln x + y^4 = C$

1.7 EQUATIONS UNRESOLVED FOR DERIVATIVE (INTRODUCTORY LOOK)

Consider first-order equation in its general form, unresolved for y':

$$F(x, y, y') = 0. \tag{1.44}$$

If this equation can be solved for derivative y', we obtain one or several equations

$$y_i' = f_i(x, y), \quad i = \overline{1, m}. \tag{1.45}$$

Integrating these equations we obtain the solution of equation (1.44). The complication comparing to the equation (1.44) is that equations (1.45) gives several direction fields.

The Cauchy problem for equation (1.44) is formulated in the same way as for equation $y' = f(x, y)$: find a particular solution of equation (1.44) satisfying the initial condition

$$y(x_0) = y_0.$$

If the number of solutions of this problem is the same as the number of functions, $f_i(x, y)$, one say that the Cauchy problem has the unique solution. In other words, through point (x_0, y_0) in the given direction (determining by each of the values $f_i(x_0, y_0)$) passes not more than one integral curve of equation (1.44). Such solutions are called *regular*. If in each point of the solution the uniqueness is not valid, such a solution is called *irregular*.

Consider equations quadratic with respect of y':

$$(y')^2 + 2P(x,y)y' + Q(x,y) = 0. \tag{1.46}$$

Solve (1.46) for y':

$$y' = -P(x,y) \pm \sqrt{P^2(x,y) - Q(x,y)}. \tag{1.47}$$

This expression is defined for $P^2 - Q \geq 0$. Integrating (1.47) we find a general integral of equation (1.46).

An irregular solution could be only the curve

$$P^2(x,y) - Q(x,y) = 0,$$

which also should satisfy the system of equations for y':

$$\begin{cases} (y')^2 + 2P(x,y)y' + Q(x,y) = 0, \\ y' + P(x,y) = 0. \end{cases} \tag{1.48}$$

Example 1.23

Find a general solution of equation

$$(y')^2 - 2(x+y)y' + 4xy = 0.$$

Solving this quadratic equation for y', we obtain $y' = (x+y) \pm (x-y)$, thus $y' = 2x$ and $y' = 2y$. Integrating each of these equations we find

$$y = x^2 + C, \quad \text{and} \quad y = Ce^{2x}.$$

Each of these families of general solutions (shown in Figure 1.7) satisfy the initial equation. Obviously, these solutions are regular. A irregular solution could arise if $(x-y) = 0$, but $y = x$ does not satisfy the given differential equation.

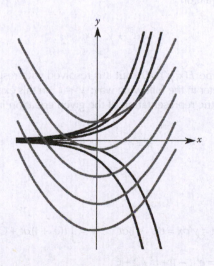

FIGURE 1.7   Two families of solutions for Example 1.23.

Example 1.24

Find a general solution of equation

$$(y')^2 - 4x^2 = 0$$

and find integral curves passing through points a) $M(1,1)$, b) $O(0,0)$.

From equation one finds $y' = 2x$, $y' = -2x$. Integration gives

$$y = x^2 + C, \ y = -x^2 + C.$$

These general integrals are two families of parabolas and there are no irregular solutions. Then solve the Cauchy problem:

a) Substituting the initial condition $x_0 = 1$, $y_0 = 1$ in general solution $y = x^2 + C$ we obtain $C = 0$, thus $y = x^2$; substituting the initial condition in $y = -x^2 + C$ we obtain $C = -2$, thus $y = -x^2 + 2$. Therefore, through point $M(1,1)$ pass *two* integral curves: $y = x^2$ and $y = -x^2 + 2$. Because at this point these solutions belong to different direction fields, the uniqueness of Cauchy problem is not violated.

b) Substituting the initial condition $x_0 = 1$, $y_0 = 1$ in general solutions gives $y = x^2$ and $y = -x^2$. Besides that, the solutions are

$$y = \begin{cases} x^2, & x \le 0, \\ -x^2, & x \ge 0 \end{cases} \quad \text{and} \quad y = \begin{cases} -x^2, & x \le 0, \\ x^2, & x \ge 0. \end{cases}$$

The uniqueness of solution in point $O(0,0)$ is violated because the direction field in this through which two integral curves pass, is the same: $y_0' = 0$.

Example 1.25

Find a general solution of equation

$$e^{y'} + y' = x.$$

This is the equation of the type $F(x, y') = 0$, but it is resolved with respect of x. In such cases it is useful to introduce a parameter in the following way: $y' = t$. In this example it gives $x = e^t + t$. As the result we have a parametric representation of the given equation in the form

$$x = e^t + t, \ y' = t.$$

This gives

$$dy = y'dx = t(e^t + 1)dt, \quad y = \int t(e^t + 1)dt + C,$$

$$y = e^t(t - 1) + t^2/2 + C,$$

from there

$$x = e^t + t, \quad y = e^t(t-1) + t^2/2 + C.$$

Problems	Answers
1. $(y')^2 - 2xy' = 8x^2$	$y = 2x^2 + C; \quad y = -x^2 + C$
2. $y'(2y - y') = y^2 \sin^2 x$	$\ln Cy = x \pm \sin x; y = 0$
3. $xy'(xy' + y) = 2y^2$	$y = Cx; \quad y = \dfrac{C}{x^2}$
4. $(y')^2 + 2yy' \cot x - y^2 = 0$	$y = C/(1 \pm \cos x)$
5. $(y')^3 + 1 = 0$	$y = -x + C$
6. $(y')^4 - 1 = 0$	$y = x + C, \quad y = -x + C$
7. $x = (y')^3 + 1 = 0$	$x = t^3 + 1, \quad y = \dfrac{3}{4}t^4 + C$
8. $x(y')^3 = 1 + y'$	$x = \dfrac{1+t}{t^3}, \quad y = \dfrac{3}{2t^2} + \dfrac{2}{t} + C$

1.8 EXAMPLES OF PROBLEMS LEADING TO DIFFERENTIAL EQUATIONS

Example 1.26: Motion of a body under the action of a time-dependent force: slowing down by friction force proportional to velocity.

Newton's law of a linear motion of a body of mass m is the second-order ordinary differential equation for the coordinate $x(t)$:

$$m\frac{dx^2(t)}{dt^2} = F(t, x, x'). \tag{1.49}$$

Consider situation when F is a friction force proportional to the velocity, i.e., $F = -kv(t)$, where k is a (positive) coefficient. Negative sign in this relation indicates that force and velocity have opposite directions.

Since $x'(t) = v(t)$, equation (1.49) can be written as the differential equation of first-order for $v(t)$:

$$m\frac{dv(t)}{dt} = -kv(t).$$

Velocity decreases; thus the derivative $dv/dt < 0$. In this equation variables are separated:

$$\frac{dv}{v} = -\frac{k}{m}dt.$$

Integration gives the general solution:

$$v(t) = Ce^{-kt/m}.$$

Let $t_0 = 0$ be the time instant when the action of the force started, and let v_0 be the velocity at $t = 0$. Then the initial condition is $v(0) = v_0$. Substitution in the general solution gives $C = v_0$, and the particular solution thus is

$$v(t) = v_0 e^{-kt/m}, \tag{1.50}$$

i.e., velocity decreases exponentially with time, and does so faster if k is larger and m is smaller. The coordinate can be obtained integrating $v(t)$:

$$x(t) = \int_0^t v(t)\,dt = \frac{v_0 m}{k}(1 - e^{-kt/m}). \tag{1.51}$$

From there it is seen that the distance covered before the mass comes to rest is $v_0 m / k$.

Example 1.27: Cooling of an object.

Let the initial object temperature be T_0, and the temperature of the surrounding medium be $T_1 = \text{const}$ (Figure 1.8). The task is to determine the dependence of the object's temperature $T(t)$ on time.

From the experiments it is known that the speed of cooling, dT/dt, is proportional to a difference $T - T_1$ of the object's and the medium's temperatures:

$$\frac{dT}{dt} = -k(T - T_1). \tag{1.52}$$

Here $k > 0$ is a coefficient, and the negative sign in (1.52) is due to the decreasing temperature—the derivative $dT/dt < 0$.

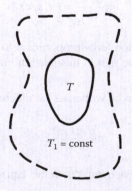

T

$T_1 = \text{const}$

FIGURE 1.8   An object cooling down in the surrounding medium.

In equation (1.52) variables are separated:

$$\frac{dT}{(T - T_1)} = -kdt.$$

Integration gives the general solution:

$$T - T_1 = Ce^{-kt}.$$

Substituting the initial condition $T(0) = T_0$, we find $C = T_0 - T_1$. Substituting this constant in the general solution gives the particular solution:

$$T(t) = T_1 + e^{-kt}(T_0 - T_1).\tag{1.53}$$

Now suppose that the object is heated up. Then $T < T_1$ and the right-hand side of (1.52) has the positive sign. The solution is still given by formula (1.53) where $T_0 - T_1 < 0$.

Equation (1.51) also describes diffusion into a surrounding medium, where concentration of the diffusing substance is constant.

Example 1.28: Radioactive decay.

In radioactive decay, the mass m of a radioactive substance decreases with time, and the speed of decay, $dm(t)/dt$, is proportional to a mass that has not yet disintegrated. Thus, the radioactive decay is described by an equation

$$\frac{dm(t)}{dt} = -km.\tag{1.54}$$

Coefficient $k > 0$ depends on the type of radioactive material, and the negative sign corresponds to the decrease of a mass with time.

Separating variables in (1.54) and integrating, we find the general solution of this equation:

$$m(t) = Ce^{-kt}.$$

Let the initial condition be $m(0) = m_0$, where m_0 is the initial mass of a material. Substitution in the general solution gives $C = m_0$, thus the particular solution is

$$m(t) = m_0 e^{-kt}.\tag{1.55}$$

Half-life is the period of time it takes to decrease the mass by half in radioactive decay. Denoting this time as $T_{1/2}$, we obtain from (1.55)

$$\frac{1}{2} m_0 = m_0 e^{-kT_{1/2}}, \quad \text{and} \quad T_{1/2} = \frac{1}{k}\ln 2.$$

If $T_{1/2}$ is known (or another, shorter time during which a certain quantity of a material decays), one can find the decay constant k.

If equation (1.54) has a positive sign at the right-hand side, then its solution is the expression (1.55) with the positive sign in the exponent. This solution describes, for instance, the exponential growth of the number of neutrons in the nuclear reactions, or the exponential growth of a bacterial colony in the situation when the growth rate of a colony is proportional to its size and also the bacteria do not die.

Problems

1. For a particular radioactive substance 50% of atoms decays in the course of 30 days. When will only 1% of the initial atoms remain?

   Answer: $t \approx 200$ days.

2. According to experiments, during a year from each gram of radium a portion of 0.44 *mg* decays. After what time will one-half of a gram decay?

   Answer: $T_{1/2} = \dfrac{\ln 0.5}{\ln(1 - 0.00044)} \approx 1570$ years.

3. A boat begins to slow down due to water resistance, which is proportional to the boat's velocity. Initial velocity of the boat is 1.5 m/s, after 4 s velocity decreased to 1 m/s. After it begins to slow down (a) when will the boat's velocity become 0.5 m/s? (b) What distance will the boat travel during 4 s? (*Hint:* find first the coefficient of friction.)

   Answer: (a) 10.8 s, 4.9 m.

4. The object cooled down from 100°C to 60°C in 10 min. Temperature of the surrounding air is constant 20°C. When will the object cool down to 25°C?

   Answer: $t = 40$ min.

5. Let m_0 be the initial mass of salt placed in water of mass M. The rate of solution, $\frac{dm(t)}{dt}$, is proportional to a mass of the salt that has not yet dissolved at time t, and the difference between the saturation concentration, $\bar{m}/M$, and the actual concentration, $(m_0 - m)/M$. Thus,

$$\frac{dm}{dt} = -km\left(\frac{\bar{m}}{M} - \frac{m_0 - m}{M}\right).$$

   Find a solution to this Cauchy problem.

6. In bimolecular chemical reactions substances A and B form molecules of type C. If a and b are the original concentrations of A and B respectively, and x is the concentration of C at time t, then

$$\frac{dx}{dt} = k(a-x)(b-x).$$

Find a solution to this Cauchy problem.

1.9 SOLUTION OF FIRST-ORDER DIFFERENTIAL EQUATIONS WITH ACCOMPANYING SOFTWARE

In this section we describe the solution of Examples 1.13, Chapter 1 and 5.1, Chapter 3 using the program **ODE 1st order** (first-order ODE).

Example 1.29 Solve $y' + 5y = e^x$.

Figure 1.9 shows the first screen of the program **ODE 1st order**.

To do this example start with *"Ordinary Differential Equations of the 1st Order"* and then click on *"Data"* and choose *"New Problem."* Then, as shown in Figure 1.10, you should enter the equation, initial conditions, and interval where you want to obtain the solution of IVP. Also, you can enter the analytical solution of IVP to compare with the numerical.

The program solves equation numerically (the solution can be presented in the form of a graph, Figure 1.11, or table, Figure 1.12) and compares with the reader's analytical solution, $y = (5e^{-5x} + e^x)/6$ (in the interface it is denoted as $y^*(x)$ and should be input as shown in the interface).

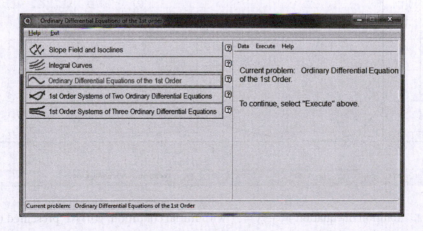

FIGURE 1.9   The first screen of the program ODE 1st order.

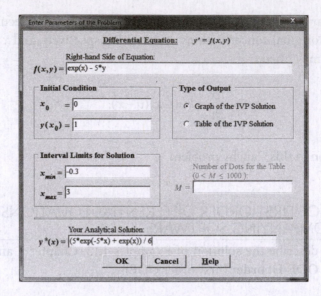

FIGURE 1.10   Equation $y' + 5y = e^x$ with initial condition $y(0) = 1$ in program interface.

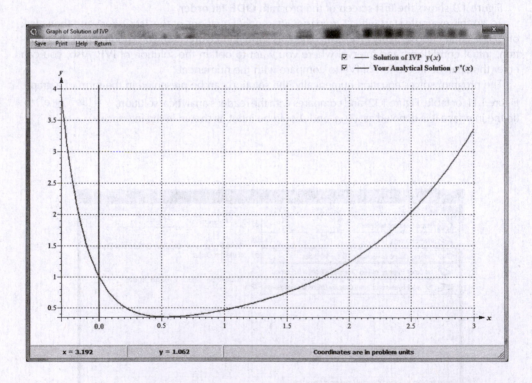

FIGURE 1.11   Solution of equation $y' + 5y = e^x$ with initial condition $y(0) = 1$ presented in graphical form.

k	x_k	$y(x_k)$	$y^*(x_k)$
0	-0.3	3.85821	3.85821
1	-0.2	2.40169	2.40169
2	-0.1	1.52474	1.52474
3	0	1	1
4	0.1	0.689637	0.689637
5	0.2	0.510133	0.510133
6	0.3	0.410918	0.410918
7	0.4	0.361417	0.361417
8	0.5	0.343191	0.343191
9	0.6	0.345176	0.345176
10	0.7	0.36079	0.36079
11	0.8	0.386187	0.386187
12	0.9	0.419191	0.419191
13	1	0.458662	0.458662
14	1.1	0.5041	0.5041
15	1.2	0.555418	0.555418

Close   Help   Save

FIGURE 1.12   Solution of equation $y'+5y=e^x$ with initial condition $y(0)=1$ presented in table form.

Example 1.30

Solve system of two linear nonhomogeneous equations

$$\begin{cases} \dfrac{dx}{dt} = -x + 8y + 1, \\ \dfrac{dy}{dt} = x + y + t \end{cases}$$

with initial conditions $x(0)=1$, $y(0)=-1$

To do this example begin with *"First Order System of Two Ordinary Differential Equations"* in the starting interface of the program **ODE 1st order** and then click on *"Data"* and choose *"New Problem."* Then, as shown in Figure 1.13, you should enter the equations, initial conditions, and interval where you want to obtain the solution of IVP. Also, you can enter the analytical solution of IVP to compare with the numerical.

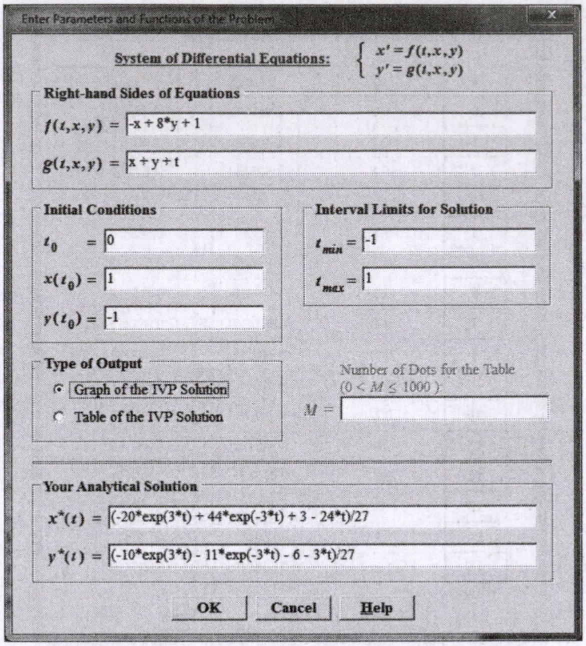

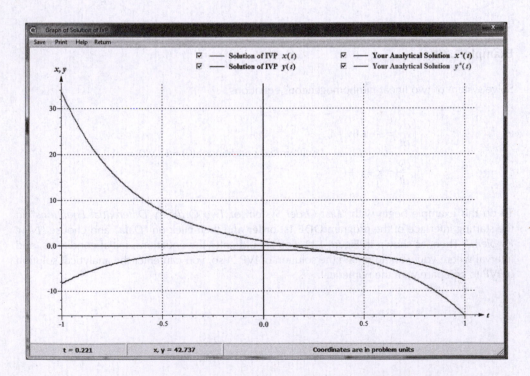

FIGURE 1.13   System of two equations with initial conditions in program interface.

FIGURE 1.14   Solution of IVP problem of Example 5.1 presented in graphical form.

Table of IVP Solution						
k	t_k	$x(t_k)$	$y(t_k)$	$x^*(t_k)$	$y^*(t_k)$	
0	-1	33.6951	-8.31255	33.6951	-8.31255	
1	-0.9	25.1098	-6.20922	25.1098	-6.20923	
2	-0.8	18.7187	-4.65786	18.7187	-4.65786	
3	-0.7	13.9505	-3.51676	13.9505	-3.51676	
4	-0.6	10.3807	-2.68145	10.3807	-2.68145	
5	-0.5	7.69377	-2.07518	7.69377	-2.07518	
6	-0.4	5.65412	-1.64197	5.65412	-1.64197	
7	-0.3	4.08486	-1.34153	4.08486	-1.34153	
8	-0.2	2.85174	-1.14561	2.85174	-1.14561	
9	-0.1	1.85102	-1.03543	1.85102	-1.03543	
10	0	1	-1	1	-1	
11	0.1	0.229586	-1.0351	0.229586	-1.0351	
12	0.2	-0.522025	-1.14289	-0.522025	-1.14289	
13	0.3	-1.31493	-1.33216	-1.31493	-1.33216	
14	0.4	-2.21296	-1.61905	-2.21296	-1.61905	
15	0.5	-3.28948	-2.02857	-3.28948	-2.02857	

Close   Help   Save

FIGURE 1.15   Solution of IVP of Example 5.1 presented in the table form.

The program solves the system numerically (the solution can be presented in the form of a graph, Figure 1.14 or table, Figure 1.15) and compares with the reader's analytical solution (in the interface it is denoted as $x^*(t)$ and $y^*(t)$, and should be input as shown in the interface).

Second-Order Differential Equations

2.1 GENERAL CONSIDERATION: NTH -ORDER DIFFERENTIAL EQUATIONS

We begin with a general case of the nth order differential equation, which is written in the form resolved for the highest derivative of the unknown function $y(x)$:

$$y^{(n)} = f(x, y, y', \ldots, y^{(n-1)}), \qquad (2.1)$$

where $f(x, y, y', \ldots, y^{(n-1)})$ is a given function.

Obviously, a general solution (general integral) of equation (2.1) depends on n arbitrary constants. For instance, for the simplest case of equation (2.1):

$$y^{(n)} = f(x),$$

a general solution obtained by consequent integration of the equation n times contains n arbitrary constants as the coefficients in the polynomial of order $n-1$:

$$y(x) = \int dx \int dx \ldots \int f(x)dx + \sum_{i=0}^{n-1} C_i x^i.$$

When equation (2.1) describes some phenomena, these constants are related to a concrete situation. For instance, consider Newton's second law for a one-dimensional motion of a body of mass m moving under the action of a force F: $m \frac{d^2 x}{dt^2} = F$. Integrating, we obtain $x'(t) = \int (F/m)dt + C_1$, then assuming constant values of F and m gives $x(t) = Ft^2/2m + C_1 t + C_2$. This is a *general solution* which gives the answer to the problem: $x(t)$ depends quadratically on time. This general solution can be presented as $x(t) = at^2/2 + v_0 t + x_0$, where $a = d^2 x/dt^2 = F/m$ is the acceleration and two arbitrary constants are denoted as v_0 and x_0.

These constants are the location, $x_0 = x(0)$, and velocity, $v_0 = x'(0)$, at time $t = 0$. For the particular values of $x(0)$ and $x'(0)$ given in the concrete situation, we have the *particular solution* of the problem. Most often, the problems given by differential equations demand a concrete particular solution, but general solutions, like in this example, are often necessary and describe general features of the solution.

The Cauchy problem (or IVP) for the nth order equation (2.1) assumes n *initial conditions* at some point x_0:

$$y(t_0) = y_0, \quad y'(t_0) = y_0', \quad y''(t_0) = y_0'', \dots, y^{(n-1)}(x_0) = y_0^{(n-1)}, \tag{2.2}$$

where $y_0, y_0', y_0'', \dots, y_0^{(n-1)}$ are n real numbers - the values of the function $y(x)$ and its $n-1$ derivatives at x_0.

Thus, equation (2.1), along with the initial conditions (2.2), constitute the IVP. The following theorem plays the key role.

Picard's Theorem (Existence and Uniqueness of a Solution to the nth-order differential equation): If function f is a continuous function of all its arguments in the vicinity of the point $(x_0, y_0, y_0', \dots, y_0^{(n-1)})$, and its partial derivatives with respect to $y, y', \dots, y^{(n-1)}$ are finite in this vicinity, then the unique solution of the Cauchy problem (2.1), (2.2) exists.

2.2 SECOND-ORDER DIFFERENTIAL EQUATIONS

For the second-order equation

$$y'' = f(x, y, y') \tag{2.3}$$

the Cauchy problem has two initial conditions:

$$y(x_0) = y_0, \quad y'(x_0) = y_0', \tag{2.4}$$

where y_0 and y_0' are the given values of $y(x)$ and $y'(x)$ at x_0. The value $y'(x_0)$ gives the angle of the tangent line (the slope) to the solution curve (integral curve) at the point (x_0, y_0): $\alpha = \arctan y_0'$ (see Figure 2.1).

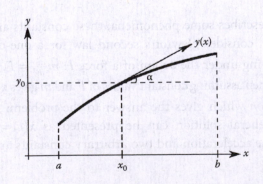

FIGURE 2.1   Solution to the Cauchy problem (2.3), (2.4).

According to the existence and uniqueness theorem, two initial conditions (2.4) are necessary and sufficient to construct a particular solution on some interval $[a,b]$ containing x_0, in which $f(x,y,y')$ is continuous and $\partial f/\partial y$, $\partial f/\partial y'$ are finite. All these restrictions guarantee that the values of y and y' at the point $x_0 + h$ (where h is small) do not differ drastically from y_0 and y'_0, respectively.

General solution of (2.3) contains two arbitrary constants, which are usually denoted as C_1 and C_2. The number of initial conditions (2.4) matches the number of arbitrary constants, and plugging the general solution in the initial conditions results in concrete values of C_1 and C_2. Plugging these values in a general solution gives a particular solution of IVP (2.3) and (2.4). The arbitrary constants are often termed parameters, and the general solution is termed *the two-parameter family of solutions*; a particular solution is just one member of this family.

Below we show several simple examples.

Example 2.1

Find the particular solution of IVP $y'' = 1$, $y(1) = 0$, $y'(1) = 1$.

Solution: Integrating the equation twice, we obtain a general solution: $y(x) = x^2/2 + C_1 x + C_2$. Substituting this into initial conditions gives $y(1) = 1/2 + C_1 + C_2 = 0$, $y'(1) = 1 + C_1 = 1$, which gives $C_1 = 0, C_2 = -1/2$. Thus, $y(x) = x^2/2 + x$ is the particular solution of IVP for given initial conditions.

Example 2.2

Find the particular solution of IVP $y''' = e^x$, $y(0) = 1$, $y'(0) = 0$, $y''(0) = 2$.

Solution. Integrating the equation three times, we obtain the general solution: $y(x) = e^x + C_1 x^2/2 + C_2 x + C_3$. Substituting this into the initial conditions gives $y(0) = 1 + C_3 = 1$, $y'(0) = 1 + C_2 + C_3 = 0$, $y''(0) = 1 + C_1 = 2$. From these three equations we have $C_1 = 1$, $C_2 = -1$, $C_3 = 0$; thus $y(x) = e^x + x^2/2 - x$ is the particular solution of the IVP.

Example 2.3

Find the particular solution of $y'' = 1/(x+2)^2$ for the initial conditions $y(-1) = 1$, $y'(-1) = 2$.

Solution. General solution is $y(x) = -\ln|x+2| + C_1 x + C_2$. Notice, that at $x = -2$ it diverges (no solution) since function $f = 1/(x+2)$ is not continuous at $x = -2$, which violates the conditions of the existence and uniqueness theorem. On the other hand, $x = -1$ is a regular point for function f and a particular solution in the vicinity of this point does exist. Substitution of the general solution in the initial conditions gives $y(-1) = -C_1 + C_2 = 1$, $y'(-1) = -1/(-1+2) + C_1 = 2$. Thus, $C_1 = 3$, $C_2 = 4$ and the particular solution is $y(x) = -\ln|x+2| + 3x + 4$ (except the line $x = -2$).

Example 2.4

Find the particular solution of $yy'' - 2y'^2 = 0$ for two sets of initial conditions, (a) $y(0) = 0$, $y'(0) = 2$; (b) $y(0) = 1$, $y'(0) = 0$.

Solution: General solution, $y(x) = 1/(C_1 x + C_2)$ will be obtained in Example 2.8—here its correctness can be checked just by substituting this function into equation. The initial condition $y(0) = 0$ obviously contradicts this general solution. This is because when the function $f(x,y,y') = 2y'^2/y$ is written as $y'' = f(x,y,y')$, it has a partial derivative $f'_y = -2y'^2/y^2$, which is infinite at $y(0) = 0$. For the second set of initial conditions f'_y is finite, and for the constants C_1 and C_2 we obtain $C_1 = 0$, $C_2 = 1$; thus $y(x) = 1$.

2.3 REDUCTION OF ORDER

In some cases the order of a differential equation can be reduced—usually this makes the solution easier. In this section we discuss several simple situations that are often encountered in practice.

1. Left side of equation

$$F(x,y,y',\ldots,y^{(n)})=0 \tag{2.5}$$

is a total derivative of some function G, that is, equation (2.5) can be written as

$$\frac{d}{dx}G(x,y,y',\ldots,y^{(n-1)})=0.$$

This means that function G is a constant:

$$G(x,y,y',\ldots,y^{(n-1)})=C_1. \tag{2.6}$$

This is the *first integral* of equation (2.5).

Example 2.5

Solve equation $yy'' + (y')^2 = 0$ using the reduction of order.

Solution. This equation can be written as $\frac{d}{dx}(yy')=0$, which gives $yy' = C_1$, or $ydy = C_1dx$. Thus, $y^2 = C_1x + C_2$ is the general solution.

2. Equation does not contain the unknown function y and its first $k-1$ derivatives:

$$F\left(x,y^{(k)},y^{(k+1)},\ldots,y^{(n)}\right)=0. \tag{2.7}$$

Substitution

$$y^{(k)} = z(x) \tag{2.8}$$

reduces the order of the equation to $(n-k)$.

Example 2.6

Solve $y^{(5)} - y^{(4)}/x = 0$ using the reduction of order.

Solution. Substitution $y^{(4)} = z(x)$ gives $z' - z/x = 0$, thus $z = C_1x$ and $y^{(4)} = C_1x$. Integrating four times gives $y = C_1x^5 + C_2x^3 + C_3x^2 + C_4x + C_5$.

As further illustration, consider the second-order equation

$$F(x, y', y'') = 0. \tag{2.9}$$

Note that y is absent from F. Substitution (2.8) in this case ($k = 1$) is

$$y' = z(x). \tag{2.10}$$

Then $y'' = z'$ and (2.9) reduces to the first-order equation:

$$F(x, z, z') = 0.$$

Its general solution contains one arbitrary constant, and we can write this solution as $z(x, C_1) = y'$. Integrating this first-order equation, the general solution of (2.9) is $y(x) = \int z(x, C_1) dx + C_2$.

Example 2.7

Solve $y'' - \frac{y'}{x} = x$ using the reduction of order.

Solution. With $y' = z(x)$, we have $y'' = z'(x)$ and the equation becomes $z' - z/x = x$. This is the linear nonhomogeneous first-order equation. Its solution is the sum of the general solution of the homogeneous equation and the particular solution of the non-homogeneous equation. The homogeneous equation is $z' - \frac{z}{x} = 0$, or $\frac{dz}{z} = \frac{dx}{x}$. Its general solution is $z = C_1 x$. The particular solution of the nonhomogeneous equation can be found using the method of the variation of a parameter: $z = C_1(x)x$. Substituting this into equation $z' - z/x = x$ gives $C_1'x + C_1 - C_1 x/x = x$, $C_1' = 1$, and $C_1(x) = x$. Thus

$$z = C_1 x + x^2,$$

and finally $y(x) = \int z(x) dx = \int (x^2 + C_1 x) dx = \frac{x^3}{3} + C_1 \frac{x^2}{2} + C_2$.

3. **Equation does not contain the independent variable (x).** For example, the second-order equation is

$$F(y, y', y'') = 0. \tag{2.11}$$

To reduce order, we make the substitution

$$y' = z(y) \tag{2.12}$$

Notice that here z is a function of y, not x. Differentiation using the chain rule gives $y'' = \frac{dz}{dy}\frac{dy}{dx} = \frac{dz}{dy}z(y)$. Equation (2.11) becomes $F(y, z(y), z'(y)) = 0$, which has a general solution $z(y, C_1)$. Thus, we arrive to the first-order equation $y'(x) = z(y, C_1)$, where variables can be separated. Then, $\int \frac{dy}{z(y, C_1)} = x + C_2$, which gives the general solution.

Example 2.8

Solve $yy'' - 2y'^2 = 0$ using the reduction of order.

Solution. Differentiation of $z(y) = y'$ gives $y'' = \frac{dz}{dy}\frac{dy}{dx}$, and replacing here y' by $z(y)$ gives $y'' = z(y)\frac{dz}{dy}$. Equation now reads $yz\frac{dz}{dy} - 2z^2 = 0$. Its solutions are (a) $z(y) = 0$, or $y'=0$, thus $y = C$; (b) $\frac{dz}{z} = 2\frac{dy}{y}$, then $z = C_1 y^2$, which gives $y' = C_1 y^2$, $dy/y^2 = C_1 dx$, $-1/y = C_1 x + C_2$, $y(x) = -\frac{1}{C_1 x + C_2}$ (sign minus can be omitted).

Example 2.9

Solve IVP, $yy'' - 2y'^2 = 0$, $y(x_0) = y_0$, $y'(x_0) = y'_0$.

Solution. The general solution (a) $y = C$ of the previous example is a constant continuous function. This solution corresponds to the initial conditions $y(x_0) = C$ (including the value $C = 0$) and $y'(x_0) = 0$.

General solution (b) $y(x) = -1/(C_1 x + C_2)$ describes hyperbolas with two asymptotes: vertical, $x = -C_2/C_1$, and horizontal, $y = 0$. If we rewrite the equation in the form $y'' = 2y'^2/y = f$ and find partial derivatives $f_y = -2y'^2/y^2$, $f_{y'} = 4y'/y$, we can notice that f, f_y, $f_{y'}$ are discontinuous at $y = 0$. At this point the conditions of the existence and uniqueness theorem are not valid.

Treating C_1 and C_2 as parameters, we can think of the solution as a two-parameter set of integral curves, $y(x, C_1, C_2)$. The sets separated by the asymptotes are unrelated, and two initial conditions pick up a particular curve from one of these four sets. If $x_0 < -C_2/C_1$ and $y(x_0) = y_0 > 0$, then we have the solution of IVP in the domain $-\infty < x < -C_2/C_1$, $y > 0$. There are also three other possibilities.

To illustrate this example, we plot four integral curves, $y(x, C_1, C_2)$, for the case of the vertical asymptote at $x = 1.5$ (that is, values of C_1 and C_2 in $y(x) = -1/(C_1 x + C_2)$ are chosen such that $-C_2/C_1 = 1.5$). In Figure 2.2, which is obtained using the program ODE 2nd order (second-order ordinary differential equations), the four particular solutions are shown.

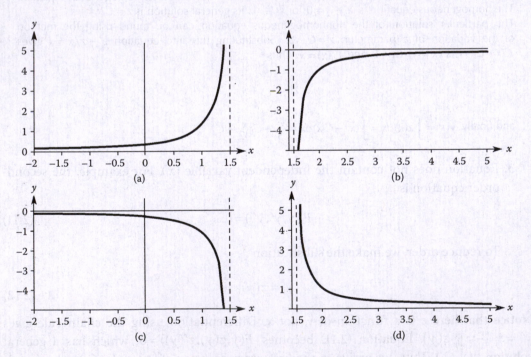

FIGURE 2.2 Four particular solutions for different initial conditions. Dashed line is the vertical asymptote $x = -C_2/C_1 = 1.5$.

1. Using the initial conditions $x_0 = 1$, $y_0 = 1$, $y'_0 = 2$ we obtain $C_1 = 2$, $C_2 = -3$, thus the particular solution is $y = -\frac{1}{2x-3}$. This solution exists on the interval $x \subset (-\infty, 1.5)$; Similarly:

2. $x_0 = 3$, $y_0 = -1/3$, $y'_0 = 2/9$, $C_1 = 2$, $C_2 = -3$, $y' = -\frac{1}{2x-3}$, $x \subset (1.5, \infty)$;

3. $x_0 = 1$, $y_0 = -1$, $y'_0 = -2$, $C_1 = -2$, $C_2 = 3$, $y = \frac{1}{2x-3}$, $x \subset (-\infty, 1.5)$;

4. $x_0 = 3$, $y_0 = 1/3$, $y'_0 = -2/9$, $C_1 = -2$, $C_2 = 3$, $y = \frac{1}{2x-3}$, $x \subset (1.5, \infty)$.

It is obvious from this example how important is to know the analytical solution: without it a numerical procedure may not "detect" the vertical asymptote at $x = -C_2/C_1$, the numerical solution may "slip" to an adjacent integral curve, and thus the obtained solution may be completely unrelated to the initial conditions.

Problems	Answers
1. $(x-3)y'' + y' = 0$	$y = C_1 \ln|x-3| + C_2$
2. $y^3 y'' = 1$	$C_1 y^2 - 1 = (C_1 x + C_2)^2$
3. $y'^2 + 2yy'' = 0$	$y^3 = C_1(x + C_2)^2$
4. $y''(e^x + 1) + y' = 0$	$y = C_1(x - e^{-x}) + C_2$
5. $yy'' = y'^2 - y'^3$	$y + C_1 \ln|y| = x + C_2$
6. $x(y'' + 1) + y' = 0$	$y = C_1 \ln|x| - \dfrac{x^2}{4} + C_2$
7. $y'' = \sqrt{1 - (y')^2}$	$y = C_2 - \cos(x + C_1)$
8. $y''' + 2xy'' = 0$	$y = C_1\left(\dfrac{1}{2}e^{-x^2} + x\int e^{-x^2}\,dx\right) + C_2 x + C_3$
9. $y'' + 2y' = e^x y'^2$	$y = -e^{-x} + C_1 \ln|e^{-x} + C_1| + C_2,\ y = 0$
10. $yy'' = (y')^3$	$y \ln|y| + x + C_1 y + C_2 = 0,\ y = C$
11. $(1-x^2)y'' + (y')^2 + 1 = 0$	$y = x + C_1(x + 2\ln|x-1|) + C_2$
12. $xy''' - y'' = 0$	$y = C_1 x^3 + C_2 x + C_3$

2.4 LINEAR SECOND-ORDER DIFFERENTIAL EQUATIONS

Linear differential equations arise frequently in applications. We discuss first second-order linear differential equations:

$$y'' + p(x)y' + q(x)y = f(x), \qquad (2.13)$$

where $p(x)$, $q(x)$ and $f(x)$ are given functions. Linearity of (2.13) with respect to $y'(x)$ and $y(x)$ stems from the observation that this equation does not contain any transcendental functions of y and y' (such as, say, $e^y \sin y$, or $y' \ln y$), and nonlinear combinations of y and y' (such as their products or powers, say y^2, $y^{3/2}$, or yy'). Equation (2.13) is *inhomogeneous*; but if $f(x) = 0$, it is *homogeneous*:

$$y'' + p(x)y' + q(x)y = 0. \tag{2.14}$$

If the functions $p(x)$, $q(x)$ and $f(x)$ are continuous in the vicinity of $x_0 \in [a,b]$, then the function $F(x, y, y', y'') = y'' + p(x)y' + q(x)y - f(x)$ and its derivatives $F_y = q(x)$ and $F_{y'} = p(x)$ are also continuous and bounded on $[a,b]$. Then, the requirements of the existence and uniqueness theorem are satisfied, and a particular solution of (2.13) that corresponds to the initial conditions $y(x_0) = y_0$, $y'(x_0) = y'_0$ (where y_0 and y'_0 are any values) exists and is unique.

2.4.1 Homogeneous Equations

Here, we consider the homogeneous equation (2.14) and prove several simple theorems. Any function that does not contain arbitrary constants and satisfies equation (2.14) is called the particular solution—this terminology overlaps with the one that is used to denote the solution of an IVP; which particular solution is in question is usually clear from the context.

Theorem 1.

Let $y_1(x)$ and $y_2(x)$ be any two particular solutions of the homogeneous equation (2.14). Then

$$y(x) = C_1 y_1(x) + C_2 y_2(x), \tag{2.15}$$

where C_1 and C_2 are two arbitrary constants, is also a solution of this equation.

Proof:
Substitute $y(x)$ and its derivatives, $y' = C_1 y'_1 + C_2 y'_2$ and $y'' = C_1 y''_1 + C_2 y''_2$, into equation (2.14) and combine separately the terms with C_1 and C_2. Then the left-hand side becomes $C_1 (y''_1 + py'_1 + qy_1) + C_2(y''_2 + py'_2 + qy_2)$. Since $y_1(x)$ and $y_2(x)$ are solutions of (2.14), the expressions in both parentheses are zero. Thus $y(x) = C_1 y_1(x) + C_2 y_2(x)$ is the solution of (2.14).

The expression $y(x) = C_1 y_1(x) + C_2 y_2(x)$ is called the *linear combination* of functions $y_1(x)$ and $y_2(x)$. More generally,

$$y(x) = C_1 y_1(x) + C_2 y_2(x) + \cdots + C_n y_n(x) \tag{2.16}$$

is called the linear combination of n functions. Thus, Theorem 1 states that the linear combination of the particular solutions of a homogeneous equation (2.14) is also its solution.

Functions $y_1(x), y_2(x), \ldots, y_n(x)$ are called *linearly dependent* on the interval $[a,b]$, if $C_1 y_1(x) + C_2 y_2(x) + \cdots + C_n y_n(x) = 0$ for all $x \in [a,b]$, and not all constants C_i are zero. Otherwise, functions are called *linearly independent*. Linear dependence of two functions, $C_1 y_1(x) + C_2 y_2(x) = 0$, simply means that functions are proportional: $y_2(x) = -C_1 y_1(x) / C_2$.

Theorem 2.

Let $y_1(x)$ and $y_2(x)$ be any two solutions of (2.14) on the interval $[a,b]$, finite or infinite, and consider the determinant, which is called *Wronskian*:

$$W(y_1(x), y_2(x)) = \begin{vmatrix} y_1(x) & y_2(x) \\ y_1'(x) & y_2'(x) \end{vmatrix}.$$

(2.17)

(a) If $y_1(x)$ and $y_2(x)$ are linearly dependent on $[a,b]$, then $W(y_1, y_2) = 0$ for at least on point x in this interval, and vice versa.

(b) If $y_1(x)$ and $y_2(x)$ are linearly independent on $[a,b]$ (and also $p(x)$ and $q(x)$ are continuous on $[a,b]$), then on this interval $W(y_1, y_2) \neq 0$, and vice versa.

Proof:

(a) Substituting $y_2(x) = -C_1 y_1(x) / C_2$ in (2.17) gives immediately $W(y_1, y_2) = 0$.

(b) Assume that $W(y_1, y_2) = 0$ at some point x_0. Then two constants, α and β, (not both equal to zero) can be chosen, such that the system of equations for α and β:

$$\alpha y_1(x_0) + \beta y_2(x_0) = 0$$

$$\alpha y_1'(x_0) + \beta y_2'(x_0) = 0$$

has a nontrivial solution. This follows from the fact that the determinant of this system of homogeneous algebraic equations is zero according to our assumption: $W(y_1(x_0), y_2(x_0)) = 0$. With such choice of α and β, a linear combination $y = \alpha y_1(x) + \beta y_2(x)$ is the solution of IVP comprised of equation (2.14) and zero initial conditions $y(x_0) = 0$, $y'(x_0) = 0$. Clearly, with such initial conditions, equation (2.14) has only a trivial solution: $y(x) = 0$. This solution is unique; thus $\alpha y_1(x) + \beta y_2(x) \equiv 0$, which contradicts the assumed linear independence of functions $y_1(x)$ and $y_2(x)$.

Theorem 2 can be also stated as follows: $y_1(x)$ and $y_2(x)$ are linearly independent if and only if their Wronskian is not zero.

Examples of linear independence on any finite or infinite interval give the following pairs of functions: $y_1 = x$ and $y_2 = e^x$, $y_1 = e^x$ and $y_2 = e^{-x}$, $y_1(x) = \cos ax$ and $y_2(x) = \sin ax$. The Wronskian for each pair is not zero and there are no nonzero coefficients C_1 and C_2, such that the relation $C_1 y_1(x) + C_2 y_2(x) = 0$ holds on any interval. On the other hand, functions $y_1 = e^x$ and $y_2 = 5e^x$ are linearly dependent, since $W(y_1, y_2) = 0$.

Theorem 3.

Let $y_1(x)$ and $y_2(x)$ be two linearly independent particular solutions of the homogeneous equation (2.14) on $[a,b]$. Then its general solution on $[a,b]$ is

$$Y(x) = C_1 y_1(x) + C_2 y_2(x),$$

(2.18)

where C_1 and C_2 are two arbitrary constants (we will use capital $Y(x)$ for a general solution of a homogeneous equation).

Proof:

To be a general solution, $Y(x)$ must satisfy equation (2.14) and allow to unambiguously determine constants C_1 and C_2 from the initial conditions

$$y(x_0) = y_0, \quad y'(x_0) = y'_0 \tag{2.19}$$

with $x_0 \in [a, b]$ and arbitrary values of y_0 and y'_0.

From Theorem 1 it follows that $y(x)$ is a solution of equation (2.14). Initial conditions (2.19) give

$$C_1 y_1(x_0) + C_2 y_2(x_0) = y_0,$$
$$C_1 y'_1(x_0) + C_2 y'_2(x_0) = y'_0. \tag{2.20}$$

The determinant of this system of equations for C_1 and C_2 is $W(y_1(x_0), y_2(x_0)) \neq 0$, thus these coefficients can be unambiguously determined from the system (2.20).

Corollary: The maximum number of linearly independent solutions of the linear second-order homogeneous equation (2.14) is two.

Any two linearly independent solutions of a second-order linear homogeneous equation are called its *fundamental systems* of solutions.

Example 2.10

Find the general solution of equation $y'' - y = 0$.

Solution. One can easily check that $y_1 = e^x$ and $y_2(x) = e^{-x}$ are the particular solutions of this equation. The Wronskian $W = \begin{vmatrix} e^x & e^{-x} \\ e^x & -e^{-x} \end{vmatrix} = -2 \neq 0$; thus $y_1(x) = e^x$ and $y_2(x) = e^{-x}$ are linearly independent and their linear combination, $Y(x) = C_1 e^x + C_2 e^{-x}$, is the general solution.

Example 2.11

For equation $y''' - y' = 0$ consider the following two sets of the particular solutions—$y_1 = e^x$, $y_2(x) = \cosh x$, $y_3(x) = \sinh x$ and $y_1(x) = 1$, $y_2(x) = \cosh x$, $y_3(x) = \sinh x$—and find a general solution.

Solution. It is easy to check that each of these six functions is a particular solution of this equation. For the first set the Wronskian $W(y_1, y_2, y_3) = 0$; for the second, $W(y_1, y_2, y_3) \neq 0$. Thus the latter three functions are linearly independent and the general solution is $Y(x) = C_1 + C_2 \cosh x + C_3 \sinh x$.

2.4.2 Inhomogeneous Equations

Next, we consider the inhomogeneous equation (2.13) and prove several useful theorems.

Theorem 4.

Let $Y(x) = C_1 y_1(x) + C_2 y_2(x)$ be the general solution of (2.14) on $[a, b]$, and let $\bar{y}(x)$ be the particular solution of the inhomogeneous equation (2.13). Then

$$y(x) = Y(x) + \bar{y}(x) \tag{2.21}$$

is the general solution of (2.13) on $[a,b]$.

Proof:

Twice differentiate relation (2.21), $y' = \bar{y}' + Y'$, $y'' = \bar{y}'' + Y''$, and then substitution of y, y', y'' in (2.13) gives: $(Y'' + p(x)Y' + q(x)Y) + (\bar{y}'' + p(x)\bar{y}' + q(x)\bar{y}) = f(x)$.

Next, notice that $Y'' + p(x)Y' + q(x)Y = 0$, since $Y(x)$ is the general solution of (2.14). Also, $\bar{y}'' + p(x)\bar{y}' + q(x)\bar{y} = f(x)$, since $y(x)$ is the particular solution of (2.13). Thus, $y(x) = Y(x) + \bar{y}(x) = C_1 y_1(x) + C_2 y_2(x) + \bar{y}(x)$ is the general solution of (2.13).

Theorem 5.

If the general solution $Y(x)$ of the homogeneous equation (2.14) has been determined, then the particular solution $\bar{y}(x)$ of the inhomogeneous equation (2.13) can be found by the method of variation of the parameters.

Proof:

Since $Y = C_1 y_1 + C_2 y_2$, then the method of variation of the parameters implies that we look for the particular solution of the nonhomogeneous equation (2.13) in the form $\bar{y}(x) = C_1(x)y_1 + C_2(x)y_2$. To find functions $C_1(x)$ and $C_2(x)$ we substitute $\bar{y}(x)$, $\bar{y}'(x)$ and $\bar{y}''(x)$ in (2.13). The first derivative is

$$\bar{y}' = C_1' y_1 + C_1 y_1' + C_2' y_2 + C_2 y_2'.$$

Choose C_1 and C_2 such that $C_1' y_1 + C_2' y_2 = 0$, since we are going to use this equation as the first in the system of equations for the unknowns C_1' and C_2'. Differentiating $\bar{y}' = C_1 y_1' + C_2 y_2'$ gives $\bar{y}'' = C_1 y_1'' + C_1' y_1' + C_2 y_2'' + C_2' y_2'$. Substitution of $\bar{y}(x)$, $\bar{y}'(x)$ and $\bar{y}''(x)$ in (2.13) gives:

$$C_1\left[y_1'' + p(x)y_1' + q(x)y_1\right] + C_2\left[y_2'' + p(x)y_2' + q(x)y_2\right] + C_1' y_1' + C_2' y_2' = f(x).$$

Since $y_1(x)$ and $y_2(x)$ are the particular solutions of the homogeneous equation (2.14), $y_1'' + p(x)y_1' + q(x)y_1 = 0$ and $y_2'' + p(x)y_2' + q(x)y_2 = 0$. Thus, we arrive to the system of algebraic equations for C_1' and C_2':

$$\begin{aligned} y_1 C_1' + y_2 C_2' &= 0, \\ y_1' C_1' + y_2' C_2' &= f(x). \end{aligned} \tag{2.22}$$

The determinant of this system is the Wronskian $W(y_1, y_2)$. Functions $y_1(x)$ and $y_2(x)$ are linearly independent since they are the two particular solutions of the homogeneous equation (2.14); thus $W \neq 0$ and the system (2.22) for C_1' and C_2' is uniquely solvable. After C_1' and C_2' have been found, the functions $C_1(x)$ and $C_2(x)$ can be determined by integration. (Notice that because we are searching for a particular solution, the constants of integrations must not be added.) This completes the construction of $\bar{y}(x)$.

Example 2.12

Solve the inhomogeneous equation $y'' - y = 1$. Then find the solution of IVP with initial conditions $y(0) = 0$, $y'(0) = 1$.

Solution. The general solution of the homogeneous equation $y'' - y = 0$ is $Y(x) = C_1 y_1(x) + C_2 y_2(x) = C_1 e^x + C_2 e^{-x}$ (Example 2.11). Thus, the system (2.22) is

$$e^x C_1' + e^{-x} C_2' = 0$$

$$e^x C_1' - e^{-x} C_2' = 1.$$

Subtracting the second equation from the first gives $C_2' = -e^x/2$, and after integration, $C_2(x) = -e^x/2$. Next, adding equations results in $C_1' = e^x/2$, and integrating, $C_1(x) = -e^{-x}/2$. Thus, the particular solution of inhomogeneous equation is $\bar{y}(x) = C_1(x)y_1 + C_2(x)y_2 = -1$ (plug $\bar{y}(x) = -1$ in $y'' - y = 1$ to verify that). Finally, $y(x) = Y(x) + \bar{y}(x) = C_1 e^x + C_2 e^{-x} - 1$ is the general solution of the inhomogeneous equation $y'' - y = 1$. To find the solution of IVP, substitute this $y(x)$ and $y'(x) = Y'(x) + \bar{y}'(x) = C_1 e^x - C_2 e^{-x}$ in the initial conditions $y(0) = 0$, $y'(0) = 1$, which gives $C_1 + C_2 - 1 = 0$, $C_1 - C_2 = 1$. Solution of this algebraic system is $C_1 = 1$, $C_2 = 0$, and the solution of IVP is $y(x) = e^x - 1$ (to verify, plug this in the equation and the initial conditions).

Theorem 6 (Principle of Superposition).

Let $y_1(x)$ be the solution of the linear nonhomogeneous equation

$$y_1'' + p(x)y_1' + q(x)y_1 = f(x) \tag{2.23}$$

and $y_2(x)$ be the solution of the linear nonhomogeneous equation

$$y_2'' + p(x)y_2' + q(x)y_2 = \varphi(x) \tag{2.24}$$

(with different right side). Then the sum,

$$y(x) = y_1(x) + y_2(x) \tag{2.25}$$

is the solution of

$$y'' + p(x)y' + q(x)\,y = f(x) + \varphi(x). \tag{2.26}$$

Proof:
Adding equations (2.23) and (2.24) immediately gives (2.26).
  Notice that the principle of superposition is valid only for linear equations.

2.5 LINEAR NTH-ORDER DIFFERENTIAL EQUATIONS

Everything that has been said above about linear second-order equations extends straightforwardly to linear equations of higher orders. The linear equation of order n has the form

$$y^{(n)} + p_{n-1}(x)y^{(n-1)} + \cdots + p_1(x)y' + p_0(x)y = f(x). \tag{2.27}$$

Its general solution is

$$y(x) = Y(x) + \bar{y}(x), \tag{2.28}$$

where $Y(x)$ is a general solution of the *homogeneous equation*

$$y^{(n)} + p_{n-1}(x)y^{(n-1)} + \cdots + p_1(x)y' + p_0(x)y = 0 \tag{2.29}$$

and $\bar{y}(x)$. is a particular solution of the *inhomogeneous equation* (2.27).
  Function $Y(x)$ is a linear combination of n linearly independent particular solutions (fundamental solutions) of the linear homogeneous equation (2.29)—the maximum number of such solutions is equal to the order of the equation, n:

$$Y(x) = C_1 y_1(x) + C_2 y_2(x) + \cdots + C_n y_n(x). \tag{2.30}$$

Functions $y_i(x)$ are linearly independent on some interval if and only if the Wronskian, which is the $n \times n$ determinant, is not zero on this interval:

$$W(y_1(x),\ldots,y_n(x)) = \begin{vmatrix} y_1(x) & y_2(x)\ldots\ldots\ldots y_n(x) \\ y_1'(x) & y_2'(x)\ldots\ldots\ldots y_n'(x) \\ \ldots\ldots\ldots\ldots\ldots\ldots\ldots\ldots\ldots\ldots\ldots\ldots\ldots\ldots \\ y_1^{(n-1)}(x) & y_2^{(n-1)}(x)\ldots\ldots y_n^{(n-1)}(x) \end{vmatrix} \neq 0. \tag{2.31}$$

Any n linearly independent solutions of an nth-order linear homogeneous equation are called its *fundamental systems of solutions*.

The method of variation of the parameters implies that we search for a particular solution of the inhomogeneous equation in the form of a function $Y(x)$, where the constants C_i are replaced by the functions $C_i(x)$:

$$\bar{y}(x) = C_1(x)y_1(x) + C_2(x)y_2(x) + \cdots + C_n(x)y_n(x). \tag{2.32}$$

The system of equations for the functions $C_i'(x)$ is

$$y_1 C_1' + y_2 C_2' + \cdots + y_n C_n' = 0$$

$$y_1' C_1' + y_2' C_2' + \cdots + y_n' C_n' = 0 \tag{2.33}$$

$$\ldots\ldots\ldots\ldots\ldots\ldots\ldots\ldots\ldots\ldots\ldots\ldots\ldots\ldots\ldots$$

$$y_1^{(n-1)} C_1' + y_2^{(n-1)} C_2' + \cdots + y_n^{(n-1)} C_n' = f(x).$$

The determinant of this system is $W(y_1, y_2 \ldots, y_n) \neq 0$; thus the solution of the system (2.33) exists and is unique. After the functions $C_i'(x)$ have been found from this system, the functions $C_i(x)$ can be determined by integration. Because we are searching for a particular solution, the constants of integrations should not be added when we obtain $C_i(x)$ from $C_i'(x)$.

Equation (2.27) is often conveniently written in a symbolic form

$$L(y) = f(x), \tag{2.34}$$

where

$$L(y) = y^{(n)} + p_{n-1}(x)y^{(n-1)} + \cdots + p_1(x)y' + p_0(x)y \tag{2.35}$$

is called a *linear differential operator*. Its linearity means that

$$L(Cy) = CL(y), \tag{2.36}$$

and for several functions

$$L(C_1 y_1 + C_2 y_2 + \cdots + C_m y_m) = C_1 L(y_1) + C_2 L(y_2) + \cdots + C_m L(y_m). \tag{2.37}$$

Next, we formulate the *principle of superposition*. This is useful when the right side of a linear inhomogeneous equation is the sum of several functions

$$f(x) = f_1(x) + f_2(x) + \cdots + f_k(x).$$

Then the particular solution of a inhomogeneous equation

$$L(y) = \sum_{i=1}^{k} f_i(x) \tag{2.38}$$

is

$$y(x) = \sum_{i=1}^{k} y_i(x), \tag{2.39}$$

where $y_i(x)$ is a particular solution of equation

$$L(y_i) = f_i(x). \tag{2.40}$$

Using the principle of superposition, the solution of a linear differential equation with a complicated right side can be reduced to a solution of several more simple equations.

Let discuss another useful result:

Theorem 7.

If $y = U + iV$ is the complex solution of a linear homogeneous equation (2.29) with the real coefficients $p_i(x)$, then the imaginary and the real parts of this complex solution (i.e., the real functions $U(x)$ and $V(x)$) are the solutions of this equation.

Proof:

The proof is simple making use of a linear operator $L(y)$. From $L(U + iV) = 0$ it follows $L(U) + iL(V) = 0$ and to be equal to zero this complex expression has to have its real and imaginary parts both be zero: $L(U) = L(V) = 0$. Thus U and V are solutions of (2.29).

Finally consider an IVP for the nth order linear equation. If the functions $p_i(x)$ ($i = 0, \ldots, n-1$) and $f(x)$ are continuous on the interval $[a, b]$, then an IVP solution of (2.27) or (2.29) with n *initial conditions* at some point $x_0 \in [a, b]$

$$y(x_0) = y_0, y'(x_0) = y'_0, y''(x_0) = y''_0, \cdots, y^{(n-1)}(x_0) = y_0^{(n-1)}, \tag{2.41}$$

exists and is unique. Here $y_0, y'_0, y''_0, \ldots, y_0^{(n-1)}$ are n real arbitrary numbers—the values of the function y and its first $n-1$ derivatives at x_0. To find the solution of IVP, substitute

$$y(x) = Y(x) + \overline{y}(x) = C_1 y_1(x) + C_2 y_2(x) + \cdots + C_n y_n(x) + \overline{y}(x)$$

(take here $\bar{y}(x) = 0$ for the homogeneous equation (2.29)) into the initial conditions (2.41). The coefficients C_i can be unambiguously determined from the system of algebraic linear equations

$$C_1 y_1(x_0) + C_2 y_2(x_0) + \cdots + C_n y_n(x_0) = y_0 - \bar{y}(x_0)$$

$$C_1 y_1'(x_0) + C_2 y_2'(x_0) + \cdots + C_n y_n'(x_0) = y_0' - \bar{y}'(x_0) \tag{2.42}$$

....

$$C_1 y_1^{(n-1)}(x_0) + C_2 y_2^{(n-1)}(x_0) + \cdots + C_n y_n^{(n-1)}(x_0) = y_0^{(n-1)} - \bar{y}^{(n-1)}(x_0)$$

with nonzero determinant $W(y_1(x_0), y_2(x_0) \ldots, y_n(x_0)) \neq 0$.

Example 2.13

Find the general solution of the linear inhomogeneous equation $y''' - y' = 3e^{2x}$. Then find the solution of IVP with the initial conditions $y(0) = 0$, $y'(0) = 1$, $y''(0) = 2$.

Solution. First, notice that $y_1(x) = e^x$, $y_2(x) = e^{-x}$ and $y_3(x) = 1$ (or any constant) are the particular solutions of the homogeneous equation $y''' - y' = 0$. These solutions are linearly independent, since the Wronskian

$$W\big(y_1(x), y_2(x), y_3(x)\big) = \begin{vmatrix} e^x & e^{-x} & 1 \\ e^x & -e^{-x} & 0 \\ e^x & e^{-x} & 0 \end{vmatrix} = 1 \cdot \begin{vmatrix} e^x & e^{-x} \\ e^x & -e^{-x} \end{vmatrix} = -2 \neq 0.$$

Thus, the general solution of the homogeneous equation is

$$Y(x) = C_1 e^x + C_2 e^{-x} + C_3.$$

Next, using variation of the parameters we look for the particular solution of the inhomogeneous equation in the form $\bar{y}(x) = C_1(x)e^x + C_2(x)e^{-x} + C_3(x)$. The system of algebraic equations (2.33) for C_1', C_2', and C_3' (with the determinant $W(y_1, y_2, y_3) = -2$) is

$$e^x C_1' + e^{-x} C_2' + C_3' = 0,$$

$$e^x C_1' - e^{-x} C_2' = 0,$$

$$e^x C_1' + e^{-x} C_2' = 3e^{2x}.$$

Solving the system (add and subtract the second and the third equations to find C_1' and C_2', then C_3' is found from the first equation), and then integrating C_1', C_2', and C_3', gives $C_1(x) = 3e^x/2$, $C_2(x) = e^{3x}/2$, and $C_3(x) = -3e^{2x}/2$. This gives $\bar{y}(x) = e^{2x}/2$. Finally, the general solution of the inhomogeneous equation is

$$y(x) = Y(x) + \bar{y}(x) = C_1 e^x + C_2 e^{-x} + C_3 + e^{2x}/2.$$

Next, plug this general solution into the initial conditions, which gives $C_1 = C_2 = 0$, $C_3 = -1/2$. Thus the solution of IVP is $y(x) = -1/2 + e^{2x}/2$.

Example 2.14

Find the general solution of equation $y''' - y' = 3e^{2x} + 5x$.

Solution. The right side of this equation is the sum of two functions, $f_1(x) = 3e^{2x}$ and $f_2(x) = 5x$. The particular solution of the inhomogeneous equation $y''' - y' = 3e^{2x}$ was found in the previous example ($\bar{y}_1(x) = e^{2x}/2$), the particular solution of the inhomogeneous equation $y''' - y' = 5x$ is easy to guess: $\bar{y}_2(x) = -5x^2/2$. Using the principle of superposition we conclude that the particular solution of the inhomogeneous equation $y''' - y' = 3e^{2x} + 5x$ is $\bar{y}(x) = \bar{y}_1(x) + \bar{y}_2(x) = e^{2x}/2 - 5x^2/2$. Thus, the general solution of this equation is

$$y(x) = Y(x) + \bar{y}(x) = C_1 e^x + C_2 e^{-x} + C_3 + e^{2x}/2 - 5x^2/2.$$

Equations in Examples 2.13–2.15 have constant coefficients. The right side is constant, exponential, and polynomial in Example 2.13, 2.14, and 2.15, respectively. We will see in Section 2.6 that for these and similar cases, $\bar{y}$ can be found more easily using the method of undermined coefficients.

Problems	Answers

Check if the following functions are linearly independent.

1. e^x, e^{x-1} no

2. $\cos x, \sin x$ yes

3. $1, x, x^2$ yes

4. $4-x, 2x + 3, 6x + 8$ no

5. e^x, e^{2x}, e^{3x} yes

6. x, e^x, xe^x yes

7. $2^x, 3^x, 6^x$ no

8. $\cos x, \sin x, \sin 2x$ yes

2.6 LINEAR SECOND-ORDER EQUATIONS WITH CONSTANT COEFFICIENTS

In this section we discuss equation (2.5), where the coefficients $p(x)$ and $q(x)$ are constants:

$$y'' + py' + qy = f(x). \tag{2.43}$$

Such constant-coefficient linear equation always can be solved analytically.

2.6.1 Homogeneous Equations

First, consider the corresponding homogeneous equation,

$$y'' + py' + qy = 0. \tag{2.44}$$

Let search for the particular solutions of (2.44) in the form $y = e^{kx}$. Substitution of this exponent in (2.44) gives $(k^2 + pk + q)e^{kx} = 0$, and then division by $e^{kx} \neq 0$ gives the *characteristic equation*

$$k^2 + pk + q = 0. \tag{2.45}$$

This quadratic equation has two roots,

$$k_{1,2} = -\frac{p}{2} \pm \sqrt{\frac{p^2}{4} - q}$$

(2.46)

which can be real or complex.

Theorem 8.

1. If the roots of the characteristic equation are real and distinct, then $y_1(x) = e^{k_1 x}$ and $y_2(x) = e^{k_2 x}$ are two linearly independent particular solutions of (2.44), and thus

$$Y(x) = C_1 e^{k_1 x} + C_2 e^{k_2 x}.$$

(2.47)

   is the general solution of (2.44).
2. If the roots of the characteristic equation are real and repeated, $k_1 = k_2 \equiv k$, then $y_1(x) = e^{k_1 x}$ and $y_2(x) = xe^{k_2 x}$ are two linearly independent particular solutions of (2.44), and thus

$$y(x) = C_1 e^{kx} + C_2 xe^{kx}$$

(2.48)

   is the general solution of (2.44).
3. If the roots of the characteristic equation are complex, $k_{1,2} = \alpha \pm i\beta$, then the *complex* form of two linearly independent particular solutions of (2.44) is $y_1(x) = e^{k_1 x}$ and $y_2(x) = e^{k_2 x}$; a pair of *real* linearly independent particular solutions is $y_1(x) = e^{\alpha x} \sin \beta x$ and $y_2(x) = e^{\alpha x} \cos \beta x$. Thus, the general solution of (2.44) can be written in two ways:
   complex

$$y(x) = C_1 e^{(\alpha + i\beta)x} + C_2 e^{(\alpha - i\beta)x},$$

(2.49)

   or real

$$y(x) = e^{\alpha x}(C_1 \cos \beta x + C_2 \sin \beta x)$$

(2.50)

   (the coefficients C_1 and C_2 are real).

Proof:

1. When real $k_1 \neq k_2$, the functions $y_1(x) = e^{k_1 x}$ and $y_2(x) = e^{k_2 x}$ are the particular solutions of (2.44) by construction. They are linearly independent because

$$W(y_1, y_2) = \begin{vmatrix} y_1 & y_2 \\ y_1' & y_2' \end{vmatrix} = \begin{vmatrix} e^{k_1 x} & e^{k_2 x} \\ k_1 e^{k_1 x} & k_2 e^{k_2 x} \end{vmatrix} = e^{(k_1 + k_2)x}(k_2 - k_1) \neq 0$$

for all x. Thus $y(x) = C_1 e^{k_1 x} + C_2 e^{k_2 x}$ is the general solution of (2.44).
2. When $k_1 = k_2 = -p/2$, then $y_1(x) = e^{k_1 x}$ is a particular solution of (2.44). Because the functions $e^{k_1 x}$ and $e^{k_2 x}$ are the same, we need another particular solution that is distinct from y_1. Next we show that $y_2 = xe^{kx}$ is such a particular solution. Plugging it in (2.44) and canceling

$e^{kx} \neq 0$, we obtain $x\underbrace{(k^2 + pk + q)}_{0} + \underbrace{2k + p}_{0} = 0$, meaning that $y_2 = xe^{kx}$ is the particular solution. Linear independence of $y_1 = e^{kx}$ and $y_2 = xe^{kx}$ follows from

$$W = \begin{vmatrix} y_1 & y_2 \\ y_1' & y_2' \end{vmatrix} = \begin{vmatrix} e^{kx} & xe^{kx} \\ ke^{kx} & e^{kx}(1 + xk) \end{vmatrix} = e^{2kx} \neq 0 \text{ for all } x.$$

Thus, we proved that $y(x) = C_1 e^{kx} + C_2 xe^{kx}$ is the general solution of (2.44) when $k = k_1 = k_2$.

3. When $k_{1,2} = \alpha \pm i\beta$, where $\alpha = -\frac{p}{2}$, $\beta = \sqrt{\frac{p^2}{4} - q}$, then $y_1 = e^{(\alpha + i\beta)x}$ and $y_2 = e^{(\alpha - i\beta)x}$ are complex-valued solutions by construction. These two functions are linearly independent, since

$$W = \begin{vmatrix} y_1 & y_2 \\ y_1' & y_2' \end{vmatrix} = \begin{vmatrix} e^{(\alpha + i\beta)x} & e^{(\alpha - i\beta)x} \\ (\alpha + i\beta)e^{(\alpha + i\beta)x} & (\alpha - i\beta)e^{(\alpha - i\beta)x} \end{vmatrix}$$

$$= (\alpha - i\beta)e^{2\alpha x} - (\alpha + i\beta)e^{2\alpha x} = -2i\beta e^{2\alpha x} \neq 0$$

for all x. Thus, we can take

$$y(x) = C_1 y_1(x) + C_2 y_2(x) = C_1 e^{(\alpha + i\beta)x} + C_2 e^{(\alpha - i\beta)x}$$

for the general solution of (2.44).

This solution is complex-valued. When one needs to find a solution of the Cauchy or the boundary value problem with real initial or boundary conditions, certainly real-valued solution is preferred. From Theorem 7 it follows that the real and imaginary parts of $y_1 = e^{(\alpha + i\beta)x}$, that is, $U = e^{\alpha x} \cos\beta x$ and $V = e^{\alpha x} \sin\beta x$ (function $y_2 = e^{(\alpha - i\beta)x}$ has the same real part U, and the imaginary part $-V$), are the solutions of equation (2.44).

The linear independence of these two functions follows from

$$W = \begin{vmatrix} y_1 & y_2 \\ y_1' & y_2' \end{vmatrix} = \begin{vmatrix} e^{\alpha x} \sin\beta x & e^{\alpha x} \cos\beta x \\ \alpha e^{\alpha x} \sin\beta x + \beta e^{\alpha x} \cos\beta x & \alpha e^{\alpha x} \cos\beta x - \beta e^{\alpha x} \sin\beta x \end{vmatrix}$$

$$= \beta e^{2\alpha x} \neq 0$$

for all x. Thus,

$$y(x) = e^{\alpha x}\left(C_1 \cos\beta x + C_2 \sin\beta x\right)$$

is the general solution of (2.44).

Obviously, the combination $C_1 \cos\beta x + C_2 \sin\beta x$ can be presented in more "physical" form:

$$A\cos(\beta x + \varphi) \text{ or } A\sin(\beta x + \varphi),$$

where constants A (amplitude) and φ (phase) play role of two arbitrary constants (instead of C_1 and C_2). Thus, the solution in case of complex roots of characteristic equation, $k_{1,2} = \alpha \pm i\beta$, can be presented in the form

$$y(x) = Ae^{\alpha x} \cos(\beta x + \varphi),$$

or

$$y(x) = Ae^{\alpha x}\sin(\beta x + \varphi).$$

Below we show several examples based on Theorem 8.

Example 2.15

Solve IVP $y'' + y' - 2y = 0$, $y(0) = 1$, $y'(0) = 2$.
  Solution. The characteristic equation $k^2 + k - 2 = 0$ has two real and distinct roots, $k_1 = 1$, $k_2 = -2$. Thus, the general solution is $y(x) = C_1e^x + C_2e^{-2x}$. In order to solve IVP, we first calculate $y'(x) = C_1e^x - 2C_2e^{-2x}$, then substitute $y(x)$ and $y'(x)$ in the initial conditions, which gives:

$$\begin{cases} C_1 + C_2 = 1 \\ C_1 - 2C_2 = 2. \end{cases}$$

The solution of this system is $C_1 = 4/3$ and $C_2 = -1/3$. Thus, the solution of IVP is $y(x) = (4e^x - e^{-2x})/3$.

Example 2.16

Solve IVP $y'' - 4y' + 4y = 0$, $y(1) = 1$, $y'(1) = 1$.
  Solution. The characteristic equation $k^2 - 4k + 4 = 0$ has repeated real root, $k_{1,2} = 2$; thus the general solution is $y(x) = C_1e^{2x} + C_2xe^{2x}$. In order to solve IVP, substitute $y(x)$ and $y'(x) = 2C_1e^{2x} + C_2e^{2x} + 2C_2xe^{2x}$ in the initial conditions, which gives

$$\begin{cases} C_1e^2 + C_2e^2 = 1, \\ 2C_1e^2 + 3C_2e^2 = 1. \end{cases}$$

The solution of this system is $C_1 = 2e^{-2}$ and $C_2 = -e^{-2}$. Thus, the solution of IVP is $y(x) = (2 - x)e^{2(x-1)}$.

Example 2.17

Solve IVP $y'' - 4y' + 13y = 0$, $y(0) = 1$, $y'(0) = 0$.
  Solution. The characteristic equation $k^2 - 4k + 13 = 0$ has complex conjugated roots $k_{1,2} = 2 \pm 3i$. The general solution in real form is $y(x) = e^{2x}(C_1\cos 3x + C_2\sin 3x)$. Substitution of $y(x)$ and $y'(x)$ in the initial conditions gives

$$\begin{cases} C_1 = 1, \\ 2C_1 + 3C_2 = 0, \end{cases}$$

which has the solution $C_1 = 1$, $C_2 = -2/3$.
  Thus, $y(x) = e^{2x}\left(\cos 3x - \frac{2}{3}\sin 3x\right)$ is the solution of IVP.

Problems	Answers
1. $y'' - 5y' - 6y = 0$	$y = C_1e^{6x} + C_2e^{-x}$
2. $y''' - 6y'' + 13y' = 0$	$y = C_1 + e^{3x}(C_2\cos 2x + C_3\sin 2x)$
3. $y^{(4)} - y = 0$	$y = C_1e^{-x} + C_2e^x + C_3\cos x + C_4\sin x$

4. $y^{(4)} + 13y^{(2)} + 36y = 0$ $\qquad$ $y = C_1 \cos 2x + C_2 \sin 2x + C_3 \cos 3x + C_4 \sin 3x.$

5. $y'' - 5y' + 6y = 0$ $\qquad$ $y = C_1 e^{2x} + C_2 e^{3x}$

6. $y''' - 4y'' + 3y' = 0$ $\qquad$ $y = C_1 + C_2 e^x + C_3 e^{3x}$

7. $y''' + 6y'' + 25y' = 0$ $\qquad$ $y = C_1 + e^{-3x}(C_2 \cos 4x + C_3 \sin 4x)$

8. $y''' + 5y'' = 0$ $\qquad$ $y = C_1 + C_2 x + C_3 e^{-5x}$

9. $y''' - 3y'' + 3y' - y = 0$ $\qquad$ $y = e^x(C_1 + C_2 x + C_3 x^2)$

10. $y^{(4)} - 8y^{(2)} - 9y = 0$ $\qquad$ $y = C_1 e^{-3x} + C_2 e^{3x} + C_3 \cos x + C_4 \sin x$

11. $y'' + 4y' + 3y = 0$ $\qquad$ $y = C_1 e^{-x} + C_2 e^{-3x}$

12. $y'' - 2y' = 0$ $\qquad$ $y = C_1 + C_2 e^{2x}$

13. $2y'' - 5y' + 2y = 0$ $\qquad$ $y = C_1 e^{2x} + C_2 e^{x/2}$

14. $y'' - 4y' + 5y = 0$ $\qquad$ $y = e^{2x}(C_1 \cos x + C_2 \sin x)$

15. $y''' - 8y = 0$ $\qquad$ $y = C_1 e^{2x} + e^{-x}(C_2 \cos x\sqrt{3} + C_3 \sin x\sqrt{3})$

16. $y^{(4)} + 4y = 0$ $\qquad$ $y = e^x(C_1 \cos x + C_2 \sin x)$
$\qquad\qquad\qquad\qquad\qquad\quad + e^{-x}(C_3 \cos x + C_4 \sin x)$

17. $y'' - 2y' + y = 0$ $\qquad$ $y = e^x(C_1 + C_2 x)$

18. $4y'' + 4y' + y = 0$ $\qquad$ $y = e^{-x/2}(C_1 + C_2 x)$

19. $y^{(5)} - 6y^{(4)} + 9y''' = 0$ $\qquad$ $y = C_1 + C_2 x + C_3 x^2 + e^{3x}(C_4 + C_5 x)$

20. $y''' - 5y'' + 6y' = 0$ $\qquad$ $y = C_1 + C_2 e^{2x} + C_3 e^{3x}$

21. $y'' + 2y = 0,\ y(0) = 0,\ y'(0) = 1$ $\qquad$ $y = \dfrac{1}{\sqrt{2}} \sin \sqrt{2} x$

22. $y''' - y' = 0,\ y(2) = 1,$ $\qquad$ $y = 1$
$\quad$ $y'(2) = 0,\ y''(2) = 0$

23. $y'' + 4y' + 5y = 0,$ $\qquad$ $y = -3e^{-2x}(\cos x + 2\sin x)$
$\quad$ $y(0) = -3,\ y'(0) = 0$

2.6.2 Inhomogeneous Equations: Method of Undetermined Coefficients

A general solution of inhomogeneous equation (2.43), $y(x)$, is the sum of a general solution of the corresponding homogeneous equation (2.44), which we denote as $Y(x)$, and a particular solution of inhomogeneous equation (2.43), denoted as $\bar{y}(x)$; thus $y(x) = Y(x) + \bar{y}(x)$. A particular solution $\bar{y}(x)$ can be found using the method of variation of the parameters discussed in Section 3.4. This method requires integrations that are often complicated. For special cases of the right-hand side of equation (2.43), which are very common in applications, this equation can be solved using simpler method, called the *method of undetermined coefficients*.

The idea of the method is to choose the particular solution $\bar{y}(x)$ of the inhomogeneous equation (2.43) in the form similar to the "inhomogeneity" $f(x)$, and then determine the coefficients in $\bar{y}(x)$ by substituting $\bar{y}(x)$ in (2.43). Below we consider functions $f(x)$ for which the method is applicable, and in each case give the form of a particular solution $\bar{y}(x)$. We will not prove the method rigorously and simply provide the solution scheme and examples.

2.6.2.1 Function f(x) Is a Polynomial

Let $f(x)$ be a polynomial of order n:

$$f(x) = P_n(x). \tag{2.51}$$

Function $\bar{y}(x)$ has to be taken in the form

$$\bar{y}(x) = x^r Q_n(x), \tag{2.52}$$

where $Q_n(x)$ is a polynomial that has same order as $P_n(x)$ (with the coefficients to be determined), and the value of r equals the number of zero roots of the characteristic equation.

Example 2.18

Solve $y'' - 2y' + y = x$.

Solution. First, solve the homogeneous equation $y'' - 2y' + y = 0$. The characteristic equation $k^2 - 2k + 1 = 0$ has the roots $k_1 = k_2 = 1$. Thus the general solution of the homogeneous equation is $Y(x) = C_1 e^x + C_2 x e^x$.

The right side of the given equation is the first-degree polynomial; thus we search the particular solution as $\bar{y} = x^r(Ax + B)$ with $r = 0$ (there is no roots $k = 0$ of the characteristic equation) and A, B will be determined by substituting this $\bar{y}$ and its derivatives, $\bar{y}' = A$, $\bar{y}'' = 0$ into equation. This gives $-2A + Ax + B = x$. Next, equating the coefficients of the like terms (zero and first power of x) at both sides of this equation gives

$$\begin{cases} x^0: \ -2A + B = 0 \\ x^1: \ A = 1. \end{cases}$$

From this system, $A = 1$, $B = 2$; thus $\bar{y} = x + 2$. Finally,

$$y(x) = Y(x) + \bar{y}(x) = e^x (C_1 + C_2 x) + x + 2$$

is the general solution of the given equation.

Example 2.19

Solve $y'' - y' = 1$.

Solution. The characteristic equation $k^2 - k = k(k - 1) = 0$ has the roots $k_1 = 0$, $k_2 = 1$; thus the general solution of the homogeneous equation is $Y(x) = C_1 + C_2 e^x$. The right side of the equation is zero-degree polynomial, also $r = 1$ (there is one zero root, $k = 0$, of the characteristic equation);

thus we search for the particular solution in the form $\bar{y} = Ax$. Substituting this $\bar{y}$ and its derivatives, $\bar{y}' = A$, $\bar{y}'' = 0$ into equation results in $A = -1$, thus $\bar{y} = -x$. Finally,

$$y(x) = Y(x) + \bar{y}(x) = C_1 + C_2 e^x - x$$

is the general solution of the given equation.

2.6.2.2 Function f(x) Is a Product of a Polynomial and Exponent

Let $f(x)$ be a product of a polynomial of order n and $e^{\gamma x}$:

$$f(x) = e^{\gamma x} P_n(x) \tag{2.53}$$

The previous situation (2.51) is obviously a particular case of (2.53) with $\gamma = 0$.

Take $\bar{y}(x)$ in the form

$$\bar{y}(x) = x^r e^{\gamma x} Q_n(x), \tag{2.54}$$

where $Q_n(x)$ is a polynomial of the same order as $P_n(x)$ (with the coefficients to be determined), and the value of r equals the number of roots of the characteristic equation that are equal to γ.

Example 2.20

Solve $y'' - 4y' + 3y = xe^x$.

Solution. Solve the homogeneous equation first: $y'' - 4y' + 3y = 0$, $k^2 - 4k + 3 = 0$, $k_1 = 3$, $k_2 = 1$, thus $Y(x) = C_1 e^{3x} + C_2 e^x$.

As can be seen, $\gamma = 1$ and since $k_2 = 1 = \gamma$ is a simple root of the characteristic equation, we must take $r = 1$; therefore, $\bar{y} = xe^x(Ax + B)$. Differentiation of $\bar{y}$ gives

$$\bar{y}' = e^x(Ax^2 + Bx) + e^x(2Ax + B) = e^x(Ax^2 + 2Ax + B + Bx),$$

$$\bar{y}'' = e^x(Ax^2 + 4Ax + Bx + 2A + 2B).$$

Next, substitution of $\bar{y}$, $\bar{y}'$ and $\bar{y}''$ in the inhomogeneous equation results in:

$$e^x\left[(Ax^2 + 4Ax + Bx + 2A + 2B) - 4(Ax^2 + 2Ax + B + Bx) + 3(Ax^2 + Bx)\right] = xe^x.$$

Canceling $e^x \neq 0$ and equating coefficients of the same powers of x at both sides gives the system

$$\begin{cases} 2A - 2B = 0 \\ -4A = 1 \end{cases}$$

which has the solution $A = B = -1/4$. Thus the particular solution is $\bar{y} = -xe^x(x + 1)/4$. Finally, the general solution is

$$y(x) = Y(x) + \bar{y}(x) = C_1 e^{3x} + C_2 e^x - x(x + 1)e^x/4.$$

For comparison, we next solve this problem using the method of variation of the parameters. Varying parameters in the general solution $Y(x) = C_1 e^{3x} + C_2 e^x$ gives $\bar{y}(x) = C_1(x)e^{3x} + C_2(x)e^x$ and then the system of algebraic equations to find C_1' and C_2' is (see Section 2.4):

$$\begin{cases} e^{3x}C_1' + e^x C_2' = 0 \\ 3e^{3x}C_1' + e^x C_2' = xe^x. \end{cases}$$

Solving the system gives $C_1' = xe^{-2x}/2$, $C_2' = -x/2$. Then, $C_2(x) = -x^2/4$, and integration by parts for C_1 gives $C_1(x) = -xe^{-2x}/4 - e^{-2x}/8$. Finally, substitution of $C_1(x)$ and $C_2(x)$ in the equation for $\bar{y}$ gives $\bar{y} = -x(x+1)e^x/4$; then $y(x) = Y(x) + \bar{y}(x) = C_1 e^{3x} + C_2 e^x - x(x+1)e^x/4$. This is the same result that was obtained using the method of undetermined coefficients.

2.6.2.3 Function f(x) Contains Sine and (or) Cosine Functions

Let $f(x)$ be a form

$$f(x) = a\cos\delta x + b\sin\delta x. \tag{2.55}$$

Take $\bar{y}(x)$ as

$$\bar{y} = x^r(A\cos\delta x + B\sin\delta x), \tag{2.56}$$

where A and B are the coefficients to be determined, and the value of r equals the number of roots of the characteristic equation equal to $i\delta$.

Note that in (2.56) for $\bar{y}(x)$ we have to keep both terms even if either a or b is zero in $f(x)$.

Example 2.21

Solve $y'' - 3y' + 2y = \sin x$.
  Solution. The characteristic equation $k^2 - 3k + 2 = 0$ has the roots $k_1 = 1$, $k_2 = 2$. Thus $Y(x) = C_1 e^x + C_2 e^{2x}$. Next, we find the particular solution of inhomogeneous equation, $\bar{y}$. Since $\delta = 1$ and there is no root equal to $i\delta = i$, $r = 0$ and we search for $\bar{y}$ in the form $\bar{y} = A\cos x + B\sin x$. Substituting $\bar{y}$, $\bar{y}'$, $\bar{y}''$ in the given equation and equating the coefficients of the like terms at the both sides gives $A = 0.3$, $B = 0.1$; thus $y(x) = Y(x) + \bar{y} = C_1 e^x + C_2 e^{2x} + 0.3\cos x + 0.1\sin x$.

Example 2.22

Solve $y'' + y = \sin x$.
  Solution. The characteristic equation $k^2 + 1 = 0$ has the roots $k_1 = i$, $k_2 = -i$. Thus $Y = C_1 \cos x + C_2 \sin x$. Next, we find the particular solution, $\bar{y}$, of the inhomogeneous equation. Since $\delta = 1$ and $k_1 = i = i\delta$, then $r = 1$; thus $\bar{y} = x(A\cos x + B\sin x)$. Substituting $\bar{y}$, $\bar{y}'$ and $\bar{y}''$:

$$\bar{y}' = A\cos x + B\sin x + x(-A\sin x + B\cos x),$$
$$\bar{y}'' = -2A\sin x + 2B\cos x + x(-A\cos x - B\sin x),$$

in the equation and equating the coefficients of the like terms at the both sides gives

$$\begin{cases} \sin x: \ -2A = 1 \\ \cos x: \ 2B = 0. \end{cases}$$

Thus, $\bar{y} = -\frac{x}{2}\cos x$ and $y(x) = C_1 \cos x + C_2 \sin x - \frac{x}{2}\cos x$ is the general solution of the given equation.

Notice that often function $f(x)$ has a form

$$f(x) = c\sin(\delta x + \varphi), \text{ or } f(x) = c\cos(\delta x + \varphi)$$

—the second can be obtained from the first one by adding $\pi/2$ to φ. Both these forms clearly are equivalent to (2.55) with constants c and φ instead of a and b. In this case a particular solution of inhomogeneous equation $\bar{y}(x)$ can be found as

$$\bar{y} = x^r[A\cos(\delta x + \varphi) + B\sin(\delta x + \varphi)],$$

where A and B are the coefficients to be determined, and the value of r equals the number of roots of the characteristic equation equal to $i\delta$. Another form of $\bar{y}(x)$ can be

$$\bar{y} = Ax^r\cos(\delta x + \gamma),$$

where the coefficients to be determined are A and γ.

Example 2.23

To illustrate, consider a physical example of forced oscillation. The linear differential equation

$$L(y) \equiv y'' + 2\lambda y' + \omega_0^2 y = f_0 \sin\omega t$$

describes oscillations generated by a source $f(t) = f_0 \sin\omega t$ with frequency ω (physical examples might include mechanical systems such as a body attached to a string, electrical circuits driven by an oscillating voltage, etc.). Here ω_0 and λ are constants: the first one is the so-called *natural frequency* (of a system), and the second is the *damping coefficient*.
The solution to this problem can be expressed as the sum of functions

$$y(t) = \bar{y}(t) + C_1 y_1(t) + C_2 y_2(t),$$

where $\bar{y}(t)$ is a particular solution to the inhomogeneous problem and $y_1(t)$, $y_2(t)$ are the fundamental solutions (two linearly independent solutions) for the respective homogeneous equation, $L(y) \equiv 0$. It is easy to check the following:

a) If $\lambda < \omega_0$ then $y_1(t) = e^{-\lambda t}\cos\tilde{\omega}t$, $y_2(t) = e^{-\lambda t}\sin\tilde{\omega}t$.
b) If $\lambda = \omega_0$ then $y_1(t) = e^{-\lambda t}$, $y_2(t) = te^{-\lambda t}$.
c) If $\lambda > \omega_0$ then $y_1(t) = e^{k_1 t}$, $y_2(t) = e^{k_2 t}$.

Here

$$\tilde{\omega} = \sqrt{\omega_0^2 - \lambda^2}, \ k_1 = -\lambda - \sqrt{\lambda^2 - \omega_0^2}, \ k_2 = -\lambda + \sqrt{\lambda^2 - \omega_0^2} \ \text{(clearly } k_{1,2} < 0\text{)}.$$

Lets seek the particular solution of the inhomogeneous equation as

$$\bar{y}(t) = a\cos\omega t + b\sin\omega t.$$

Comparing the coefficients of $\cos\omega t$ and $\sin\omega t$ in both sides of equation $\bar{y}'' + 2\lambda\bar{y}' + \omega_0^2\bar{y} = f(t)$, find that

$$\begin{cases} -b\omega^2 - 2\lambda a\omega + b\omega_0^2 = f_0 \\ -a\omega^2 + 2\lambda b\omega + a\omega_0^2 = 0. \end{cases}$$

From there

$$a = -\frac{2f_0\lambda\omega}{\left(\omega_0^2 - \omega^2\right)^2 + (2\lambda\omega)^2}, \quad b = \frac{f_0\left(\omega_0^2 - \omega^2\right)}{\left(\omega_0^2 - \omega^2\right)^2 + (2\lambda\omega)^2}.$$

The particular solution $\bar{y}(t)$ can also be written as

$$\bar{y} = A\cos(\omega t + \delta).$$

With $\cos(\omega t + \delta) = \cos\omega t \cos\delta - \sin\omega t \sin\delta$ we see that $a = A\cos\delta$, $b = -A\sin\delta$; thus the amplitude A and the phase shift δ are

$$A = \sqrt{a^2 + b^2} = \frac{f_0}{\left(\omega_0^2 - \omega^2\right)^2 + (2\lambda\omega)^2}, \quad \delta = \arctan\left(\frac{2\lambda\omega}{\omega^2 - \omega_0^2}\right).$$

The initial conditions $y(0)$ and $y'(0)$ determine the constants C_1 and C_2. Obviously for big t a general solution $C_1 y_1(t) + C_2 y_2(t)$ vanishes because of dissipation and only a particular solution $\bar{y}(t)$ describing the forced oscillations survives. For ω close to ω_0 we have resonance with the maximum amplitude $A = f_0/(2\lambda\omega)^2$.

2.6.2.4 Function f(x) Is a Product of Polynomial, Exponent, Sine and (or) Cosine Functions

Let $f(x)$ be a form

$$f(x) = e^{\gamma x}(P_m(x)\cos\delta x + Q_n(x)\sin\delta x), \qquad (2.57)$$

where $P_m(x)$ and $Q_n(x)$ are polynomials of orders m and n, respectively.

Take $\bar{y}(x)$ in the form

$$\bar{y} = e^{\gamma x}(M_p(x)\cos\delta x + N_p(x)\sin\delta x)x^r, \qquad (2.58)$$

where M_p and N_p are polynomials of degree p, with p being the largest of m and n, and the value of r equals the number of roots of the characteristic equation that are equal to $\gamma + i\delta$. Note that in (2.58) for $\bar{y}(x)$ we have to keep both terms even if either $P_m(x)$ or $Q_n(x)$ is zero in $f(x)$.

Obviously (2.57) includes the cases (2.51), (2.53), and (2.56).

Example 2.24

Solve $y'' - 2y' + 5y = e^x(\cos 2x - 3\sin 2x)$.

Solution. The characteristic equation $k^2 - 2k + 5 = 0$ has the roots $k_1 = 1 + 2i$, $k_2 = 1 - 2i$; thus the general solution of the homogeneous equation $y'' - 2y' + 5y = 0$ is $Y(x) = e^x(C_1\cos 2x + C_2\sin 2x)$.

Next, we find the particular solution $\bar{y}(x)$. Since $\gamma + i\delta = 1 + i2 = k_1$, then $r = 1$ and $\bar{y} = xe^x(A\cos 2x + B\sin 2x)$. Substituting $\bar{y}$, $\bar{y}'$, $\bar{y}''$ in the equation and equating the coefficients of the like terms, $xe^x\cos 2x$ and $xe^x\sin 2x$, at the both sides gives $A = 3/4, B = 1/4$. Thus $\bar{y} = xe^x\left(\frac{3}{4}\cos 2x + \frac{1}{4}\sin 2x\right)$, and the general solution of the equation is

$$y(x) = e^x(C_1\cos 2x + C_2\sin 2x) + xe^x\left(\frac{3}{4}\cos 2x + \frac{1}{4}\sin 2x\right).$$

The principle of superposition helps to handle the situations where $f(x)$ is a sum of two or more functions considered above.

Example 2.25

Solve IVP $y'' + y = \sin x + x$. $y(0) = 1$, $y'(0) = 0$.

Solution. Here $f(x) = f_1(x) + f_2(x) = \sin x + x$. The general solution of the homogeneous equation is $Y(x) = C_1 \cos x + C_2 \sin x$. The particular solution of equation $y'' + y = \sin x$ was obtained in Example 2.21: $\bar{y}_1 = -\frac{x}{2} \cos x$. It remains to find the particular solution of $y'' + y = x$. According to (2.52), we take the particular solution in the form $\bar{y}_2 = Ax + B$. Substitution of $\bar{y}_2$ and $\bar{y}_2'' = 0$ in $y'' + y = x$ gives $A = 1$ and $B = 0$. Thus the general solution of the equation $y'' + y = x + \sin x$ is the sum, $y(x) = Y(x) + \bar{y}_1(x) + \bar{y}_2(x) = C_1 \cos x + C_2 \sin x - \frac{x}{2} \cos x + x$.

The initial conditions give $C_1 = 1$, $C_2 = -1/2$; thus the solution of IVP is $y(x) = \cos x - \frac{1}{2} \sin x - \frac{x}{2} \cos x + x$.

2.7 LINEAR NTH-ORDER EQUATIONS WITH CONSTANT COEFFICIENTS

Similar to the case of linear equations with general coefficients (Sections 2.4 and 2.5), everything said above about linear second-order equations with constant coefficients extends in the obvious manner to equations of higher orders. The linear equation of order n with constant coefficients has the form

$$y^{(n)} + a_{n-1} y^{(n-1)} + \cdots + a_1 y' + a_0 y = f(x). \tag{2.59}$$

The *homogeneous* equation is

$$y^{(n)} + a_{n-1} y^{(n-1)} + \cdots + a_1 y' + a_0 y = 0. \tag{2.60}$$

Solution of (2.60) is taken in the form $y = e^{kx}$, and after substitution in (2.60) and cancellation of $e^{kx} \neq 0$ we obtain the *characteristic equation*

$$k^n + a_{n-1} k^{n-1} + \cdots + a_1 k + a_0 = 0. \tag{2.61}$$

This algebraic equation of nth order has n roots.

1. For each *simple* real root k of (2.61), the corresponding particular solution is

$$y = e^{kx}. \tag{2.62}$$

2. For real root k of *multiplicity* $r \geq 2$, there are r linearly independent particular solutions:

$$y_1 = e^{kx}, \; y_2 = xe^{kx}, \; y_3 = x^2 e^{kx}, \; \ldots, \; y_r = x^{r-1} e^{kx}. \tag{2.63}$$

3. For a *simple pair* of complex-conjugate solutions $k_{1,2} = \alpha \pm i\beta$, there are two linearly independent, real particular solutions:

$$y_1(x) = e^{\alpha x} \cos \beta x,$$

$$y_2(x) = e^{\alpha x} \sin \beta x. \tag{2.64}$$

4. For a *pair of complex-conjugate* solutions $k_{1,2} = \alpha \pm i\beta$ of *multiplicity* $r \geq 2$, there are $2r$ linearly independent, real particular solutions:

$$y_1(x) = e^{\alpha x}\cos\beta x, \quad y_2(x) = xe^{\alpha x}\cos\beta x, \dots, \quad y_r(x) = x^{r-1}e^{\alpha x}\cos\beta x,$$

$$y_{r+1}(x) = e^{\alpha x}\sin\beta x, \quad y_{r+2}(x) = xe^{\alpha x}\sin\beta x, \dots, \quad y_{2r}(x) = x^{r-1}e^{\alpha x}\sin\beta x. \tag{2.65}$$

A general solution of homogeneous equation (2.60) is a linear combination of linearly independent particular (fundamental) solutions $y_i(x)$:

$$Y(x) = \sum_{i=1}^{n} C_i y_1(x). \tag{2.66}$$

Example 2.26

Solve IVP $y''' + 25y' = 0$, $y(\pi) = 0$, $y'(\pi) = 1$, $y''(\pi) = 0$.

Solution. The characteristic equation $k^3 + 25k = 0$ has the roots $k_1 = 0$, $k_{2,3} = \pm 5i$. These roots are simple (not repeating); thus the general solution is $Y(x) = C_1 + C_2\cos 5x + C_3\sin 5x$. Substituting it and $Y'(x)$, $Y''(x)$ in initial conditions gives $C_1 = C_2 = 0$, $C_3 = -1/5$; thus the solution of the IVP problem is $y(x) = -\frac{1}{5}\sin 5x$.

Example 2.27

Solve IVP $y^{(4)} = 8y'' - 16y$, $y(0) = 1$, $y'(0) = 0$, $y''(0) = 2$, $y'''(0) = 1$.

Solution. The characteristic equation corresponding to this homogeneous equation is $k^4 - 8k^2 +$, $16 = 0$ or $(k^2 - 4)^2 = (k-2)^2(k+2)^2 = 0$. Its solutions, $k_{1,2} = 2$, $k_{3,4} = -2$, are real roots, each having multiplicity two; thus the general solution of the equation is $Y(x) = C_1 e^{2x} + C_2 xe^{2x} + C_3 e^{-2x} + C_4 xe^{-2x}$. Calculating the derivatives and substituting initial conditions gives the system

$$C_1 + C_3 = 1,$$

$$2C_1 + C_2 - 2C_3 + C_4 = 0,$$

$$4C_1 + 4C_2 + 4C_3 - 4C_4 = 2,$$

$$8C_1 + 12C_2 - 8C_3 + 12C_4 = 1.$$

The solution of the system is $C_1 = 17/12$, $C_2 = -49/12$, $C_3 = -5/12$, $C_4 = 5/12$. Thus, $y(x) = \frac{17}{12}e^{2x} - \frac{49}{12}xe^{2x} - \frac{5}{12}e^{-2x} + \frac{5}{12}xe^{-2x}$ is the solution of IVP.

Now consider *inhomogeneous* equations. The method of variation of constants discussed in Section 2.5 gives a particular solution of inhomogeneous equation (2.59). This method is universal, but often it needs rather lengthy integrations to solve the system of equations for functions $C_i'(x)$ and then to integrate them to find functions $C_i(x)$.

In some cases the method of undetermined coefficients allows to obtain a particular solution of inhomogeneous equations substantially easier. The method given by equations (2.51)–(2.58) is also applicable to higher-order equations.

Example 2.28

Solve $y^{(4)} - 16y = \cos 3x$.

Solution. The characteristic equation is $k^4 - 16 = 0$, or $(k^2 - 4)(k^2 + 4) = 0$, and its roots are $k_{1,2} = \pm 2, k_{2,3} = \pm 2i$ Thus, the solution of the homogeneous equation is $Y = C_1 e^{2x} + C_2 e^{-2x} + C_3 \cos 2x + C_4 \sin 2x$. Since there is no root $k = 3i$, $r = 0$ and we take the particular solution in the form $\bar{y} = A\cos 3x + B\sin 3x$. Substituting $\bar{y}$ and $\bar{y}^{(4)} = 81A\cos 3x + 81B\sin 3x$ in the given equation and equating the coefficients of sine and cosine terms at both sides gives $A = 1/65$, $B = 0$. Thus, the general solution of the nonhomogeneous equation is

$$y = C_1 e^{2x} + C_2 e^{-2x} + C_3 \cos 2x + C_4 \sin 2x + \frac{1}{65} \cos 3x.$$

Finally, consider the **Euler equation**:

$$a_0 x^n y^{(n)} + a_1 x^{n-1} y^{(n-1)} + \cdots + a_{n-1} xy' + a_n y = 0. \tag{2.67}$$

It solution can be searched as

$$y = x^k.$$

Substituting in the equation (2.67) and canceling x^k, we obtain the equation for k:

$$a_0 k(k-1)\cdots(k-n+1) + a_1 k(k-1)\cdots(k-n+2) + \cdots + a_n = 0. \tag{2.68}$$

It can be checked that (for $x > 0$) a real root k_i multiplicity α_i of this equation gives the following particular solutions of equation (2.67):

$$x^{k_i}, x^{k_i} \ln x, x^{k_i} \ln^2 x, \ldots, x^{k_i} \ln^{\alpha_i - 1} x; \tag{2.69}$$

complex roots $p \pm qi$ multiplicity α give the particular solutions

$$x^p \cos(q\ln x), \; x^p \ln x \cos(q\ln x), \ldots, \; x^p \ln^{\alpha-1} x \cos(q\ln x),$$

$$x^p \sin(q\ln x), \; x^p \ln x \sin(q\ln x), \ldots, \; x^p \ln^{\alpha-1} x \sin(q\ln x). \tag{2.70}$$

Example 2.29

Solve $x^2 y'' + \frac{5}{2} xy' - y = 0$.

Solution. Search solution in the form $y = x^k$. It gives equation $k(k-1) + \frac{5}{2}k - 1 = 0$, which has two roots: $k_1 = 1/2$, $k_2 = -2$. Thus, the general solution (for $x > 0$) is

$$y = C_1 x^{\frac{1}{2}} + C_2 x^{-2}.$$

Example 2.30

Solve $x^3 y''' - x^2 y'' + 2xy' - 2y = 0$.

Solution. Equation for k is $k(k-1)(k-2) - k(k-1) + 2k - 2 = 0$, or $(k-1)(k^2 - 3k + 2) = 0$. The roots are $k_1 = k_2 = 1$, $k_3 = 2$; thus a general solution (for $x > 0$) is

$$y = (C_1 + C_2 \ln x)x + C_3 x^2.$$

Example 2.31

Solve $x^2 y'' + xy' + y = 0$.

   Solution. Equation for k is $k(k-1) + k + 1 = 0$, its roots are $k_{1,2} = \pm i$, and the general solution for $x > 0$ is

$$y = C_1 \cos \ln x + C_2 \sin \ln x.$$

Problems	Answers
1. $y'' + 6y' + 5y = 25x^2 - 2$	$y = C_1 e^{-5x} + C_2 e^{-x} + 5x^2 - 12x + 12$
2. $y^{(4)} + 3y'' = 9x^2$	$y = C_1 + C_2 x + C_3 \cos\sqrt{3}x + C_4 \sin\sqrt{3}x$ $+ x^4/4 - x^2$
3. $y'' + 6y = 5e^x$	$y = C_1 \cos\sqrt{6}x + C_2 \sin\sqrt{6}x + \dfrac{5}{7}e^x$
4. $y'' + 6y' + 9y = 10\sin x$	$y = e^{-3x}(C_1 + C_2 x) - 0.6\cos x + 0.8\sin x$
5. $\dfrac{d^2 x}{dt^2} - 2x = te^{-t}$	$x = C_1 e^{-t\sqrt{2}} + C_2 e^{t\sqrt{2}} + e^{-t}(2-t)$
6. $y'' - 2y' = 4(x+1)$	$y = C_1 + C_2 e^{2x} - x^2 - 3x$
7. $y'' - 2y' - 3y = e^{4x}$	$y = C_1 e^{-x} + C_2 e^{3x} + (1/5)e^{4x}$
8. $y'' + y = 4xe^x$	$y = C_1 \cos x + C_2 \sin x + 2e^x(x-1)$
9. $y'' - y = 2e^x - x^2$	$y = C_1 e^x + C_2 e^{-x} + xe^x + x^2 + 2$
10. $y'' + y' - 2y = 3xe^x$	$y = C_1 e^x + C_2 e^{-2x} + \left(\dfrac{x^2}{2} - \dfrac{x}{3}\right)e^x$
11. $y'' - 3y' + 2y = \sin x$	$y = C_1 e^x + C_2 e^{2x} + 0.1\sin x + 0.3\cos x$
12. $y'' + y = 4\sin x$	$y = C_1 \cos x + C_2 \sin x - 2x\cos x$
13. $y'' - 3y' + 2y = x\cos x$	$y = C_1 e^x + C_2 e^{2x} + (0.1x - 0.12)$ $\cos x - (0.3x + 0.34)\sin x.$
14. $y'' + 3y' - 4y = e^{-4x} + xe^{-x}$	$y = C_1 e^x + C_2 e^{-4x} - \dfrac{x}{5}e^{-4x} - \dfrac{1}{6}\left(x + \dfrac{1}{6}\right)e^{-x}$
15. $y'' - 4y' + 8y = e^{2x} + \sin 2x$	$y = e^{2x}(C_1 \cos 2x + C_2 \sin 2x) + 0.25e^{2x}$ $+ 0.1\cos 2x + 0.05\sin 2x$
16. $y'' + y = x\sin x$	$y = \left(C_1 - \dfrac{x^2}{4}\right)\cos x + \left(C_2 + \dfrac{x}{4}\right)\sin x$
17. $y'' + 4y' + 4y = xe^{2x}$	$y(x) = (C_1 + C_2 x)e^{-2x} + \dfrac{(2x-1)}{32}e^{2x}$

18. $y'' - 5y' = 3x^2 + \sin 5x$ $\quad$ $y = C_1 + C_2 e^{5x} - 0.2x^3 - 0.12x^2$
$$- 0.048x + 0.02(\cos 5x - \sin 5x)$$

19. $y'' + y = 4e^x$, $y(0) = 0$, $y'(0) = -3$ $\quad$ $y = -2\cos x - 5\sin x + 2e^x$

20. $y'' - 2y' = 2e^x$, $y(1) = -1$, $y'(1) = 0$ $\quad$ $y = e^{2x-1} - 2e^x + e - 1$

21. $y'' + 2y' + 2y = xe^{-x}$, $y(0) = y'(0) = 0$ $\quad$ $y = e^{-x}(x - \sin x)$

22. $y'' - y = 2\sin x - 4\cos x$ $\quad$ $y = C_1 e^x + C_2 e^{-x} - \sin x + 2\cos x$

23. $y'' + y = e^x + \cos x$ $\quad$ $y = C_1 \cos x + C_2 \sin x + \dfrac{1}{2}(e^x + x\sin x)$

24. $y'' + y = \cos x + \cos 2x$ $\quad$ $y = \dfrac{1}{2} x \sin x - \dfrac{1}{3}\cos 2x + C_1 \cos x + C_2 \sin x$

25. $y^{(4)} + 2y'' + y = \cos x$ $\quad$ $y = -\dfrac{x^2}{8}\cos x + (C_1 + C_2 x)\cos x$
$$+ (C_3 + C_4 x)\sin x$$

26. $y'' - 4y = \sin 2x$, $y(0) = 0$, $y'(0) = 0$ $\quad$ $y = \dfrac{1}{16}(e^{2x} - e^{-2x} - 2\sin 2x)$

27. $y'' - y = x$, $y(0) = 1$, $y'(0) = -1$ $\quad$ $y = -x + \cosh x$

28. $y'' + 4y' + 4y = 3e^{-2x}$, $\quad$ $y = \dfrac{3}{2} x^2 e^{-2x}$
$y(0) = 0$, $y'(0) = 0$

29. $x^2 y'' - xy' + y = 0$ $\quad$ $y = (C_1 + C_2 \ln|x|)x$

30. $x^2 y'' - 4xy' + 6y = 0$ $\quad$ $y = C_1 x^2 + C_2 x^3$

31. $x^2 y'' - xy' - 3y = 0$ $\quad$ $y = C_1 x^{-1} + C_2 x^3$

32. $x^3 y''' + xy' - y = 0$ $\quad$ $y = x\left(C_1 + C_2 \ln|x| + C_3 \ln^2|x|\right)$

2.8 SOLUTION OF IVP FOR SECOND-ORDER EQUATIONS WITH ACCOMPANYING SOFTWARE

In this appendix we describe solutions of Examples 2.9 and 2.25 with the program **ODE 2nd order**.

Figure 2.3 shows the starting interface of the program **ODE 2nd order**.

To do Example 2.9 start with "*Ordinary Differential Equations of the 2nd Order*" and then click on "*Data*" and choose "*New Problem.*" Then, as shown in Figure 2.4, you should enter the equation, initial conditions, and interval where you want to obtain the solution of IVP. Also, you can enter the analytical solution of IVP to compare with the numerical. The solution can be presented in graphical or table form (Figures 2.5 and 2.6).

The program solves it numerically and compares with the reader's analytical solution $y = 1/(1-2x)$. In the interface this solution is denoted as y^* and should be input in the form shown in the interface.

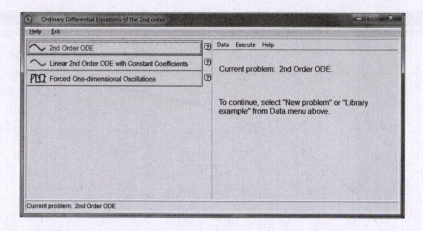

FIGURE 2.3   Starting interface of the program **ODE 2nd order**.

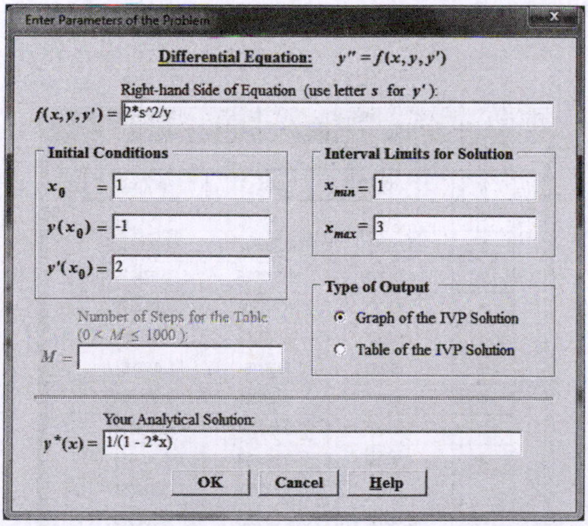

FIGURE 2.4   Equation $yy'' - 2y'^2 = 0$ with initial conditions $y(1) = -1$, $y'(1) = 2$.

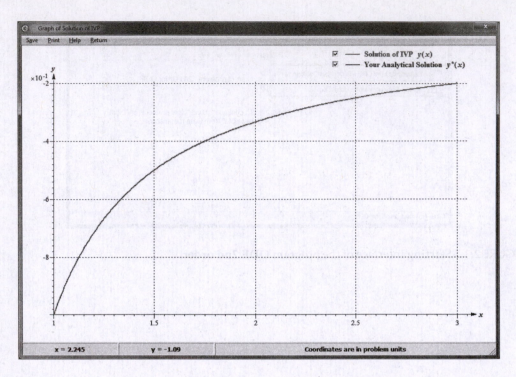

FIGURE 2.5 Graph of the solution of equation $yy'' - 2y'^2 = 0$ with initial conditions $y(1) = -1$, $y'(1) = 2$.

k	x_k	$y(x_k)$	$y^*(x_k)$
0	1	-1	-1
1	1.00667	-0.986842	-0.986842
2	1.01333	-0.974026	-0.974026
3	1.02	-0.961538	-0.961538
4	1.02667	-0.949367	-0.949367
5	1.03333	-0.9375	-0.9375
6	1.04	-0.925926	-0.925926
7	1.04667	-0.914634	-0.914634
8	1.05333	-0.903614	-0.903614
9	1.06	-0.892857	-0.892857
10	1.06667	-0.882353	-0.882353
11	1.07333	-0.872093	-0.872093
12	1.08	-0.862069	-0.862069
13	1.08667	-0.852273	-0.852273
14	1.09333	-0.842697	-0.842697
15	1.1	-0.833333	-0.833333

FIGURE 2.6 Solution of equation $yy'' - 2y'^2 = 0$ with initial conditions $y(1) = -1$, $y'(1) = 2$ presented in the table form.

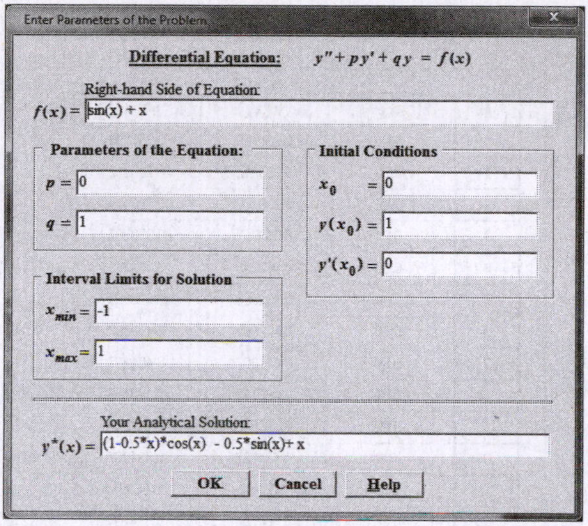

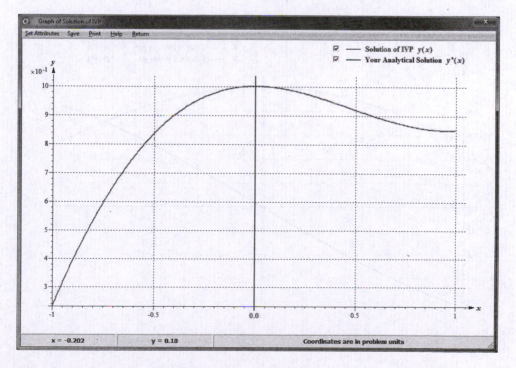

FIGURE 2.7 Equation $y'' + y = \sin x + x$ with initial conditions $y(0) = 1$, $y'(0) = 0$.

Now do Example 2.25. Start with *"Linear 2nd Order DE with constant coefficients"* and then click on *"Data"* and choose *"New Problem."* Figure 2.7 shows the data related to this example. In Figure 2.8 the solution is presented in graphical form; in Figure 2.9, in table form.

The program solves it numerically and compares with the reader's analytical solution (denoted as $y^*(x)$).

FIGURE 2.8 Example 2.25. IVP solution: analytical and numerical.

Table of IVP Solution

k	x_k	$y(x_k)$	$y^*(x_k)$
0	-1	0.231189	0.231189
1	-0.98	0.265212	0.265212
2	-0.96	0.298405	0.298405
3	-0.94	0.330767	0.330767
4	-0.92	0.362298	0.362298
5	-0.9	0.392998	0.392998
6	-0.88	0.422867	0.422867
7	-0.86	0.451907	0.451907
8	-0.84	0.480119	0.480119
9	-0.82	0.507505	0.507505
10	-0.8	0.534067	0.534067
11	-0.78	0.55981	0.55981
12	-0.76	0.584734	0.584734
13	-0.74	0.608846	0.608846
14	-0.72	0.632148	0.632148
15	-0.7	0.654646	0.654646

Parameters...

Close Help Save

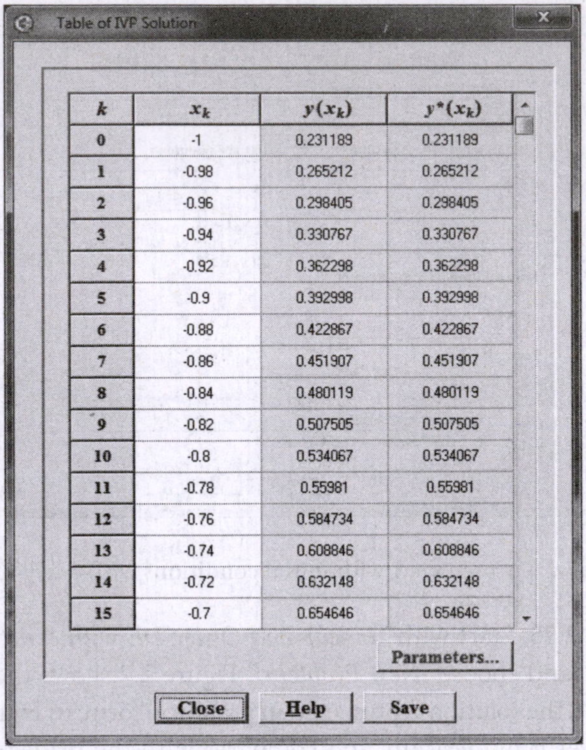

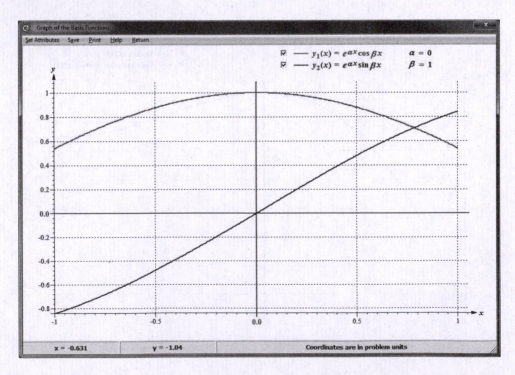

FIGURE 2.9 Example 2.25. Solution of IVP presented in the table form.

FIGURE 2.10 Example 2.25. Graphs of the fundamental solutions of equation $y'' + y = \sin x + x$.

CHAPTER 3

Systems of Differential Equations

3.1 INTRODUCTION

A system of ordinary differential equations consists of several equations containing derivatives of unknown functions of one variable. The general form of such system of n equations is

$$F_j(t, y_1, y_1', \ldots, y_1^{(n_1)}, y_2, y_2', \ldots, y_2^{(n_2)}, \ldots, y_{n_k}, y_{n_k}', \ldots, y_{n_k}^{(n_k)}) = 0, \quad (j = \overline{1,n}), \tag{3.1}$$

where $y_i(t)$ are unknown functions, t is independent variable, and F_j are given functions. A *solution* of a system is composed of all the functions $y_i(t)$ such that when they are substituted for the corresponding unknown functions in the system, each equation becomes the identity $(0 = 0)$.

As the example let consider Newton's equation describing the motion of a mass m under the influence of a force $\vec{f}(t, \vec{r}, \vec{r}')$:

$$m\frac{d^2\vec{r}}{dt^2} = \vec{f}(t, \vec{r}, \vec{r}'), \tag{3.2}$$

where $\vec{r}(t) = (x(t), y(t), z(t))$ is a vector pointing from the origin of the Cartesian coordinate system to the current position of the mass. Thus, equation (3.2) reduces to three second-order differential equations

$$\begin{cases} m\dfrac{d^2x}{dt^2} = f_x(t, x, y, z, x', y', z'), \\[2mm] m\dfrac{d^2y}{dt^2} = f_y(t, x, y, z, x', y', z'), \\[2mm] m\dfrac{d^2z}{dt^2} = f_z(t, x, y, z, x', y', z'). \end{cases} \tag{3.3}$$

To find the unknown functions $x(t), y(t), z(t)$ one has to specify (at $t = t_0$) the initial position of the mass, $(x(t_0), y(t_0), z(t_0))$, and the three projections of the velocity on the coordinate axes, $(x'(t_0), y'(t_0), z'(t_0))$. This is the total of six initial conditions, that is, two conditions (position plus velocity) for each of the three second-order equations:

$$\begin{cases} x(t_0) = x_0, \\ y(t_0) = y_0, \\ z(t_0) = z_0, \\ x'(t_0) = x_0', \\ y'(t_0) = y_0', \\ z'(t_0) = z_0'. \end{cases} \tag{3.4}$$

Equations (3.3) together with the initial conditions (3.4) form the Cauchy problem, or the IVP. If the force $\vec{f}(t, \vec{r}, \vec{r}')$ is sufficiently smooth, this IVP has a unique solution that describes a specific trajectory of the mass.

It is important that this problem can be reduced to a system of six *first-order* equations by introducing new functions. This is done as follows. Let rename the functions x, y, z as x_1, x_2, x_3 and introduce three new functions x_4, x_5, x_6 defined as

$$\begin{cases} x_4 = x', \\ x_5 = y', \\ x_6 = z'. \end{cases}$$

System (3.3) now is rewritten as

$$\begin{cases} mx_4' = f_x, \\ mx_5' = f_y, \\ mx_6' = f_z, \\ x_1' = x_4, \\ x_2' = x_5, \\ x_3' = x_6. \end{cases}$$

Thus we arrived to a system of six first-order differential equations for the functions $x_1, x_2, x_3, x_4, x_5, x_6$. Initial conditions for this system have the form

$$\begin{cases} x_1(t_0) = x_0, \\ x_2(t_0) = y_0, \\ x_3(t_0) = z_0, \\ x_4(t_0) = x_0', \\ x_5(t_0) = y_0', \\ x_6(t_0) = z_0'. \end{cases}$$

The first three conditions give the initial coordinates of the mass; the last three, the projections of the initial velocity on the coordinate axes.

Example 1.1: Rewrite the system of two second-order differential equations

$$\begin{cases} x'' = a_1x + b_1y + c_1y' + f_1(t), \\ y'' = a_2x + b_2y + c_2x' + f_2(t) \end{cases}$$

as a system of four first-order equations.

Solution. Let introduce the notations:

$$x_1 = x, \quad x_2 = y, \quad x_3 = x', \quad x_4 = y'.$$

Then the system takes the form

$$x_3' = a_1x_1 + b_1x_2 + c_1x_4 + f_1(t),$$
$$x_4' = a_2x_1 + b_2x_2 + c_2x_3 + f_2(t).$$

Note that these equations are first-order. Two additional first-order equations are (see the notations):

$$x_3 = x_1',$$
$$x_4 = x_2'.$$

Finally, the first-order system that is equivalent to the original second-order system is

$$\begin{cases} x_1' = x_3, \\ x_2' = x_4, \\ x_3' = a_1x_1 + b_1x_2 + c_1x_4 + f_1(t), \\ x_4' = a_2x_1 + b_2x_2 + c_2x_3 + f_2(t). \end{cases}$$

These examples illustrate the general principle that any system of differential equations can be reduced to a larger system of first-order equations. Thus it is sufficient to study only the systems of first-order differential equations. In the next section we introduce such systems.

3.2 SYSTEMS OF FIRST-ORDER DIFFERENTIAL EQUATIONS

A general system of n first-order differential equations can be written as

$$\begin{cases} \dfrac{dx_1}{dt} = f_1(t, x_1, x_2, ..., x_n), \\[2mm] \dfrac{dx_2}{dt} = f_2(t, x_1, x_2, ..., x_n), \\ \quad \cdots\cdots\cdots\cdots\cdots \\ \dfrac{dx_n}{dt} = f_n(t, x_1, x_2, ..., x_n), \end{cases} \tag{3.5}$$

where $x_i(t)$ are unknown functions, and f_i are given functions. The *general solution* of system (3.5) contains n arbitrary constants, and it has the form

$$x_i = x_i(t, C_1, C_2, \ldots, C_n), \quad i = \overline{1, n}.$$

When solving real-world problems using system (3.5), the solution of IVP (called the *particular solution*) is determined from the general solution by specifying some *initial condition* for each function $x_i(t)$:

$$x_i(t_0) = x_{i0} \quad i = \overline{1, n}. \tag{3.6}$$

Solutions $x_i(t)$ determine a solution curve (called an *integral curve*) in the space $(t, x_1, x_2, \ldots, x_n)$.

Next, we state the existence and uniqueness theorem for the solution of system (3.5).

Cauchy's Theorem (Existence and Uniqueness of a Solution)

System (3.5) together with the initial conditions (3.6) has a unique solution (which is determined by a single set of constants C_j), if the functions f_i and their partial derivatives with respect to all arguments, $\partial f_i / \partial x_k$ $(i, k = \overline{1, n})$, are bounded in the vicinity of the initial conditions (that is, in some domain D in the space $(t, x_1, x_2, \ldots, x_n)$ that contains the initial point $(t_0, x_{10}, x_{20}, \ldots, x_{n0})$.

Solution $x_i(t)$ $(i = \overline{1, n})$ of system (3.5) can be represented by a vector function X, whose components are $x_i(t)$. Similarly, the set of functions f_i can be represented by a vector function F:

$$X = \begin{pmatrix} x_1 \\ x_2 \\ \vdots \\ x_n \end{pmatrix}, \quad F = \begin{pmatrix} f_1 \\ f_2 \\ \vdots \\ f_n \end{pmatrix}.$$

Using this, we can compactly rewrite the original system (3.5) and the initial conditions (3.6) in the *vector form*:

$$\frac{dX}{dt} = F, \tag{3.5a}$$

$$X(t_0) = X_0. \tag{3.6a}$$

One can solve system (3.5) by reducing it to a single high-order equation. Using equations (3.5), as well as using equations that are obtained by differentiating (3.5), we can obtain, and then solve, a single nth-order equation for any one of the unknown functions $x_i(t)$. The remaining unknown functions are obtained from the equations of the original system (3.5) and from the intermediate equations that have been obtained by differentiation. Examples 2.1 and 2.2 illustrate this method.

Example 2.1

(a) Find the general solution of the system of two first-order equations

$$\begin{cases} \dfrac{dx}{dt} = -x - 5y, \\[2mm] \dfrac{dy}{dt} = x + y, \end{cases}$$

by reducing it to a single second-order equation. (b) Solve IVP, given $x(\pi/2) = 0$, $y(\pi/2) = 1$.
Solution. Differentiate the second equation:

$$\frac{d^2y}{dt^2} = \frac{dx}{dt} + \frac{dy}{dt}.$$

Substitute the derivative $\frac{dx}{dt}$ from the first equation:

$$\frac{d^2y}{dt^2} = -x - 5y + \frac{dy}{dt}.$$

The variable x can be taken from the second equation:

$$x = \frac{dy}{dt} - y,$$

and then the second-order equation for $y(t)$ is

$$\frac{d^2y}{dt^2} + 4y = 0.$$

We obtained a homogeneous linear differential equation of order two with constant coefficients. Its characteristic equation is

$$k^2 + 4 = 0,$$

solving which gives $k_{1,2} = \pm 2i$. Then the general solution is

$$y(t) = C_1 \cos 2t + C_2 \sin 2t.$$

We have determined one of the unknown functions of the system. Next, using $x = dy/dt - y$, we find $x(t)$:

$$x(t) = (2C_2 - C_1)\cos 2t - (2C_1 + C_2)\sin 2t.$$

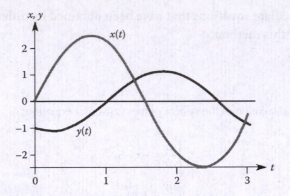

FIGURE 3.1   The particular solution of the system in Example 2.1 (on the interval [1,3]).

To solve IVP, we substitute the initial conditions in the general solution:

$$\begin{cases} 0 = C_1 - 2C_2, \\ 1 = -C_1 \end{cases}$$

and obtain the integration constants: $C_1 = -1$, $C_2 = -\frac{1}{2}$.
   Finally, the solution of the IVP is

$$\begin{cases} x(t) = \dfrac{5}{2}\sin 2t, \\[2mm] y(t) = -\cos 2t - \dfrac{1}{2}\sin 2t. \end{cases}$$

Functions $x(t)$ and $y(t)$ are shown in Figure 3.1.

Example 2.2

Find general solution of the system

$$\begin{cases} \dfrac{dx}{dt} = y, \\[2mm] \dfrac{dy}{dt} = -x, \end{cases}$$

by reducing it to a single second-order equation.
   Solution. Differentiate the first equation:

$$\frac{d^2x}{dt^2} = \frac{dy}{dt}.$$

Using the second equation, we arrive to a second-order equation for x:

$$x'' + x = 0.$$

The solution of this equation is $x(t) = C_1 \cos t + C_2 \sin t$. Substituting this solution into $y = \frac{dx}{dt}$ and differentiating gives $y(t)$. Thus we finally have

$$\begin{cases} x(t) = C_1 \cos t + C_2 \sin t, \\ y(t) = -C_1 \sin t + C_2 \cos t. \end{cases}$$

Remark. We found $y(t)$ from the equation $y = \frac{dx}{dt}$. It seems at first glance that same solution can be obtained by substituting $x(t) = C_1 \cos t + C_2 \sin t$ in the second equation of the system:

$$\frac{dy}{dt} = -x = -C_1 \cos t - C_2 \sin t,$$

following by integration. However, upon integration there appears an additional constant:

$$y(t) = -C_1 \sin t + C_2 \cos t + C_3.$$

This function $y(t)$ cannot be the solution of the first-order system, since the number of constants exceeds the number of equations in the system. Indeed, when this $y(t)$ (and $x(t)$ that we have found) are substituted in the system, it becomes clear that $y(t)$ is solution only when $C_3 = 0$ (check). Thus the second function should preferably be determined without integration.

To visualize our solution, let square $x(t)$ and $y(t)$, and add:

$$x^2(t) + y^2(t) = C_1^2 + C_2^2.$$

This equation describes a family of concentric circles in the plane XOY centered at the origin (see Figure 3.2). The equations $x(t) = C_1 \cos t + C_2 \sin t$ and $y(t) = -C_1 \sin t + C_2 \cos t$ are the parametric equations in the parameter t, and the xy-plane (in which a plane curve C: $\{x(t), y(t)\}$ is located) is called the *phase plane*. The curve C: $\{x(t), y(t)\}$ is a *phase curve*, and a set of two or more phase curves is a *phase portait* of the system. Substituting some initial conditions in the original equation, we can obtain certain values of the integration constants C_1, C_2, and thus obtain a circle of a certain radius $\sqrt{C_1^2 + C_2^2}$ in the phase plane. Thus, to every set of initial conditions there corresponds a certain particular phase curve. Take, for instance, the initial conditions $x(0) = 0$, $y(0) = 1$. Substitution in the general solution results in $C_1 = 0$, $C_2 = 1$, and thus the IVP solution has the form $x(t) = \sin t$, $y(t) = \cos t$. When t increases from zero on the interval $[0, 2\pi]$, the circular phase curve is traced clockwise: point $(0,1)$ on the y-axis corresponds to value $t = 0$, point $(1,0)$ on the

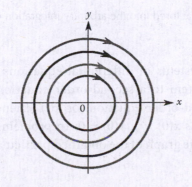

FIGURE 3.2   Phase plane in Example 2.2.

x-axis corresponds to $t = \pi/2$, point $(0,-1)$ – to $t = \pi$, $(-1,0)$ – to $t = 3\pi/2$, and finally, at $t = 2\pi$ the phase curve returns to the initial point $(0,1)$.

Sometimes a system of differential equations can be easily solved by finding *integrating combinations* of the unknown functions. Next we consider the example of using this solution method.

Example 2.3

Solve the system

$$\begin{cases} \dfrac{dx}{dt} = 3x + y, \\[3mm] \dfrac{dy}{dt} = x + 3y, \end{cases}$$

by finding integrating combination.

Solution. Addition of two equations (first integrating combination) gives separable equation for *x+y*:

$$\frac{d(x+y)}{dt} = 4(x+y).$$

Separating variables and integrating we obtain

$$\ln|x+y| = 4t + \ln C_1, \quad \text{or} \quad x + y = C_1 e^{4t}.$$

Next, subtract second equation from the first one (second integrating combination). This results in the separable equation for *x − y*:

$$\frac{d(x-y)}{dt} = 2(x-y).$$

Separation of variables and integration gives $x - y = C_2 e^{2t}$.
From the two linear algebraic equations for *x* and *y* one can easily obtain the solution of the system:

$$\begin{cases} x(t) = C_1 e^{4t} + C_2 e^{2t}, \\[2mm] y(t) = C_1 e^{4t} - C_2 e^{2t} \end{cases}$$

(here, the multiplier 1/2 is factored into the arbitrary integration constants).

Problems

Find general solutions of systems of differential equations by increasing the order. In problems 1–6 convert a system to a second-order equation for $y(t)$; in problems 7–10 convert a system to a second-order equation for $x(t)$. Then solve the IVP. In problems 1–5 use the initial conditions $x(0) = 1$, $y(0) = 0$. In problems 6–10 use the initial conditions $x(0) = 0$, $y(0) = 1$. Plot the graph of the obtained particular solution with the program **ODE_1st_order**.

1. $\begin{cases} x' = -2x + 4y, \\ y' = -x + 3y. \end{cases}$

6. $\begin{cases} x' = 5x + 2y, \\ y' = -4x - y. \end{cases}$

2. $\begin{cases} x' - 2y + 3x = 0, \\ y' - y + 2x = 0. \end{cases}$

7. $\begin{cases} x' = 3x + y, \\ y' = -x + y. \end{cases}$

3. $\begin{cases} x' + 2x + 3y = 0, \\ y' + x = 1. \end{cases}$

8. $\begin{cases} x' = 3x + 2y, \\ y' = -x - y. \end{cases}$

4. $\begin{cases} x' = -3x - y - 1, \\ y' = x - y. \end{cases}$

9. $\begin{cases} x' = -x - 2y, \\ y' = 3x + 4y. \end{cases}$

5. $\begin{cases} x' = x - 3y, \\ y' = 3x + y. \end{cases}$

10. $\begin{cases} x' = 2x - y, \\ y' = x + 2y. \end{cases}$

Answers

1. $\begin{cases} x = C_1 e^{2t} + 4C_2 e^{-t}, \\ y = C_1 e^{2t} + C_2 e^{-t}. \end{cases}$

6. $\begin{cases} x(t) = -C_1 e^{3t} - \dfrac{1}{2}C_2 e^{t}, \\ y(t) = C_1 e^{3t} + C_2 e^{t}. \end{cases}$

2. $\begin{cases} x(t) = e^{-t}\left[C_1 + \left(t - \dfrac{1}{2} \right)C_2 \right], \\ y(t) = e^{-t}\left(C_1 + C_2 t \right). \end{cases}$

7. $\begin{cases} x(t) = e^{2t}\left(C_1 + C_2 t \right), \\ y(t) = e^{2t}\left[-C_1 + C_2(1 - t) \right]. \end{cases}$

3. $\begin{cases} x(t) = -C_1 e^{t} + 3C_2 e^{-3t} + 1, \\ y(t) = C_1 e^{t} + C_2 e^{2t} - 2/3. \end{cases}$

8. $\begin{cases} x(t) = C_1 e^{(1+\sqrt{2})t} + C_2 e^{(1-\sqrt{2})t}, \\ y(t) = \dfrac{1}{2}\Big[C_1(-2 + \sqrt{2})e^{(1+\sqrt{2})t} - \\ \qquad\qquad - C_2(2 + \sqrt{2})e^{(1-\sqrt{2})t} \Big]. \end{cases}$

4. $\begin{cases} x(t) = e^{-2t}\left[C_2(1 - t) - C_1 \right] - 1/4, \\ y(t) = e^{-2t}\left(C_1 + C_2 t \right) - 1/4. \end{cases}$

9. $\begin{cases} x(t) = C_1 e^{2t} + C_2 e^{t}, \\ y(t) = -\dfrac{3}{2}C_1 e^{2t} - C_2 e^{t}. \end{cases}$

5. $\begin{cases} x(t) = e^{t}\left(C_2 \cos 3t - C_1 \sin 3t \right), \\ y(t) = e^{t}\left(C_1 \cos 3t + C_2 \sin 3t \right). \end{cases}$

10. $\begin{cases} x(t) = e^{2t}\left(C_1 \cos t + C_2 \sin t \right), \\ y(t) = e^{2t}\left(C_1 \sin t - C_2 \cos t \right). \end{cases}$

3.3 SYSTEMS OF FIRST-ORDER LINEAR DIFFERENTIAL EQUATIONS

A system of ordinary differential equations is *linear* if it is linear with respect to all unknown functions and their derivatives. Consider system of n linear differential equations:

$$\begin{cases} \dfrac{dx_1}{dt} = a_{11}(t)x_1 + a_{12}(t)x_2 + \ldots + a_{1n}(t)x_n + f_1(t), \\[2mm] \dfrac{dx_2}{dt} = a_{21}(t)x_1 + a_{22}(t)x_2 + \ldots + a_{2n}(t)x_n + f_2(t), \\[2mm] \cdots\cdots\cdots\cdots\cdots\cdots\cdots\cdots\cdots\cdots\cdots \\[2mm] \dfrac{dx_n}{dt} = a_{n1}(t)x_1 + a_{n2}(t)x_2 + \ldots + a_{nn}(t)x_n + f_n(t), \end{cases} \tag{3.7}$$

where $a_{ij}(t)$ $(i, j = \overline{1,n})$ and $f_i(t)$ $(i = \overline{1,n})$ are given functions. System (3.7) also can be written in vector form:

$$\frac{dX}{dt} = AX + F, \tag{3.7a}$$

where $A = [a_{ij}(t)]$ $(i, j = \overline{1,n})$ is the coefficients matrix, and F is the n-dimensional vector with components $f_1(t), f_2(t), \ldots, f_n(t)$:

$$A = \begin{pmatrix} a_{11}(t) & \cdots & a_{1n}(t) \\ \vdots & \ddots & \vdots \\ a_{n1}(t) & \cdots & a_{nn}(t) \end{pmatrix}, \quad F = \begin{pmatrix} f_1(t) \\ \vdots \\ f_n(t) \end{pmatrix}. \tag{3.8}$$

As before, X is a vector with components $x_1, x_2, \ldots, x_n$. We assume that functions $a_{ij}(t)$ and $f_i(t)$ are continuous on the interval $a < t < b$; thus by the Cauchy's theorem the system (3.7) has a unique solution for any given initial condition.

The properties of first-order linear differential systems are generally analogous to the properties of second- and higher-order linear differential equations. Below we briefly state some of these properties.

Consider *homogeneous system* ($F = 0$):

$$\frac{dX}{dt} = AX \tag{3.9}$$

and let us assume that we know n solutions of this system:

$$X^{(1)} = \begin{pmatrix} x_1^{(1)}(t) \\ x_2^{(1)}(t) \\ \vdots \\ x_n^{(1)}(t) \end{pmatrix}, \quad X^{(2)} = \begin{pmatrix} x_1^{(2)}(t) \\ x_2^{(2)}(t) \\ \vdots \\ x_n^{(2)}(t) \end{pmatrix}, \quad \ldots, \quad X^{(n)} = \begin{pmatrix} x_1^{(n)}(t) \\ x_2^{(n)}(t) \\ \vdots \\ x_n^{(n)}(t) \end{pmatrix}. \tag{3.10}$$

The combination of these vectors,

$$\sum_{k=1}^{n} C_k X^{(k)}, \tag{3.11}$$

is called the *linear combination* of solutions $X^{(k)}$. Here C_1, C_2, ..., C_n are some constants. This vector function has the following properties:

Property 1

A linear combination of the solutions of the homogeneous system is also the solution of the system.

This property is easily verified by substitution of $\sum_{k=1}^{n} C_k X^{(k)}$ into the homogeneous system (3.9).

We say that particular solutions (3.10) are linearly dependent on the interval (a,b), if there exists a nonzero set of n coefficients C_k, such that at every point in (a,b)

$$\sum_{k=1}^{n} C_k X^{(k)} = 0.$$

If no such set of coefficients exists, then the vector-functions (3.10) are linearly independent. System of n linearly independent solutions (3.10) of system (3.9) is called *the fundamental solution set* of this system.

Property 2

For linear system of differential equations (3.9) a fundamental solution set always exists, and the linear combination of the solutions in the fundamental set is the *general solution* of the linear homogeneous system.

By this property, the values of the coefficients C_1, C_2, ..., C_n in the general solution

$$X(t) = \sum_{k=1}^{n} C_k X^{(k)}(t) \tag{3.12}$$

of the homogeneous system (3.9) can be uniquely determined from this general solution if n initial conditions

$$x_i(t_0) = x_{i0}, \quad i = \overline{1, n}$$

are given. Substitution of the coefficients values in the general solution gives IVP solution for system (3.9).

3.4 SYSTEMS OF FIRST-ORDER LINEAR HOMOGENEOUS DIFFERENTIAL EQUATIONS WITH CONSTANT COEFFICIENTS

The discussion in Section 3.3 provides the solution scheme for a system of linear equations with constant coefficients a_{ij}. For such a system,

$$\begin{cases} \dfrac{dx_1}{dt} = a_{11}x_1 + a_{12}x_2 + \cdots + a_{1n}x_n + f_1(t), \\[2mm] \dfrac{dx_2}{dt} = a_{21}x_1 + a_{22}x_2 + \cdots + a_{2n}x_n + f_2(t), \\[2mm] \cdots \cdots \cdots \cdots \cdots \cdots \cdots \cdots \cdots \\[2mm] \dfrac{dx_n}{dt} = a_{n1}x_1 + a_{n2}x_2 + \cdots + a_{nn}x_n + f_n(t), \end{cases} \tag{3.13}$$

an analytical solution is always possible.

Solution of this system of inhomogeneous equations starts from determination of the general solution of the corresponding homogeneous system:

$$\begin{cases} \dfrac{dx_1}{dt} = a_{11}x_1 + a_{12}x_2 + \cdots + a_{1n}x_n, \\[2mm] \dfrac{dx_2}{dt} = a_{21}x_1 + a_{22}x_2 + \cdots + a_{2n}x_n, \\[2mm] \cdots \cdots \cdots \cdots \cdots \cdots \cdots \cdots \\[2mm] \dfrac{dx_n}{dt} = a_{n1}x_1 + a_{n2}x_2 + \cdots + a_{nn}x_n. \end{cases} \tag{3.14}$$

The vector form of the system is

$$\frac{dX}{dt} = AX, \tag{3.14a}$$

where A is the constant coefficients matrix.

For linear systems with constant coefficients we describe a standard solution method, which does not involve raising the system's order. The method is based on determination of the n linearly independent particular solutions (the fundamental set) and then forming the linear combination of these solutions—which is the general solution of the system by Property 2. Below we introduce this method.

We look for solutions of the system (3.14) in the form

$$X = \bar{\alpha}e^{kt}, \tag{3.15}$$

where

$$\bar{\alpha} = \begin{pmatrix} \alpha_1 \\ \alpha_2 \\ \vdots \\ \alpha_n \end{pmatrix}$$

is the column-vector of the coefficients that must be determined (below we write this vector in the form $[\alpha_1, \alpha_2, ..., \alpha_n]$). Let us substitute (3.15) in (3.14), move all terms to the right side, and cancel the exponent e^{kt}. This results in the system of n linear homogeneous algebraic equations:

$$\begin{cases} (a_{11} - k)\alpha_1 + a_{12}\alpha_2 + \cdots + a_{1n}\alpha_n = 0, \\ a_{21}\alpha_1 + (a_{22} - k)\alpha_2 + \cdots + a_{2n}\alpha_n = 0, \\ \cdots \cdots \cdots \cdots \cdots \cdots \cdots \cdots \\ a_{n1}\alpha_1 + a_{n2}\alpha_2 + \cdots + (a_{nn} - k)\alpha_n = 0. \end{cases} \tag{3.16}$$

This system is compactly written as follows (I is the identity $n \times n$ matrix):

$$(A - kI)\bar{\alpha} = 0. \tag{3.16a}$$

As it is well known from linear algebra, the system (3.16) has a nontrivial solution only when its determinant is zero:

$$\Delta = \begin{vmatrix} a_{11} - k & a_{12} & \cdots & a_{1n} \\ a_{21} & a_{22} - k & \cdots & a_{2n} \\ \cdots & \cdots & \cdots & \cdots \\ a_{n1} & a_{n2} & \cdots & a_{nn} - k \end{vmatrix} = 0, \tag{3.17}$$

or, in the vector notation,

$$\det|A - kI| = 0. \tag{3.17a}$$

Equation (3.17) (or 3.17a) is called the *characteristic equation* and it is the algebraic equation of order n; thus it has n roots. For each root determined from the characteristic equation one finds the set of coefficients α_i from system (3.16). Then, after substitution in (3.15), the solution is complete. The particulars depend on whether the roots of the characteristic equation, k, are real and distinct, real and repeated, or complex. We consider these three cases separately.

3.4.1 Roots of the Characteristic Equation Are Real and Distinct

System (3.16) is solved for $\bar{\alpha} = [\alpha_1, \alpha_2, ..., \alpha_n]$ with each root k_i. Since we are looking for the solution of system (3.14) in the form (3.15), then for k_1 we write the set of particular solutions:

$$x_1^{(1)} = \alpha_1^{(1)}e^{k_1 t}, \quad x_2^{(1)} = \alpha_2^{(1)}e^{k_1 t}, \quad ..., \quad x_n^{(1)} = \alpha_n^{(1)}e^{k_1 t}.$$

Superscript indicates that the corresponding particular solutions and the set of coefficients α_i are associated with the root k_1. Similarly, for the root k_2 we write

$$x_1^{(2)} = \alpha_1^{(2)}e^{k_2 t}, \quad x_2^{(2)} = \alpha_2^{(2)}e^{k_2 t}, \quad ..., \quad x_n^{(2)} = \alpha_n^{(2)}e^{k_2 t}.$$

For k_n

$$x_1^{(n)} = \alpha_1^{(n)}e^{k_n t}, \quad x_2^{(n)} = \alpha_2^{(n)}e^{k_n t}, \quad ..., \quad x_n^{(n)} = \alpha_n^{(n)}e^{k_n t}.$$

Note that the coefficients $\alpha_i^{(j)}$ found from system (3.16) are not unique, since by construction this system is homogeneous and has zero determinant. This means that one of the coefficients $\alpha_i^{(j)}$ for each j can be set equal to an arbitrary constant.

Recall that exponents with different arguments are linearly independent functions and they form a fundamental set. Linear combination of functions $x_k^{(1)}(t)$ (with n different k values) is the general solution for the function $x_1(t)$, the linear combination of functions $x_k^{(2)}(t)$ is the general solution for the function $x_2(t)$, and so on. Thus, the general solution of homogeneous system (3.14) has the form

$$\begin{cases} x_1(t) = C_1\alpha_1^{(1)}e^{k_1 t} + C_2\alpha_1^{(2)}e^{k_2 t} + \cdots + C_n\alpha_1^{(n)}e^{k_n t}, \\ x_2(t) = C_1\alpha_2^{(1)}e^{k_1 t} + C_2\alpha_2^{(2)}e^{k_2 t} + \cdots + C_n\alpha_2^{(n)}e^{k_n t}, \\ \cdots \cdots \cdots \cdots \cdots \cdots \cdots \cdots \cdots \cdots \\ x_n(t) = C_1\alpha_n^{(1)}e^{k_1 t} + C_2\alpha_n^{(2)}e^{k_2 t} + \cdots + C_n\alpha_n^{(n)}e^{k_n t}. \end{cases} \tag{3.18}$$

Example 4.1

Solve the system

$$\begin{cases} \dfrac{dx}{dt} = -x + 8y, \\ \dfrac{dy}{dt} = x + y. \end{cases}$$

Solution. We look for solutions in the form

$$x = \alpha_1 e^{kt},$$

$$y = \alpha_2 e^{kt}.$$

Next, substitute these expressions into the system, move all terms to the right side, cancel the exponent e^{kt}, and equate the determinant of the resulting system of homogeneous algebraic equations to zero:

$$\begin{vmatrix} -1-k & 8 \\ 1 & 1-k \end{vmatrix} = 0, \quad \text{or} \quad k^2 - 9 = 0.$$

This characteristic equation has the roots $k_1 = 3$; $k_2 = -3$. Particular solutions are

$$\begin{cases} x^{(1)} = \alpha_1^{(1)} e^{3t}, \\ y^{(1)} = \alpha_2^{(1)} e^{3t}, \end{cases} \quad \text{and} \quad \begin{cases} x^{(2)} = \alpha_1^{(2)} e^{-3t}, \\ y^{(2)} = \alpha_2^{(2)} e^{-3t}. \end{cases}$$

Next, we find coefficients $\alpha_i^{(j)}$. System (3.16) for $k = 3$ is

$$\begin{cases} -4\alpha_1^{(1)} + 8\alpha_2^{(1)} = 0, \\ \alpha_1^{(1)} - 2\alpha_2^{(1)} = 0. \end{cases}$$

As can be seen, these two equations are linearly dependent (the first one is -4 times the second one); thus one of the unknown coefficients can be chosen arbitrarily.
Since $\alpha_1^{(1)} = 2\alpha_2^{(1)}$, it is convenient to set $\alpha_2^{(1)}$ equal to one. Then $\alpha_1^{(1)} = 2$.
Similarly we find the second pair of coefficients (for $k = -3$). System (3.16) is

$$\begin{cases} 2\alpha_1^{(2)} + 8\alpha_2^{(2)} = 0, \\ \alpha_1^{(2)} + 4\alpha_2^{(2)} = 0. \end{cases}$$

One of its solutions is $\alpha_1^{(2)} = -4$, $\alpha_2^{(2)} = 1$.
The particular solutions are

$$\begin{cases} x^{(1)} = 2e^{3t}, \\ y^{(1)} = e^{3t} \end{cases} \quad \begin{cases} x^{(2)} = -4e^{-3t}, \\ y^{(2)} = e^{-3t} \end{cases}$$

and they form a fundamental set. The general solution of the original system of differential equations is the linear combination of the solutions in the fundamental set:

$$\begin{cases} x(t) = 2C_1 e^{3t} - 4C_2 e^{-3t}, \\ y(t) = C_1 e^{3t} + C_2 e^{-3t}. \end{cases}$$

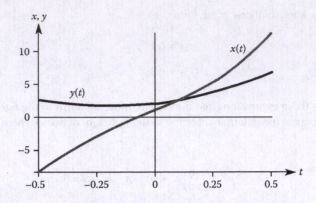

FIGURE 3.3   Solution of IVP in Example 4.1.

Let the initial conditions for the IVP be

$$\begin{cases} x(0) = 1, \\ y(0) = 2. \end{cases}$$

Substitution in the general solution gives

$$\begin{cases} 1 = 2C_1 - 4C_2, \\ 2 = C_1 + C_2. \end{cases}$$

The solution of this system is $C_1 = 3/2$, $C_2 = 1/2$. Thus, the solution of the IVP is

$$\begin{cases} x(t) = 3e^{3t} - 2e^{-3t}, \\ y(t) = \dfrac{3}{2}e^{3t} + \dfrac{1}{2}e^{-3t}. \end{cases}$$

In Figure 3.3 the functions $x(t)$ and $y(t)$ are shown on the interval $[-0.5, 0.5]$.

3.4.2 Roots of the Characteristic Equation Are Complex and Distinct

Note first that for a system of differential equations with real-valued coefficients complex roots occur only in conjugate pairs. To the pair of complex conjugate roots $k_{1,2} = p \pm iq$ there correspond $2n$ particular solutions:

$$x_1^{(1)} = \alpha_1^{(1)} e^{(p+iq)t}, \quad x_2^{(1)} = \alpha_2^{(1)} e^{(p+iq)t}, \quad \ldots, \quad x_n^{(1)} = \alpha_n^{(1)} e^{(p+iq)t},$$

$$x_1^{(2)} = \alpha_1^{(2)} e^{(p-iq)t}, \quad x_2^{(2)} = \alpha_2^{(2)} e^{(p-iq)t}, \quad \ldots, \quad x_n^{(2)} = \alpha_n^{(2)} e^{(p-iq)t}.$$

As in the case of real roots, in order to determine coefficients $\alpha_i^{(j)}$ one has to solve system (3.16) with k_1 and k_2. Recall again that exponents with different arguments, or their real and imaginary parts, are linearly independent functions that form a fundamental solution set. Thus the linear combination of n vectors $X = \bar{\alpha}e^{kt}$ is the general solution of the homogeneous differential system (3.14).

Example 4.2

Solve the system

$$\begin{cases} \dfrac{dx}{dt} = -7x + y, \\[2mm] \dfrac{dy}{dt} = -2x - 5y. \end{cases}$$

Solution. We look for a solution in the form

$$x = \alpha_1 e^{kt},$$

$$y = \alpha_2 e^{kt}.$$

Substitution of these x and y into the system leads to the characteristic equation:

$$\begin{vmatrix} -7-k & 1 \\ -2 & -5-k \end{vmatrix} = 0.$$

The roots are $k_{1,2} = -6 \pm i$. System (3.16) for $k_1 = -6 + i$ is

$$\begin{cases} (-1-i)\alpha_1^{(1)} + \alpha_2^{(1)} = 0, \\ -2\alpha_1^{(1)} + (1-i)\alpha_2^{(1)} = 0. \end{cases}$$

The equations in the latter system are linearly dependent: we see that multiplication of the first equation by $1-i$ gives the second equation. Then from either equation, we obtain the following relation between $\alpha_1^{(1)}$ and $\alpha_2^{(1)}$:

$$\alpha_2^{(1)} = (1+i)\alpha_1^{(1)}.$$

If we let $\alpha_1^{(1)} = 1$, then $\alpha_2^{(1)} = 1 + i$. Next, we repeat the steps for $k_2 = -6 - i$. System (3.16) in this case is

$$\begin{cases} (-1+i)\alpha_1^{(2)} + \alpha_2^{(2)} = 0, \\ -2\alpha_1^{(2)} + (1+i)\alpha_2^{(2)} = 0. \end{cases}$$

Omitting the second equation and letting $\alpha_1^{(2)} = 1$ in the first equation results in $\alpha_2^{(2)} = 1 - i$.

Now we can write two particular solutions of the original differential system:

$$x^{(1)} = e^{(-6+i)t}, \qquad x^{(2)} = e^{(-6-i)t},$$

$$y^{(1)} = (1+i)e^{(-6+i)}, \quad y^{(2)} = (1-i)e^{(-6-i)t}.$$

Their linear combination is the general solution:

$$\begin{cases} x(t) = C_1 e^{(-6+i)t} + C_2 e^{(-6-i)t}, \\ y(t) = C_1(1+i)\,e^{(-6+i)t} + C_2(1-i)e^{(-6-i)t}. \end{cases}$$

Note that the general solution containing complex exponents is often convenient for further calculations (if these are warranted), but for IVP solution (where initial conditions are real numbers), it must be transformed to the real-valued form. It is a fact that for differential equations with real coefficients the real and the imaginary parts of the complex solution are separately solutions. Thus we apply Euler's formula to complex solutions $x^{(1)}$ and $x^{(2)}$:

$$x^{(1)} = e^{-6t}(\cos t + i \sin t),$$

$$x^{(2)} = e^{-6t}(\cos t - i \sin t).$$

Thus

$$\mathrm{Re}\,x^{(1)} = e^{-6t}\cos t, \qquad \mathrm{Re}\,x^{(2)} = \mathrm{Re}\,x^{(1)},$$

$$\mathrm{Im}\,x^{(1)} = e^{-6t}\sin t. \qquad \mathrm{Im}\,x^{(2)} = -\mathrm{Im}\,x^{(1)}.$$

A linear combination of linear independent functions $\mathrm{Re}\,x^{(1)}$ and $\mathrm{Im}\,x^{(1)}$ gives a general solution for $x(t)$:

$$x(t) = C_1 \mathrm{Re}\,x^{(1)} + C_2 \mathrm{Im}\,x^{(1)}.$$

Obviously, if we use $\mathrm{Re}\,x^{(2)}$ and $\mathrm{Im}\,x^{(2)}$ we obtain the same result.
Then take $y^{(1)}$ and also isolate real and imaginary parts:

$$\mathrm{Re}\,y^{(1)} = e^{-6t}(\cos t - \sin t), \qquad \mathrm{Re}\,y^{(2)} = \mathrm{Re}\,y^{(1)},$$

$$\mathrm{Im}\,y^{(1)} = e^{-6t}(\cos t + \sin t). \qquad \mathrm{Im}\,y^{(2)} = -\mathrm{Im}\,y^{(1)}.$$

Forming a linear combination of $\mathrm{Re}\,y^{(1)}$ and $\mathrm{Im}\,y^{(1)}$ with the same coefficients C_1 and C_2 we have

$$y(t) = C_1 \mathrm{Re}\,y^{(1)} + C_2 \mathrm{Im}\,y^{(1)}.$$

Substitution of $\mathrm{Re}\,x^{(1)}$, $\mathrm{Im}\,x^{(1)}$, $\mathrm{Re}\,y^{(1)}$ and $\mathrm{Im}\,y^{(1)}$ gives

$$\begin{cases} x(t) = e^{-6t}(C_1\cos t + C_2\sin t), \\ y(t) = e^{-6t}(C_1 + C_2)\cos t + e^{-6t}(C_2 - C_1)\sin t. \end{cases}$$

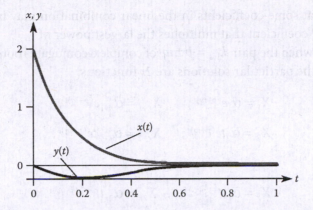

FIGURE 3.4 The particular solution of the system in Example 4.2 (on the interval [0,1]).

Thus it can be seen that for the real-valued solution the functions $x^{(2)}$ and $y^{(2)}$ are not needed, and one can omit finding them altogether—that is, the second of the complex-conjugate roots of the characteristic equation may be disregarded.

Assume that the following initial conditions are given:

$$\begin{cases} x(0) = 2, \\ y(0) = 0. \end{cases}$$

Substituting them in the general solution we find the integration constants: $C_1 = 2$, $C_2 = -2$. Thus solution of IVP is

$$\begin{cases} x(t) = 2e^{-6t}(\cos t - \sin t), \\ y(t) = -4e^{-6t}\sin t. \end{cases}$$

The graph of this particular solution is shown in Figure 3.4.

3.4.3 Roots of the Characteristic Equation Are Repeated (the Number of Different Roots Is Less Than *n*)

In this case a fundamental solution set cannot be constructed only from exponents with different arguments, since the number of such different exponents is less than *n*. We proceed as follows (note that a similar situation occurs when solving linear high-order differential equations; see Chapter 2).

First, consider the real root *k* of the characteristic equation of multiplicity *r*. Then the particular solutions corresponding to this root have the form

$$X_1 = \bar{\alpha}_1 e^{kt}, \ X_2 = \bar{\alpha}_2 t e^{kt}, ..., X_r = \bar{\alpha}_r t^{r-1} e^{kt}.$$

It is easy to check that these *r* vector functions are linearly independent and their linear combination gives the part of the general solution of system (3.14) (corresponding to this root). Note that some coefficients in the linear combination may later turn out to be zero, including the coefficient that multiplies the largest power of t.

In the situation when the pair $k_{1,2} = p \pm iq$ of complex-conjugate roots of equation (3.17) has multiplicity r, the particular solutions are $2r$ functions:

$$X_1 = \bar{\alpha}_1 e^{(p+iq)t}, \qquad X_{r+1} = \bar{\alpha}_{r+1} e^{(p-iq)t},$$

$$X_2 = \bar{\alpha}_2 t e^{(p+iq)t}, \qquad X_{r+2} = \bar{\alpha}_{r+2} t e^{(p-iq)t},$$

$$\cdots \cdots \cdots \cdots \cdots \cdots$$

$$X_r = \bar{\alpha}_r t^{r-1} e^{(p+iq)t}, \quad X_{2r} = \bar{\alpha}_{2r} t^{r-1} e^{(p-iq)t}.$$

Instead of these complex functions their real and imaginary parts can be taken.

Consider example where the characteristic equation has two repeated real roots.

Example 4.3

Solve the system

$$\begin{cases} \dfrac{dx}{dt} = 5x + 3y, \\[2mm] \dfrac{dy}{dt} = -3x - y. \end{cases}$$

Solution. The characteristic equation is

$$\begin{vmatrix} 5-k & 3 \\ -3 & -1-k \end{vmatrix} = 0.$$

The roots are $k_{1,2} = 2$. We look for solution in the form

$$\begin{cases} x(t) = (\alpha_1 + \beta_1 t)e^{2t}, \\[2mm] y(t) = (\alpha_2 + \beta_2 t)e^{2t}, \end{cases}$$

where $\alpha_1, \alpha_2, \beta_1, \beta_2$ are undetermined coefficients. These equations must be substituted in one of the system's equation, for instance in the first one (substitution in another equation gives the same result):

$$2\alpha_1 + \beta_1 + 2\beta_1 t = 5\alpha_1 + 5\beta_1 t + 3\alpha_2 + 3\beta_2 t,$$

from which

$$\alpha_2 = \frac{\beta_1}{3} - \alpha_1,$$

$$\beta_2 = -\beta_1.$$

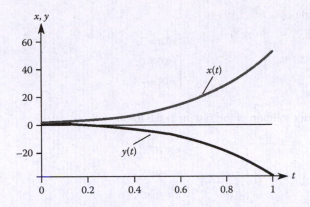

FIGURE 3.5   The particular solution of the system in Example 4.3 (on the interval [0,1]).

Coefficients α_1 and β_1 are still arbitrary. We introduce the notation C_1 and C_2 for them, respectively, and then write the general solution:

$$\begin{cases} x(t) = (C_1 + C_2 t)e^{2t}, \\ y(t) = (C_2/3 - C_1 - C_2 t)e^{2t}. \end{cases}$$

Let the initial conditions be

$$\begin{cases} x(0) = 1, \\ y(0) = 1. \end{cases}$$

Substituting these initial conditions gives $C_1 = 1$, $C_2 = 6$, thus the solution of IVP is

$$\begin{cases} x(t) = (1 + 6t)e^{2t}, \\ y(t) = (1 - 6t)e^{2t}. \end{cases}$$

The graph of this particular solution is shown in Figure 3.5.

If a system contains different kinds of roots, then the three approaches described above are combined in order to determine the fundamental system of solutions.

Next, consider examples of systems of three differential equations for all possible situations. Note that in Appendix 4 we formulate solutions in more formal way.

Example 4.4

Find the solution of the system

$$\begin{cases} \dfrac{dx}{dt} = 3x - y + z, \\[2mm] \dfrac{dy}{dt} = x + y + z, \\[2mm] \dfrac{dz}{dt} = 4x - y + 4z \end{cases}$$

with the initial conditions

$$\begin{cases} x(0) = 3, \\ y(0) = 0, \\ z(0) = -5. \end{cases}$$

Solution. Consider solution of the system in the form

$$x = \alpha_1 e^{kt},$$
$$y = \alpha_2 e^{kt},$$
$$z = \alpha_3 e^{kt}.$$

Substitution of these expressions in the system gives the characteristic equation

$$\begin{vmatrix} 3-k & -1 & 1 \\ 1 & 1-k & 1 \\ 4 & -1 & 4-k \end{vmatrix} = 0.$$

Its roots are $k_1 = 1$, $k_2 = 2$, $k_3 = 5$. For $k_1 = 1$ the system (3.16) is

$$\begin{cases} 2\alpha_1^{(1)} - \alpha_2^{(1)} + \alpha_3^{(1)} = 0, \\ \alpha_1^{(1)} + \alpha_3^{(1)} = 0, \\ 4\alpha_1^{(1)} - \alpha_2^{(1)} + 3\alpha_3^{(1)} = 0. \end{cases}$$

This is the homogeneous algebraic system with the zero determinant; thus it has infinitely many solutions. We find one of these solutions. Let $\alpha_1^{(1)} = 1$; then from any two equations of this system we obtain $\alpha_2^{(1)} = 1$, $\alpha_3^{(1)} = -1$.

For the second root $k_2 = 2$ the system (3.16) is

$$\begin{cases} \alpha_1^{(2)} - \alpha_2^{(2)} + \alpha_3^{(2)} = 0, \\ \alpha_1^{(2)} - \alpha_2^{(2)} + \alpha_3^{(2)} = 0, \\ 4\alpha_1^{(2)} - \alpha_2^{(2)} + 2\alpha_3^{(2)} = 0. \end{cases}$$

Choosing again $\alpha_1^{(2)} = 1$, gives $\alpha_2^{(2)} = -2$, $\alpha_3^{(2)} = -3$. And for the third root $k_3 = 3$ we have

$$\begin{cases} -2\alpha_1^{(3)} - \alpha_2^{(3)} + \alpha_3^{(3)} = 0, \\ \alpha_1^{(3)} - 4\alpha_2^{(3)} + \alpha_3^{(3)} = 0, \\ 4\alpha_1^{(3)} - \alpha_2^{(3)} - \alpha_3^{(3)} = 0, \end{cases}$$

and $\alpha_1^{(3)} = 1$, $\alpha_2^{(3)} = 1$, $\alpha_3^{(3)} = 3$. Now we can write the general solution of the system:

$$\begin{cases} x(t) = C_1 e^t + C_2 e^{2t} + C_3 e^{5t}, \\ y(t) = C_1 e^t - 2C_2 e^{2t} + C_3 e^{5t}, \\ z(t) = -C_1 e^t - 3C_2 e^{2t} + 3C_3 e^{5t}. \end{cases}$$

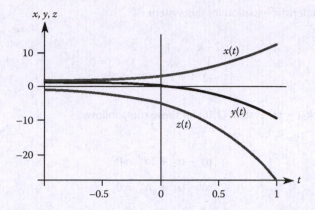

FIGURE 3.6 The particular solution of the system in Example 4.4 (on the interval [0,1]).

Finally, substitution of the initial conditions gives the algebraic system:

$$\begin{cases} 3 = C_1 + C_2 + C_3, \\ 0 = C_1 - 2C_2 + C_3, \\ -5 = -C_1 - 3C_2 + 3C_3. \end{cases}$$

Its solution is $C_1 = 2$, $C_2 = 1$, $C_3 = 0$. Thus the solution of IVP is

$$\begin{cases} x(t) = 2e^t + e^{2t}, \\ y(t) = 2e^t - 2e^{2t}, \\ z(t) = -2e^t - 3e^{2t}. \end{cases}$$

Graphs of $x(t)$, $y(t)$, and $z(t)$ are shown in Figure 3.6.

Example 4.5

Find the general solution of the system

$$\begin{cases} \dfrac{dx}{dt} = 2x - y + 2z, \\[2mm] \dfrac{dy}{dt} = x + 2z, \\[2mm] \dfrac{dz}{dt} = -2x + y - z. \end{cases}$$

Solution. Characteristic equation for this system is

$$\begin{vmatrix} 2-k & -1 & 2 \\ 1 & -k & 2 \\ -2 & 1 & -1-k \end{vmatrix} = 0,$$

with roots $k_1 = 1$, $k_{2,3} = \pm i$. System (3.16) for these roots follows.

For $k_1 = 1$:

$$\begin{cases} \alpha_1^{(1)} - \alpha_2^{(1)} + 2\alpha_3^{(1)} = 0, \\ \alpha_1^{(1)} - \alpha_2^{(1)} + 2\alpha_3^{(1)} = 0, \\ -2\alpha_1^{(1)} + \alpha_2^{(1)} - 2\alpha_3^{(1)} = 0. \end{cases}$$

Choosing $\alpha_3^{(1)} = 1$, gives $\alpha_1^{(1)} = 0$, $\alpha_2^{(1)} = 2$.

For $k_2 = i$:

$$\begin{cases} (2-i)\alpha_1^{(2)} - \alpha_2^{(2)} + 2\alpha_3^{(2)} = 0, \\ \alpha_1^{(2)} - i\alpha_2^{(2)} + 2\alpha_3^{(2)} = 0, \\ -2\alpha_1^{(2)} + \alpha_2^{(2)} - (1+i)\alpha_3^{(2)} = 0. \end{cases}$$

Choosing $\alpha_1^{(2)} = 1$ gives $\alpha_2^{(2)} = 1$, $\alpha_3^{(2)} = \frac{1}{2}(i-1)$; thus

$$x^{(2)} = e^{it}, \ y^{(2)} = e^{it}, z^{(2)} = \frac{1}{2}(i-1)e^{it}.$$

Instead of these complex functions one can use their real and imaginary parts (below the indices 2 and 3 denote these linearly independent functions):

$$x^{(2)} = \cos t, \qquad\qquad x^{(3)} = \sin t,$$

$$y^{(2)} = \cos t, \qquad\qquad y^{(3)} = \sin t,$$

$$z^{(2)} = -\frac{1}{2}(\cos t + \sin t), \ \ z^{(3)} = \frac{1}{2}(\cos t - \sin t).$$

In this approach the solution for the complex-conjugate root $k_3 = -i$ is not needed.

One can directly check that the triplet of particular solutions, $x^{(1)}$, $y^{(1)}$, $z^{(1)}$, as well as the other two triplets, satisfy the original differential system.

Linear combinations of these functions give the general solution:

$$\begin{cases} x(t) = C_2 \cos t + C_3 \sin t, \\ y(t) = 2C_1 e^t + C_2 \cos t + C_3 \sin t, \\ z(t) = C_1 e^t - \dfrac{C_2}{2}(\cos t + \sin t) + \dfrac{C_3}{2}(\cos t - \sin t). \end{cases}$$

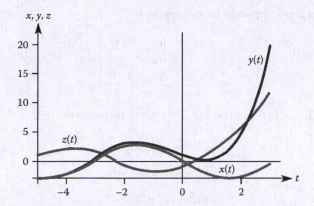

FIGURE 3.7 The particular solution of the system in Example 4.5 (on the interval $[-5,3]$).

Using the initial conditions

$$\begin{cases} x(0) = 0, \\ y(0) = 1, \\ z(0) = -1, \end{cases}$$

results in the integration constants $C_1 = 0.5$, $C_2 = 0$, $C_3 = -3$, and then the solution of the IVP is

$$\begin{cases} x(t) = -3\sin t, \\ y(t) = e^t - 3\sin t, \\ z(t) = \dfrac{1}{2}e^t - \dfrac{3}{2}(\cos t - \sin t). \end{cases}$$

Graphs of this solution are shown in Figure 3.7.

Example 4.6

Find the general solution of the system

$$\begin{cases} \dfrac{dx}{dt} = -2x + y - 2z, \\[2mm] \dfrac{dy}{dt} = x - 2y + 2z, \\[2mm] \dfrac{dz}{dt} = 3x - 3y + 5z. \end{cases}$$

Solution. Characteristic equation for this system is

$$\begin{vmatrix} -2-k & 1 & -2 \\ 1 & -2-k & 2 \\ 3 & -3 & 5-k \end{vmatrix} = 0.$$

Roots are $k_1 = 3$, $k_2 = k_3 = -1$. For $k_1 = 3$ system (3.16) becomes

$$\begin{cases} -5\alpha_1^{(1)} + \alpha_2^{(1)} - 2\alpha_3^{(1)} = 0, \\ \alpha_1^{(1)} - 5\alpha_2^{(1)} + 2\alpha_3^{(1)} = 0, \\ 3\alpha_1^{(1)} - 3\alpha_2^{(1)} + 2\alpha_3^{(1)} = 0. \end{cases}$$

As always, this system of homogeneous linear algebraic equations with zero determinant has infinitely many solutions. Choosing $\alpha_1^{(1)} = 1$, we get $\alpha_2^{(1)} = -1$, $\alpha_3^{(1)} = -3$. This gives the first triplet of particular solutions:

$$x^{(1)} = e^{3t}, \ y^{(1)} = -e^{3t}, \ z^{(1)} = -3e^{3t}.$$

Each of these solutions will enter the general solution multiplied by the coefficient C_1.
The solution for the repeated root $k_{2,3} = -1$ has the form

$$\begin{cases} x(t) = (\alpha_1 + \beta_1 t)e^{-t}, \\ y(t) = (\alpha_2 + \beta_2 t)e^{-t}, \\ z(t) = (\alpha_3 + \beta_3 t)e^{-t}. \end{cases}$$

To determine the coefficients, these expressions must be substituted in the equations of the differential system.
Substitution in the first equation gives

$$-\alpha_1 + \beta_1 - \beta_1 t = -2\alpha_1 - 2\beta_1 t + \alpha_2 + \beta_2 t - 2\alpha_3 - 2\beta_3 t \ .$$

Equating coefficients of the equal powers of t, we find

$$\alpha_3 = \frac{1}{2}(\alpha_2 - \alpha_1 - \beta_1),$$

$$\beta_3 = \frac{1}{2}(\beta_2 - \beta_1).$$

Substitution in the second equation gives

$$\alpha_3 = \frac{1}{2}(\alpha_2 - \alpha_1 + \beta_2),$$

$$\beta_3 = \frac{1}{2}(\beta_2 - \beta_1).$$

Substitution in the third equation gives

$$\alpha_3 = \frac{1}{2}(\alpha_2 - \alpha_1 + \beta_3 / 3),$$

$$\beta_3 = \frac{1}{2}(\beta_2 - \beta_1).$$

Thus we obtained four independent equations for six constants α_i, β_i. Comparing expressions for α_3, it becomes clear that they are satisfied for $\beta_1 = \beta_2 = \beta_3 = 0$. Let $\alpha_1 \equiv C_2$, $\alpha_3 \equiv C_3$; then from the equation $\alpha_3 = (\alpha_2 - \alpha_1)/2$ we obtain $\alpha_2 = 2\alpha_3 + \alpha_1 = C_1 + 2C_3$. Thus the general solution can be written as

$$\begin{cases} x(t) = C_1 e^{3t} + C_2 e^{-t}, \\ y(t) = -C_1 e^{3t} + e^{-t}(C_2 + 2C_3), \\ z(t) = -3C_1 e^{3t} + C_3 e^{-t}. \end{cases}$$

The initial conditions

$$\begin{cases} x(0) = 3, \\ y(0) = 0, \\ z(0) = -5 \end{cases}$$

give $C_1 = 1.75$, $C_2 = 1.25$, $C_3 = 0.25$, and the solution of IVP is

$$\begin{cases} x(t) = 1.75 e^{3t} + 1.25 e^{-t}, \\ y(t) = -1.75 e^{3t} + 1.75 e^{-t}, \\ z(t) = -5.25 e^{3t} + 0.25 e^{-t}. \end{cases}$$

Graphs of this solution are shown in Figure 3.8.

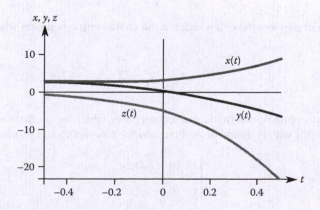

FIGURE 3.8   The particular solution of the system in Example 4.6 (on the interval $[-0.5, 0.5]$).

A choice of coefficients α_1 и α_3 for C_2 and C_3 is only one a possibility. If one were to choose $\alpha_2 \equiv C_2$, $\alpha_3 \equiv C_3$, this gives $\alpha_1 = C_2 - 2C_3$. In this case a general solution looks different:

$$\begin{cases} x(t) = C_1 e^{3t} + (C_2 - 2C_3)e^{-t}, \\ y(t) = -C_1 e^{3t} + C_2 e^{-t}, \\ z(t) = -3C_1 e^{3t} + C_3 e^{-t}. \end{cases}$$

But clearly the particular solution for the given initial conditions is the same, as obtained above.

Example 4.7

Find the general solution of the system

$$\begin{cases} \dfrac{dx}{dt} = 2x - y - z, \\[2mm] \dfrac{dy}{dt} = 2x - y - 2z, \\[2mm] \dfrac{dz}{dt} = -x + y + 2z. \end{cases}$$

Solution. Characteristic equation for this system is

$$\begin{vmatrix} 2-k & -1 & -1 \\ 2 & -1-k & -2 \\ -1 & -1 & 2-k \end{vmatrix} = 0.$$

It has three coinciding roots: $k_1 = k_2 = k_3 = 1$. In general, a solution of a system of differential equations when a root of characteristic equation is real and three times repeating is seeking in the form:

$$\begin{cases} x(t) = (\alpha_1 + \beta_1 t + \gamma_1 t^2)e^{kt}, \\ y(t) = (\alpha_2 + \beta_2 t + \gamma_2 t^2)e^{kt}, \\ z(t) = (\alpha_3 + \beta_3 t + \gamma_3 t^2)e^{kt}. \end{cases}$$

But in the current problem the determinant of the characteristic equation when $k = 1$,

$$\Delta = \begin{vmatrix} 1 & -1 & -1 \\ 2 & -2 & -2 \\ -1 & -1 & 1 \end{vmatrix}$$

does not contain any nonzero determinant of second order, which means that the rank of Δ equals one. In this case terms with t^2 should be omitted and the solution should be searched as

$$\begin{cases} x = (\alpha_1 + \beta_1 t)e^t, \\ y = (\alpha_2 + \beta_2 t)e^t, \\ z = (\alpha_3 + \beta_3 t)e^t. \end{cases}$$

Substitute these expressions in the first equation of the system of DE. Dividing the result by e^t, we have

$$\beta_1 + \alpha_1 + \beta_1 t = 2\alpha_1 + 2\beta_1 t - \alpha_2 - \beta_2 t - 2\alpha_3 - 2\beta_3 t.$$

Compare the coefficients at the same degrees of t:

$$\beta_1 = \alpha_1 - \alpha_2 - \alpha_3,$$
$$0 = \beta_1 - \beta_2 - \beta_3.$$

Similarly, substituting $x(t)$, $y(t)$, $z(t)$ in the second equation of the system of DE, we have

$$\beta_2 = 2(\alpha_1 - \alpha_2 - \alpha_3),$$
$$0 = \beta_1 - \beta_2 - \beta_3.$$

Substitution in the third equation gives

$$\beta_3 = -\alpha_1 + \alpha_2 + \alpha_3,$$
$$0 = \beta_1 - \beta_2 - \beta_3.$$

Three out of six equations for coefficients α_i, β_i coincide. Choosing $\alpha_1 \equiv C_1$, $\alpha_2 \equiv C_2$, $\beta_1 \equiv C_3$, we can get the other coefficients from them:

$$\alpha_3 = C_1 - C_2 - C_3,$$
$$\beta_2 = 2C_3,$$
$$\beta_3 = -C_3.$$

Thus, a general solution of the system of differential equations is

$$\begin{cases} x = (C_1 + C_2 t)e^t, \\ y = (C_2 + 2C_3 t)e^t, \\ z = (C_1 - C_2 - C_3 - C_3 t)e^t. \end{cases}$$

Problems

Solve differential systems without resorting to increase of the system's order. Then solve the IVP. In problems 1–5 use the initial conditions $x(0) = 1$, $y(0) = 1$; in problems 6–12, the initial conditions $x(0) = 1$, $y(0) = 2$; in problems 13–17, the initial conditions $x(0) = 1$, $y(0) = 2$, $z(0) = 3$. Plot the graph of the obtained particular solution with the program **ODE_1st_order**.

To simplify the task, the roots of the characteristic equation are stated for systems of three equations

1. $\begin{cases} x' + 2x + 3y = 0, \\ y' + x = 0. \end{cases}$

2. $\begin{cases} x' = -3x - y, \\ y' = x - y. \end{cases}$

3. $\begin{cases} x' = x - 3y, \\ y' = 3x + y. \end{cases}$

4. $\begin{cases} x' = 3x - 2y, \\ y' = 4x - y. \end{cases}$

5. $\begin{cases} x' = -x - 2y, \\ y' = 3x + 4y. \end{cases}$

6. $\begin{cases} x' = x + y, \\ y' = 2x. \end{cases}$

7. $\begin{cases} x' = x - 2y, \\ y' = 2x - 3y. \end{cases}$

8. $\begin{cases} x' = 4x - y, \\ y' = x + 2y. \end{cases}$

9. $\begin{cases} x' = 2x - y, \\ y' = x + 2y. \end{cases}$

10. $\begin{cases} x' = x - 2y, \\ y' = x - y. \end{cases}$

11. $\begin{cases} x' = x - y, \\ y' = -4x + 4y. \end{cases}$

12. $\begin{cases} x' = 2x - 3y, \\ y' = 3x + 2y. \end{cases}$

13. $\begin{cases} x' = 4y - 2z - 3x, \\ y' = z + x, \\ z' = 6x - 6y + 5z. \end{cases}$

14. $\begin{cases} x' = x - y - z, \\ y' = x + y, \\ z' = 3x + z. \end{cases}$

$k_1 = 1,\ k_2 = 2,\ k_3 = -1.$

$k_1 = 1,\ k_{2,3} = 1 \pm 2i.$

15. $\begin{cases} x' = 2x - y - z, \\ y' = 3x - 2y - 3z, \\ z' = -x + y + 2z. \end{cases}$

16. $\begin{cases} x' = 2x + y, \\ y' = x + 3y - z, \\ z' = -x + 2y + 3z. \end{cases}$

$k_1 = 0,\ k_2 = k_3 = 1.$

$k_1 = 2,\ k_{2,3} = 3 \pm i.$

17. $\begin{cases} x' = 3x + 12y - 4z, \\ y' = -x - 3y + z, \\ z' = -x - 12y + 6z. \end{cases}$

$k_1 = 1,\ k_2 = 2,\ k_3 = 3.$

Answers

1. $\begin{cases} x = C_1 e^{-3t} + C_2 e^t, \\ y = C_1 e^{-3t} / 3 - C_2 e^t. \end{cases}$

2. $\begin{cases} x = e^{-2t} \left[C_1(t-1) - C_2 \right], \\ y = e^{-2t} \left(-C_1 t + C_2 \right). \end{cases}$

3. $\begin{cases} x = e^t \left(-C_1 \sin 3t + C_2 \cos 3t \right), \\ y = e^t \left(C_1 \cos 3t + C_2 \sin 3t \right). \end{cases}$

4. $\begin{cases} x(t) = e^t \left[C_1 (\cos 2t - \sin 2t) \right. \\ \qquad\qquad \left. + C_2 (\cos 2t + \sin 2t) \right], \\ y(t) = 2e^t \left(C_1 \cos 2t + C_2 \sin 2t \right). \end{cases}$

5. $\begin{cases} x = 2C_1 e^{2t} + C_2 e^t, \\ y = -3C_1 e^{2t} - C_2 e^t. \end{cases}$

6. $\begin{cases} x = C_1 e^{2t} + C_2 e^{-t}, \\ y = C_1 e^{2t} - 2C_2 e^{-t}. \end{cases}$

7. $\begin{cases} x = e^{-t} \left[C_1 + C_2 \left(\dfrac{1}{2} + t \right) \right], \\ y = e^{-t} \left(C_1 + C_2 t \right). \end{cases}$

8. $\begin{cases} x = e^{3t} \left(C_1 + C_2 t \right), \\ y = e^{3t} \left[C_1 + C_2 (t-1) \right]. \end{cases}$

9. $\begin{cases} x = e^{2t} \left(-C_1 \sin t + C_2 \cos t \right), \\ y = e^{2t} \left(C_1 \cos t + C_2 \sin t \right). \end{cases}$

10. $\begin{cases} x = C_1 (\cos t - \sin t) \\ \qquad\quad + C_2 (\cos t + \sin t), \\ y = C_1 \cos t + C_2 \sin t. \end{cases}$

11. $\begin{cases} x = C_1 - C_2 e^{5t}, \\ y = C_1 + 4C_2 e^{5t}. \end{cases}$

12. $\begin{cases} x = e^{2t} \left(C_1 \cos 3t + C_2 \sin 3t \right), \\ y = e^{2t} \left(C_1 \sin 3t - C_2 \cos 3t \right). \end{cases}$

13. $\begin{cases} x = C_1 e^t - C_2 e^{-t}, \\ y = C_1 e^t + C_3 e^{2t} / 2, \\ z = C_2 e^{-t} + C_3 e^{2t}. \end{cases}$

14. $\begin{cases} x(t) = e^t \left(-C_2 \sin 2t + C_3 \cos 2t \right), \\ y(t) = \dfrac{1}{2} e^t \left(-2C_1 + C_2 \cos 2t + C_3 \sin 2t \right), \\ z(t) = \dfrac{1}{2} e^t \left(2C_1 + 3C_2 \cos 2t + 3C_3 \sin 2t \right). \end{cases}$

15. $\begin{cases} x(t) = -C_1 + C_2 e^t, \\ y(t) = -3C_1 + C_3 e^t, \\ z(t) = C_1 + (C_2 - C_3) e^t. \end{cases}$

16.
$$\begin{cases} x = C_1 e^{2t} + C_2(2\cos t - \sin t)e^{3t} + C_3(\cos t + 2\sin t)e^{3t}, \\ y = C_2(\cos t - 3\sin t)e^{3t} + C_3(\sin t + 3\cos t)e^{3t}, \\ z = C_1 e^{2t} + 5C_2 e^{3t}\cos t + 5C_3 e^{3t}\sin t. \end{cases}$$

17.
$$\begin{cases} x = -2C_1 e^t - 8C_2 e^{2t} - 3C_3 e^{3t}, \\ y = C_1 e^t + 3C_2 e^{2t} + C_3 e^{3t}, \\ z = 2C_1 e^t + 7C_2 e^{2t} + 3C_3 e^{3t}. \end{cases}$$

3.5 SYSTEMS OF FIRST-ORDER LINEAR INHOMOGENEOUS DIFFERENTIAL EQUATIONS WITH CONSTANT COEFFICIENTS

When solving inhomogeneous systems of the form

$$\frac{dX}{dt} = AX + F \tag{3.19}$$

we first determine the general solution of the corresponding homogeneous system:

$$\frac{dX}{dt} = AX.$$

Assume that this solution is known: $x_i(t) = \sum_{k=1}^{n} C_k x_i^{(k)}, (i = \overline{1,n})$. Then, we need to determine one particular solution of the nonhomogeneous system, using, for instance, *variation of parameters*. The idea of this method is as follows. Particular solution is sought in the form

$$\overline{x}_i(t) = \sum_{k=1}^{n} C_k(t) x_i^{(k)}, \quad (i = \overline{1,n}), \tag{3.20}$$

where $C_k(t)$ are unknown functions. Substitution of this particular solution in the original nonhomogeneous system gives the system of equations for $C_k'(t)$. Integration gives functions $C_k(t)$. Then, from (3.20) one finds the particular solution $\overline{x}_i(t)$ $(i = \overline{1,n})$, and then the general solution of the inhomogeneous system (3.19) is the sum of general solution of homogeneous equation and a particular solution of inhomogeneous equation:

$$x_i(t) = X_i(t) + \overline{x}_i(t), \quad (i = \overline{1,n}).$$

Example 5.1

Solve inhomogeneous differential system

$$\begin{cases} \dfrac{dx}{dt} = -x + 8y + 1, \\[2mm] \dfrac{dy}{dt} = x + y + t. \end{cases}$$

Solution. The corresponding homogeneous system is

$$\begin{cases} \dfrac{dx}{dt} = -x + 8y, \\[2mm] \dfrac{dy}{dt} = x + y. \end{cases}$$

Its general solution was obtained in Example 4.1:

$$\begin{cases} x(t) = 2C_1 e^{3t} - 4C_2 e^{-3t}, \\[2mm] y(t) = C_1 e^{3t} + C_2 e^{-3t}. \end{cases}$$

(Recall that for convenience, when solving inhomogeneous systems of linear differential equations, we use capital letters to denote the general solution of the homogeneous system.) Next, we find the particular solution of the inhomogeneous differential system. Varying the parameters,

$$\begin{cases} \bar{x}(t) = 2C_1(t) e^{3t} - 4C_2(t) e^{-3t}, \\[2mm] \bar{y}(t) = C_1(t) e^{3t} + C_2(t) e^{-3t}. \end{cases}$$

Substituting these particular solutions in the first equation of the inhomogeneous system gives

$$2C_1' e^{3t} + 6C_1 e^{3t} - 4C_2' e^{-3t} + 12C_2 e^{-3t} = -2C_1 e^{3t} + 4C_2 e^{-3t} + 8C_1 e^{3t} + C_2 e^{-3t} + 1.$$

The terms that contain functions $C_1(t)$ and $C_2(t)$ cancel (this always happens by the design of the method), and we get

$$2C_1' e^{3t} - 4C_2' e^{-3t} = 1.$$

Repeating for the second equation of the system gives

$$C_1' e^{3t} + C_2' e^{-3t} = t.$$

Thus we arrive to the algebraic system of two equations for two unknowns C_1' and C_2':

$$\begin{cases} 2C_1'e^{3t} - 4C_2'e^{-3t} = 1, \\ C_1'e^{3t} + C_2'e^{-3t} = t. \end{cases}$$

Its solution is

$$C_1' = \frac{1}{6}(1+4t)e^{-3t},$$

$$C_2' = \frac{1}{6}(2t-1)e^{3t}.$$

Integration gives $C_1(t)$ и $C_2(t)$ (arbitrary integration constants are omitted since we are interested in the particular solution):

$$C_1(t) = -\frac{1}{18}e^{-3t}\left(\frac{7}{3} + 4t\right),$$

$$C_2(t) = \frac{1}{18}e^{3t}\left(2t - \frac{5}{3}\right).$$

Substituting these functions in the expressions for $x(t)$ and $y(t)$ gives the particular solution of the inhomogeneous differential system:

$$\begin{cases} \overline{x}(t) = \frac{1}{9}(1-8t), \\ \overline{y}(t) = -\frac{1}{9}(2+t). \end{cases}$$

The general solution of the inhomogeneous system is the sum of the general solution of the homogeneous system and the particular solution of the inhomogeneous one:

$$\begin{cases} x(t) = 2C_1e^{3t} - 4C_2e^{-3t} + \frac{1}{9}(1-8t), \\ y(t) = C_1e^{3t} + C_2e^{-3t} - \frac{1}{9}(2+t). \end{cases}$$

Let the initial conditions be

$$\begin{cases} x(0) = 1, \\ y(0) = -1. \end{cases}$$

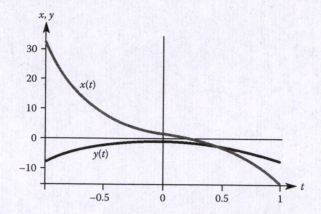

FIGURE 3.9 The particular solution of the system in Example 5.1 (on the interval [−1,1]).

Substitution in the general solution gives

$$\begin{cases} 1 = 2C_1 - 4C_2 + \dfrac{1}{9}, \\[2mm] -1 = C_1 + C_2 - \dfrac{2}{9}. \end{cases}$$

Solving gives $C_1 = -\frac{10}{27}$, $C_2 = -\frac{11}{27}$ and then the solution of the IVP is

$$\begin{cases} x(t) = -\dfrac{20}{27}e^{3t} + \dfrac{44}{27}e^{-3t} + \dfrac{1}{9}(1 - 8t), \\[3mm] y(t) = -\dfrac{10}{27}e^{3t} - \dfrac{11}{27}e^{-3t} - \dfrac{1}{9}(2 + t). \end{cases}$$

The graph of this solution is shown in Figure 3.9.

Problems

Solve inhomogeneous systems of differential equations. Then solve the IVP for the initial conditions $x(0) = 1$, $y(0) = 0$. Plot the graph of the obtained particular solution with the program **ODE_1st_order**.

1. $\begin{cases} x' = 2x - y, \\ y' = 2y - x - 5e^t \sin t \end{cases}$

2. $\begin{cases} x' + 4y = \cos 2t, \\ y' + 4x = \sin 2t. \end{cases}$

3. $\begin{cases} x' = 5x - 3y + 2e^{3t}, \\ y' = x + y + 5e^{-t}. \end{cases}$

4. $\begin{cases} x' = -2x + 4y, \\ y' = -x + 3y + 3t^2. \end{cases}$

5. $\begin{cases} x' = 2x + y + 2e^t, \\ y' = x + 2y - 3e^{4t}. \end{cases}$

6. $\begin{cases} x' = y - 5\cos t, \\ y' = 2x + y. \end{cases}$

7. $\begin{cases} x' = 2x + y - 7te^{-t} - 3, \\ y' = -x + 2y - 1. \end{cases}$

Answers

1. $\begin{cases} x = -C_1 e^{3t} + C_2 e^t + e^t(2\cos t - \sin t), \\ x = C_1 e^{3t} + C_2 e^t + e^t(3\cos t + \sin t). \end{cases}$

2. $\begin{cases} x = C_1 e^{-4t} + C_2 e^{4t} + 0.3\sin 2t, \\ y = C_1 e^{-4t} - C_2 e^{4t} + 0.1\cos 2t. \end{cases}$

3. $\begin{cases} x = 3C_1 e^{4t} + C_2 e^{2t} - 4e^{3t} - e^{-t}, \\ y = C_1 e^{4t} + C_2 e^{2t} - 2e^{3t} - 2e^{-t}. \end{cases}$

4. $\begin{cases} x = C_1 e^{2t} + 4C_2 e^{-t} - 6t^2 + 6t - 9, \\ y = C_1 e^{2t} + C_2 e^{-t} - 3t^2 - 3. \end{cases}$

5. $\begin{cases} x = C_1 e^t + C_2 e^{3t} + \left(t - \dfrac{1}{2}\right)e^t - e^{4t}, \\ y = C_1 e^t + C_2 e^{3t} - \left(t + \dfrac{1}{2}\right)e^t - 2e^{4t}. \end{cases}$

6. $\begin{cases} x = C_1 e^{-t} + C_2 e^{2t} - 2\sin t - \cos t, \\ y = -C_1 e^{-t} + 2C_2 e^{2t} + \sin t + 3\cos t. \end{cases}$

7. $\begin{cases} x = e^{2t}(C_1 \cos t + C_2 \sin t) + \dfrac{7}{50}e^{-t}(4 + 15t) + 1, \\ y = e^{2t}(-C_1 \sin t + C_2 \cos t) + \dfrac{7}{50}e^{-t}(3 + 5t) + 1. \end{cases}$

CHAPTER 4

Boundary Value Problems for Second-Order ODE and Sturm-Liouville Theory

4.1 INTRODUCTION TO BVP

The concept of a boundary value problem can be illustrated considering a projectile motion for a body of mass m launched at point $\vec{r}(t_0)$ with the goal to reach point $\vec{r}(t_1)$. The body is moving under the action of a force $\vec{F}$. To find the trajectory we need to solve the differential equation $m\frac{d^2\vec{r}}{dt^2} = \vec{F}(t, \vec{r}, \dot{\vec{r}})$ for the position vector $\vec{r}(t)$ (actually this is a system of two equations for $x(t)$ and $y(t)$ for the components of vector $\vec{r}(x, y)$ if we consider motion in x, y -plane).

From the physical point of view it is clear that such a problem with the fixed initial and final points of the trajectory may not always have a solution; also it may have multiple solutions describing different ways to deliver a body from initial to final point.

The problem consisting of a differential equation and boundary conditions is called a *boundary value problem* (BVP).

Consider several simple BVP for linear second-order differential equations.

Example 1.1: Solve the BVP

$$y'' + y = 0, \quad x \in [0, x_1]$$

$$\begin{cases} y(0) = 0, \\ y(x_1) = y_1 \end{cases} \quad \text{– the boundary conditions.}$$

The general solution is $y(x) = C_1 \cos x + C_2 \sin x$.

Substitutions into the boundary condition at $x = 0$ gives $C_1 \cos 0 + C_2 \sin 0 = 0$, from which $C_1 = 0$. Thus $y(x) = C_2 \sin x$ and we have to identify C_2 from the boundary condition at x_1.

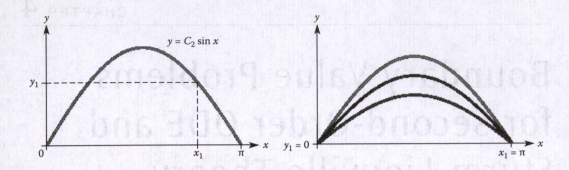

FIGURE 4.1 Integral curve for the BVP (Example 1.1). (a) $x_1 \neq n\pi$ (the unique solution), (b) $x_1 = n\pi$ $y_1 = 0$ (infinite number of solutions).

Three situations are possible:

(a) $x_1 \neq n\pi$; thus $\sin x_1 \neq 0$. It allows one to find $C_2 = y_1/\sin x_1$; thus the solution of the BVP exists and is unique (it is shown in Figure 4.1a for $x_1 < \pi$):

$$y(x) = \frac{y_1}{\sin x_1} \sin x.$$

(b) $x_1 = n\pi$ and $y_1 = 0$. Because $\sin x_1 = 0$ the identity $y(x_1) = C_2 \sin x_1$ is satisfied for an arbitrary value of C_2. Thus there are an infinite number of solutions of the BVP (it is shown in Figure 4.1b for $x_1 = \pi$ and $C_2 > 0$).

(c) $x_1 = n\pi$ and $y_1 \neq 0$. Because $\sin x_1 = 0$ the identity $y(x_1) = C_2 \sin x_1$ cannot be satisfied and the value of C_2 cannot be identified. Thus, the BVP has no solution.

The boundary conditions do not necessarily imply the values of a function at the ends of the interval, there can be other options. Consider several examples.

Example 1.2: Solve the BVP

$$y'' + y' = 1, \quad x \in [0,1]$$

$$\begin{cases} y'(0) = 0, \\ y(1) = 1 \end{cases} \quad \text{– the boundary conditions.}$$

In this problem the value of the function is given at one end of the interval; the value of the function's derivative, at another end.

The solution of the homogeneous equation $y'' + y' = 0$ (the roots of the characteristic equation are $k_1 = 0$, $k_2 = -1$) is $Y(x) = C_1 + C_2 e^{-x}$. Searching a particular solution of the nonhomogeneous equation in the form $\bar{y}(x) = ax$, we obtain $a = 1$, thus $\bar{y}(x) = x$. Therefore, a general solution of the given equation is

$$y(x) = C_1 + C_2 e^{-x} + x.$$

From there, $y'(x) = -C_2 e^{-x} + 1$. Substitution it in the boundary conditions gives

$$\begin{cases} -C_2 + 1 = 0, \\ C_1 + C_2 e^{-1} + 1 = 1, \end{cases}$$

and we obtain $C_2 = 1$, $C_1 = -1/e$, thus the solution of the BVP is

$$y(x) = -e^{-1} + e^{-x} + x.$$

Example 1.3: Solve the BVP

$$y'' + y = 2x - \pi, \quad x \in [0, \pi]$$

$$\begin{cases} y(0) = 0, \\ y(\pi) = 0. \end{cases}$$

The general solution of the given equation is

$$y(x) = C_1 \cos x + C_2 \sin x + 2x - \pi.$$

Either boundary condition gives $C_1 = \pi$, but the value of C_2 cannot be determined. Thus, the solution of the BVP is

$$y(x) = \pi \cos x + C \sin x + 2x - \pi$$

with an arbitrary C.

Example 1.4: Solve the BVP

$$y'' - y' = 0, \quad x \in [0, 1]$$

$$\begin{cases} y(0) = -1, \\ y'(1) - y(1) = 2. \end{cases}$$

The second condition, combining the value of the function and its derivative, is called the *mixed boundary conditions*.

The general solution of the equation (the roots of the characteristic equation are $k_1 = 0$, $k_2 = 1$) is $y(x) = C_1 + C_2 e^x$; thus $y'(x) = C_2 e^x$. Substitution it in the boundary conditions gives

$$\begin{cases} C_1 + C_2 = -1, \\ C_2 e - C_1 - C_2 e = 2, \end{cases}$$

and we obtain $C_1 = -2$, $C_2 = 1$, thus the solution of the BVP is

$$y(x) = -2 + e^x.$$

Notice that for the function $y(x)$ defined on an infinite or semi-infinite interval, boundary conditions may not be specified and are often replaced by the condition of regularity or physically reasonable behavior as $x \to \pm\infty$. For instance on the interval $[a,\infty)$, it can be the condition of regularity at $x \to \infty$, for example that $y(\infty)$ be finite.

Example 1.5: Solve the BVP

$$y'' - y' - 2y = 0, \quad x \in [0, \infty)$$
$$\begin{cases} y(0) = 0, \\ y(\infty) = 0. \end{cases}$$

The second boundary determines the asymptotic behavior at $x \to \infty$.

The general solution of the equation (the roots of the characteristic equation are $k_1 = 2$, $k_2 = -1$) is $y(x) = C_1 e^{-x} + C_2 e^{2x}$.

The second boundary condition can be satisfied only if $C_2 = 0$; thus $y(x) = C_1 e^{-x}$. Substitution it into first boundary condition gives $y(0) = C_1 = 0$.

Thus, the BVP has only a trivial solution $y(x) = 0$.

Example 1.6: Solve the BVP

$$y'' - y = 1, \quad x \in [0, \infty)$$
$$\begin{cases} y(0) = 0, \\ y \text{ is bounded at } x \to \infty. \end{cases}$$

The second boundary condition describes the behavior of the function we search at $x \to \infty$.

The solution of the homogeneous equation $y'' - y = 0$ (the roots of the characteristic equation are $k_1 = 1$, $k_2 = -1$) is $Y(x) = C_1 e^x + C_2 e^{-x}$. Searching a particular solution of the nonhomogeneous equation in the form $\bar{y}(x) = a$, we obtain $a = -1$; thus $\bar{y}(x) = -1$. Therefore, a general solution of the given equation is $y(x) = C_1 e^x + C_2 e^{-x} - 1$.

The second boundary condition can be satisfied only if $C_1 = 0$. Then the condition $y(0) = 0$ gives $C_2 - 1 = 0$. The solution of the BVP is

$$y(x) = e^{-x} - 1.$$

Example 1.7: Solve the BVP

$$y'' - y = 1, \quad x \in (-\infty, \infty)$$

with the boundary conditions formulated in the following way: function $y(x)$ is bounded when $x \to \infty$ and $x \to -\infty$.

From a general solution $y(x) = C_1 e^x + C_2 e^{-x} - 1$ we see that both $C_1 = C_2 = 0$; thus the solution of the BVP is $y(x) = -1$.

Example 1.8: Solve the BVP

$$y'' + p(x)y' + q(x)y = f(x), \quad x \in [a,b],$$

$$\begin{cases} y(a) = \alpha, \\ y(b) = \beta \end{cases} \quad - \text{ the boundary conditions.}$$

If two fundamental solutions of the homogeneous equations are $y_1(x)$ and $y_2(x)$, a particular solution of the nonhomogeneous equation is $\bar{y}(x)$; then a general solution of nonhomogeneous equation is

$$y(x) = C_1 y_1(x) + C_2 y_2(x) + \bar{y}(x).$$

Substitution into the boundary conditions gives the system of equations for coefficients C_1 и C_2

$$\begin{cases} C_1 y_1(a) + C_2 y_2(a) = \alpha - \bar{y}(a), \\ C_1 y_1(b) + C_2 y_2(b) = \beta - \bar{y}(b). \end{cases}$$

The determinant of this system is

$$\Delta = \begin{vmatrix} y_1(a) & y_2(a) \\ y_1(b) & y_2(b) \end{vmatrix}.$$

When $\Delta \neq 0$, the system (and the BVP) has the unique solution; when $\Delta = 0$, the system can have an infinite number of solutions if both supplemental determinants Δ_1, Δ_2 are equal to zero; no solutions, if one of these determinants is not equal to zero.

Problems

1. $y'' + y = 1$, $y(0) = 0$, $y(\pi / 2) = 0$

2. $y'' + 2y' = 1$, $y'(0) = 0$, $y(1) = 1$

3. $y'' - y' = 0$, $y(0) = -1$, $y'(1) - y(1) = 2$

4. $y'' + y = 1$, $y(0) = 0$, $y(\pi) = 0$

5. $y'' + 4y = x$, $y(0) = 0$, $y(\pi / 4) = 1$

6. $y'' - y' - 2y = 0$, $y'(0) = 2$, $y(+\infty) = 0$

7. $y'' - y = 1$, $y(0) = 0$, $y(x)$ limited at $x \to +\infty$

8. $y'' - y' = 0$, $y(0) = 3$, $y(1) - y'(1) = 1$

9. $y'' + y = 1$, $y(0) = 0$, $y'(\pi / 2) = 1$

10. $y'' - 2y' - 3y = 0$,

   a) $y(0) = 1$, $\lim\limits_{x \to \infty} y(x) = 0$

   b) $y'(0) = 2$, $\lim\limits_{x \to \infty} y'(x) = 0$

Answers

$y = 1 - \sin x - \cos x$

$y(x) = (x + 1) / 2 + (e^{-2x} - e^{-2}) / 4$

$y = e^x - 2$

No solution.

$y = (1 - \pi/16)\sin 2x + x/4$

$y = -2e^{-x}$

$y = e^{-x} - 1$

$y = 1 + 2e^x$

$y = 1 + C\sin x - \cos x$

a) $y = e^{-x}$

b) $y = -2e^{-x}$

4.2 THE STURM-LIOUVILLE PROBLEM

In many applications in physics and engineering we must solve linear, second-order differential equations that fall into a class of problems known as Sturm-Liouville eigenvalue problems. Such problems consist of a linear, second-order, ordinary differential equation containing a parameter whose value is determined so that the solution to the equation satisfies a given *boundary condition*. Thus, Sturm-Liouville problems are special kinds of *boundary value problems* for certain types of ordinary differential equations. The set of orthogonal functions generated by the solution to such problems we will use later as the basis functions for the Fourier expansion method of solving partial differential equations. The special functions we briefly introduce in this chapter arise from one or another Sturm-Liouville problem and will be discussed in more detail later in the book.

Notice that the material of this and the following sections is most needed for Part II of the book and can be studied just before it.

To begin we notice that a general linear second-order ordinary differential equation with a parameter λ multiplied by the function $y(x)$

$$a(x)y''(x)+b(x)y'(x)+c(x)y(x)+\lambda d(x)y(x)=0, \tag{4.1}$$

$(a(x) \neq 0)$ can be written in *the Sturm-Liouville form*

$$\frac{d}{dx}[p(x)y'(x)]+[q(x)+\lambda r(x)]y(x)=0, \tag{4.2}$$

where

$$p(x)=e^{\int \frac{b(x)}{a(x)}dx}, \quad q(x)=\frac{p(x)c(x)}{a(x)}, \quad r(x)=\frac{p(x)d(x)}{a(x)}. \tag{4.3}$$

Here $y(x)$ may represent some physical quantity in which we are interested, such as the amplitude of a wave at a particular location or the temperature at a particular time or location.

Reading Exercise

Verify that substitution of equations (4.3) into equation (4.2) gives equation (4.1).

As we will see studying partial differential equations, many physical problems result in the linear ordinary equation (4.2) where the function $y(x)$ is defined on an interval $[a,b]$ and obeys *homogeneous boundary conditions* of the form

$$\alpha_1 y' + \beta_1 y\big|_{x=a} = 0,$$

$$\alpha_2 y' + \beta_2 y\big|_{x=b} = 0. \tag{4.4}$$

Clearly, the constants α_1 and β_1 cannot both be zero simultaneously, nor the constants α_2 and β_2. The constants α_k and β_k are real since boundary conditions represent physical restrictions. Therefore, for BVP it is more convenient to use real solutions $y(x)$. We note also that the relative signs for a_k and β_k are not arbitrary; we generally must have $\beta_1/\alpha_1 < 0$ and $\beta_2/\alpha_2 > 0$. This choice of signs (details of which are discussed in book [1]) is necessary in setting up boundary conditions for various classes of physical problems. The very rare cases when the signs are different occur in problems where there is explosive behavior, such as an exponential temperature increase. Everywhere in the book we consider "normal" physical situations in which processes occur smoothly and thus the parameters in boundary conditions are restricted as above.

Equations (4.2) and (4.4) define a Sturm-Liouville problem. Solving this problem involves determining the values of the constant λ for which nontrivial solutions $y(x)$ exist. If $\alpha_k = 0$, the boundary condition simplifies to $y = 0$ (known as the *Dirichlet boundary condition*), and if $\beta_k = 0$, the boundary condition is $y' = 0$ (called the *Neumann boundary condition*); otherwise the boundary condition is referred to as a *mixed boundary condition*.

Notice that for the function $y(x)$ defined on an infinite or semi-infinite interval, the conditions (4.4) may not be specified and are often replaced by the condition of regularity or physically reasonable behavior as $x \to \pm\infty$—for example, that $y(\infty)$ be finite.

For the following we let $p(x)$, $q(x)$, $r(x)$, and $p'(x)$ be continuous, real functions on an interval $[a,b]$ and let $p(x) \geq 0$ and $r(x) \geq 0$ on the interval $[a,b]$. The coefficients α_k and β_k in equations (4.4) are assumed to be real and independent of λ.

The differential equation (4.2) and boundary conditions (4.4) are *homogeneous*, which is essential for the subsequent development. The trivial solution $y(x) = 0$ is always possible for homogeneous equations but we seek special values of λ (called *eigenvalues*) for which there are nontrivial solutions (called *eigenfunctions*) that depend on λ.

If we introduce the differential operator (called the *Sturm-Liouville operator*)

$$Ly(x) = -\frac{d}{dx}\left[p(x)y'(x)\right] - q(x)y(x) = -p(x)y''(x) - p'(x)y'(x) - q(x)y(x) \qquad (4.5)$$

then equation (4.2) becomes

$$Ly(x) = \lambda r(x)y(x). \qquad (4.6)$$

As can be seen from equation (4.5), $Ly(x)$ is a linear operator. When $r(x) = 1$ this equation appears as an ordinary *eigenvalue problem*, $Ly(x) = \lambda y(x)$, for which we have to determine λ and $y(x)$. For $r(x) \neq 1$ we have a modified problem where the function $r(x)$ is called a *weight function*. As we stated above, the only requirement on $r(x)$ is that it is real and non-negative. Equations (4.2) and (4.6) are different ways to specify the same boundary value problem.

Now we discuss the properties of eigenvalues and eigenfunctions of the Sturm-Liouville problem. Let us write equation (4.6) for two eigenfunctions, $y_n(x)$ and $y_m(x)$, and take the complex conjugate of the equation for $y_m(x)$. Notice that in spite of the fact that $p(x)$, $q(x)$, and $r(x)$ are real, λ and $y(x)$ can be complex. We have

$$Ly_n(x) = \lambda_n r(x)y_n(x),$$

and

$$Ly_m^*(x) = \lambda_m^* r(x) y_m^*(x).$$

Multiplying the first of these equations by $y_m^*(x)$ and the second by $y_n(x)$, we then integrate both from a to b and subtract the two results to obtain

$$\int_a^b y_m^*(x)Ly_n(x)dx - \int_a^b y_n(x)Ly_m^*(x)dx = (\lambda_n - \lambda_m^*)\int_a^b r(x)y_m^*(x)y_n(x)dx. \qquad (4.7)$$

Using the definition of L given by equation (4.5), the left side of equation (4.7) is

$$\left\{ p(x)\left[\frac{dy_m^*}{dx} y_n(x) - y_m^*(x)\frac{dy_n}{dx} \right] \right\}_a^b. \qquad (4.8)$$

Reading Exercise

Verify the previous statement.

Then, using the boundary conditions (4.4), it can be easily proved that the expression (4.8) equals to zero.

Reading Exercise

Verify that the expression in equation (4.8) equals zero.

Thus we are left with

$$\int_a^b y_m^*(x)Ly_n(x)dx = \int_a^b y_n(x)Ly_m^*(x)dx. \qquad (4.9)$$

An operator, L, that satisfies equation (4.9) is named a *Hermitian* or *self-adjoint operator*. Thus we may say that *the Sturm-Liouville linear operator* satisfying homogeneous boundary conditions is *Hermitian*. Many important operators in physics, especially in quantum mechanics, are Hermitian.

Let us show that Hermitian operators have *real eigenvalues* and their eigenfunctions are *orthogonal*. Using equation (4.9), equation (4.7) may be written as

$$(\lambda_n - \lambda_m^*)\int_a^b r(x)y_m^*(x)y_n(x)dx = 0. \qquad (4.10)$$

When $m = n$ the integral cannot be zero (recall that $r(x) > 0$); thus $\lambda_n^* = \lambda_n$ and we have proved that *the eigenvalues of a Sturm-Liouville problem are real*.

Then, for $\lambda_m \neq \lambda_n$, equation (4.10) is

$$\int_a^b r(x)y_m^*(x)y_n(x)dx = 0 \qquad (4.11)$$

and we conclude that the eigenfunctions corresponding to different eigenvalues of a Sturm-Liouville problem are orthogonal (with the weight function $r(x)$).

The squared *norm* of the eigenfunction $y_n(x)$ is defined to be

$$\|y_n\|^2 = \int_a^b r(x)|y_n(x)|^2 \, dx. \qquad (4.12)$$

Note that the eigenfunctions of Hermitian operators always *can be chosen to be real*. This can be done by using some linear combinations of the functions $y_n(x)$—for example, choosing $\sin x$ and $\cos x$ instead of $\exp(\pm ix)$ for the solutions of the equation $y'' + y = 0$. Real eigenfunctions are more convenient to work with because it is easier to match them to boundary conditions that are intrinsically real since they represent physical restrictions.

The above proof fails if $\lambda_m = \lambda_n$ for some $m \neq n$ (in other words, there exist different eigenfunctions belonging to the same eigenvalues) in which case we cannot conclude that the corresponding eigenfunctions, $y_m(x)$ and $y_n(x)$, are orthogonal (although in some cases they are). If there are f eigenfunctions that have the same eigenvalues, we have an *f-fold degeneracy* of the eigenvalue. In general a degeneracy reflects a symmetry of the underlying physical system (examples will be given below). For a Hermitian operator it is always possible to construct linear combinations of the eigenfunctions belonging to the same eigenvalue so that these new functions are orthogonal.

If $p(a) \neq 0$ and $p(b) \neq 0$, then $p(x) > 0$ on the closed interval $[a,b]$ (which follows from $p(x) \geq 0$) and we have the so-called *regular Sturm-Liouville problem*. If $p(a) = 0$, then we do not have the first of the boundary conditions in equation (4.4); instead we require $y(x)$ and $y'(x)$ to be finite at $x = a$. Similar situations occur if $p(b) = 0$, or if both $p(a) = 0$ and $p(b) = 0$. All these cases correspond to the so-called *singular Sturm-Liouville problem*.

If $p(a)=p(b)$ and also instead of equations (4.4) we have *periodic boundary conditions* $y(a) = y(b)$ and $y'(a) = y'(b)$, we have what is referred to as the *periodic Sturm-Liouville problem*.

The following summarizes the types of Sturm-Liouville problems:

1) For $p(x) > 0$ and $r(x) > 0$ we have the *regular* problem;

2) For $p(x) \geq 0$ and $r(x) \geq 0$ we have the *singular* problem;

3) For $p(a) = p(b)$ and $r(x) > 0$ we have the *periodic* problem.

Notice that the interval (a,b) can be infinite in which case the Sturm-Liouville problem is also classified as singular.

The following theorem gives a list of several important *properties of the Sturm-Liouville problem*:

Theorem

1) Each regular and each periodic Sturm-Liouville problem has an infinite number of non-negative, discrete eigenvalues $0 \leq \lambda_1 < \lambda_2 < ... < \lambda_n < ...$ such that $\lambda_n \to \infty$ as $n \to \infty$. All eigenvalues are real numbers.
2) For each eigenvalue of a regular Sturm-Liouville problem there is only one eigenfunction; for a periodic Sturm-Liouville problem this property does not hold.
3) For each of the types of Sturm-Liouville problems the eigenfunctions corresponding to different eigenvalues are linearly independent.
4) For each of the types of Sturm-Liouville problems the set of eigenfunctions is orthogonal with respect to the weight function $r(x)$ on the interval $[a,b]$.
5) If $q(x) \leq 0$ on $[a,b]$ and $\beta_1/\alpha_1 < 0$ and $\beta_2/\alpha_2 > 0$, then all $\lambda_n \geq 0$.

Some of these properties have been proven previously, such as property 4 and part of property 1. The remaining part of property 1 should be obvious and will be shown in several examples below, as well as property 5. Property 2 can be proved by postulating that there are two eigenfunctions corresponding to the same eigenvalue. We then apply equations (4.2) and (4.4) to show that these two eigenfunctions coincide or differ at most by some multiplicative constant. We leave this proof to the reader as a Reading Exercise. Similarly, property 3 is easily proven.

Bessel functions and the orthogonal polynomials, such as Legendre, arise from singular Sturm-Liouville problems; thus the first statement in the above theorem is not directly applicable to these important cases. In spite of that, singular Sturm-Liouville problems may also have an infinite sequence of discrete eigenvalues, which we will later verify directly for Bessel functions and for the orthogonal polynomials. It is interesting to notice that there is the possibility for a singular Sturm-Liouville problem to have a continuous range of eigenvalues, in other words a continuous spectrum; however, we will not meet such situations in the problems we study in this book.

When equation (4.11) is satisfied, eigenfunctions $y_n(x)$ form a *complete orthogonal set* on $[a,b]$. This means that any reasonable well-behaved function, $f(x)$, defined on $[a,b]$ can be expressed as a series of eigenfunctions (called a *generalized Fourier series*) of a Sturm-Liouville problem in which case we may write

$$f(x) = \sum_n^\infty a_n y_n(x), \tag{4.13}$$

where it is convenient, in some cases, to start the summation with $n = 1$, in other cases with $n = 0$. An expression for the coefficients a_n can be found by multiplying both sides of equation (4.13) by $r(x)y_n^*(x)$ and integrating over $[a,b]$ to give

$$a_n = \frac{\int_a^b r(x)f(x)y_n^*(x)dx}{\|y_n\|^2}. \tag{4.14}$$

The Sturm-Liouville theory provides a theorem for convergence of the series in equation (4.13) at every point x of $[a,b]$:

Theorem

Let $\{y_n(x)\}$ be the set of eigenfunctions of a regular Sturm-Liouville problem and let $f(x)$ and $f'(x)$ be piecewise continuous on a closed interval. Then the series expansion (4.13) converges to $f(x)$ at every point where $f(x)$ is continuous and to the value $[f(x_0 + 0) + f(x_0 - 0)]/2$ if x_0 is a point of discontinuity.

The theorem is also valid for the orthogonal polynomials and Bessel functions related to singular Sturm-Liouville problems. This theorem, which is extremely important for applications, is similar to the theorem for trigonometric Fourier series.

Formulas in equations (4.11) through (4.14) can be written in a more convenient way if we define a *scalar product of* eigenfunctions φ and ψ (for simplicity consider real functions) as the number given by

$$\varphi \cdot \psi = \int_a^b r(x)\varphi(x)\psi(x)\,dx. \tag{4.15}$$

This definition of the scalar product has properties identical to those for vectors in linear Euclidian space, a result which can be easily proved:

$$\varphi \cdot \psi = \psi \cdot \varphi,$$

$$(a\varphi) \cdot \psi = a\varphi \cdot \psi, \qquad \text{(where } a \text{ is a number)}$$

$$\varphi \cdot (a\psi) = a\varphi \cdot \psi, \tag{4.16}$$

$$\varphi \cdot (\psi + \phi) = \varphi \cdot \psi + \varphi \cdot \phi,$$

$$\varphi \cdot \varphi \geq 0.$$

The last property relies on the assumption made for the Sturm-Liouville equation that $r(x) \geq 0$. If φ is continuous on $[a,b]$, then $\varphi \cdot \varphi = 0$ only if φ is zero.

Reading Exercise

Prove the relations given in equations (4.18).

In terms of the scalar product, the orthogonality of eigenfunctions (defined by equation (4.11)) means that

$$y_n \cdot y_m = 0 \quad \text{if } n \neq m \tag{4.17}$$

and the formula for the Fourier coefficients in equation (4.14) becomes

$$a_n = \frac{f \cdot y_n}{y_n \cdot y_n}.$$
(4.18)

Functions satisfying the condition

$$\varphi \cdot \varphi = \int_a^b r(x) |\varphi(x)|^2 \, dx < \infty$$
(4.19)

belong to a Hilbert space, L^2, having infinite dimensionality. The complete orthogonal set of functions $\{y_n(x)\}$ serves as the orthogonal basis in L^2, where here, completeness means that series in equation (4.13) converges to $f(x)$ in the mean.

A discussion of Green's functions for the Sturm-Liouville problem can be found in book [1].

4.3 EXAMPLES OF STURM-LIOUVILLE PROBLEMS

Example 4.1

Solve the equation

$$y''(x) + \lambda y(x) = 0$$
(4.20)

on the interval $[0,l]$ with boundary conditions

$$y(0) = 0 \text{ and } y(l) = 0.$$
(4.21)

Solution. First, comparing equation (4.22) with equations (4.5) and (4.6), it is clear that we have a Sturm-Liouville problem with linear operator $L = -d^2/dx^2$ and functions $q(x) = 0$ and $p(x) = r(x) = 1$. (Note that L can be taken with either negative or positive sign as is clear from equations (4.2), (4.5), and (4.6).) As a Reading Exercise, verify that L is Hermitian.

Let us discuss the cases $\lambda = 0$, $\lambda < 0$, and $\lambda > 0$ separately. If $\lambda = 0$, then a general solution to equation (4.22) is

$$y(x) = C_1 x + C_2$$

and from boundary conditions (4.23) we have $C_1 = C_2 = 0$; that is, there exists only the trivial solution $y(x) = 0$. If $\lambda < 0$, then

$$y(x) = C_1 e^{\sqrt{-\lambda} x} + C_2 e^{-\sqrt{-\lambda} x}$$

and the boundary conditions (4.23) again give $C_1 = C_2 = 0$ and therefore the trivial solution $y(x) = 0$. Thus we have only the possibility $\lambda > 0$, in which case we write $\lambda = \mu^2$ with μ real and we have a general solution of equation (4.22) given by

$$y(x) = C_1 \sin \mu x + C_2 \cos \mu x.$$

The boundary condition $y(0) = 0$ requires that $C_2 = 0$ and the boundary condition $y(l) = 0$ gives $C_1 \sin \mu l = 0$. From this we must have $\sin \mu l = 0$ and $\mu_n = n\pi / l$ since the choice $C_1 = 0$ again gives the trivial solution. Thus, the eigenvalues are

$$\lambda_n = \mu_n^2 = \left(\frac{n\pi}{l} \right)^2, \quad n = 1, 2, \ldots, \tag{4.22}$$

and the eigenfunctions are $y_n(x) = C_n \sin \frac{n\pi x}{l}$ where for $n = 0$ we have the trivial solution $y_0(x) = 0$. It is obvious that we can restrict ourselves to positive values only of n since negative values do not give new solutions in the case that the constants C_n are arbitrarily. These eigenfunctions are orthogonal over the interval $[0, l]$ since we can easily show that

$$\int_0^l \sin \frac{n\pi x}{l} \cdot \sin \frac{m\pi x}{l} dx = 0 \quad \text{for } m \neq n. \tag{4.23}$$

The orthogonality of eigenfunctions follows from the fact that the Sturm-Liouville operator, L, is Hermitian for the boundary conditions given in equation (4.23).

The eigenfunctions may be normalized by writing

$$C_n^2 \int_0^1 \sin^2 \frac{n\pi x}{l} dx = C_n^2 \cdot \frac{l}{2} = 1$$

which results in the orthonormal eigenfunctions

$$y_n(x) = \sqrt{\frac{2}{l}} \sin n\pi x, \quad n = 1, 2, \ldots. \tag{4.24}$$

Thus we have shown that the boundary value problem consisting of equations (4.22) and (4.23) has eigenfunctions that are sine functions. It means that *the expansion in eigenfunctions of the Sturm-Liouville problem for solutions to equations (4.22) and (4.23) is equivalent to the trigonometric Fourier sine series.*

Reading Exercise: Suggest alternatives to boundary conditions (4.23) that will result in cosine functions as the eigenfunctions for equation (4.22).

The system of these eigenfunctions and eigenvalues is used in library problems 1,2,4,5,6,7,8,11,12 of program **Waves** *(Vibrations of a Finite String)* and in the library problems 1,4,5,6,10,11,12 of program **Heat** *(Heat Conduction within a Finite Uniform Rod)*.

Example 4.2

Determine the eigenvalues and corresponding eigenfunctions for the Sturm-Liouville problem

$$y''(x) + \lambda y(x) = 0 \tag{4.25}$$

$$y'(0) = 0, \quad y(l) = 0. \tag{4.26}$$

Solution. As in Example 1, the reader may check as a Reading Exercise that the parameter λ must be positive in order to have nontrivial solutions. Thus we may write $\lambda = \mu^2$, so that we have oscillating solutions given by

$$y(x) = C_1 \sin\mu x + C_2 \cos\mu x.$$

The boundary condition $y'(0) = 0$ gives $C_1 = 0$ and the boundary condition $y(l) = 0$ gives $C_2 \cos\mu l = 0$. If $C_2 = 0$, we have a trivial solution; otherwise we have $\mu_n = (2n+1)\pi/2l$, for $n = 0,1,2,....$ Therefore the eigenvalues are

$$\lambda_n = \mu_n^2 = \left[\frac{(2n-1)\pi}{2l}\right]^2, \quad n = 1,2,..., \tag{4.27}$$

and the eigenfunctions are

$$y_n(x) = C_n \cos\frac{(2n-1)\pi x}{2l}, \quad n = 1,2,.... \tag{4.28}$$

We leave it to the reader to prove that the eigenfunctions in equation (4.30) are orthogonal on the interval [0,1]. The reader may also normalize these eigenfunctions to find the normalization constant C_n, which is equal to $\sqrt{2/l}$.

Example 4.3

Determine the eigenvalues and eigenfunctions for the Sturm-Liouville problem

$$y''(x) + \lambda y(x) = 0 \tag{4.29}$$

$$y(0) = 0, \quad y'(l) + hy(l) = 0. \tag{4.30}$$

Solution. As in the previous examples, nontrivial solutions exist only when $\lambda > 0$ (the reader should verify this as a Reading Exercise). Letting $\lambda = \mu^2$ we obtain a general solution as

$$y(x) = C_1 \sin\mu x + C_2 \cos\mu x.$$

From the boundary condition $y(0) = 0$ we have $C_2 = 0$. The other boundary condition gives $\mu\cos\mu l + h\sin\mu l = 0$. Thus, the eigenvalues are given by the equation

$$\tan\mu_n l = -\mu_n / h. \tag{4.31}$$

We can obtain these eigenvalues by plotting $\tan\mu_n l$ and $-\mu_n / h$ on the same graph as in Figure 4.2. The graph is plotted for positive μ, because negative μ do not bring new solutions. From the figure it is directly seen that there is an infinite number of discrete eigenvalues. The eigenfunctions

$$y_n(x) = C_n \sin\mu_n x, \quad n = 1,2,... \tag{4.32}$$

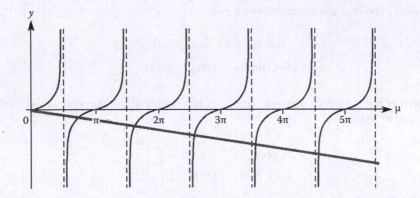

FIGURE 4.2　The functions $\tan\mu_n l$ and $-\mu_n/h$ (for $h = 5$) plotted against μ. The eigenvalues of the Sturm-Liouville problem in Example 4.3 are given by the intersections of these lines.

are orthogonal so that

$$\int_0^l \sin\mu_n x \cdot \sin\mu_m x\, dx = 0 \quad \text{for } m \neq n. \tag{4.33}$$

The orthogonality condition shown in equation (4.35) follows from the general theory as a direct consequence of the fact that the operator $L = -d^2/dx^2$ is Hermitian for the boundary conditions (4.32). We leave it to the reader to verify the previous statement as a Reading Exercise.

The normalized eigenfunctions are

$$y_n(x) = \sqrt{\frac{2\left(\mu_n^2 + h^2\right)}{l\left(\mu_n^2 + h^2\right) + h}}\ \sin\mu_n x. \tag{4.34}$$

Example 4.4

Solve the Sturm-Liouville problem

$$y'' + \lambda y = 0, \quad 0 < x < l, \tag{4.35}$$

on the interval $[0, l]$ with periodic boundary conditions

$$y(0) = y(l), \quad y'(0) = y'(l). \tag{4.36}$$

Solution. Again, verify as a Reading Exercise that nontrivial solutions exist only when $\lambda > 0$ (for which we will have oscillating solutions as before). Letting $\lambda = \mu^2$ we can write a general solution in the form

$$y(x) = C_1 \cos\mu x + C_2 \sin\mu x.$$

The boundary conditions in equations (4.38) give

$$\begin{cases} C_1(\cos\mu l - 1) + C_2\sin\mu l = 0, \\ -C_1\sin\mu l + C_2(\cos\mu l - 1) = 0. \end{cases} \tag{4.37}$$

This system of homogeneous algebraic equations has nontrivial solution only when its determinant is equal to zero:

$$\begin{vmatrix} \cos\mu l - 1 & \sin\mu l \\ -\sin\mu l & \cos\mu l - 1 \end{vmatrix} = 0, \tag{4.38}$$

which yields

$$\cos\mu l = 1. \tag{4.39}$$

The roots of equation (4.41) are

$$\lambda_n = \left(\frac{2\pi n}{l}\right)^2, \quad n = 0,1,2,\dots \tag{4.40}$$

With these values of λ_n, equations (4.39) for C_1 and C_2 have two linearly independent nontrivial solutions given by

$$C_1 = \begin{pmatrix} 1 \\ 0 \end{pmatrix} \quad \text{and} \quad C_2 = \begin{pmatrix} 0 \\ 1 \end{pmatrix}. \tag{4.41}$$

Substituting each set into the general solution we obtain the eigenfunctions

$$y_n^{(1)}(x) = \cos\sqrt{\lambda_n}\,x, \quad \text{and} \quad y_n^{(2)}(x) = \sin\sqrt{\lambda_n}\,x. \tag{4.42}$$

Therefore for the eigenvalue $\lambda_0 = 0$ we have the eigenfunction $y_0(x) = 1$ and a trivial solution $y(x) \equiv 0$. Each nonzero eigenvalue, λ_n, has two linearly independent eigenfunction so that for this example we have twofold degeneracy.

Collecting the above results we have that this boundary value problem with periodic boundary conditions has the following eigenvalues and eigenfunctions:

$$\lambda_n = \left(\frac{2\pi n}{l}\right)^2, \quad n = 0,1,2,\dots \tag{4.43}$$

$$y_0(x) \equiv 1, \quad y_n(x) = \begin{cases} \cos(2\pi nx/l), \\ \sin(2\pi nx/l), \end{cases} \tag{4.44}$$

$$\|y_0\|^2 = l, \quad \|y_n\|^2 = \frac{l}{2} \quad \text{(for } n=1,2,\dots\text{)}.$$

In particular, when $l = 2\pi$ we have

$$\lambda_n = n^2, \quad y_0(x) \equiv 1, \quad y_n(x) = \begin{cases} \cos nx \\ \sin nx . \end{cases}$$

From this we see that the boundary value problem consisting of equations (4.37) and (4.38) results in eigenfunctions for this Sturm-Liouville problem which allows *an expansion of the solution equivalent to the complete trigonometric Fourier series expansion.*

Example 4.5

In the following chapters of this book we will meet a number of two-dimensional Sturm-Liouville problems. Here we present a simple example for future reference.
  Consider the equation

$$\frac{\partial^2 u}{\partial x^2} + \frac{\partial^2 u}{\partial y^2} + k^2 u = 0 \tag{4.45}$$

where k is a real constant that determines the function $u(x,y)$ with independent variables in domains $0 \le x \le l$, $0 \le y \le h$. Define the two-dimensional Sturm-Liouville operator in a similar fashion as was done for the one-dimensional case.
  Solution. We have

$$L = -\frac{d^2}{dx^2} - \frac{d^2}{dy^2}. \tag{4.46}$$

Let the boundary conditions be Dirichlet type so that we have

$$u(0,y) = u(l,y) = u(x,0) = u(x,h) = 0. \tag{4.47}$$

Reading Exercise

By direct substitution into equation (4.47) and using boundary conditions (4.49), check that this Sturm-Liouville problem has the eigenvalues

$$k_{nm}^2 = \frac{1}{\pi^2}\left(\frac{n^2}{l^2} + \frac{m^2}{h^2}\right) \tag{4.48}$$

and the corresponding eigenfunctions

$$u_{nm} = \sin\frac{n\pi x}{l}\sin\frac{m\pi y}{h}, \quad n,m = 1,2,\dots \tag{4.49}$$

In the case of a square domain where $l = h$, the eigenfunctions u_{nm} and u_{mn} have the same eigenvalues, $k_{nm} = k_{mn}$, which is a degeneracy reflecting the symmetry of the problem with respect to x and y.

The system of these eigenfunctions and eigenvalues is used in library problems 1,5,6 of program **Waves** (*Vibrations of a Flexible Rectangular Membrane*) and in the library problems 1,3,7 of program **Heat** (*Heat Conduction within a Thin Uniform Rectangular Membrane*).

Example 4.6

Obtain the eigenfunction expansion (*generalized* Fourier series expansion) of the function $f(x) = x^2(1 - x)$ using the eigenfunctions of the Sturm-Liouville problem

$$y'' + \lambda y = 0, \quad 0 \leq x \leq \pi/2, \tag{4.50}$$

$$y'(0) = y'(\pi/2) = 0. \tag{4.51}$$

Solution. First, prove as a Reading Exercise that the eigenvalues and eigenfunctions of this boundary value problem are

$$\lambda = 4n^2, \quad y_n(x) = \cos 2nx, \ n = 0,1,2,... \tag{4.52}$$

A Fourier series expansion, given in equation (4.13), of the function $f(x)$ using the eigenfunctions above is

$$x^2(1 - x) = \sum_{n=0}^{\infty} c_n y_n(x) = \sum_{n=0}^{\infty} c_n \cos 2nx. \tag{4.53}$$

Since f and f' are continuous functions, this expansion will converge to $x^2(1 - x)$ for $0 < x < \pi/2$ as was shown previously. In equation (4.52) we see that the function $r(x) = 1$; thus the coefficients of this expansion obtained from equation (4.14) are

$$c_0 = \frac{\int_0^{\pi/2} x^2(1 - x)\,dx}{\int_0^{\pi/2} dx} = \frac{\pi^2}{4}\left(\frac{1}{3} - \frac{\pi}{8}\right),$$

$$c_n = \frac{\int_0^{\pi/2} x^2(1 - x)\cos 2nx \ dx}{\int_0^{\pi/2} \cos^2 2nx \ dx} = \frac{(-1)^n}{n^4}\left[1 - \frac{3\pi}{4} + \frac{3}{2\pi n^2}\right] - \frac{3}{2\pi n^4}, \quad n=1,2,3,...$$

Figure 4.3 shows the partial sum ($n = 10$) of this series, compared with the original function $f(x) = x^2(1 - x)$.

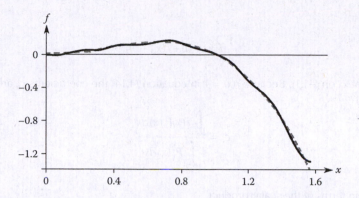

FIGURE 4.3   Graphs of the function $f(x) = x^2(1 - x)$ (dashed line) and partial sum with $n = 10$ of the Fourier expansion of $f(x)$ (solid line).

The system of these eigenfunctions and eigenvalues is used in library problems 14 of program **Waves** (*Vibrations of a Finite String*) and in the library problems 7 of program **Heat** (*Heat Conduction within a Finite Uniform Rod*).

Two important *special functions*, Legendre and Bessel functions, will be discuss later in detail. In the two following examples they serve simply as illustrations of Sturm-Liouville problems.

Example 4.7: (Fourier-Legendre Series).

The Legendre equation is

$$\frac{d}{dx}\left[(1-x^2)y'\right]+\lambda y = 0 \tag{4.54}$$

for x on the closed interval $[-1,1]$. There are no boundary conditions in a straight form because $p(x) = 1 - x^2$ vanishes at the endpoints. However, we seek a finite solution, a condition which in this case acts as a boundary condition.

   Solution. The Legendre polynomials, $P_n(x)$, are the only solutions of Legendre's equation that are bounded on the closed interval $[-1,1]$. The set of functions $\{P_n(x)\}$, where $n = 0,1,2,...$, is orthogonal with respect to the weight function $r(x) = 1$ on the interval $[-1,1]$ in which case the orthogonality relation is

$$\int_{-1}^{1} P_n(x)P_m(x)\,dx = 0 \quad \text{for } m \neq n. \tag{4.55}$$

The eigenfunctions for this problem are thus $P_n(x)$ with, as we will see in Chapter 8, eigenvalues $\lambda = n(n + 1)$ for $n = 0,1,2,...$
   If $f(x)$ is a piecewise smooth on $[-1,1]$, the series

$$\sum_{n=0}^{\infty} c_n P_n(x) \tag{4.56}$$

converges to

$$\frac{1}{2}[f(x_0 + 0) + f(x_0 - 0)] \tag{4.57}$$

at any point x_0 on $(-1,1)$. Because $r(x) = 1$ in equation (4.14) the coefficients c_n are

$$c_n = \frac{\int_{-1}^{1} f(x)P_n(x)dx}{\int_{-1}^{1} P_n^2(x)dx}, \tag{4.58}$$

or written in terms of the scalar product,

$$c_n = \frac{f(x) \cdot P_n(x)}{P_n(x) \cdot P_n(x)}. \tag{4.59}$$

Example 4.8: (Fourier-Bessel Series).

Consider the Sturm-Liouville problem

$$(xy')' + \left(\lambda x - \frac{v^2}{x}\right)y = 0, \quad 0 \leq x \leq 1 \tag{4.60}$$

with boundary conditions such that $y(0)$ is finite and $y(1) = 0$. Here v is a constant.

Solution. The eigenvalues for this problem are $\lambda = j_n^2$ for $n = 1,2,...$, where j_1, j_2, j_3, ... are the positive zeros of the functions $J_v(x)$, which are Bessel functions of order v. If $f(x)$ is a piecewise smooth function on the interval $[0,1]$, then for $0 < x < 1$ it can be resolved in the series

$$\sum_{n=1}^{\infty} c_n J_v(j_n x), \tag{4.61}$$

which converges to

$$\frac{1}{2}[f(x_0 + 0) + f(x_0 - 0)]. \tag{4.62}$$

Since, in this Sturm-Liouville problem, $r(x) = x$, the coefficients c_n are

$$c_n = \frac{\int_0^1 xf(x)J_v(j_n x)dx}{\int_0^1 xJ_v^2(j_n x)dx}, \tag{4.63}$$

or in terms of the scalar product,

$$c_n = \frac{f(x) \cdot J_v(j_n x)}{J_v(j_n x) \cdot J_v(j_n x)}. \tag{4.64}$$

Problems

In problems 1 through 7, find eigenvalues and eigenfunctions of the Sturm-Liouville problem for the equation

$$y''(x) + \lambda y(x) = 0$$

with the following boundary conditions:

1. $y(0)=0$, $y'(l)=0$

2. $y'(0)=0$, $y(\pi) = 0$

3. $y'(0)=0$, $y'(1)=0$

4. $y'(0)=0$, $y'(1)+y(1)=0$

5. $y'(0)+y(0)=0$, $y(\pi) = 0$

6. $y(-l) = y(l)$, $y'(-l) = y'(l)$ (periodic boundary conditions). As was noted above, if the boundary conditions are periodic, then the eigenvalues can be degenerate. Show that in this problem two linearly independent eigenfunctions exist for each eigenvalue.

7. For the operator $L = -d^2/dx^2$ acting on functions $y(x)$ defined on the interval $[0,1]$, find their eigenvalues and eigenfunctions (assume $r(x) = 1$):

   a) $y(0) = 0$, $y'(1)+y(1)=0$

   b) $y'(0)-y(0)=0$, $y(1) = 0$

   c) $y(0) = y(1)$, $y'(0) = y'(1)$

CHAPTER 5

Qualitative Methods and Stability of ODE Solutions

5.1 QUALITATIVE APPROACH FOR AUTONOMOUS FIRST-ORDER EQUATIONS: EQUILIBRIUM SOLUTIONS AND PHASE LINES

Very often we would like to obtain some rough understanding of the solutions of an IVP without actually solving it. In this section we describe one such method, called *the phase line*. This method works equally well for linear and nonlinear autonomous equations. However, it must be always remembered that it is not a full substitute for the rigorously calculated solution. The method does not result in a solution formula, and it is the latter that is the most valuable.

The phase line is most useful, and even indispensable, for the analysis of nonlinear equations, since such equations rarely permit analytical solution. Thus one often resorts to solving the equation numerically, but this has a disadvantage of computing only one solution at a time. In the numerical approach there is always an inherent risk of not choosing the initial condition that results in the solution which is qualitatively different from other solutions, and thus failing to completely determine the solution set. The phase line, on the other hand, allows one to determine (qualitatively) all solutions in a single step.

Certainly, the qualitative method such as phase line should not be used for the equations that can be solved analytically, but for the purpose of instruction, in this section we are applying this method to rather simple equations. We compare to analytical solutions where such comparison is warranted.

We first introduce the notions of an autonomous equation (AE) and its equilibrium solution. An autonomous first-order equation has the form $y' = f(y)$; that is, the right-hand side is t- independent. (The independent variable is denoted as t throughout this chapter and it is ascribed the meaning of time.) The equilibrium solution of AE is the solution that does not change (is a constant). Thus if $y = C$ is such solution, then $y' = 0$, and the equation $y' = f(y)$ now states $f(y) = 0$. Thus to find all equilibrium solutions means to solve the algebraic equation $f(y) = 0$. Equilibrium solutions are also often called equilibrium points or critical points.

As a side note, an autonomous first-order equation is separable—its solutions (in fact, the inverse solution function $t = F(y)$) are given by the integral of $1/f(y)$. However, for many nonlinear functions $f(y)$ there is no analytical method to calculate the integral in elementary functions ($f(y) = e^{\pm y^2}$ is one such example, and one which is most often cited). The phase line comes to the rescue here.

Why are equilibrium solutions useful? Since they are constants, we can plot them on the real number line. Then, we mark the sign of $f(y)$ in the intervals between equilibrium solutions. It is convenient to position the line vertically and use the up (down) arrow for the positive (negative) sign. Notice that since $y' = f(y)$, we are actually marking the sign of y'. Thus, if $f(y) > 0$ (<0) on an interval, then the solution is increasing (decreasing) on this interval. Now, the completed phase line allows one to determine the qualitative behavior of time-dependent (non-equilibrium) solutions. The three examples below demonstrate how this works.

Example 1.1

For IVP $\dfrac{dy}{dt} = ky, \; y(t_0) = y_0$:

(a) Find equilibrium solutions.
(b) Use a phase line to determine qualitative behavior of nontrivial solutions.

Equilibrium solutions are determined from the equation $f(y) = 0$. Thus $ky = 0$, and the equilibrium solution is $y = 0$. Clearly, for $y < 0$, $f(y) < 0$ (>0) if $k > 0$(<0). But, for $y > 0$, $f(y) > 0$(<0) if $k > 0$(<0). The phase lines for $k > 0$ and $k < 0$ are shown in Figure 5.1. Next to the phase lines we show the sketches of some solutions. The first phase line indicates that a positive initial condition y_0 gives rise to an increasing solution, and a negative initial condition leads to a decreasing one. This is certainly correct, since the exact solution of this IVP by separation of variables is $y = y_0 e^{kt}$, which is increasing and positive (decreasing and negative) for $k > 0$, $y_0 > 0$ ($k > 0$, $y_0 < 0$). Similarly, the second line indicates that a positive initial condition y_0 gives rise to a decreasing solution, and a negative initial condition yields an increasing one. Again this is correct, since the analytical solution $y = y_0 e^{kt}$ is decreasing and positive (increasing and negative) for $k < 0$, $y_0 > 0$ ($k < 0$, $y_0 < 0$).

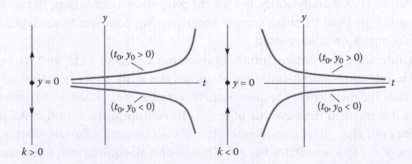

FIGURE 5.1 Phase lines and two characteristic solutions of IVP $y' = ky$, $y(t_0) = y_0$. Upward-pointing arrows mark increasing solutions ($y' > 0$); downward-pointing arrows, decreasing solutions ($y' < 0$). Solutions that correspond to other values of y_0 have same trends (are qualitatively same) as the two solutions shown.

Next, notice that the uniqueness theorem prohibits crossing or merging of the solution curves that correspond to different initial conditions; this includes intersections or merging with the equilibrium solution line $y = 0$. Taken together with the solution increasing/decreasing analysis just performed, this leaves only one option for the shape of a solution curve when $y > 0$ or $y < 0$, as shown next to the phase line. That is, the time-dependent solutions of $y' = ky$ (integral curves) *are asymptotic to the equilibrium solution*. Certainly, this asymptotic property matches the behavior of the exact solution $y = y_0 e^{kt}, y_0 \neq 0$ as $t \rightarrow \infty$ ($k < 0$), or $t \rightarrow -\infty$ ($k > 0$).

Since the method is qualitative, we are not concerned with the exact shapes of the solution curves. It is sufficient here to determine that the solutions are increasing or decreasing in the certain intervals of y values shown on the phase line. We can sketch more solution curves in Figure 5.1 that follow the solution trends already shown. In the sketch we should only care that the solution curves do not intersect and do not show a tendency toward intersecting, since intersections are not allowed due to the uniqueness property of solutions. Saying this differently, while the phase line succeeds in establishing the solution trend (increasing vs. decreasing), it is incapable of predicting the *rate* of the solution change, that is, the steepness of the solution curve. Of course, the growth or decay rate of an exact solution is defined precisely at all times; it is $y' = y_0 k e^{kt}$.

Example 1.2

For IVP $\dfrac{dy}{dt} = y(y - 1)$, $y(t_0) = y_0$:

(a) Find equilibrium solutions.
(b) Use a phase line to determine qualitative behavior of nontrivial solutions.

Equilibrium solutions are determined from the equation $f(y) = 0$. Thus $y(y - 1) = 0$, and the equilibrium solutions are $y = 0$ and $y = 1$. The phase line and the sketch of solutions are shown in Figure 5.2. According to the phase line, all initial conditions above the line $y = 1$ in the sketch give rise to increasing solutions; all initial conditions between this line and the line $y = 0$ give rise to the decreasing solutions; and all initial conditions below the line $y = 0$ give rise to increasing solutions. The time-dependent integral curves must be asymptotic to the equilibrium solutions due to the uniqueness theorem, which results in the unique trend of IVP solution curves in the sketch. Note that two vertical asymptotes (dashed lines in Figure 5.2) are the result of the analytical solution of the IVP; they can't be obtained from the qualitative reasoning.

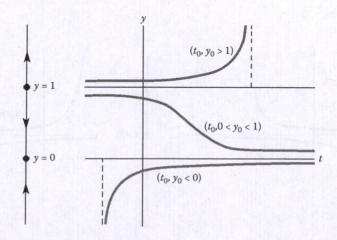

FIGURE 5.2 Phase line and three characteristic solutions of IVP $y' = y(y - 1)$, $y(t_0) = y_0$.

In Example 1.2, the equilibrium solution $y = 1$ is unstable, because all nearby initial conditions produce solutions that diverge (tend away) from $y = 1$ as t increases. It is commonplace to call such equilibrium solution *a source, or unstable node*. Similarly, the equilibrium solution $y = 0$ is called *a sink, or stable node* since it is stable—all nearby initial conditions produce solutions that converge to $y = 0$ as t increases. In Example 1.1, $y = 0$ is a source for $k > 0$, and a sink for $k < 0$. We will discuss stability of solutions in Section 5.3.

Exact analytical solution of IVP 1.2 is $y = 1/(1 \pm Ce^t)$, where $C = e^{-t_0} |1 - 1/y_0|$. It can be easily obtained using separation of variables, partial fractions decomposition, and integration. To make sure that the phase line gives correct qualitative information about the solutions, you can plot a few analytical solution curves for the initial conditions as in Figure 5.2.

Example 1.3

For IVP $\dfrac{dy}{dt} = ky^2$, $y(t_0) = y_0$:

- (a) Find equilibrium solutions.
- (b) Use a phase line to determine qualitative behavior of nontrivial solutions.
- (c) Confirm your analysis by solving IVP exactly using separation of variables.

Equilibrium solutions are determined from the equation $f(y) = 0$. Thus $ky^2 = 0$, and the equilibrium solution is $y = 0$. The phase line and the sketch of solutions are shown in Figure 5.3. According to the phase line, all initial conditions for $k > 0(< 0)$ give rise to increasing (decreasing) solutions. The time-dependent integral curves must be asymptotic to the equilibrium solutions due to the uniqueness theorem, which results in the unique trend of the IVP solution curves in the sketch.

The exact solution of the IVP using separation of variables is $y = 1/[(\frac{1}{y_0} + kt_0) - kt]$. Let $k, y_0, t_0 > 0$. As t increases from t_0 the denominator decreases, and as long as it is positive, the solution is positive and increases. This matches the solution trend in Figure 5.3 (left sketch) for $y > 0$. However, the exact solution ceases to exist at $t = \frac{1}{ky_0} + t_0$ (vertical asymptote there, solution growth rate tends to infinity as $t \to (ky_0)^{-1} + t_0$), which the qualitative analysis is incapable of predicting. Similar to this case, other cases of k, y_0, t_0 can be analyzed.

The equilibrium solution $y = 0$ of Example 1.3 is called *a degenerate node*, and it is unstable. Lastly, we provide the following analytical criteria for a sink, a source, and a degenerate node. Let $y = y_*$ be

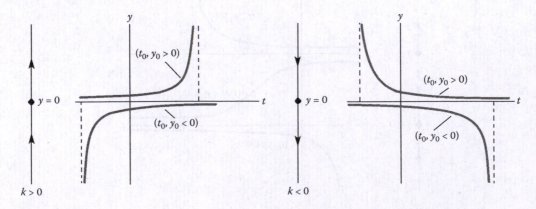

FIGURE 5.3   Phase line and two characteristic solutions of IVP $y' = ky^2$, $y(t_0) = y_0$.

$$a \text{ degenerate node, then } f'(y_*) = 0;$$
$$a \text{ sink, then } f'(y_*) < 0; \tag{1.1}$$
$$a \text{ source, then } f'(y_*) > 0.$$

In cases when the derivative $f'(y)$ can be easily calculated, these conditions are convenient for confirming the equilibrium point type. For instance, in Example 1.3 $f'(y) = 2ky$, $y_* = 0$, and $f'(y_*) = 0$. Importantly, the condition $f'(y_*) = 0$ can be useful when one wishes to *find* degenerate nodes. However, caution is needed, as the following example shows.

Example 1.4

For IVP $\dfrac{dy}{dt} = ky^3$, $y(t_0) = y_0$:

(a) Find equilibrium solutions;
(b) Use a phase line to determine qualitative behavior of nontrivial solutions.

The equilibrium solution is again $y = 0$. The phase line and the sketch of solutions coincide with the one in Figure 5.1. The equilibrium solution $y = 0$ is the source for $k > 0$ and sink for $k < 0$, but notice that $f'(0) = 0$. Thus it can be concluded that the converse of the statements (1.1) is not true, and one has to always supplement the calculation of the derivative at the equilibrium point with the analysis of the sign of $f(y)$ at the neighboring points. (As an exercise, obtain the analytical solution: $y = \pm 1/\sqrt{C - 2kt}$, where $C = y_0^{-4} + 2kt_0$.)

A phase line helps in the studies of *bifurcations* in autonomous equations containing a parameter. A bifurcation is an event, such as the parameter attaining a critical value, which results in a drastic (qualitative) change in the solution behavior. For instance, when the parameter k in the Example 1.1 changes from negative to positive—that is, passes through zero—the sink bifurcates into the source. (Value $k = 0$ is called the bifurcation value.)

Problems | Answers

1. $y' = ky^3$, $y(t_0) = y_0$
Eq. sol.: $y = 0$ (source). Exact sol.: $y = \pm (C - 2kt)^{-1/2}$

2. $y' = y^3(y-2)(y+1)$, $y(t_0) = y_0$
Eq. sol.: $y = -1$ (source), $y = 0$ (sink), $y = 2$ (source)

3. $y' = (y+3)\ln|y|$, $y(t_0) = y_0$
Eq. sol.: $y = -3$ (source), $y = -1$ (sink), $y = 1$ (source)

4. $y' = 2ye^y \sqrt[3]{y}$, $y(t_0) = y_0$
Eq. sol.: $y = 0$ (degenerate node)

5. $y' = \dfrac{-3y}{(y-1)(y-9)}$, $y(t_0) = y_0$
Eq. sol.: $y = 0$ (sink)

6. $y' = \sin y$, $y(t_0) = y_0$
Eq. sol.: $y = 2n\pi$ (source), $y = (2n + 1)\pi$ (sink), $n = 0$, ± 1, ± 2,...

7. $y' = -1 + \cos y$, $y(t_0) = y_0$
Eq. sol.: $y = 2n\pi$ (degenerate node), $n = 0$, ± 1, ± 2,...

8. $y' = -y\sin y$, $y(t_0) = y_0$
Eq. sol.: $y = 0$ (degenerate node), $y = (2n + 1)\pi$, $n = -1, -2,...$ (sink), $y = (2n + 1)\pi$, $n = 0, 1, 2,...$ (source), $y = 2n\pi$, $n = -1, -2,...$ (source), $y = 2n\pi$, $n = 1, 2,...$ (sink).

We will not dwell further on this topic; interested readers should consult other sources, for instance Chapter 3 in *Nonlinear Dynamics and Chaos* by S.H. Strogatz, or the more advanced *Nonlinear Systems* by P.G. Drazin.

In problems 1–8, draw a phase line, sketch solutions next to a line, and classify equilibrium points into sources, sinks, and degenerate nodes. Also, in problem 1 obtain the exact solutions using separation of variables and compare the qualitative analysis to these solutions.

5.2 QUALITATIVE APPROACH FOR AUTONOMOUS SYSTEMS OF FIRST-ORDER EQUATIONS: EQUILIBRIUM SOLUTIONS AND PHASE PLANES

The phase plane method is the useful qualitative approach to study some important properties of solutions of systems of autonomous ordinary differential equations. (*Autonomous* again means that the right-hand sides of the system's equations do not depend explicitly on the independent variable.) This method has all the advantages (and disadvantages) over the numerical and analytical solution methods for systems that the phase line method has over the corresponding methods for a single first-order equation. The phase plane method can be also applied to high-order equations because they can be reduced to a system of first-order equations. Let first show how to do this reduction of any nth-order equation

$$y^{(n)} = f\left(t, y, y', ..., y^{(n-1)}\right). \tag{5.1}$$

(Also see Section 3.1.) The equation does not need to be autonomous to allow reduction to a system of equations; thus here, for extra generality, we assume that f depends explicitly on t, that is, (2.1) is non-autonomous. It is convenient to change the notation of $y(t)$ to $x_1(t)$, and then denote $y'(t)$ as $x_2(t)$, $y''(t)$ as $x_3(t)$, and so on, and finally, $y^{(n-1)}(t)$ as $x_n(t)$. Then, equation (5.1) can be considered as a system of n first-order equations:

$$\begin{cases} x_1' = x_2(t), \\ x_2' = x_3(t), \\ x_3' = x_4(t), \\ \qquad \text{.....................} \\ x_n' = f\left(t, x_1(t), x_2(t), ..., x_n(t)\right). \end{cases} \tag{5.2}$$

As a side note, such reduction to a system of the first-order equations, besides its importance as a possible general scheme to solve the nth-order differential equation, also allows discussion of the existence and uniqueness of a particular solution of a problem consisting of equation (5.1) and initial conditions.

For instance, the second-order autonomous equation

$$y'' = f\left(y, y'\right) \tag{5.3}$$

is equivalent to the system

$$x_1' = x_2$$
$$x_2' = f(x_1, x_2).$$

(5.4)

In vector form the general autonomous system of two first-order equations is written as

$$X' = F(X),$$

(5.5)

where

$$X = \begin{pmatrix} x_1(t) \\ x_2(t) \end{pmatrix}, \quad F(X) = \begin{pmatrix} f_1(x_1, x_2) \\ f_2(x_1, x_2) \end{pmatrix}$$

(5.6)

are the column-vectors whose components are the unknown functions $x_1(t), x_2(t)$ and the known functions f_1 and f_2.

How can we visualize solutions of system (5.5)? Once the pair of solutions $(x_1(t), x_2(t))$ is known, it can be thought of as a point that sweeps out a solution curve in the $x_1 x_2$-plane as time varies. The $x_1 x_2$-plane in this context is called a phase plane, and the solution curve is called a *phase trajectory*, or *orbit*. (In other words, a phase trajectory is a parametric curve, where the parameter is t). The shape of a phase trajectory is determined by the functions f_1 and f_2 and the initial conditions for $x_1(t)$ and $x_2(t)$. A few trajectories that correspond to different initial conditions comprise a *phase portrait* of the system, from which one can see the behavior of all solutions. Isolated points in the phase portrait mark equilibrium solutions, also called the critical points (more on this terminology below). Since the derivative of a constant solution is zero, from system (5.5) it follows that equilibrium points are the solutions of an algebraic system

$$f_1(x_1, x_2) = 0,$$
$$f_2(x_1, x_2) = 0.$$

(5.7)

A trajectory that is a closed curve corresponds to periodic solutions, since the solution that starts at any point on a curve returns to this point after some time (this time is, of course, the period of the solutions $x_1(t)$ and $x_2(t)$).

A vector function $F(X)$ defines a *vector field* (also called *direction plot*) in the phase plane. That is, every point (x_1, x_2) in the plane could be a base of a vector with the components $f_1(x_1, x_2) / \sqrt{f_1(x_1, x_2)^2 + f_2(x_1, x_2)^2}$ and $f_2(x_1, x_2) / \sqrt{f_1(x_1, x_2)^2 + f_2(x_1, x_2)^2}$; the radical ensures that vectors at different points have unit length, otherwise they will overlap when plotted. Picking up at random (or according to a smart rule) a sufficient number of points and plotting a vector at each point is a way of obtaining the information on solutions, *since a vector at a point is tangent to a trajectory there.* Thus trajectories are easy to visualize from a vector field. (Of course, the trajectories always can be found numerically if high solution accuracy is needed, but this lies beyond the scope of a qualitative technique.) Construction of a vector field is rarely done by hand, since this would be very tedious for even a few tens of vectors—the number that usually is not sufficient for plotting all representative trajectories.

To understand why equilibrium points are also called critical points, note that an equilibrium point is equivalently defined as such point, where the tangent to the solution curve does not exist, that is, $dx_2/dx_1 = f_2(x_1,x_2)/f_1(x_1,x_2) = 0/0$ (see (5.4) and (5.7)). For a *linear* system, $f_1(x_1,x_2) = ax_1 + bx_2$, and $f_2(x_1,x_2) = cx_1 + dx_2$ (where a,b,c,d are constants) and when the determinant $\begin{vmatrix} a & b \\ c & d \end{vmatrix} \neq 0$ the indeterminate form $dx_2/dx_1 = 0/0$ is possible only at the origin $(x_1,x_2) = (0,0)$. *Thus for a linear system the origin is the only equilibrium point when the coefficient matrix is nonsingular.* This is also stated in Theorem I below.

If a system (linear or nonlinear) is autonomous, then it can be proved that there is a unique trajectory through each point in the phase plane—in other words, trajectories do not cross or merge.

Example 2.1

Consider the nonlinear second-order equation $y'' = y^2 - (y')^2$. The corresponding system of two first-order equations is

$$x_1' = f_1(x_2) = x_2$$

$$x_2' = f_2(x_1, x_2) = x_1^2 - x_2^2.$$

Equilibrium points are found from the system

$$x_2 = 0$$

$$x_1^2 - x_2^2 = 0.$$

The only solution of this system is $(x_1, x_2) = (0,0)$, which is the origin of a phase plane. The vector field is

$$F\begin{pmatrix} x_1 \\ x_2 \end{pmatrix} = \begin{pmatrix} x_2 \\ x_1^2 - x_2^2 \end{pmatrix}.$$

Figure 5.4 shows that all initial conditions in the upper half-plane give rise to trajectories that after some time approach the line $y = x$ in the first quadrant and both x_1 and x_2 tend to infinity as $t \to \infty$. All initial conditions in the lower half-plane give rise to trajectories that after some time approach the line $y = -x$ in the fourth quadrant and $x_1 \to \infty$, $x_2 \to -\infty$ as $t \to \infty$. Solutions that start in the left half-plane are first attracted and then are repelled by the equilibrium. The system never settles at the equilibrium.

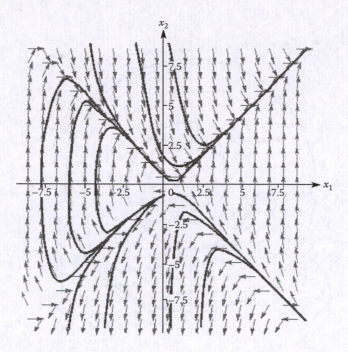

FIGURE 5.4 Phase portrait and the vector field of the system in Example 2.1.

Example 2.2 (Mass-Spring)

Consider a mass-spring system—a one-dimensional harmonic oscillator with zero damping. The equation of motion for the mass m is $y'' + \omega^2 y = 0$. Here $y(t)$ is the coordinate of an oscillator as a function of time, $\omega = \sqrt{k/m}$ is the frequency of the oscillation, and k is a spring constant. The corresponding system of two first-order equations is

$$x_1' = f_1(x_2) = x_2$$

$$x_2' = f_2(x_1, x_2) = -\omega^2 x_1.$$

The only equilibrium point is the origin, $(x_1, x_2) = (0,0)$. Now let $\omega = 1$. Solving the homogeneous equation $y'' + y = 0$ gives the general solution $y = C_1 \cos t + C_2 \sin t$. Next, suppose the initial condition for the oscillator is $(y, y') = (1,0)$, which translates into the initial condition $(x_1, x_2) = (1,0)$ in the phase plane. The solution of the IVP is $y = \cos t$. Its derivative is $y' = -\sin t$. Since $y^2 + (y')^2 = 1$, the corresponding trajectory in the phase plane is $x_1^2 + x_2^2 = 1$, which is the circle of radius one centered at the origin. As t increases from zero, the point moves clockwise along the trajectory and returns to $(1,0)$ at $t = 2\pi$ (the period of the solution $y = \cos t$). Clearly, the trajectories corresponding to other initial conditions are also circles centered at the origin, and the period of motion along all trajectories is constant and equal to 2π when $\omega = 1$ (the period is $2\pi/\omega$). Saying this a little differently, in the phase plot the signature of a periodic oscillation of solutions can be seen in that x_1 and x_2 change the sign periodically and for as long as the trajectory is traced (i.e., forever); also the trajectory returns to where it started after a fixed and constant time elapsed (the period).

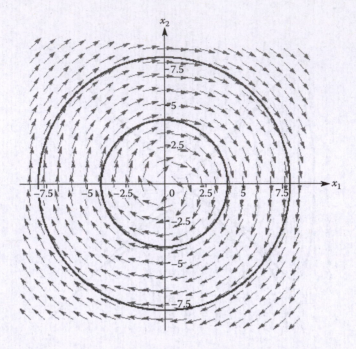

FIGURE 5.5   Phase portrait and the vector field of the system in Example 2.2.

5.2.1 Phase Planes of Linear Autonomous Equations

Consider first the important case of a linear, homogeneous second-order equation with constant coefficients:

$$y'' + py' + qy = 0. \tag{5.8}$$

The corresponding planar system is

$$
\begin{aligned}
x_1' &= f_1(x_2) = x_2 \\
x_2' &= f_2(x_1, x_2) = -qx_1 - px_2.
\end{aligned}
\tag{5.9}
$$

Clearly, the only equilibrium point of this system is (0,0). The vector form of the system is

$$X' = F(X) = AX, \tag{5.10}$$

where

$$X = \begin{pmatrix} x_1 \\ x_2 \end{pmatrix} = \begin{pmatrix} y \\ y' \end{pmatrix}$$

is the column-vector of unknowns, and

$$A = \begin{pmatrix} 0 & 1 \\ -q & -p \end{pmatrix}$$

is the two-by-two matrix of coefficients. Equations (5.9) and (5.10) are the particular case of a more general system

$$x_1' = f_1(x_1, x_2) = ax_1 + bx_2$$
$$x_2' = f_2(x_1, x_2) = cx_1 + dx_2, \tag{5.11}$$

where a,b,c,d are constants; it is assumed that at least one of these constants is nonzero. In vector form,

$$X' = AX, \ A = \begin{pmatrix} a & b \\ c & d \end{pmatrix}. \tag{5.12}$$

Next, we focus attention on the system (5.11) ((5.12)). The analysis that we present applies to the system (5.9) ((5.10)) as well, if in the formulas one replaces (a,b,c,d) by $(0,1,-q,-p)$. Without presenting the proof, we state the following theorem.

Theorem I

The planar linear system $X' = AX$ has a unique equilibrium point if $\det(A) \neq 0$, and a straight line of (infinitely many) equilibrium points if $\det(A) = 0$.

We will further consider only the first case; that is, we assume that matrix A is nonsingular. The topology of a phase portrait and thus the behavior of solutions to system (5.11) depends at large on *eigenvalues* of A (see Appendix 4 and also Section 3.4, where eigenvalues are denoted by k). Eigenvalues λ are the solutions of the equation

$$\det(A - \lambda I) = 0, \tag{5.13}$$

where I is the identity matrix, $I = \begin{pmatrix} 1 & 0 \\ 0 & 1 \end{pmatrix}$. Equation (5.13) is called the characteristic equation for matrix A. In expanded form, it reads

$$(a-\lambda)(d-\lambda) - bc = \lambda^2 - \lambda(a+d) + (ad-bc) = \lambda^2 - T(A)\lambda + \det(A) = 0, \tag{5.14}$$

where $T(A)$ is trace of A. We can immediately notice that the assumption we made above, $\det(A) \neq 0$, is equivalent to $\lambda \neq 0$, $\lambda \neq T(A)$. Thus the system has the unique equilibrium point $(0,0)$ only if there is no zero eigenvalue(s). Being second-order, the characteristic equation (5.14) has two solutions. These solutions can be real, or two complex-conjugate solutions may occur. Since our interest in this section is in determining the topology of a phase portrait, we briefly consider all possible cases for eigenvalues (excluding zero eigenvalue). The reader is referred to Section 3.4 and Appendix 4 for further details.

5.2.2 Real and Distinct Eigenvalues

Obviously, there are three cases to consider:

1. $\lambda_1 < 0 < \lambda_2$; the (unstable) equilibrium point $(0,0)$ is called *a saddle*.

2. $\lambda_1 < \lambda_2 < 0$; the (stable) equilibrium point $(0,0)$ is called *a sink, or stable node*.

3. $0 < \lambda_1 < \lambda_2$; the (unstable) equilibrium point $(0,0)$ is called *a source, or unstable node*.

We discuss one example for each case.

Example 2.3 (Sink)

Consider (5.12), where $A = \begin{pmatrix} -2 & -2 \\ -1 & -3 \end{pmatrix}$. The characteristic equation (5.14) is

$\lambda^2 + 5\lambda + 4 = 0$, which has solutions $\lambda_1 = -1$, $\lambda_2 = -4$. The corresponding eigenvectors are

$X_1 = \begin{pmatrix} -2 \\ 1 \end{pmatrix}$, $X_2 = \begin{pmatrix} 1 \\ 1 \end{pmatrix}$ (see Appendix 4). The general solution of the system is

$$X = C_1 X_1 e^{\lambda_1 t} + C_2 X_2 e^{\lambda_2 t} = C_1 \begin{pmatrix} -2 \\ 1 \end{pmatrix} e^{-t} + C_2 \begin{pmatrix} 1 \\ 1 \end{pmatrix} e^{-4t}.$$

Linearly independent eigenvectors X_1 and X_2 define two distinct lines through the origin, given by equations $x_2 = x_1, x_2 = -x_1/2$. In Figure 5.6 the vector field and several phase trajectories are

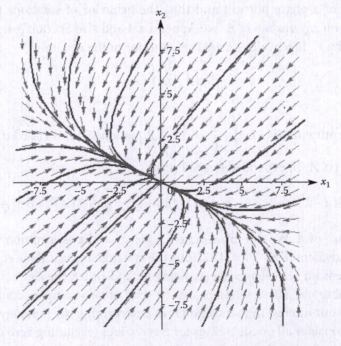

FIGURE 5.6 Phase portrait and the vector field of the system in Example 2.3.

plotted. Solutions curves come from infinity along the line $x_2 = x_1$, corresponding to the largest (in magnitude) negative eigenvalue, bend near the equilibrium point so that they become tangent to the line $x_2 = -x_1/2$, and then "sink into" the equilibrium.

Example 2.4 (Saddle)

Let in (5.12) $A = \begin{pmatrix} 0 & 1 \\ 4 & 3 \end{pmatrix}$. Comparing this matrix to the matrix in (5.10), we see that $q = -4$, $p = -3$, and therefore the system (5.12) corresponds to the second-order equation $y'' - 3y' - 4y = 0$. The characteristic equation for this homogeneous equation is $k^2 - 3k - 4 = 0$, which has solutions $k_1 = -1$, $k_2 = 4$. Thus the general solution is $y = C_1 e^{-t} + C_2 e^{4t}$. Now we can write the vector form of the general solution using $y' = -C_1 e^{-t} + 4C_2 e^{4t}$ as the second component:

$$X = \begin{pmatrix} y \\ y' \end{pmatrix} = \begin{pmatrix} x_1 \\ x_2 \end{pmatrix} = C_1 X_1 e^{k_1 t} + C_2 X_2 e^{k_2 t} = C_1 \begin{pmatrix} 1 \\ -1 \end{pmatrix} e^{-t} + C_2 \begin{pmatrix} 1 \\ 4 \end{pmatrix} e^{4t}.$$

This is the same general solution that we would have obtained if we decided instead to solve (5.12) directly. Indeed, the characteristic equation for A is $\lambda^2 - 3\lambda - 4 = 0$ (see (5.14)), which is the same equation as the characteristic equation we just solved. Thus the eigenvalues are the same two numbers, $\lambda_1 = -1$, $\lambda_2 = 4$. The eigenvectors are $\begin{pmatrix} 1 \\ -1 \end{pmatrix}$ and $\begin{pmatrix} 1 \\ 4 \end{pmatrix}$, that is, they coincide with X_1 and X_2 found above. The phase portrait is shown in Figure 5.7. Note that trajectories approach the origin along the line of eigenvectors $x_2 = -x_1$, corresponding to the negative eigenvalue (stable axis), and tend to infinity along the line of eigenvectors $x_2 = 4x_1$, corresponding to the positive eigenvalue (unstable axis).

Example 2.5 (Source)

Let in (5.12) $A = \begin{pmatrix} 2 & 0 \\ 0 & 3 \end{pmatrix}$. The characteristic equation for A is $(2 - \lambda)(3 - \lambda) = 0$. Its solutions are $\lambda_1 = 2$, $\lambda_2 = 3$. As in Example 2.4, eigenvectors can be easily determined. We notice that $a = 2, d = 3, b = c = 0$, and thus equations in system (5.11) are completely decoupled:

$$x_1' = 2x_1$$

$$x_2' = 3x_2.$$

Thus each equation can be solved independently of another one by using, say, separation of variables. The solutions are $x_1 = C_1 e^{2t}$, $x_2 = C_2 e^{3t}$. The vector form of the solution to the system is

$$X = \begin{pmatrix} x_1 \\ x_2 \end{pmatrix} = C_1 X_1 e^{2t} + C_2 X_2 e^{3t} = C_1 \begin{pmatrix} 1 \\ 0 \end{pmatrix} e^{2t} + C_2 \begin{pmatrix} 0 \\ 1 \end{pmatrix} e^{3t},$$

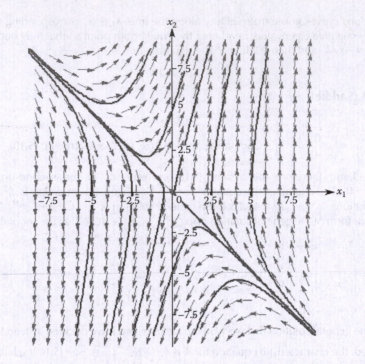

FIGURE 5.7 Phase portrait and the vector field of the system in Example 2.4.

where $\begin{pmatrix} 1 \\ 0 \end{pmatrix}$ and $\begin{pmatrix} 0 \\ 1 \end{pmatrix}$ are the eigenvectors. The lines of eigenvectors are $x_2 = 0$ and $x_1 = 0$.

The solutions leave the equilibrium tangent to the first line (smallest eigenvalue) and tend to infinity along the second line (largest eigenvalue).

The phase portrait is shown in Figure 5.8. Notice that eigenvectors are aligned with coordinate axes.

5.2.3 Complex Eigenvalues

There are three cases to consider:

1. $\lambda_1 = \lambda_r + i\lambda_i$, $\lambda_2 = \lambda_r - i\lambda_i$, $\lambda_r < 0$; the (stable) equilibrium point $(0, 0)$ is called a **spiral sink**.

2. $\lambda_1 = \lambda_r + i\lambda_i$, $\lambda_2 = \lambda_r - i\lambda_i$, $\lambda_r > 0$; the (unstable) equilibrium point $(0, 0)$ is called a **spiral source**.

3. $\lambda_1 = i\lambda_i$, $\lambda_2 = -i\lambda_i$, $\lambda_r = 0$; the (stable) equilibrium point $(0, 0)$ is called a **center**.

We discuss one example for each case.

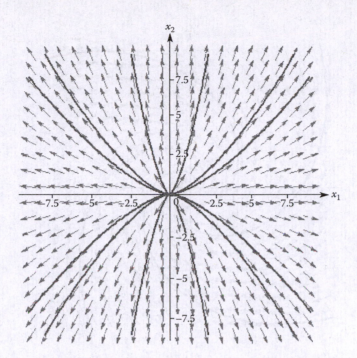

FIGURE 5.8   Phase portrait and the vector field of the system in Example 2.5.

Example 2.6 (Spiral Sink)

Let in (5.12) $A = \begin{pmatrix} -2 & -3 \\ 3 & -2 \end{pmatrix}$. The characteristic equation for A is $\lambda^2 + 4\lambda + 13 = 0$. Its solutions

are $\lambda_1 = -2 + 3i$, $\lambda_2 = -2 - 3i$. The direction field and the phase portrait are shown in Figure 5.9. The trajectories spiral toward the origin. The period of rotation about the origin is $2\pi/|\lambda_i| = 2\pi/3$. Note that the eigenvectors are complex-valued and thus they cannot be plotted. Clearly, the phase plane points to both the oscillatory and decaying behavior of solutions. Indeed (see Appendix 4), the general solution of the system is

$$X = \begin{pmatrix} x_1 \\ x_2 \end{pmatrix} = C_1 \begin{pmatrix} -\sin 3t \\ \cos 3t \end{pmatrix} e^{-2t} + C_2 \begin{pmatrix} \cos 3t \\ \sin 3t \end{pmatrix} e^{-2t},$$

from which it is obvious that solutions $x_1(t)$ and $x_2(t)$ oscillate with the period $2\pi/3$ and their amplitude is exponentially decaying.

Example 2.7 (Spiral Source)

Let in (2.12) $A = \begin{pmatrix} 0 & 1 \\ -5 & 2 \end{pmatrix}$. The characteristic equation is $\lambda^2 - 2\lambda + 5 = 0$ and the eigenvalues

are $\lambda_1 = 1 + 2i$, $\lambda_2 = 1 - 2i$. The direction field and the phase portrait are shown in Figure 5.10. The trajectories spiral away from the origin. The period of the rotation about the origin is

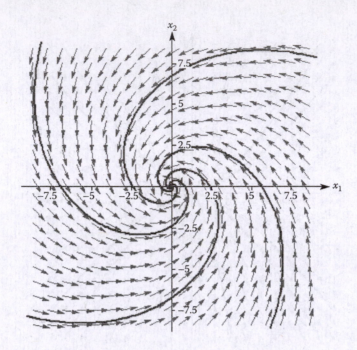

FIGURE 5.9 Phase portrait and the vector field of the system in Example 2.6.

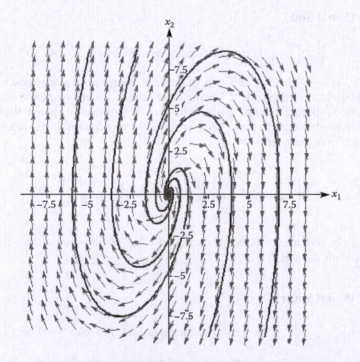

FIGURE 5.10 Phase portrait and the vector field of the system in Example 2.7.

$2\pi / |\lambda_i| = 2\pi / 2 = \pi$. Unlike in Example 2.6, here the solution of the system can be easily found. Comparing matrix A to the matrix in (5.10), we see that $q=5$, $p=-2$, and therefore the system (5.11) ((5.12)) corresponds to the second-order equation $y'' - 2y' + 5y = 0$. This is the equation of harmonic oscillator with negative damping. Thus from the form of the equation only, we expect oscillatory solutions with increasing amplitude, and the phase portrait shows precisely that. Solving the oscillator give $r_1 = \lambda_1$, $r_2 = \lambda_2$ and the general solution is $y = C_1 e^t \cos 2t + C_2 e^t \sin 2t$, which has period π.

Example 2.8 (Center)

Let in (2.12) $A = \begin{pmatrix} 0 & 1 \\ -2 & 0 \end{pmatrix}$. The characteristic equation is $\lambda^2 + 2 = 0$ and the eigenvalues are $\lambda_1 = i\sqrt{2}$, $\lambda_2 = -i\sqrt{2}$. The direction field and the phase portrait are shown in Figure 5.11. The trajectories are circles centered the origin. The period of the rotation about the origin is $2\pi / |\lambda_i| = 2\pi / \sqrt{2} = \pi\sqrt{2}$. Comparing matrix A to the matrix in (5.10), we see that $q = 2$, $p = 0$, and therefore the system (5.11) ((5.12)) corresponds to the second-order equation $y'' + 2y = 0$. This is the equation of an undamped harmonic oscillator. Solving the oscillator yields $r_1 = \lambda_1, r_2 = \lambda_2$ and the general solution $y = C_1 \cos \sqrt{2}t + C_2 \sin \sqrt{2}t$. A careful reader may notice that this example coincides with Example 2.2, if in Example 2.2 one takes $\omega = \sqrt{2}$. Then the system's matrix in Example 2.2 is precisely the matrix A from this example, and the phase portraits are identical. In general, all solution curves in a system with purely imaginary eigenvalues lie on ellipses that enclose the origin.

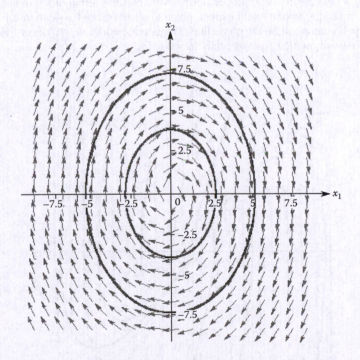

FIGURE 5.11   Phase portrait and the vector field of the system in Example 2.8.

5.2.4 Repeated Real Eigenvalues

Example 2.9

Let in (5.12) $A = \begin{pmatrix} 0 & 1 \\ -4 & -4 \end{pmatrix}$. The characteristic equation is $\lambda^2 + 4\lambda + 4 = 0$ and the eigen-

values are $\lambda_1 = \lambda_2 = \lambda = -2$. Thus we expect the origin to be a sink. Comparing matrix A to the matrix in (5.10), we see that $q = p = 4$, and therefore the system (5.11) ((5.12)) corresponds to the second-order equation $y'' + 4y' + 4y = 0$. This is the equation of the critically damped oscillator. Solution gives $r_1 = r_2 = \lambda = -2$, and the general solution is $y = C_1 e^{-2t} + C_2 t e^{-2t}$. The vector form of the solution is obtained by differentiating this and writing the two formulas as a single vector

formula. This gives $X = C_1 \begin{pmatrix} 1 \\ -2 \end{pmatrix} e^{-2t} + C_2 \left[\begin{pmatrix} 1 \\ -2 \end{pmatrix} t + \begin{pmatrix} 0 \\ 1 \end{pmatrix} \right] e^{-2t}$. The direction field and the

phase portrait are shown in Figure 5.12. Notice how the phase trajectories try to spiral around the origin but the line of eigenvectors $x_2 = -2x_1$ terminates the spiraling, indicating attempted but not completed oscillation.

If the eigenvalue is repeated ($\lambda_1 = \lambda_2$) and its geometric multiplicity is two (which means that the number of linearly independent eigenvectors is two; see Appendix 4), then the equilibrium point (sink if $\lambda_1 = \lambda_2 < 0$, or source if $\lambda_1 = \lambda_2 > 0$) is called a *proper node*. In this case the general solution is $X = C_1 X_1 e^{\lambda t} + C_2 X_2 e^{\lambda t}$, where $\lambda = \lambda_1 = \lambda_2$, and X_1, X_2 are two linearly independent eigenvectors corresponding to eigenvalue λ. If there is only one linearly independent eigenvector that corresponds to the repeated eigenvalue (geometric multiplicity is one), then the equilibrium point is called a *degenerate*, or *improper node*. (Note that the terminology *degenerate node* is not applied to a saddle, as one might expect, given that saddle is a two-dimensional analog of a degenerate equilibrium point on the phase line). Degenerate nodes do not occur if A is symmetric ($a_{ij} = a_{ji}$) or skew-symmetric ($a_{ij} = -a_{ji}$, thus $a_{ii} = 0$).

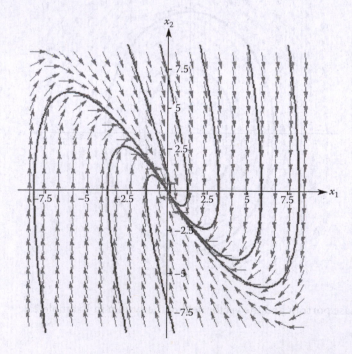

FIGURE 5.12 Phase portrait and the vector field of the system in Example 2.9.

In Examples 2.4, 2.7–2.9, rather than solving the systems directly, we chose instead to reduce the system to the second-order linear equation with constant coefficients and then solve that equation using the method described in Section 2.6. The general reduction technique is described in Section 3.2; see Examples 2.1 and 2.2 there. Solution by reduction to a single equation is easy in the planar case, but its complexity grows fast as the dimension of the system increases. Certainly, Examples 2.4, 2.7–2.9 can be also solved using the techniques specially developed for systems, as described in Section 3.4 and in Appendix 4. In particular, in Appendix 4 we show that the equilibrium point in Example 2.9 is the degenerate node and demonstrate in detail the method of calculation of the general solution. This method is applicable to any autonomous linear system with a degenerate node.

Phase portraits of all planar autonomous systems retain the typical topology demonstrated in the examples. That is, depending on eigenvectors, all saddles differ only in the angle of rotation of the axes of a saddle with respect to the coordinate axes x_1 and x_2, and also in the angle that the saddle's axes make. The same is true for phase portraits near other equilibrium points.

Problems   Answers

In exercises 1–3, write a system of first-order equations, corresponding to the given equation or the IVP.

1. $2y'' - 3y' + 11y = 2.$

$$x_1' = x_2$$
$$x_2' = \frac{-11}{2} x_1 + \frac{3}{2} x_2 + 1$$

2. $\begin{aligned} y'' + (t+1)y' - 5t &= 0, \\ y(0) &= 0, \quad y'(0) = 1. \end{aligned}$

$$x_1' = x_2$$
$$x_2' = -(t+1)x_2 + 5t, \quad x_1(0) = 0, x_2(0) = 1$$

3. $y^{(4)} = y + y''.$

$$x_1' = x_2, x_2' = x_3, x_3' = x_4, x_4' = x_1 + x_3.$$

In exercises 4 and 5, replace the planar ODE system by a second-order equation.

4. $X' = \begin{pmatrix} 0 & 1 \\ -2 & 3 \end{pmatrix} X.$

$$y'' - 3y' + 2y = 0$$

5. $X' = \begin{pmatrix} 0 & 1 \\ t-4 & 1/t \end{pmatrix} X.$

$$y'' - (1/t)y' + (4-t)y = 0$$

In exercises 6–17, solve the linear system, state the equilibrium point type, and, if the system is planar, use CAS to plot the vector field and phase portrait.

6. System in exercise 2.4.

$$X = C_1 \begin{pmatrix} 1 \\ 1 \end{pmatrix} e^t + C_2 \begin{pmatrix} 1 \\ 2 \end{pmatrix} e^{2t}, \ (0,0) \text{ is source.}$$

7. $X' = \begin{pmatrix} 8 & -3 \\ 16 & -8 \end{pmatrix} X.$

$$X = C_1 \begin{pmatrix} 3 \\ 4 \end{pmatrix} e^{4t} + C_2 \begin{pmatrix} 1 \\ 4 \end{pmatrix} e^{-4t}, \ (0,0) \text{ is saddle.}$$

8. $X' = \begin{pmatrix} 12 & -15 \\ 4 & -4 \end{pmatrix} X.$

$$X = C_1 \begin{pmatrix} 3 \\ 2 \end{pmatrix} e^{2t} + C_2 \begin{pmatrix} 5 \\ 2 \end{pmatrix} e^{6t}, \ (0,0) \text{ is source.}$$

9. $X' = \begin{pmatrix} -1 & 0 \\ 0 & -4 \end{pmatrix} X.$

$$X = C_1 \begin{pmatrix} 1 \\ 0 \end{pmatrix} e^{-t} + C_2 \begin{pmatrix} 0 \\ 1 \end{pmatrix} e^{-4t}, \ (0,0) \text{ is sink.}$$

10. $X' = \begin{pmatrix} 1 & -1 & -1 \\ 0 & 1 & 3 \\ 0 & 3 & 1 \end{pmatrix} X,$

$$X = C_1 \begin{pmatrix} 1 \\ 0 \\ 0 \end{pmatrix} e^t + C_2 \begin{pmatrix} 2 \\ -3 \\ -3 \end{pmatrix} e^{4t}$$

$\lambda_{1,2,3} = 1, 4, -2.$

$$+ C_3 \begin{pmatrix} 0 \\ 1 \\ -1 \end{pmatrix} e^{-2t}, \ (0,0) \text{ is 3D saddle.}$$

11. $X' = \begin{pmatrix} 2 & -5 \\ 2 & -4 \end{pmatrix} X.$

$$X = C_1 \begin{pmatrix} 5\cos t \\ 3\cos t + \sin t \end{pmatrix} e^{-t}$$

$$+ C_2 \begin{pmatrix} 5\sin t \\ -\cos t + 3\sin t \end{pmatrix} e^{-t}, (0,0) \text{ is spiral sink.}$$

12. $X' = \begin{pmatrix} 4 & 1 \\ -8 & 8 \end{pmatrix} X.$

$$X = C_1 \begin{pmatrix} \cos 2t \\ 2\cos 2t - 2\sin 2t \end{pmatrix} e^{6t}$$

$$+ C_2 \begin{pmatrix} \sin 2t \\ 2\cos 2t + 2\sin 2t \end{pmatrix} e^{6t},$$

(0,0) is spiral source.

13. $X' = \begin{pmatrix} 4 & 5 \\ -4 & -4 \end{pmatrix} X.$ $X = C_1 \begin{pmatrix} 5\cos 2t \\ -4\cos 2t - 2\sin 2t \end{pmatrix}$

$$+ C_2 \begin{pmatrix} 5\sin 2t \\ 2\cos 2t - 4\sin 2t \end{pmatrix},$$

(0,0) is center.

14. $X' = \begin{pmatrix} 4 & -13 \\ 2 & -6 \end{pmatrix} X.$ $X = C_1 \begin{pmatrix} 13\cos t \\ 5\cos t + \sin t \end{pmatrix} e^{-t}$

$$+ C_2 \begin{pmatrix} 13\sin t \\ -\cos t + 5\sin t \end{pmatrix} e^{-t},$$

(0,0) is spiral sink.

15. $X' = \begin{pmatrix} 1 & 2 & -1 \\ 0 & 1 & 1 \\ 0 & -1 & 1 \end{pmatrix} X,$ $X = C_1 \begin{pmatrix} 1 \\ 0 \\ 0 \end{pmatrix} e^t + C_2 \begin{pmatrix} 2\cos t - \sin t \\ -\sin t \\ -\cos t \end{pmatrix} e^t$

$\lambda_{1,2,3} = 1, 1+i, 1-i.$

$$+ C_3 \begin{pmatrix} -\cos t + 2\sin t \\ \cos t \\ -\sin t \end{pmatrix} e^t,$$

(0,0) is 3D spiral source,

16. $X' = \begin{pmatrix} 8 & -1 \\ 4 & 12 \end{pmatrix} X.$ $X = C_1 \begin{pmatrix} 1 \\ -2 \end{pmatrix} e^{10t}$

$$+ C_2 \left[\begin{pmatrix} 1 \\ -2 \end{pmatrix} t + \begin{pmatrix} -1 \\ 1 \end{pmatrix} \right] e^{10t},$$

(0,0) is degenerate node.

17. $X' = \begin{pmatrix} 4 & 1 \\ -1 & 2 \end{pmatrix} X.$ $X = C_1 \begin{pmatrix} 1 \\ -1 \end{pmatrix} e^{3t}$

$$+ C_2 \left[\begin{pmatrix} 1 \\ -1 \end{pmatrix} t + \begin{pmatrix} 0 \\ 1 \end{pmatrix} \right] e^{3t},$$

(0,0) is degenerate node.

5.3 STABILITY OF SOLUTIONS

In this section we take a close look at stability. In order to be able to introduce all necessary concepts, we omit the discussion of the simplest case, that is, of a single first-order equation, and start with the higher-order equations, or equivalently by virtue of a reduction method, with systems.

5.3.1 Lyapunov and Orbital Stability

The following Definition I is for *linear*, autonomous systems of n first-order equations with a nonsingular matrix A. In particular, when $n = 2$ the Definition I is valid for equations (5.11) (or (5.12)). Note that the dimension of vectors is n.

Definition I (Lyapunov stability of the equilibrium solution of a linear system)

Let $\bar{X} = 0$ be the equilibrium solution of $X' = AX$. Also let X be the perturbation, at $t = t_0$ (where t_0 is any time), of the equilibrium solution. Then the equilibrium solution is said to be:

   (i) **Stable**, if for all $\varepsilon > 0$ there exists a $\delta(\varepsilon, t_0)$ such that $|X(t_0)| < \delta$ implies that $|X(t)| < \varepsilon$ for all $t > t_0$;
       This means that solution that is inside the "disc" of radius δ (with the center at the origin, $\bar{X} = 0$) at time t_0, stays within the disc of radius ε (which depends on δ and t_0) for all $t > t_0$.
   (ii) **Uniformly (neutrally) stable**, if stable and $\delta = \delta(\varepsilon)$ is independent of t_0;
   (iii) **Asymptotically stable**, if $|X(t_0)| < \delta$ implies that $|X(t)| \to 0$ as $t \to \infty$.

This means that if the solution is inside the disc of radius δ (with the center at the origin, $\bar{X} = 0$) at time t_0, then it approaches the origin as $t \to \infty$.

If (i) does not hold, the solution $\bar{X} = 0$ is said to be **unstable**.

Clearly, asymptotic stability is a stronger property than stability, since the former requires the latter. However, it is important to understand that asymptotic stability does not *imply* stability. Indeed, the solution trajectory may leave the disc of radius ε at some $t > t_0$ (and thus the solution is unstable according to part (i)), but then at a later time reenter the disc and approach the origin asymptotically as $t \to \infty$. Such solution is asymptotically stable according to part (iii).

The examples in Section 5.2 demonstrate unstable (Examples 2.4, 2.5, 2.7), asymptotically stable (Examples 2.3, 2.6, 2.9), and uniformly stable equilibrium solutions (Example 2.8) in the Lyapunov sense. In principle, each of these examples can be analyzed in the spirit of the formal Definition I, and the instability or stability proved, but in this text we will not demonstrate such direct proofs.

It is reasonable to be concerned also about stability of other solutions, that is, non-equilibrium (time-dependent) solutions $\bar{X}(t)$. For such solutions we will now demonstrate that Definition I is sufficient, that is, Lyapunov stability analysis of any solution reduces to the analysis of the equilibrium solution. Indeed, consider another solution,

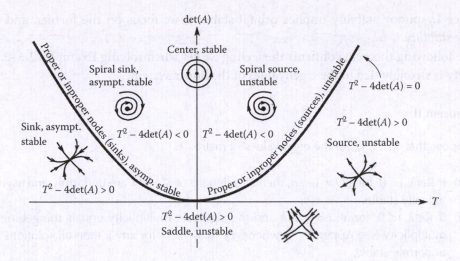

FIGURE 5.13 Stability diagram.

$X(t)$, and let $Y(t) = X(t) - \bar{X}(t)$, that is, $Y(t)$ is thought of as a perturbation (not necessarily a small one) of the solution whose stability is in question, $\bar{X}(t)$. Differentiation gives $Y'(t) = X'(t) - \bar{X}'(t) = AX(t) - A\bar{X}(t) = A(X(t) - \bar{X}(t)) = AY(t)$. This shows that the perturbation must be determined from the same ODE system. If the equilibrium point of this system (this point is, of course, the good old $Y=0$) is stable (according to the Definition I), then the perturbation decays ($|X(t) - \bar{X}(t)| \to 0$) and the solution $\bar{X}(t)$ is stable; if the equilibrium point is unstable, then the perturbation grows and $\bar{X}(t)$ is unstable, and so on.

There is also another notion of stability, called orbital (trajectorial), or Poincare, stability. Roughly speaking, a solution that has an orbit C is called orbitally stable if a small perturbation to C remains small. In other words, if a solution that is close to C at some $t = t_0$ remains close to C at $t > t_0$, then the orbit is stable. It can be shown that Lyapunov stability implies orbital stability, but the converse is not true—an orbit may be stable simultaneously with the solution, defining the orbit, being unstable in the Lyapunov sense. The following definition of orbital stability is for periodic solution, that is, $X(t + T) = X(t)$ for all t, defining a closed orbit C in the phase plane, enclosing the equilibrium point at the origin (a center; see for instance Figure 5.11).

Definition II (Orbital stability)

(i) Suppose that $Y(t)$ is a solution of (5.13) such that, for all $\varepsilon > 0$ and any t_0, there is a $\delta(\varepsilon) > 0$ such that $d(Y(t_0), C) < \delta$ implies that $d(Y(t), C) < \varepsilon$ for all $t > t_0$. Then C is said to be orbitally, or Poincare, stable.

(ii) If C is orbitally stable, and $d(Y(t), C) \to 0$ as $t \to \infty$, then C is said to be asymptotically orbitally, or Poincare, stable.

Here $d(Y(t), C) = \min_{X \in C} |X - Y(t)|$ is the distance of $Y(t)$ from the orbit C.

Since Lyapunov stability implies orbital stability, we focus on the former, and call it simply *stability*.

The following theorem confirms the feeling we get when solving Examples 2.3–2.9 that stability is strongly tied to the eigenvalues of the linear system.

Theorem II

Suppose that $\lambda_1,...,\lambda_n$ are the eigenvalues of matrix A.

(i) If $\text{Re}(\lambda_i) < 0$ for all $i = 1,...,n$, then all solutions of $X' = AX$ are uniformly and asymptotically stable.

(ii) If $\text{Re}(\lambda_i) \leq 0$ for all $i = 1,...,n$ and the algebraic multiplicity equals the geometric multiplicity (see Appendix 4) whenever $\text{Re}(\lambda_j) = 0$ for any j, then all solutions are uniformly stable.

(iii) If $\text{Re}(\lambda_j) > 0$ for at least one j, or the algebraic multiplicity is greater than the geometric multiplicity should $\text{Re}(\lambda_j) = 0$, then all solutions are unstable.

Rigorous proof of Theorem II is difficult, but it is easy once again to get the feeling that Theorem II is correct. Consider, for instance, the simplest case of a system that has n distinct eigenvalues and the corresponding n different, linearly independent eigenvectors X_i. Then the general solution is the linear combination $X = X_1 e^{\lambda_1 t} + X_2 e^{\lambda_2 t} + \cdots + X_n e^{\lambda_n t}$. If all eigenvalues have negative real parts, then X approaches zero as $t \to \infty$; indeed, by the Euler's formula $e^{\lambda_i t} = e^{(r_i \pm i\theta_i)t} = e^{r_i t} e^{\pm i\theta_i t} \to 0$ as $t \to \infty$. Such solution is asymptotically stable. If at least one eigenvalue has a positive real part, then the corresponding term in the general solution grows exponentially, while the terms that correspond to eigenvalues with negative real parts decay to zero. Thus such a solution is unstable, et cetera.

Applying Theorem II to the case $n = 2$ (system (5.12) with $\det(A) \neq 0$), we arrive to the following summary of stability:

Eigenvalues	Type of Equilibrium Point	Stability
$\lambda_1 > \lambda_2 > 0$	Source (unstable node)	Unstable
$\lambda_1 < \lambda_2 < 0$	Sink (stable node)	Asymptotically stable
$\lambda_2 < 0 < \lambda_1$	Saddle	Unstable
$\lambda_1 = \lambda_2 > 0$	Source (proper or improper (degenerate) node)	Unstable
$\lambda_1 = \lambda_2 < 0$	Sink (proper or improper (degenerate) node)	Asymptotically stable
$\lambda_1, \lambda_2 = r \pm i\theta$	Spiral	
$r > 0$	Spiral source	Unstable
$r < 0$	Spiral sink	Asymptotically stable
$r_1 = i\theta, r_2 = -i\theta$	Center	Uniformly stable

Sink and spiral sink are not only asymptotically stable equilibrium points but also *glob-ally* asymptotically stable since all nonzero initial conditions give rise to phase trajectories that asymptotically approach the equilibrium. (Note that for nonlinear systems, as a rule only the initial conditions in the vicinity of these equilibrium points result in trajectories that asymptotically approach equilibrium; thus nonlinear sink and spiral sink are *locally* asymptotically stable.)

5.3.2 Sensitivity to Small Perturbations

Eigenvalues λ_1 and λ_2 that determine the type and stability of the equilibrium point of course depend on the values of the matrix elements. These values usually result from a measurement or a computation. Thus it is natural to ask how sensitive are the eigenvalues to small changes in the matrix element's values. Recall that the eigenvalues are the roots of the characteristic polynomial. *It can be shown that small perturbations in some or all matrix elements result in small perturbations in some or all polynomial coefficients and thus the type and stability of the equilibrium point is not altered by small perturbations.* As far as stability (but not the type) is the only concern, there is one exception to this rule, which occurs when the eigenvalues are purely imaginary (equilibrium point is the center). In this case a small perturbation leads to emergence of a nonzero real part (same number) in λ_1 and λ_2, and the center is transformed into a spiral point. This sensitivity to small perturbations is reflected in a "weaker" stability of a center, that is, uniform but not asymptotic stability. All asymptotically stable equilibrium points (see the table) are not sensitive to small perturbations of the matrix elements; that is, the real parts of eigenvalues stay negative after a small perturbation is imposed. Another exception to the above stated rule occurs when $\lambda_1 = \lambda_2$. In this case the type of the equilibrium point is affected by small perturbations, but not its stability.

5.3.3 Stability Diagram in the Trace-Determinant Plane

There is another (graphical) way of summarizing stability results for planar linear systems. Referring to equation (5.14), we observe that eigenvalues λ_1 and λ_2 can be determined in terms of trace and determinant of A. One obtains $\lambda_1\lambda_2 = \det(A), \lambda_1 + \lambda_2 = T(A)$, and $\lambda_{1,2} = \frac{T(A) \pm \sqrt{T(A)^2 - 4\det(A)}}{2}$. Using, say, the last equation one may plot all equilibrium point types in the trace-determinant plane. Such a plot is called the stability diagram. The curve $\Delta(A) \equiv T(A)^2 - 4\det(A) = 0$ (equivalently, the curve $\det(A) = T(A)^2/4$) is the *repeated roots parabola*, since whenever $\Delta(A) = 0$, $\lambda_1 = \lambda_2 = T(A)/2$. For all points above the repeated root parabola $\Delta(A) < 0$, and thus λ_1 and λ_2 are complex (and conjugate). Thus in this region only the spiral equilibrium is possible. Additionally, if $T(A) < 0$ (left of the vertical axis) $\mathrm{Re}(\lambda_{1,2}) < 0$ (spiral sink), if $T(A) > 0$ (right of the vertical axis) $\mathrm{Re}(\lambda_{1,2}) > 0$ (spiral source), and if $T(A) = 0$ (on the vertical axis) $\mathrm{Re}(\lambda_{1,2}) = 0$ (center). Similar analysis demarcates other regions of the trace-determinant plane.

Example 3.1

In Example 2.3, $A = \begin{pmatrix} -2 & -2 \\ -1 & -3 \end{pmatrix}$. Thus $T(A) = -5, det(A) = 4$. The point $(-5,4)$ is below the repeated

root parabola (since the point on the repeated root parabola that corresponds to $T = -5$ is $(-5,25/4)$) and it is in the second quadrant. The stability diagram indicates that the equilibrium point is a sink (a stable node). Indeed it is, since $\lambda_1 = -1$, $\lambda_2 = -4$; we determined this in Example 2.3.

Example 3.2

In Example 2.9, $A = \begin{pmatrix} 0 & 1 \\ -4 & -4 \end{pmatrix}$. Thus $T(A) = -4, det(A) = 4$. The point $(-4,4)$ is on the repeated

root parabola. The stability diagram indicates that the equilibrium point is stable proper or improper node. Using only the trace and determinant it is not possible to further conclude whether it is proper or improper, but indeed it is stable (a sink), since $\lambda_1 = \lambda_2 = -2$.

5.3.4 Lyapunov Stability of a General System

Next, consider *general (nonautonomous and nonlinear) system of n first-order ODEs (5.1) in Section 3.2.* Let all f_i and $\partial f_i / \partial x_k$ be continuous for all $t \geq t_0$. Lyapunov stability of any solution of this system is established by Theorem III, which is very similar to Definition I for linear system and the remarks that follow it.

Theorem III (Lyapunov stability of any solution of a general first-order system)

Let $X = \Phi(t)$ be a solution of (5.1). Then solution $\Phi(t)$ is:

   (i) **Stable**, if for all $\varepsilon > 0$ and any t_0 there exists a $\delta(\varepsilon, t_0) > 0$ such that for any solution $X(t)$ of (5.1) that satisfies $|X(t_0) - \Phi(t_0)| < \delta$ it is true that $|X(t) - \Phi(t)| < \varepsilon$ for all $t > t_0$;
   (ii) **Uniformly (neutrally) stable**, if $\delta = \delta(\varepsilon)$ is independent of t_0;
   (iii) **Asymptotically stable**, if $|X(t_0) - \Phi(t_0)| < \delta$ implies that $|X(t) - \Phi(t)| \to 0$ as $t \to \infty$.

If (i) does not hold, the solution $\Phi(t)$ is said to be **unstable**. Of course when (5.1) defines a linear system and $\Phi(t) = 0$ (equilibrium solution), Theorem III is equivalent to Definition I.

5.3.5 Stability Analysis of Equilibria in Nonlinear Autonomous Systems

As an application, we now consider a frequently encountered issue of determining stability of the critical points of *a planar nonlinear autonomous system* (5.5). All critical points are determined by solving equations (5.7). Let $f_1(\alpha, \beta) = f_2(\alpha, \beta) = 0$; that is, let the critical point be $x_1(t) = \alpha$, $x_2(t) = \beta$, where α and β are any real values (including zero). We assume that $f_1, f_2, \partial f_i / \partial x_j, i, j = 1..2$ are continuous in a neighborhood of the critical point, and also that the critical point is isolated; that is, there is no other critical points in the immediate vicinity (in fact, in the infinitesimally small disc centered at (α,β)). To determine stability, introduce new variables $u(t)$ and $v(t)$ by the translation $u = x_1 - \alpha$, $v = x_2 - \beta$. Plugging in (5.5) gives

$$u' = f_1(u+\alpha, v+\beta)$$
$$v' = f_2(u+\alpha, v+\beta). \tag{5.15}$$

Clearly, system (5.15) has the critical point $u(t) = v(t) = 0$; thus the translation moved the critical point (α,β) to the origin. Next, we expand f_1 and f_2 in a double Taylor series about (α,β):

$$u' = f_1(\alpha,\beta) + \frac{\partial f_1}{\partial u}u + \frac{\partial f_1}{\partial v}v + h_1(u,v) = \frac{\partial f_1}{\partial x_1}u + \frac{\partial f_1}{\partial x_2}v + h_1(u,v),$$

$$v' = f_2(\alpha,\beta) + \frac{\partial f_2}{\partial u}u + \frac{\partial f_2}{\partial v}v + h_2(u,v) = \frac{\partial f_2}{\partial x_1}u + \frac{\partial f_2}{\partial x_2}v + h_2(u,v), \tag{5.16}$$

where the partial derivatives are evaluated at $(x_1, x_2) = (\alpha,\beta)$, and nonlinear functions $h_1(u,v)$, $h_2(u,v)$ are the remainders of the expansions. The nonlinear system (5.16) can be compactly written in a vector form

$$\Psi' = J\Psi + H(\Psi), \tag{5.17}$$

where

$$\Psi = \begin{pmatrix} u(t) \\ v(t) \end{pmatrix}, \quad J = \begin{pmatrix} a_{11} & a_{12} \\ a_{21} & a_{22} \end{pmatrix}, \quad H(\Psi) = \begin{pmatrix} h_1(u,v) \\ h_2(u,v) \end{pmatrix}, \tag{5.18}$$

and $a_{ij} = \partial f_i / \partial x_j, i, j = 1..2$ (at (α,β)). Matrix J is called the *Jacobian* of the system (5.5) at (α,β). We now assume that (i) $|H(\Psi)|/|\Psi| \to 0$ as $|\Psi| \to 0$, that is, $|H(\Psi)|$ is small in comparison with $|\Psi|$ itself near the origin.

The linearized system is obtained by dropping $H(\Psi)$:

$$\Psi' = J\Psi. \tag{5.19}$$

Next, we also assume, as usual, that (ii) $\det(J) \neq 0$, that is, the origin is the only critical point of the linear system (5.19). When conditions (i) and (ii) are satisfied, the system (5.17) is said to be an *almost linear system* in the neighborhood of the critical point (α,β). (It can be shown that condition (i) is equivalent to the requirement that not only $f_1, f_2, \partial f_i / \partial x_j, i, j = 1..2$ are continuous in the neighborhood of (α,β), but also the second partial derivatives are continuous there. This makes condition (i) very easy to check.) Finally, we can state the following theorem.

Theorem IV

If (5.17) is an almost linear system, then the type and stability of the critical point $(x_1, x_2) = (\alpha, \beta)$ of the nonlinear system (5.5) are the same as those of the zero critical point of the linear system (5.19). Also the trajectories of the linear system (5.19) near the origin are a good approximation to the trajectories of the nonlinear system (5.5) near (α, β). Exceptions occur if the zero critical point of (5.19) is a center (J has pure imaginary eigenvalues), or if the eigenvalues of J are equal. Then (5.5) may have the same type of critical point as (5.19) or a spiral point.

This theorem is reminiscent of the sensitivity analysis to small perturbations that was discussed above for a linear system, in that it turns out again that the center and the degenerate case $\lambda_1 = \lambda_2$ are special. *The reason* they are special is their sensitivity to a small perturbation $H(\Psi)$. For all other combinations of eigenvalues from the table, the corresponding critical point is insensitive to a small perturbation $H(\Psi)$. Next, if a critical point of (5.5) is asymptotically stable, then it is only *locally* asymptotically stable; that is, it is asymptotically stable in some finite domain of the phase space, called basin of attraction, that includes the critical point (see the comment after the table). And finally, the above ideas and analysis carry over, with only some minor modifications, to higher-dimensional autonomous nonlinear systems ($n > 2$). Of course, the phase portraits of such systems are not easy to visualize.

Problems	Answers

1. Using the trace and determinant of matrix A, confirm the equilibrium point type in problems 2.6–2.9, 2.11–2.14, 2.16, 2.17.

In problems 2–10, show that the system is almost linear and find the location and type of all critical points by linearization. Show the details of your work. Use CAS to plot phase portraits of nonlinear and linearized systems near the equilibrium point(s) and study differences and similarities.

2. System in Example 2.1

3.
$$x_1' = -x_1 + x_2 - x_2^2$$
$$x_2' = -x_1 - x_2$$

Spiral sink at (0,0) and saddle at (–2,2)

4.
$$x_1' = x_1^2 - x_2^2 - 4$$
$$x_2' = x_2$$

Saddle at (–2,0) and source at (2,0)

5.
$$x_1' = x_1 + x_2 + x_2^2$$
$$x_2' = 2x_1 + x_2$$

Saddle at (0,0), unstable improper node at (1/4,–1/2)

6.
$$x_1' = 4 - 2x_2$$
$$x_2' = 12 - 3x_1^2$$

Saddle at $(2,2)$, center at $(-2,2)$

7.
$$x_1' = -(x_1 - x_2)(1 - x_1 - x_2)$$
$$x_2' = x_1(2 + x_2)$$

Saddle at $(0,0)$, spiral sink at $(0,1)$, sink at $(-2,-2)$, source at $(3,-2)$

8.
$$x_1' = x_1 - x_1^2 - x_1 x_2$$
$$x_2' = 3x_2 - x_1 x_2 - 2x_2^2$$

Source at $(0,0)$, saddle at $(1,0)$, sink at $(0,3/2)$, saddle at $(-1,2)$

9.
$$x_1' = -2x_2 - x_1 \sin(x_1 x_2)$$
$$x_2' = -2x_1 - x_2 \sin(x_1 x_2)$$

Saddle at $(0,0)$

10.
$$x_1' = (1 + x_1)\sin x_2$$
$$x_2' = 1 - x_1 - \cos x_2$$

Center at $(0, \pm 2n\pi), n=0,1,2,...$, saddle at $(2, \pm(2n-1)\pi), n=1,2,3,...$

In problems 11–13 find and classify all critical points by first converting the ODE to a system and then linearizing it. Note that (i) the equation in 12 is one of free undamped pendulum, where $k=g/L>0$, and g is the acceleration of gravity, L is the pendulum length; (ii) the equation in 13 is one of damped pendulum, where a nonzero $\gamma>0$ is the damping coefficient.

11. $y'' - 9y + y^3 = 0$

Saddle at $(0,0)$ and centers at $(-3,0)$ and $(3,0)$

12. $\theta'' + k\sin\theta = 0$

Centers at $(n\pi,0), n = 0, \pm 2, \pm 4,...$ and saddles at $(n\pi,0), n = \pm 1, \pm 3,...$

13. $\theta'' + \gamma\theta' + k\sin\theta = 0$

$(n\pi,0), n=0, \pm 2, \pm 4,...$ are spiral sinks if $\gamma < 2k$, sinks if $\gamma > 2k$, and stable proper or improper nodes, or spiral sinks, if $\gamma = 2k$. $(n\pi,0), n = \pm 1, \pm 3,...$ are saddles.

14. The predator-prey population model, also called the Lotka-Volterra model, is given by a nonlinear system
$$x_1' = ax_1 - bx_1 x_2$$
, where all parameters
$$x_2' = -cx_2 + dx_1 x_2$$
are positive. Find and classify all equilibrium points.

Saddle at $(0,0)$, center at $(c/d, a/b)$

CHAPTER 6

Method of Laplace Transforms for ODE

Laplace transforms are widely used in scientific research. In this chapter we introduce the main concepts and show how to use Laplace transform (LT) to solve ordinary differential equations.

6.1 INTRODUCTION

The Laplace transform $L[f(x)]$ of a real function of a real variable $f(x)$ is defined as follows:

$$\hat{f}(p) = L[f(x)] = \int_0^\infty e^{-px} f(x)\,dx, \qquad (6.1)$$

where p is, generally, a complex parameter. Function $\hat{f}(p)$ is often called the *image* of the *original* function $f(x)$.

The right side of equation (6.1) is called the *Laplace integral*. For its convergence in the case where p is real, it is necessary that $p > 0$; when p is complex, then for convergence it is necessary that $\operatorname{Re} p > 0$. Also for the convergence a growth of function $f(x)$ should be restricted. We will consider only functions $f(x)$ that are increasing slower than some exponential function as $x \to \infty$. This means that for any x there exist positive constants M and a, such that

$$|f(x)| \le Me^{ax}.$$

Clearly, for the convergence of the integral, it should be $\operatorname{Re} p > a$. And obviously, function $f(x)$ should be regular on $(0, \infty)$; for instance, for $f(x) = 1/x^\beta$, $\beta > 0$, the LT clear not exist.

To determine the original function from the image, $\hat{f}(p)$, one has to apply the *inverse Laplace transform*, which is denoted as

$$f(x) = L^{-1}[\hat{f}(p)]. \qquad (6.2)$$

For example, below we find Laplace transforms of two functions:

Let $f(x) = 1$, then

$$L[f(x)] = \int_0^\infty e^{-px} dx = -\frac{1}{p} e^{-px}\Big|_0^\infty = \frac{1}{p};$$

Let $f(x) = e^{ax}$, then

$$L[f(x)] = \int_0^\infty e^{-px} e^{ax} dx = \int_0^\infty e^{(a-p)x} dx = \frac{1}{a-p} e^{(a-p)x}\Big|_0^\infty = \frac{1}{p-a}.$$

Clearly for the convergence of the integral, Re p should be greater than a.

Laplace transforms of some functions can be found in Table 6.1.

6.2 PROPERTIES OF THE LAPLACE TRANSFORM

Property 1

The Laplace transform is linear:

$$L[C_1 f_1(x) + C_2 f_2(x)] = C_1 L[f_1(x)] + C_2 L[f_2(x)]. \tag{6.3}$$

This follows from the linearity of the integral (1.1).

The inverse transform is also linear:

$$L^{-1}[C_1 \hat{f}_1(p) + C_2 \hat{f}_2(p)] = C_1 L^{-1}[\hat{f}_1(p)] + C_2 L^{-1}[\hat{f}_2(p)]. \tag{6.4}$$

Property 2

Let $L[f(x)] = \hat{f}(p)$. Then

$$L[f(kx)] = \frac{1}{k}\hat{f}\left(\frac{p}{k}\right). \tag{6.5}$$

This can be proven using the change of a variable, $x = t/k$:

$$\int_0^\infty e^{-px} f(kx) dx = \frac{1}{k}\int_0^\infty e^{-\frac{p}{k}t} f(t) dt = \frac{1}{k}\hat{f}\left(\frac{p}{k}\right).$$

TABLE 6.1   Laplace Transforms of Some Functions

$f(x)$	$L[f(x)]$	Convergence Condition
1	$\dfrac{1}{p}$	$\operatorname{Re} p > 0$
x^n	$\dfrac{n!}{p^{n+1}}$	$n \geq 0$ is integer, $\operatorname{Re} p > 0$
x^a	$\dfrac{\Gamma(a+1)}{p^{a+1}}$	$a > -1$, $\operatorname{Re} p > 0$
e^{ax}	$\dfrac{1}{p-a}$	$\operatorname{Re} p > \operatorname{Re} a$
$\sin cx$	$\dfrac{c}{p^2+c^2}$	$\operatorname{Re} p > |\operatorname{Im} c|$
$\cos cx$	$\dfrac{p}{p^2+c^2}$	$\operatorname{Re} p > |\operatorname{Im} c|$
$\sinh bx$	$\dfrac{b}{p^2-b^2}$	$\operatorname{Re} p > |\operatorname{Re} b|$
$\cosh bx$	$\dfrac{p}{p^2-b^2}$	$\operatorname{Re} p > |\operatorname{Re} b|$
$x^n e^{ax}$	$\dfrac{n!}{(p-a)^{n+1}}$	$\operatorname{Re} p > \operatorname{Re} a$
$x \sin cx$	$\dfrac{2pc}{(p^2+c^2)^2}$	$\operatorname{Re} p > |\operatorname{Im} c|$
$x \cos cx$	$\dfrac{p^2-c^2}{(p^2+c^2)^2}$	$\operatorname{Re} p > |\operatorname{Im} c|$
$e^{ax} \sin cx$	$\dfrac{c}{(p-a)^2+c^2}$	$\operatorname{Re} p > (\operatorname{Re} a + |\operatorname{Im} c|)$
$e^{ax} \cos cx$	$\dfrac{p-a}{(p-a)^2+c^2}$	$\operatorname{Re} p > (\operatorname{Re} a + |\operatorname{Im} c|)$

Property 3

Let $L[f(x)] = \hat{f}(p)$, and

$$f_a(x) = \begin{cases} 0, & x < a, \\ f(x-a), & x \geq a, \end{cases}$$

where $a > 0$. Then

$$\hat{f}_a(p) = e^{-pa} \hat{f}(p). \tag{6.6}$$

This property is also known as the *Delay Theorem*.

Property 4

Let $L[f(x)] = \hat{f}(p)$. Then for any complex constant c,

$$L\left[e^{cx} f(x)\right] = \int_0^\infty e^{-px} e^{cx} f(x) dx = \hat{f}(p-c).$$

This property is also known as the *Shift Theorem*.

Property 5

First define *convolution* of functions $f_1(x)$ and $f_2(x)$ as function $f(x)$ given by

$$f(x) = \int_0^x f_1(x-t) f_2(t) dt. \tag{6.7}$$

Convolution is symbolically denoted as

$$f(x) = f_1(x) * f_2(x).$$

The property states that if $f(x)$ is a convolution of functions $f_1(x)$ and $f_2(x)$, then

$$\hat{f}(p) = \hat{f}_1(p) \hat{f}_2(p), \tag{6.8}$$

in other words,

$$L[f(x)] = L\left[\int_0^x f_1(x-t) f_2(t) dt\right] = \hat{f}_1(p) \hat{f}_2(p), \tag{6.9}$$

where (the integrals should converge absolutely)

$$\hat{f}_1(p) = \int_0^\infty f_1(x) e^{-px} dx, \quad \hat{f}_2(p) = \int_0^\infty f_2(x) e^{-px} dx, \quad \hat{f}(p) = \int_0^\infty f(x) e^{-px} dx.$$

Property 6

The Laplace transform of a derivative is

$$L[f'(x)] = p\hat{f}(p) - f(0). \tag{6.10}$$

Indeed, integration by parts gives

$$\int_0^\infty e^{-px} f'(x)dx = f(x)e^{-px}\Big|_0^\infty + p\int_0^\infty e^{-px} f(x)dx = -f(0) + p\hat{f}(p).$$

Analogously, for the second derivative we obtain

$$L[f''(x)] = p^2 \hat{f}(p) - pf(0) - f'(0). \tag{6.11}$$

And for the nth derivative,

$$L[f^{(n)}(x)] = p^n \hat{f}(p) - p^{n-1} f(0) - p^{n-2} f'(0) - \cdots - f^{(n-1)}(0). \tag{6.12}$$

Property 7

If $L[f(x)] = \hat{f}(p)$, then the Laplace transform of the integral can be represented as

$$L\left[\int_0^t f(t)dt\right] = \frac{1}{p}\hat{f}(p). \tag{6.13}$$

This property can be proven by writing the Laplace transform of the integral as a double integral and then interchanging the order of the integrations.

Property 8

Laplace transform of the delta function, $\delta(x)$, is

$$L[\delta(x)] = \int_{-0}^{+0} e^{-px}\delta(x)dx + \int_{+0}^{\infty} e^{-px}\delta(x)dx = \int_{-0}^{+0} \delta(x)dx + 0 = 1. \tag{6.14}$$

Recall the main properties of $\delta(x)$:

$$\int_{-\infty}^{\infty} f(x')\delta(x-x')dx' = f(x), \qquad \int_{-\infty}^{\infty} \delta(x-x')dx' = 1. \tag{6.15}$$

6.3 APPLICATIONS OF LAPLACE TRANSFORM FOR ODE

One of the primary applications of the Laplace transform is the solution of IVPs for differential equations, since it often turns out that an equation that involves the transform of the unknown function can be solved much easier than the original equation.

After the transform has been found, it must be *inverted*, that is, the original function (the solution of a differential equation) must be obtained. Often this step needs a calculation of integrals of a complex variable, but in many situations, such as the case of linear equations with constant coefficients, the inverted transforms can be found in Table 6.1. Some partial differential equations also can be solved using the method of Laplace transform. In this section we consider solutions of ordinary differential equations and systems of ODEs.

Let the task be to solve the IVP for the second-order linear equation with constant coefficients:

$$ax'' + bx' + cx = f(t), \tag{6.16}$$

$$x(0) = \beta, \quad x'(0) = \gamma. \tag{6.17}$$

Let us apply the Laplace transform to both sides of equation (6.16). Using linearity of the transform and Property 6 gives the *algebraic* equation for the transform function $\hat{x}(p)$:

$$a(p^2\hat{x} - \beta p - \gamma) + b(p\hat{x} - \beta) + c\hat{x} = \hat{f}(p). \tag{6.18}$$

We have used the initial conditions (6.17). Solving this algebraic equation, one finds

$$\hat{x}(p) = \frac{\hat{f}(p) + a\beta p + a\gamma + b\beta}{ap^2 + bp + c}. \tag{6.19}$$

Note that when the initial conditions are zero, the transform takes a simple form:

$$\hat{x}(p) = \frac{\hat{f}(p)}{ap^2 + bp + c}. \tag{6.20}$$

Next, inverting the transform, gives the function $x(t)$—the solution of the IVP.

Below we consider examples.

Example 3.1: Solve differential equation

$$x'' + 9x = 6\cos 3t$$

with zero initial conditions.

Solution. Applying the Laplace transform to both sides of the equation, and taking into account that $x(0) = 0$, $x'(0) = 0$, gives

$$p^2\hat{x}(p) + 9\hat{x}(p) = \frac{6p}{p^2 + 9}.$$

Then,

$$\hat{x}(p) = \frac{6p}{(p^2+9)^2} = \frac{2 \cdot 3p}{(p^2+3^2)^2}.$$

The original function $x(t)$ is read out directly from the Laplace transform table:

$$x(t) = t \sin 3t.$$

This function is the solution of the IVP.

Example 3.2

Solve the IVP

$$x'' + 4x = e^t, \quad x(0) = 4, \quad x'(0) = -3.$$

Solution. Applying the LT to both sides of the equation and taking into the account the initial conditions $x(0) = 4$, $x'(0) = -3$, we obtain

$$p^2 \hat{x}(p) - 4p + 3 + 4\hat{x}(p) = \frac{1}{p-1}.$$

Solving for $\hat{x}(p)$ gives

$$\hat{x}(p) = \frac{4p^2 - 7p + 4}{(p^2+4)(p-1)}.$$

Next, using the partial fractions we can write

$$\frac{4p^2 - 7p + 4}{(p^2+4)(p-1)} = \frac{A}{p-1} + \frac{Bp+C}{p^2+4}.$$

From there,

$$4p^2 - 7p + 4 = Ap^2 + 4A + Bp^2 + Cp - Bp - C$$

and equating the coefficients of the second, first, and zeroth degrees of p, we have

$$4 = A + B, \quad -7 = C - B, \quad 4 = 4A - C$$

The solution to this system of equations is

$$A = 1/5, \quad B = 19/5, \quad C = -16/5.$$

Then

$$x(t) = \frac{1}{5} L^{-1}\left[\frac{1}{p-1}\right] + \frac{19}{5} L^{-1}\left[\frac{p}{p^2+4}\right] - \frac{16}{5} L^{-1}\left[\frac{1}{p^2+4}\right].$$

Using the inverse transform from the Table gives *the solution of the IVP*:

$$x(t) = (e^t + 19 \cos 2t - 16 \sin 2t)/5.$$

Easy to check that this solution satisfies the equation and the initial conditions.

Example 3.3

Solve system of differential equations

$$\begin{cases} x' = 2y + t, \\ y' = x + y, \end{cases}$$

with the initial conditions $x(0) = 0, \ y(0) = 0$.

Solution. Applying the LT to each equation, we obtain an algebraic system for the functions $\hat{x}(p)$ and $\hat{y}(p)$:

$$\begin{cases} p\hat{x} = 2\hat{y} + \dfrac{1}{p^2}, \\ p\hat{y} = \hat{x} + \hat{y}. \end{cases}$$

From the second equation, $\hat{x} = (p - 1)\hat{y}$. Substitute this into the first equation to get

$$p(p - 1)\hat{y} - 2\hat{y} = \frac{1}{p^2},$$

from which we have

$$\hat{y} = \frac{1}{p^2(p^2 - p - 2)} = \frac{1}{p^2(p - 2)(p + 1)} = \frac{1}{3p^2}\left[\frac{1}{p - 2} - \frac{1}{p + 1}\right].$$

Now use partial fractions:
1)

$$\frac{1}{p^2(p - 2)} = \frac{A}{p} + \frac{B}{p^2} + \frac{C}{p - 2},$$

which gives

$$1 = Ap(p - 2) + B(p - 2) + Cp^2 \text{ and } A + C = 0, \ -2A + B = 0, \ -2B = 1.$$

Thus, $A = -1/4, \ B = -1/2, \ C = 1/4$.
2)

$$\frac{1}{p^2(p + 1)} = \frac{A}{p} + \frac{B}{p^2} + \frac{C}{p + 1},$$

which gives

$$1 = Ap(p + 1) + B(p + 1) + Cp^2 \quad \text{and} \quad A + C = 0, \ A + B = 0, \ B = 1.$$

Thus, $A = -1, \ B = 1, \ C = 1$.

Then

$$\hat{y} = \frac{1}{3}\left[-\frac{1}{4p} - \frac{1}{2p^2} + \frac{1}{4(p-2)} + \frac{1}{p} + \frac{1}{p^2} - \frac{1}{p+1}\right]$$

$$= \frac{1}{4p} - \frac{1}{2p^2} + \frac{1}{12(p-2)} - \frac{1}{3(p+1)}.$$

The inverse LT gives

$$y(t) = \frac{1}{4} - \frac{t}{2} + \frac{1}{12}e^{2t} - \frac{1}{3}e^{-t}.$$

Function $x(t)$ can be found from the second equation of the initial system:

$$x(t) = y' - y = -\frac{3}{4} + \frac{t}{2} + \frac{1}{12}e^{2t} + \frac{2}{3}e^{-t}.$$

Thus *the solution of IVP is*

$$\begin{cases} x(t) = -\dfrac{3}{4} + \dfrac{t}{2} + \dfrac{1}{12}e^{2t} + \dfrac{2}{3}e^{-t}, \\[3mm] y(t) = \dfrac{1}{4} - \dfrac{t}{2} + \dfrac{1}{12}e^{2t} - \dfrac{1}{3}e^{-t}. \end{cases}$$

It is easy to check that these $x(t)$ and $y(t)$ satisfy the equations and the initial conditions.

Example 3.4

Find the current in the electrical circuit consisting of sequentially connected resistor R, inductor L, capacitor C, and a voltage source $E(t)$, provided that at the initial time $t = 0$ the switch is closed, and the capacitor is not charged. Consider the cases when:

1. The voltage source is constant: $E(t) = E_0$ for $t \geq 0$;
2. At $t = 0$ the voltage source produces a short impulse: $E(t) = E_0\delta(t)$.

Solution. The equation for the current in the circuit is

$$L\frac{di}{dt} + Ri + \frac{q}{C} = E(t),$$

or, taking into account that electric charge on a capacitor is $q = \int_0^t i(\tau)d\tau$, we can write this equation as

$$L\frac{di}{dt} + Ri + \frac{1}{C}\int_0^t i(\tau)d\tau = E(t).$$

Applying the Laplace transform to both sides of the equation gives

$$Lp\hat{i} + R\hat{i} + \frac{1}{C}\frac{\hat{i}}{p} = \hat{E}(p).$$

To transform the integral, we used property 7 of the Laplace transform. From the last equation the function $\hat{i}(p)$ is

$$\hat{i}(p) = \frac{\hat{E}(p)}{R + Lp + 1/pC}.$$

1) If voltage is constant, then $\hat{E}(p) = E_0/p$ and thus

$$\hat{i}(p) = \frac{1}{p}\frac{E_0}{R + Lp + 1/pC} = \frac{E_0}{L}\frac{1}{p^2 + (R/L)p + 1/LC}.$$

Next, complete the square in the denominator and introduce notations:

$$\omega^2 = \frac{1}{LC} - \frac{R^2}{4L^2}, \quad \gamma = \frac{R}{2L}.$$

Then,

$$\hat{i}(p) = \frac{E_0}{L}\frac{1}{(p+\gamma)^2 + \omega^2}.$$

From the table of Laplace transforms we find

$$\frac{1}{\omega}L[e^{-\gamma t}\cos\omega t] = \frac{1}{(p+\gamma)^2 + \omega^2},$$

and thus the current in case of constant voltage is

$$i(t) = \frac{E_0}{\omega L}e^{-\gamma t}\sin\omega t.$$

2) If $E(t) = E_0\delta(t)$, then $\hat{E}(p) = E_0$ and

$$\hat{i}(p) = \frac{E_0}{R + Lp + 1/pC} = \frac{E_0}{L}\frac{1}{p + (R/L) + 1/pLC}.$$

As in the previous solution, we complete the square in the denominator and write the expression for $\hat{i}(p)$ in the form convenient for the application of the inverse transform:

$$\hat{i}(p) = \frac{E_0}{L}\frac{p}{(p+\gamma)^2 + \omega^2} = \frac{E_0}{L}\left[\frac{p+\gamma}{(p+\gamma)^2 + \omega^2} - \frac{\gamma}{\omega}\frac{\omega}{(p+\gamma)^2 + \omega^2}\right].$$

Now from the table of the transforms the original function $i(t)$ is

$$i(t) = \frac{E_0}{L} e^{-\gamma t} \left[\cos \omega t - \frac{\gamma}{\omega} \sin \omega t \right].$$

The current is damped (due to resistance R) periodic.

Example 3.5

The support points of two pendulums are connected by a spring that has the stiffness k. Each pendulum has length l and mass m. Find the motions of the pendulums, if both pendulums at the initial time are at equilibrium, and one of them is given the velocity v, and the second one has zero velocity.

Solution. The equations of motion and the initial conditions for the pendulums are

$$\begin{cases} m\ddot{x}_1 = -\dfrac{mg}{l} x_1 + k(x_2 - x_1), \\[2mm] m\ddot{x}_2 = -\dfrac{mg}{l} x_2 + k(x_1 - x_2), \end{cases}$$

$$x_1(0) = x_2(0) = 0, \ \dot{x}_1(0) = v, \ \dot{x}_2(0) = 0.$$

Applying the Laplace transform to each of these equations and using the formula (6.11) and the initial conditions, we obtain for the left sides of equations

$$L[m\ddot{x}_1] = m(p^2 \hat{x}_1 - v),$$

$$L[m\ddot{x}_2] = mp^2 \hat{x}_2.$$

Thus after the transformation the system takes the form

$$\begin{cases} m(p^2 \hat{x}_1 - v) = -\dfrac{mg}{l} \hat{x}_1 + k(\hat{x}_2 - \hat{x}_1), \\[2mm] mp^2 \hat{x}_2 = -\dfrac{mg}{l} \hat{x}_2 + k(\hat{x}_1 - \hat{x}_2). \end{cases}$$

From this algebraic system we find $\hat{x}_1$:

$$\hat{x}_1 = \frac{v(p^2 + g/l + k/m)}{(p^2 + g/l + 2k/m)(p^2 + g/l)} = \frac{v}{2}\left(\frac{1}{p^2 + g/l + 2k/m} + \frac{1}{p^2 + g/l} \right).$$

Inverting the transform gives

$$x_1(t) = \frac{v}{2}\left(\frac{1}{\omega_1} \sin \omega_1 t + \frac{1}{\omega_2} \sin \omega_2 t \right),$$

where $\omega_1 = \sqrt{g/l + 2k/m}$, $\omega_2 = \sqrt{g/l}$ are the frequencies. The equation for $x_2(t)$ and its solution can be found analogously (this makes a good exercise).

Problems

Solve the IVPs for differential equations and systems of differential equations using the Laplace transform.

1. $x'' + x = 4\sin t$, $x(0) = 0$, $x'(0) = 0$

2. $x'' + 16x = 3\cos 2t$, $x(0) = 0$, $x'(0) = 0$

3. $x'' + 4x = e^t$, $x(0) = 0$, $x'(0) = 0$

4. $x'' + 25x = e^{-2t}$, $x(0) = 1$, $x'(0) = 0$

5. $x'' + x' = te^t$, $x(0) = 0$, $x'(0) = 0$

6. $x'' + 2x' + 2x = te^{-t}$, $x(0) = 0$, $x'(0) = 0$

7. $x'' - 9x = e^{3t}\cos t$, $x(0) = 0$, $x'(0) = 0$

8. $x'' - x = 2e^t - t^2$, $x(0) = 0$, $x'(0) = 0$

9. $\begin{cases} x' = 5x - 3y, \\ y' = x + y + 5e^{-t}, \end{cases}$ $\qquad \begin{cases} x(0) = 0, \\ y(0) = 0. \end{cases}$

10. $\begin{cases} x' + 4y = \sin 2t, \\ y' + x = 0, \end{cases}$ $\qquad \begin{cases} x(0) = 0, \\ y(0) = 0. \end{cases}$

11. $\begin{cases} x' + 4y = \cos 2t, \\ y' + 4x = \sin 2t, \end{cases}$ $\qquad \begin{cases} x(0) = 0, \\ y(0) = 0. \end{cases}$

12. $\begin{cases} x' = 2x - y, \\ y' = 2y - x - 5e^t \sin t, \end{cases}$ $\qquad \begin{cases} x(0) = 0, \\ y(0) = 0. \end{cases}$

13. $\begin{cases} x' = 2y + \cos t, \\ y' = x + y, \end{cases}$ $\qquad \begin{cases} x(0) = 0, \\ y(0) = 0. \end{cases}$

Answers

1. $x(t) = 2(\sin t - t\cos t)$

2. $x = (\cos 2t - \cos 4t)/4$

3. $x(t) = (2e^t - 2\cos 2t - \sin 2t)/10$

4. $x(t) = (e^{-2t} + 28\cos 5t + 0.4\sin 5t)/29$

5. $x(t) = (4 - e^{-t} - 3e^t + 2te^t)/4$

6. $x(t) = e^{-t}(t - \sin t)$

7. $x(t) = \{e^{-3t} + e^{3t}(6\sin t - \cos t)\}/37$

8. $x(t) = -0.5e^{-t} - 1.5e^{t} + te^{t} + 2 + t^2$

9. $\begin{cases} x(t) = -e^{-t} + \dfrac{5}{2}e^{2t} - \dfrac{3}{2}e^{4t}, \\[3mm] y(t) = -2e^{-t} + \dfrac{5}{2}e^{2t} - \dfrac{1}{2}e^{4t}. \end{cases}$

10. $\begin{cases} x(t) = -\dfrac{1}{4}\cos 2t + \dfrac{1}{8}(e^{2t} + e^{-2t}), \\[3mm] y(t) = \dfrac{1}{8}\sin 2t - \dfrac{1}{16}(e^{2t} - e^{-2t}). \end{cases}$

11. $\begin{cases} x(t) = \dfrac{3}{10}\sin 2t + \dfrac{1}{20}(e^{4t} - e^{-4t}), \\[3mm] y(t) = -\dfrac{1}{10}\cos 2t - \dfrac{1}{20}(e^{4t} + e^{-4t}). \end{cases}$

12. $\begin{cases} x(t) = e^{t}(2\cos t - \sin t) + \dfrac{1}{2}(e^{3t} - 5e^{t}), \\[3mm] y(t) = e^{t}(3\cos t + \sin t) - \dfrac{1}{2}(e^{3t} + 5e^{t}). \end{cases}$

13. $\begin{cases} x(t) = \dfrac{1}{5}(\cos t + 2\sin t) + \dfrac{1}{15}(2e^{2t} - 5e^{-t}), \\[3mm] y(t) = -\dfrac{1}{10}(3\cos t + \sin t) + \dfrac{1}{30}(4e^{2t} + 5e^{-t}). \end{cases}$

CHAPTER 7

Integral Equations

7.1 INTRODUCTION

Integral equations appear in the descriptions of the processes in electrical circuits, in the models of materials deformation, and in the theory of radiation transport through matter. Many applications of integral equations are associated with partial differential equations of mathematical physics.

An integral equation is the equation that contains the unknown function under the integral sign (as an integrand).

Consider the following integral equation:

$$y(x) = y_0 + \int_{x_0}^{x} f(x, y)dx, \tag{7.1}$$

where $y(x)$ is the unknown function, $f(x,y)$ is the given function. From (7.1) it is clear that $y(x_0) = y_0$.

Differentiation of (7.1) in x gives the first-order ODE

$$\frac{dy}{dx} = f(x, y) \tag{7.2}$$

with the initial condition $y(x_0) = y_0$. Thus, the integral equation (7.1) reduces to the first-order Cauchy problem (first-order IVP). Conversely, the differential equation (7.2) together with the initial condition is equivalent to the integral equation (7.1).

Often it is easier to solve the integral equation (7.1) than the differential equation (7.2). Approximate solution of (7.1) can be obtained *iteratively*. Let set $y(x) = y_0$ to be initial approximation. Substituting it into (7.1), we obtain the first approximation (first iteration):

$$y_1(x) = y_0 + \int_{x_0}^{x} f(x, y_0)dx. \tag{7.3}$$

The second approximation is obtained by substituting the first approximation in the original equation (7.1):

$$y_2(x) = y_0 + \int_{x_0}^{x} f(x, y_1(x))dx. \tag{7.4}$$

Continuing the iterative process, we obtain the nth approximation (iteration), which with some precision gives the solution of equation (7.1):

$$y_n(x) = y_0 + \int_{x_0}^{x} f(x, y_{n-1}(x))dx. \tag{7.5}$$

For smooth functions $f(x,y)$, the iterative process converges rapidly to the exact solution of (7.1) in the neighborhood of the point (x_0, y_0).

Example 1.1

Solve the following equation iteratively in the vicinity of the point $x_0 = 1$:

$$y(x) = 2 + \int_{1}^{x} \frac{y(x)}{x} dx.$$

Solution. The initial approximation, as can be seen from the equation, is $y_0(1) = 2$. The first approximation is

$$y_1(x) = 2 + \int_{1}^{x} \frac{2}{x} dx = 2 + 2\ln|x|.$$

And then the second approximation is

$$y_2(x) = 2 + \int_{1}^{x} \frac{2 + 2\ln|x|}{x} dx = 2 + 2\ln|x| + \ln^2|x|.$$

The original integral equation is equivalent to the differential equation $y' = \frac{y}{x}$ with the initial condition $y(1) = 2$. This IVP has the exact solution

$$y = 2x.$$

At the point $x_0 = 1$ the exact and the approximate solutions give the same result. Let compare the exact solution and the first and second approximations at $x = 1.1$. The exact solution is $y(1.1) = 2 \cdot 1.1 = 2.2$, and the first and second approximations are

$$y_1(1.1) = 2.19062,$$

$$y_2(1.1) = 2.19970.$$

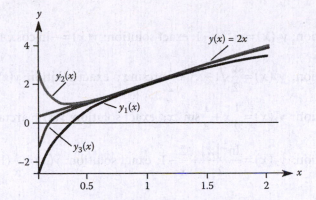

FIGURE 7.1 Iterative solution of the equation in the Example 1.1 on the interval $x \in [0.1, 2]$.

It is seen that the iterative solution is rapidly approaching the exact solution. However, as we move away from the point $x_0 = 1$, the difference of exact and approximate values increases and more iterations are required to approximate the exact solution with acceptable accuracy.

Figure 7.1 shows the graphs of the exact solution $y(x)$ and the first and second iterations $y_1(x)$, $y_2(x)$ on the interval $x \in [0.1, 2]$. These graphs allow one to compare the results of the iterative solution with the exact solution, as one moves away from the point $y(1) = 2$. Graphs are plotted using the program **ODE 1st order**. The exact solution is the straight line.

Problems

Solve the following integral equations: (a) iteratively—find the first two approximations, not counting the initial one—and (b) by reducing the integral equations to differential equations—note that in these problems the differential equations can be solved exactly.

1. $y(x) = \int\limits_{\pi/2}^{x} e^{y(x)} \sin x \, dx$

2. $y(x) = \int\limits_{0}^{x} \sqrt{1 - y^2(x)} \, dx$

3. $y(x) = \int\limits_{0}^{x} \cos^2 y(x) \, dx$

4. $y(x) = 1 - \int\limits_{1}^{x} \frac{xy(x) + 1}{x^2} \, dx$

Answers

1. Second iteration: $y_2(x) = e^{-\cos x} - 1$; exact solution: $y(x) = -\ln|\cos x + 1|$.

2. Second iteration: $y_2(x) = \dfrac{x}{2}\sqrt{1-x^2} + \dfrac{1}{2}\arcsin x$; exact solution: $y(x) = \sin x$.

3. Second iteration: $y_2(x) = \dfrac{1}{2}x + \dfrac{1}{4}\sin 2x$; exact solution: $y(x) = \arctan x$.

4. Second iteration: $y_2(x) = \dfrac{\ln^2|x|}{2} + \dfrac{2}{x} - 1$; exact solution: $y(x) = \dfrac{1}{x}(1 - \ln|x|)$.

7.2 INTRODUCTION TO FREDHOLM EQUATIONS

In physical problems integral equations often arise when one needs to find function $y(x)$ that acts on a system, when the net outcome is given by the integral over some interval $[a,b]$ and the integrand is the function $y(x)$ multiplied by some function $K(x,t)$, which describes the response of the system.

Consider the following two types of linear (with respect to unknown function) integral equations:

$$\lambda \int_a^b K(x,t)y(t)dt = g(x), \tag{7.6}$$

$$y(x) - \lambda \int_a^b K(x,t)y(t)dt = g(x). \tag{7.7}$$

Here $y(x)$ is the unknown function, $K(x,t)$ is the given continuous function (called *the kernel* of the integral equation), $g(x)$ is the given continuous function on $[a,b]$, and λ is the parameter. Variables x and t (here t is the integration variable) are real, but λ, $y(x)$, $K(x,t)$, $g(x)$ may be complex. The unknown function $y(x)$ also may be complex.

Equation (7.6) is the *Fredholm integral equation of the first kind*, and equation (7.7) is the *Fredholm integral equation of the second kind*. When $g(x) = 0$, the integral equation is said to be *homogeneous*; otherwise, it is *nonhomogeneous*. For instance, one problem that leads to equation (7.6) is one of finding the density of the force $y(x)$, acting on a string, such that the string takes on the given shape $g(x)$ on $a \le x \le b$.

Let consider first the Fredholm equations of the second kind with singular *kernels*. A kernel $K(x,t)$ is singular, if it can be written as a product, or a sum of products, of some functions of x and of t:

$$K(x,t) = \sum_{i=1}^n \varphi_i(x)\psi_i(t).$$

Then equation (7.7) reads

$$y(x) = \lambda \int_a^b K(x,t)y(t)dt + g(x) = \lambda \int_a^b \sum_{i=1}^n \varphi_i(x)\psi_i(t)y(t)dt + g(x), \qquad (7.8)$$

and it can be reduced to a system of n linear algebraic equations for n unknowns. Next we show how this reduction is carried out.

First, introduce the notation

$$A_i = \int_a^b \psi_i(t)y(t)dt \quad (i = 1, 2, \ldots, n). \qquad (7.9)$$

Then (7.8) reads

$$y(x) = \lambda \sum_{i=1}^n A_i\varphi_i(x) + g(x). \qquad (7.10)$$

Let change the summation index to k and substitute this expression for $y(x)$ in (7.9), where the integration variable t is replaced by x:

$$A_i = \lambda \sum_{k=1}^n \int_a^b A_k\varphi_k(x)\psi_i(x)dx + \int_a^b \psi_i(x)g(x)dx \quad (i = 1, 2, \ldots, n). \qquad (7.11)$$

Second, introduce the notations

$$\alpha_{ik} = \int_a^b \varphi_k(x)\psi_i(x)dx, \quad g_i = \int_a^b \psi_i(x)g(x)dx \quad (i,k = 1, 2, \ldots, n). \qquad (7.12)$$

Now we can see that (7.11) results in a system of linear algebraic equations for A_i:

$$A_i - \lambda \sum_{k=1}^n \alpha_{ik}A_k = g_i \quad (i = 1, 2, \ldots, n). \qquad (7.13)$$

Coefficients α_{ik} and right-hand sides g_i are given by (7.12).

For the nonhomogeneous equation (7.8) ($g(x) \neq 0$), (7.13) constitutes a system of non-homogeneous linear algebraic equations that has a unique solution if its determinant is not zero. If for some values of the parameter λ the determinant is zero, then the nonhomogeneous equation (7.8) has no solutions, but at the same values of λ the corresponding homogeneous equation, obviously, has infinitely many solutions. From this discussion the purpose of introducing the parameter λ becomes clear. Values of λ at which the homogeneous equation has a solution are called the *eigenvalues* of the problem. These values constitute a *spectrum* of the problem. Functions $y_i(x)$ that are obtained from (7.10) for each value of λ from the spectrum are called the *eigenfunctions*.

Example 2.1

Solve integral equation

$$y(x) - \lambda \int_0^1 (xt^2 - x^2 t) y(t) \, dt = x. \qquad (7.14)$$

Solution. Let $A = \int_0^1 t^2 y(t) \, dt$, $B = \int_0^1 ty(t) \, dt$. Then (7.14) has the form

$$y(x) = \lambda Ax - \lambda Bx^2 + x.$$

Change x to t in $y(x)$, and plug $y(t)$ in the expressions for A and B:

$$A = \int_0^1 t^2 (\lambda At - \lambda Bt^2 + t) \, dt,$$

$$B = \int_0^1 t(\lambda At - \lambda Bt^2 + t) \, dt.$$

Integration gives the system of linear nonhomogeneous algebraic equations for the unknowns A and B:

$$\begin{cases} A = \dfrac{1}{4}\lambda A - \dfrac{1}{5}\lambda B + \dfrac{1}{4}, \\[2mm] B = \dfrac{1}{3}\lambda A - \dfrac{1}{4}\lambda B + \dfrac{1}{3}. \end{cases}$$

The solution of this system is

$$A = \frac{60 - \lambda}{240 + \lambda^2}, \qquad B = \frac{80}{240 + \lambda^2}.$$

Substitution of these expressions in $y(x) = \lambda Ax - \lambda Bx^2 + x$ gives the solution of the integral equation (7.14) on [0, 1]. This can be verified as follows: when $y(x) = \lambda Ax - \lambda Bx^2 + x$ and $y(t) = \lambda At - \lambda Bt^2 + t$ with the above expressions for $A(\lambda)$ and $B(\lambda)$ are substituted in (7.14) and the integration is performed, the result is the equality $x = x$. In this solution λ is the arbitrary parameter. In Figure 7.2 the graphs of the solutions of equation (7.14) are shown for several values of λ.

For some values of λ the original nonhomogeneous equation has no solutions. For the equation of this example there are two such values—the roots of the quadratic equation:

$$240 + \lambda^2 = 0.$$

Solutions of this equation are the eigenvalues $\lambda_{1,2} = \pm i\sqrt{240}$. For these values the homogeneous equation

$$y(x) - \lambda \int_0^1 (xt^2 - x^2 t) y(t) \, dt = 0, \qquad (7.15)$$

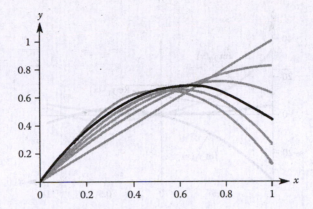

FIGURE 7.2   Solutions of equation (7.14) for $\lambda = 0, 2, 4, 6, 8, 10$. The solution corresponding to value $\lambda = 6$ is shown by the black line.

that corresponds to the original nonhomogeneous equation has nontrivial solutions—the eigenfunctions of the homogeneous equation. Next, we proceed to find these functions.

For equation (7.15) the system (7.13) is homogeneous, and for the coefficients A and B we obtain the system

$$\begin{cases} A = \dfrac{1}{4}\lambda A - \dfrac{1}{5}\lambda B, \\[2mm] B = \dfrac{1}{3}\lambda A - \dfrac{1}{4}\lambda B. \end{cases} \tag{7.16}$$

It is not difficult to see that this system is linearly dependent (its determinant is zero), and it has nontrivial solution only for $\lambda = \lambda_1$ or $\lambda = \lambda_2$. Then, assigning to A an arbitrary real number α and finding B in terms of α, we obtain from the system (7.16) $A = \alpha$, $B = \frac{5\lambda - 20}{4\lambda}\alpha$. Thus, the eigenfunctions of equation (7.15) are

$$y_1(x) = \lambda_1 \alpha x - \lambda_1 B x^2, \; y_2(x) = \lambda_2 \alpha x - \lambda_2 B x^2.$$

The spectrum of eigenvalues of the homogeneous equation, as we found above, is $\lambda_{1,2} = \pm i\sqrt{240}$. The value of α is obviously arbitrary, since both sides of equation (7.15) can be divided by this parameter.

The functions $y_1(x)$ and $y_2(x)$ are complex. Taking $\alpha = 1$, we obtain $\mathrm{Re}\, B = 5/4$, $\mathrm{Im}\, B = \pm 5/\sqrt{240}$. In Figure 7.3 we show graphs of the real and imaginary parts of these solutions:

$$\mathrm{Re}\, y_{1,2}(x) = 5x^2, \quad \mathrm{Im}\, y_{1,2}(x) = \pm\sqrt{240}\left(x - \frac{5}{4}x^2\right).$$

Example 2.2

Solve the integral equation

$$y(x) - \lambda \int_0^1 (xe^t + x^2 t)y(t)\,dt = x.$$

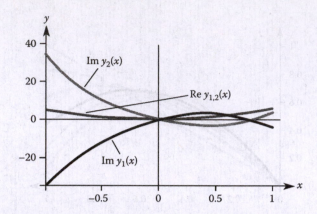

FIGURE 7.3   Eigenfunctions of equation (7.15).

Solution. First, denote $A = \int_0^1 e^t y(t)\,dt$, $B = \int_0^1 t y(t)\,dt$. Then the original integral

equation takes the form

$$y(x) = \lambda A x + \lambda B x^2 + x.$$

Changing the variable x to the variable t in this function $y(x)$, substituting $y(t)$ in the expressions for A and B, and then integrating, results in the system of linear nonhomogeneous algebraic equations:

$$\begin{cases} (1-\lambda)A + (2-e)\lambda B = 1, \\ -\dfrac{1}{3}\lambda A + \left(1 - \dfrac{\lambda}{4}\right)B = \dfrac{1}{3}. \end{cases}$$

Its solution is

$$A = \frac{12 - (11 - 4e)\lambda}{(11 - 4e)\lambda^2 - 15\lambda + 12}, \qquad B = \frac{4}{(11 - 4e)\lambda^2 - 15\lambda + 12}.$$

Substituting these A and B in $y(x) = \lambda A x + \lambda B x^2 + x$ gives the solution of the nonhomogeneous integral equation on the interval $[0, 1]$. The parameter λ is arbitrary, but for the values of λ that turn into zero the denominator of the expressions for A and B,

$$\lambda_1 = \frac{15 + \sqrt{144e - 171}}{22 - 8e} \approx 117, \qquad \lambda_2 = \frac{15 - \sqrt{144e - 171}}{22 - 8e} \approx 0.603,$$

the original nonhomogeneous integral equation does not have solutions. But in this case the corresponding homogeneous equation

$$y(x) - \lambda \int_0^1 (x e^t + x^2 t) y(t)\,dt = 0$$

has solutions. When solving the homogeneous equation, the system of equations for the coefficients A and B is also homogeneous:

$$\begin{cases} (1-\lambda)A + (2-e)\lambda B = 0, \\ -\dfrac{1}{3}\lambda A + \left(1-\dfrac{\lambda}{4}\right)B = 0. \end{cases}$$

For $\lambda = \lambda_1$ or $\lambda = \lambda_2$ this system has a nontrivial solution. Letting $A = \alpha$, from the last equation we obtain

$$B = \frac{\lambda-1}{(2-e)\lambda}\alpha.$$

For $\lambda = \lambda_1$ we have $B_1 = -1.38\alpha$, and for $\lambda = \lambda_2$ we have $B_2 = 0.917\alpha$. The eigenfunctions of the homogeneous equation are $y(x) = \lambda Ax + \lambda Bx^2$, that is,

$$y_1(x) = 117\alpha x - 161\alpha x^2, \quad y_2(x) = 0.603\alpha x + 0.553\alpha x^2,$$

where α is the arbitrary parameter, which can be taken equal to one, for simplicity.

Notice that *nonlinear* equations with degenerated kernels can be solved in a similar way.

Example 2.3

Solve homogeneous nonlinear integral equation

$$y(x) = \lambda \int_0^1 xty^2(t)\,dt.$$

Solution. Let

$$A = \int_0^1 ty^2(t)\,dt,$$

then

$$y(x) = A\lambda x.$$

Substitution $y(x) = A\lambda x$ in the expression for A gives

$$A = \int_0^1 t\lambda^2 A^2 t^2\,dt.$$

Integration results in

$$A = \frac{\lambda^2}{4}A^2.$$

This equation has two solutions:

$$A_1 = 0, \quad A_2 = \frac{4}{\lambda^3}.$$

Consequently, the original integral equation on the interval [0, 1] has two solutions (eigenfunctions) for any $\lambda \neq 0$:

$$y_1(x) = 0, \quad y_2(x) = \frac{4}{\lambda} x.$$

The first one is a trivial solution of the homogeneous equation. It is easy to verify that the substitution of these functions into the original equation transforms this equation into the identity. For the first function, the substitution obviously gives $0 = 0$. Let us substitute the second function in the equation:

$$\frac{4}{\lambda} x = \lambda \int_0^1 xt \frac{16}{\lambda^2} t^2 \, dt.$$

Canceling common factors and calculating the integral again gives the identity. The spectrum of the problem is continuous—it consists of all real, nonzero values of λ.

What if the kernel is not singular? In this situation, we can use the fact that any continuous kernel (as well as kernels with finite discontinuities) can be represented as an infinite sum (Fourier series) of singular kernels. Using this, Ivar Fredholm (1866–1927) proved several theorems. We present one of these theorems (without a proof) for the equation (7.7) (in fact, the schematic of the proof has been presented above). This theorem is known as the *Fredholm alternative*.

Theorem 2.1

Either the nonhomogeneous equation (7.7) with a continuous kernel has a unique continuous solution for any function $g(x)$ (that is, λ is not an eigenvalue), or the homogeneous equation corresponding to (7.7) has at least one nontrivial solution (an eigenfunction).

The theory of Fredholm integral equations of the first kind is much more complicated. Here we will describe only the general idea of a solution.

Solution of equation (7.6):

$$\lambda \int_a^b K(x,t) y(t) \, dt = g(x)$$

can be found in the form of a Fourier series

$$y(x) = \sum_i c_i f_i(x), \tag{7.17}$$

where $f_1(x)$, $f_2(x)$, $f_3(x)$,... is some complete set of functions on the interval (a,b), and C_i are the coefficients of the Fourier expansion. Completeness of the set means that the series (7.17) converges to $y(x)$ on (a,b), and the coefficients c_i are given by

$$c_i = \frac{\int_a^b y(x) f_i(x) dx}{\int_a^b f_i^2(x) dx} \tag{7.18}$$

(for simplicity this formula is written for the real functions $f(x)$).
    Substitution of (7.17) in (7.6) gives

$$g(x) = \lambda \sum_i c_i h_i(x), \tag{7.19}$$

where

$$h_i(x) = \int_a^b K(x,t) f_i(t) dt. \tag{7.20}$$

From equation (7.19) one can find the coefficients c_i, which finalizes the solution of equation (7.6).

Problems
Solve the following Fredholm integral equations of the second kind with singular kernels. Note that in some problems $\lambda = 1$.

1. $y(x) = \sin x + \lambda \int_0^{\pi/2} \sin x \cos t \, y(t) dt$

2. $y(x) = x + \lambda \int_0^{2\pi} (2\pi - t) y(t) \sin x \, dt$

3. $y(x) = 2 \int_0^1 xt y^3(t) dt$

4. $y(x) = \int_{-1}^{1} (xt + x^2 t^2) y^2(t) dt$

5. $y(x) = \int_{0}^{1} x^2 t^2 y^2(t) dt$

6. $y(x) = \lambda \int_{0}^{2\pi} y(t) \sin x \sin t \, dt$

7. $y(x) = \lambda \int_{0}^{1} (2xt - 4x^2) y(t) dt$

8. $y(x) = \lambda \int_{-1}^{1} (5xt^3 + 4x^2 t) y(t) dt$

9. $y(x) = \lambda \int_{-1}^{1} (5xt^3 + 4x^2 t + 3xt) y(t) dt$

10. $y(x) = \lambda \int_{0}^{\pi} x \, y(t) \sin 2t \, dt + \cos 2x$

11. $y(x) = \lambda \int_{0}^{2\pi} \sin(x+t) y(t) dt + 2$

Answers

(For homogeneous equations a solution can be multiplied by an arbitrary constant; also we do not show trivial solutions.)

1. $y(x) = \dfrac{2}{2-\lambda} \sin x; \quad \lambda \neq 2$

2. $y(x) = x + \dfrac{4\lambda\pi^3}{3(1 - 2\pi\lambda)} \sin x$

3. $y_{1,2}(x) = \pm\sqrt{\dfrac{5}{2}} x$

4. $y(x) = \pm\dfrac{15}{4\sqrt{7}} x + \dfrac{5}{4} x^2$

5. $y(x) = 7x^2$

6. $\lambda = \dfrac{1}{\pi}, \quad y(x) = \sin x$

7. $\lambda_1 = \lambda_2 = -3, \quad y(x) = \alpha_1 x + \alpha_2 x^2$

8. $\lambda = \dfrac{1}{2}, \quad y(x) = \dfrac{5}{2} x + \dfrac{10}{3} x^2$

9. $\lambda = \dfrac{1}{4}, \quad y(x) = \dfrac{5}{3} \left(x^2 + \dfrac{3}{2} \right)$

10. $\lambda = -\dfrac{2}{\pi}, \quad y(x) = -\dfrac{2}{\pi} x + \cos 2x$

11. $y(x) = C(\sin x + \cos x) + 2$, C is arbitrary

7.3 ITERATIVE METHOD FOR THE SOLUTION OF FREDHOLM EQUATIONS OF THE SECOND KIND

The iterative method allows one to obtain a sequence of functions that converges to the solution of the integral equation.

When solving equation (7.7),

$$y(x) = \lambda \int_a^b K(x,t) y(t)\, dt + g(x),$$

it is natural to take $g(x)$ to be the zeroth approximation of the solution.

Then, g is substituted in the integrand for y:

$$y(x) = g(x) + \lambda \int_a^b K(x,t) g(t)\, dt. \tag{7.21}$$

Next, (7.21) is substituted in the integrand in equation (7.7) and so on. As a result, we obtain

$$y(x) = g(x) + \lambda \int_a^b K(x,t) g(t)\, dt + \lambda^2 \int_a^b dt \int_a^b dt' K(x,t) K(t,t') g(t') + \dots \tag{7.22}$$

Series (7.22) is called the *Neumann series* of equation (7.7). If the kernel $K(x,t)$ is bounded on (a,b), then the series converges for small values of λ (as elsewhere in this chapter, we are not concentrated on the questions of mathematical rigor).

Example 3.1

Find solution of the integral equation

$$y(x) = \lambda \int_0^1 \frac{x}{1+t^2} y(t)\, dt + 1 + x^2 \tag{7.23}$$

by the iterative method. Find the range of values of λ, for which the Neumann series converges.

Solution. The zeroth approximation is $y_0(x) = 1 + x^2$. Substituting this in the integrand and integrating results in the first approximation:

$$y_1(x) = \lambda \int_0^1 \frac{x}{1+t^2}(1+t^2)\, dt + 1 + x^2 = \lambda \int_0^1 x\, dt + 1 + x^2 = \lambda x + 1 + x^2.$$

Substituting first approximation in the integrand of equation (7.23) gives the second approximation:

$$y_2(x) = \lambda \int_0^1 \frac{x}{1+t^2}(\lambda t + 1 + t^2)\, dt + 1 + x^2$$

$$= \int_0^1 \frac{\lambda x t}{1+t^2}\, dt + \lambda \int_0^1 x\, dt + 1 + x^2 = x\left(\lambda + \frac{\lambda^2 \ln 2}{2}\right) + 1 + x^2.$$

The third approximation is (check)

$$y_3(x) = x\left(\lambda + \frac{\lambda^2 \ln 2}{2} + \frac{\lambda^3 \ln^2 2}{4}\right) + 1 + x^2.$$

It is clear that the nth approximation can be written as

$$y_n(x) = x\left(\lambda + \frac{\lambda^2 \ln 2}{2} + \frac{\lambda^3 \ln^2 2}{4} + \cdots + \frac{\lambda^n \ln^{n-1} 2}{2^{n-1}}\right) + 1 + x^2.$$

For $n \to \infty$ we obtain the series

$$y(x) = x \sum_{n=1}^{\infty} \frac{\lambda^n \ln^{n-1} 2}{2^{n-1}} + 1 + x^2.$$

The series $\sum_{n=1}^{\infty} \frac{\lambda^n \ln^{n-1} 2}{2^{n-1}}$ is geometric with the common ratio $q = \frac{\lambda \ln 2}{2}$ and the first term $a_1 = \lambda$. This series converges to the sum $S = \frac{a_1}{1-q} = \frac{\lambda}{1-\lambda \frac{\ln 2}{2}}$, if $\left|\frac{\lambda \ln 2}{2}\right| < 1$. Thus, the range of values of the parameter λ, such that the solution of the integral equation on the interval [0, 1] exists, is $|\lambda| < \frac{2}{\ln 2}$. The solution is the function

$$y(x) = \frac{\lambda x}{1 - \frac{\ln 2}{2}\lambda} + 1 + x^2. \tag{7.24}$$

In Figure 7.4 the graphs of the functions $y_0(x)$, $y_1(x)$, $y_2(x)$, and $y_3(x)$ are shown for two values of λ.

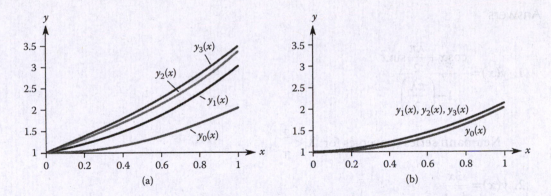

FIGURE 7.4   Solutions of equation (7.23) obtained by the iterative method: (a) $\lambda = 1$; (b) $\lambda = 0.1$.

Problems

Solve the following Fredholm integral equations of the second kind by the iterative method, and find the range of values of λ for which the Neumann series converges. Plot the graphs of the zeroth, first, and second iterations on the corresponding intervals for several values of λ.

1. $y(x) = \lambda \int_0^\pi \sin(x+t) y(t) dt + \cos x$

2. $y(x) = \lambda \int_0^1 xt y(t) dt + x$

3. $y(x) = \lambda \int_0^1 y(t) dt + \sin \pi x$

4. $y(x) = \lambda \int_0^{\pi/2} \sin x \cos x \, y(t) dt + 1$

5. $y(x) = \lambda \int_0^1 xe^t y(t) dt + e^{-x}$

6. $y(x) = \lambda \int_0^{1/2} y(t) dt + x$

7. $y(x) = \lambda \int_0^\pi x \sin 2t \, y(t) dt + \cos 2x$

8. $y(x) = \lambda \int_0^{2\pi} \sin(x+t) y(t) dt + 2$

Answers

1. $y(x) = \dfrac{\cos x + \dfrac{\lambda \pi}{2} \sin x}{1 - \left(\dfrac{\pi \lambda}{2}\right)^2}$.

Neumann series converges for $|\lambda| < \dfrac{2}{\pi}$.

2. $y(x) = \dfrac{3x}{3 - \lambda}$.

Neumann series converges for $|\lambda| < 3$.

3. $y(x) = \sin \pi x + \dfrac{2\lambda}{\pi(1 - \lambda)}$.

Neumann series converges for $|\lambda| < 1$.

4. $y(x) = 1 + \dfrac{\lambda \pi}{2(2 - \lambda)} \sin 2x$.

Neumann series converges for $|\lambda| < 2$.

5. $y(x) = e^{-x} + \dfrac{\lambda x}{1 - \lambda}$.

Neumann series converges for $|\lambda| < 1$.

6. $y(x) = x + \dfrac{\lambda}{4(2 - \lambda)}$.

Neumann series converges for $|\lambda| < 2$.

7. $y(x) = \cos 2x$.

8. $y(x) = 2$.

7.4 VOLTERRA EQUATION

Integral equations that contain the integral with a variable limit of integration are called *Volterra equations*:

$$\lambda \int_a^x K(x,t) y(t) \, dt + g(x) = 0, \qquad (7.25)$$

$$\lambda \int_a^x K(x,t)y(t)dt + g(x) = y(x). \tag{7.26}$$

Equations (7.25) and (7.26) are called Volterra equations of the first kind and of the second kind, respectively. Volterra equations are linear.

Solution of an initial value problem for linear differential equations leads to Volterra integral equations of the second kind, and vice versa—often the Volterra equation can be solved by reducing it to a differential equation. The following example illustrates this solution scheme. Note that constant λ is often omitted.

Example 4.1

Solve a Volterra equation of the second kind

$$x + \int_0^x xty(t)dt = y(x),$$

by reducing it to a differential equation.

Solution. Let

$$u(x) = \int_0^x ty(t)dt.$$

Next, we differentiate this expression and use the original equation in the form $y(x) = x + xu(x)$:

$$u' = xy(x) = x(x + xu(x)).$$

The solution of this differential equation is found using separation of variables:

$$u(x) = -1 + Ce^{x^3/3}.$$

Turning again to the equation $y(x) = x + xu(x)$, we obtain

$$y(x) = xCe^{x^3/3}.$$

The initial condition follows from the integral equation and it is $y(0) = 0$. Obviously, constant C cannot be obtained from this condition.

To find C, substitute the expression for $y(x)$ in the original equation:

$$x + \int_0^x xt^2 Ce^{t^3/3}\,dt = xCe^{x^3/3}.$$

Calculation of the integral gives

$$xCe^{x^3/3} = xCe^{x^3/3} + x(1-C),$$

from which $C = 1$. Thus the final answer is

$$y(x) = xe^{x^3/3}.$$

Problems

Solve the following integral equations by reducing them to the differential equations.

1. $y(x) = x + \dfrac{x^2}{2} - \displaystyle\int_0^x y(t)\,dt$

2. $y(x) = e^x + \displaystyle\int_0^x y(t)\,dt$

3. $y(x) = \cos x + \displaystyle\int_0^x e^{x-t} y(t)\,dt$

4. $y(x) = 2 + \displaystyle\int_0^x t y(t)\,dt$

Answers

1. $y(x) = x$

2. $y(x) = e^x(1+x)$

3. $y(x) = \dfrac{1}{5}(3\cos x + \sin x + 2e^{2x})$

4. $y(x) = 2e^{x^2/2}$

7.5 SOLUTION OF VOLTERRA EQUATIONS WITH THE DIFFERENCE KERNEL USING THE LAPLACE TRANSFORM

The Volterra equation of the second kind with the difference kernel has the form

$$\lambda \int_0^x K(x-t)y(t)\,dt + g(x) = y(x). \tag{7.27}$$

For instance, the initial value problem for linear differential equations with constant coefficients can be reduced to such equations. Equation (7.27) can be solved using the Laplace transform.

Recall that the Laplace transform $L[f(x)]$ of the real function $f(x)$ is

$$\hat{f}(p) = L[f(x)] = \int_0^\infty e^{-px} f(x)\,dx, \tag{7.28}$$

where p is, generally, a complex parameter. $\hat{f}(p)$ is the image of the original function $f(x)$.

The right side of equation (7.28) is called the *Laplace integral*. For its convergence in the case where p is complex, it is necessary that Re$p > 0$; if p is real, then for convergence $p > 0$. Also, a growth of function $f(x)$ should be restricted and $f(x)$ should be bounded on $(0, \infty)$.

Convolution of functions $f_1(x)$ and $f_2(x)$ of the real variable x is the function $f(x)$, which is defined as

$$f(x) = \int_0^x f_1(x-t) f_2(t)\, dt. \tag{7.29}$$

The integral in equation (7.27) is the convolution of functions $K(x)$ and $y(x)$; thus

$$L\left[\int_0^x K(x-t) y(t)\, dt \right] = \hat{K}(p)\hat{y}(p). \tag{7.30}$$

Also, the Laplace transform is linear:

$$L\left[C_1 f_1(x) + C_2 f_2(x) \right] = C_1 L\left[f_1(x) \right] + C_2 L\left[f_2(x) \right]. \tag{7.31}$$

Applying the Laplace transform to equation (7.27) and using these properties, we obtain

$$\hat{y}(p) = \int_0^\infty \left[g(x) + \lambda \int_0^x K(x-t)y(t)\, dt \right] e^{-px}\, dx = \hat{g}(p) + \lambda \int_0^\infty \left[\int_0^x K(x-t)y(t)\, dt \right] e^{-px}\, dx$$

$$= \hat{g}(p) + \lambda \hat{K}(p)\hat{y}(p).$$

Then,

$$\hat{y}(p) = \frac{\hat{g}(p)}{1 - \lambda \hat{K}(p)} \quad \left(\hat{K}(p) \neq \frac{1}{\lambda} \right), \tag{7.32}$$

where $\hat{y}(p)$, $\hat{g}(p)$, $\hat{K}(p)$ are the Laplace transforms of the corresponding functions.

To determine the original function from the image, $\hat{f}(p)$, one has to perform the *inverse Laplace transform*, which is denoted as

$$f(x) = L^{-1}\left[\hat{f}(p) \right]. \tag{7.33}$$

The inverse transform is also linear:

$$L^{-1}\left[C_1 \hat{f}_1(p) + C_2 \hat{f}_2(p) \right] = C_1 L^{-1}\left[\hat{f}_1(p) \right] + C_2 L^{-1}\left[\hat{f}_2(p) \right]. \tag{7.34}$$

Laplace transforms of some functions can be found in the table in Chapter 6.

Example 5.1

Solve the integral equation

$$y(x) = \sin x + 2\int_0^x \cos(x - t)y(t)\,dt.$$

Solution. Calculating directly from the definition (7.28), or using the table, we have

$$L[\sin x] = \frac{1}{p^2 + 1}, \quad L[\cos x] = \frac{p}{p^2 + 1}.$$

Applying the Laplace transform to both sides of the equation, taking into the account linearity of the transform and the convolution theorem, gives

$$\hat{y}(p) = \frac{1}{p^2 + 1} + \frac{2p}{p^2 + 1}\hat{y}(p).$$

Then,

$$\hat{y}(p)\left[1 - \frac{2p}{p^2 + 1}\right] = \frac{1}{p^2 + 1},$$

and simplifying:

$$\hat{y}(p) = \frac{1}{(p - 1)^2}.$$

Solution of the given integral equation is the inverse Laplace transform of this function. Using the table, we finally find

$$y(x) = xe^x.$$

Example 5.2

Solve the integral equation

$$y(x) = \cos x - \int_0^x (x - t)\cos(x - t)y(t)\,dt.$$

Solution. Applying the Laplace transform to this equation and using transforms

$$L[\cos x] = \frac{p}{p^2 + 1}, \quad L[x\cos x] = \frac{p^2 - 1}{(p^2 + 1)^2},$$

gives

$$\hat{y}(p) = \frac{p}{p^2 + 1} - \frac{p^2 - 1}{(p^2 + 1)^2}\hat{y}(p).$$

Thus,

$$\hat{y}(p) = \frac{p^2 + 1}{p^3 + 3p} = \frac{p}{p^2 + 3} + \frac{1}{p(p^2 + 3)}.$$

The second term in the above formula is not in the table. Using the method of partial fractions, we can decompose this term into the sum of two elementary fractions:

$$\hat{y}(p) = \frac{p}{p^2 + 3} + \frac{1}{3p} - \frac{1}{3p^2 + 9}.$$

Finally, the inverse Laplace transform gives the solution of the integral equation:

$$y(x) = \frac{2}{3}\cos\sqrt{3}x + \frac{1}{3}.$$

Problems

Solve the following integral equations using the Laplace transform:

1. $y(x) = e^x - \int_0^x e^{x-t} y(t)dt$

2. $y(x) = x - \int_0^x e^{x-t} y(t)dt$

3. $y(x) = e^{2x} + \int_0^x e^{t-x} y(t)dt$

4. $y(x) = 2 + \frac{1}{2}\int_0^x (x-t)^2 y(t)dt$

Answers

1. $y(x) = 1$

2. $y(x) = x - \dfrac{x^2}{2}$

3. $y(x) = e^{2x}(1+x)$

4. $y(x) = \dfrac{2}{3}e^x + \dfrac{4}{3}e^{-x/2} \cdot \cos\dfrac{3x}{4}$

CHAPTER 8

Series Solutions of ODE and Bessel and Legendre Equations

In this chapter we discuss the Bessel and Legendre equations and their solutions: Bessel and Legendre functions. These famous functions are widely used in the sciences; in particular, Bessel functions serve as a set of basis functions for BVP with circular or cylindrical symmetry, and the Legendre functions serve as a set of basis functions for BVP in spherical coordinates.

8.1 SERIES SOLUTIONS OF DIFFERENTIAL EQUATIONS: INTRODUCTION

Solutions of differential equations most often are not available in closed form. However, they may be found in the form of convergent power series.

Let us begin with a simple equation

$$y' = xy.$$

The analytical solution can be easily obtained by separating the variables: $y = Ce^{x^2/2}$. We will show how to solve the equation using a power series expansion for the unknown function:

$$y = a_0 + a_1 x + a_2 x^2 + a_3 x^3 + \ldots = \sum_{k=0}^{\infty} a_k x^k. \tag{8.1}$$

Assuming that this series (absolutely) converges, we can differentiate it to obtain

$$y' = a_1 + 2a_2x + 3a_3x^2 + 4a_4x^3 + \ldots = \sum_{k=1}^{\infty} ka_kx^{k-1}. \tag{8.2}$$

Substituting expansions (8.1) and (8.2) in the equation, we have

$$a_1 + 2a_2x + 3a_3x^2 + 4a_4x^3 + 5a_5x^4 + 6a_6x^6 + \cdots$$

$$= a_0x + a_1x^2 + a_2x^3 + a_3x^4 + a_4x^5 + a_5x^6 + \cdots$$

Equating the coefficients of the same powers of x we obtain

$$a_1 = 0, \quad a_2 = a_0/2, \quad a_3 = a_1/3 = 0, \quad a_4 = a_2/4 = a_0/(2 \cdot 4), \quad a_5 = a_3/5 = 0,$$

$$a_6 = a_4/6 = a_0/(2 \cdot 4 \cdot 6), \ldots$$

Thus, the expansion (8.1) is

$$y = a_0 \left(1 + \frac{x^2}{2} + \frac{1}{2}\frac{x^4}{4} + \frac{1}{2 \cdot 3}\frac{x^6}{8} + \cdots \right).$$

The expression in the brackets is the MacLaurin expansion of $e^{x^2/2}$; thus the previous formula is $y = a_0 e^{x^2/2}$. We see that with initial condition $y(0) = C = a_0$ the power series solution coincides with the one obtained using separation of variables.
Consider another equation

$$y' = 1 + xy.$$

Substituting into this equations expansions (8.1) and (8.2) we obtain

$$a_1 + 2a_2x + 3a_3x^2 + 4a_4x^3 + 5a_5x^4 + 6a_6x^6 + \cdots$$

$$= 1 + a_0x + a_1x^2 + a_2x^3 + a_3x^4 + a_4x^5 + \cdots$$

Equating the coefficients of the same powers of x we have

$$a_1 = 1, \quad a_2 = a_0/2, \quad a_3 = a_1/3 = 1/(1 \cdot 3), \quad a_4 = a_2/4 = a_0/(2 \cdot 4),$$

$$a_5 = a_3/5 = 1/(1 \cdot 3 \cdot 5), \quad a_6 = a_4/6 = a_0/(2 \cdot 4 \cdot 6), \ldots$$

Therefore, we obtain a general solution of a given equation as the expansion

$$y = a_0\left(1 + \frac{x^2}{2} + \frac{x^4}{2\cdot4} + \frac{x^6}{2\cdot4\cdot6} + \cdots\right) + \left(\frac{x}{1} + \frac{x^3}{1\cdot3} + \frac{x^5}{1\cdot3\cdot5} + \frac{x^7}{1\cdot3\cdot5\cdot7} + \cdots\right)$$

which converges for any x, as can be easily checked by the convergence test.

The value of a_0 can be obtained from the initial condition $y(0) = y_0$; thus $a_0 = y_0$.

If instead of MacLaurin expansion we use the Taylor expansion

$$y = a_0 + a_1(x - x_0) + a_2(x - x_0)^2 + a_3(x - x_0)^3 + \cdots = \sum_{k=0}^{\infty} a_k(x - x_0)^k \qquad (8.3)$$

then, to find a particular solution, the initial condition should be taken as $y(x_0) = y_0$.

If an equation has singular points x at which its coefficients become infinite, the method of integration in series may break down.

Similar approach based on expansions (8.1) or (8.3) can be applied to second-order (and higher-order) differential equations.

Example 1.1

Find the particular solution of IVP $y'' = xy^2 - y'$, $y(0) = 2$, $y'(0) = 1$.

Solution. Search for solution in the form (8.1). First, from the initial conditions we obtain $a_0 = 2$, $a_1 = 1$; thus

$$y = 2 + x + a_2x^2 + a_3x^3 + a_4x^4 + \ldots$$

Substituting this and

$$y' = 1 + 2a_2x + 3a_3x^2 + 4a_4x^3 + \cdots, \quad y'' = 2a_2 + 6a_3x + 12a_4x^2 + \cdots$$

in the equation and equating the coefficients of the same powers of x we have

$$a_2 = -1/2, \, a_3 = 5/6, a_4 = 1/8, \, \ldots;$$

therefore

$$y = 2 + x - x^2/2 + 5x^3/6 + x^4/8 + \cdots$$

Because we cannot find the sum of this series, we cannot finish up this example with the solution in a closed form.

In the problems below search for particular solutions in the form of series expansion— find first several terms up to x^4. Use programs ODE_1st_order and ODE_2st_order to plot the graphs of series and numerical solutions of IVP on some interval which includes the initial point x_0. (Input formula for series solution into Edit window "*Analytical solution.*")

Problems	Answers
1. $y' = y^2 - x$, $y(0) = 1$	$y = 1 + x + \dfrac{x^2}{2} + \dfrac{2x^3}{3} + \dfrac{7x^4}{12} + \cdots$
2. $y' = y^3 + x^2$, $y(1) = 1$	$y = 1 + 2(x-1) + 4(x-1)^2$ $+ \dfrac{25(x-1)^3}{3} + \dfrac{81(x-1)^4}{4} + \cdots$
3. $y' = y + xe^y$, $y(0) = 0$	$y = \dfrac{x^2}{2} + \dfrac{x^3}{6} + \dfrac{x^4}{6} + \cdots$
4. $y' = 2x + \cos y$, $y(0) = 0$	$y = x + x^2 - \dfrac{x^3}{6} - \dfrac{x^4}{4} + \cdots$
5. $y'' + xy = 0$, $y(0) = 1$, $y'(0) = -1$	$y = 1 - x + \dfrac{x^3}{6} + \dfrac{x^4}{12} + \cdots$
6. $y'' = xy' - y^2$, $y(0) = 1$, $y'(0) = 2$	$y = 1 + 2x - \dfrac{x^2}{2} - \dfrac{x^3}{3} - \dfrac{x^4}{3} + \cdots$
7. $y'' = y'^2 + xy$, $y(0) = 4$, $y'(0) = -2$	$y = 4 - 2x + 2x^2 - 2x^3 + \dfrac{19x^4}{6} + \cdots$

8.2 BESSEL EQUATION

In many problems one often encounters a differential equation

$$r^2 y''(r) + ry'(r) + (\lambda r^2 - p^2) y(r) = 0, \tag{8.4}$$

where p is given fixed value and λ is an additional parameter. Equation (8.4) is called the *Bessel equation of order p*. In applications of the Bessel equation we will encounter boundary conditions. For example the function $y(r)$, defined on a closed interval $[0, l]$, could be restricted to a specified behavior at the points $r = 0$ and $r = l$. For instance at $r = l$ the value of the solution, $y(l)$, could be prescribed, or its derivative $y'(l)$, or their linear combination $\alpha y'(l) + \beta y(l)$. Generally equation (8.4) has nontrivial solutions that correspond to a given set of boundary conditions only for certain values of the parameter λ, which are called *eigenvalues*. The goal then is to find eigenvalues for these boundary conditions and the corresponding solutions, $y(r)$, which are called *eigenfunctions*.

Let's make the change of variables $x = \sqrt{\lambda} r$ in equation (8.4), which, with $y'_r = y'_x \cdot x'_r = \sqrt{\lambda} y'_x$, $y''_{r^2} = \lambda y''_{x^2}$, gives

$$x^2 y''(x) + xy'(x) + (x^2 - p^2) y(x) = 0. \tag{8.5}$$

By dividing this equation by x we obtain the Bessel equation in the form

$$(xy')' + \left(x - \frac{p^2}{x} \right) y = 0, \quad \text{or} \quad y'' + \frac{y'}{x} + \left(1 - \frac{p^2}{x^2} \right) y = 0. \tag{8.6}$$

To solve equation (8.6) we first consider integer values of the parameter p. Let us try to find the solution in the form of power series in x and let a_0 be the first nonzero coefficient of the series; then

$$y = a_0 x^m + a_1 x^{m+1} + a_2 x^{m+2} + \cdots + a_n x^{m+n} + \cdots \tag{8.7}$$

where integer $m \geq 0$ and $a_0 \neq 0$.

We may differentiate equation (8.7) to find an expression for y' and, multiplying the series for y' by x and differentiating once more, we obtain

$$(xy')' = a_0 m^2 x^{m-1} + a_1 (m+1)^2 x^m + a_2 (m+2)^2 x^{m+1} + \cdots$$
$$+ a_n (m+n)^2 x^{m+n-1} + \cdots \tag{8.8}$$

Substitution of this expression for $(xy')'$ and the expressions for y into equation (8.6) results in the following equality:

$$a_0 m^2 x^{m-1} + a_1 (m+1)^2 x^m + a_2 (m+2)^2 x^{m+1} + \cdots + a_n (m+n)^2 x^{m+n-1} + \cdots$$
$$- a_0 p^2 x^{m-1} - a_1 p^2 x^m - a_2 p^2 x^{m+1} - \cdots - a_n p^2 x^{m+n-1} - \cdots$$
$$+ a_0 x^{m+1} + a_1 x^{m+2} + \cdots + a_{n-2} x^{m+n-1} + \cdots = 0.$$

Because functions x^k are linearly independent, the coefficients of each power of x^k must be zero. Thus we obtain an infinite system of equations for the coefficients, a_n:

$$a_0 (m^2 - p^2) = 0,$$

$$a_1 [(m+1)^2 - p^2] = 0,$$

$$a_2 [(m+2)^2 - p^2] + a_0 = 0,$$

$$a_3 [(m+3)^2 - p^2] + a_1 = 0,$$

....

$$a_n [(m+n)^2 - p^2] + a_{n-2} = 0,$$

....

From the first equation we obtain $m^2 - p^2 = 0$ (since $a_0 \neq 0$). Thus $m = p$ (recall that m is nonnegative; thus, $m \neq -p$).

From other equations we obtain the coefficients a_1, a_2, a_3, ... a_n,.... To simplify the calculations let us transform the expressions in the square brackets above, taking into account that $m = p$, in the following way:

$$a_1[(p+1)^2 - p^2] = 0,$$
$$a_2[(p+2)^2 - p^2] + a_0 = 0,$$
$$a_3[(p+3)^2 - p^2] + a_1 = 0,$$
$$\cdots\cdots\cdots\cdots\cdots\cdots\cdots\cdots\cdots$$
$$a_n[(p+n)^2 - p^2] + a_{n-2} = 0,$$
$$\cdots\cdots\cdots\cdots\cdots\cdots\cdots\cdots\cdots$$

or

$$a_1(2p+1) = 0,$$

$$a_2 2(2p+2) = -a_0,$$

$$a_3 3(2p+3) = -a_1,$$

$$\cdots\cdots\cdots\cdots\cdots\cdots\cdots$$

$$a_n \cdot n \cdot (2p+n) = -a_{n-2},$$

$$\cdots\cdots\cdots\cdots\cdots\cdots\cdots$$

Thus we see that $a_1 = a_3 = a_5 = \cdots = a_{2k+1} = 0$,

$$a_2 = \frac{-a_0}{2(2p+2)}, \quad a_4 = -\frac{a_2}{4(2p+4)} = \frac{a_0}{2\cdot4\cdot(2p+2)(2p+4)},$$

$$a_6 = \frac{-a_4}{6(2p+6)} = \frac{-a_0}{2\cdot4\cdot6\cdot(2p+2)(2p+4)(2p+6)}, \text{ and so on.}$$

Finally, on inspection, we obtain a recurrence relation that will generate any term in the series:

$$a_{2k} = \frac{a_0(-1)^k}{2\cdot4\cdot\cdots\cdot 2k\cdot(2p+2)(2p+4)...(2p+2k)}$$

$$= \frac{a_0(-1)^k}{2^k\cdot k!\cdot 2^k\cdot(p+1)(p+2)...(p+k)} = \frac{a_0 p!(-1)^k}{2^{2k}k!(p+k)!}$$

By substituting the coefficients that we have found in the series (8.4) and taking into account that $m = p$, we obtain

$$y = a_0 \left[x^p - \frac{p!}{2^2 1!(p+1)!} x^{p+2} + \frac{p!}{2^4 2!(p+2)!} x^{p+4} - \cdots + \frac{p!(-1)^k}{2^{2k} k!(p+k)!} x^{p+2k} + \cdots \right],$$

which may be written in a simpler form as

$$y = a_0 2^p\, p! \left[\frac{\left(\frac{x}{2}\right)^p}{p!} - \frac{\left(\frac{x}{2}\right)^{p+2}}{1!(p+1)!} + \frac{\left(\frac{x}{2}\right)^{p+4}}{2!(p+2)!} - \cdots + (-1)^k \frac{\left(\frac{x}{2}\right)^{p+2k}}{k!(p+k)!} + \cdots \right].$$

The series in the square brackets is absolutely convergent for all values of x, which is easy to confirm using the D'Alambert criteria: $\lim_{k \to \infty} |a_{k+1}/a_k| = 0$. Due to the presence of factorials in the denominator this series converges very fast; its sum is called a *Bessel function* of order p and it is denoted as $J_p(x)$:

$$J_p(x) = \frac{\left(\frac{x}{2}\right)^p}{p!} - \frac{\left(\frac{x}{2}\right)^{p+2}}{1!(p+1)!} + \frac{\left(\frac{x}{2}\right)^{p+4}}{2!(p+2)!} - \cdots + (-1)^k \frac{\left(\frac{x}{2}\right)^{p+2k}}{k!(p+k)!} + \cdots, \tag{8.9}$$

that is,

$$J_p(x) = \sum_{k=0}^{\infty} \frac{(-1)^k}{k!(p+k)!} \left(\frac{x}{2}\right)^{p+2k}. \tag{8.10}$$

We obtained the solution of equation (8.6) in the case of the function $y(x)$ finite at $x = 0$. (The constant coefficient $a_0 2^p\, p!$ in the series for $y(x)$ can be omitted because the Bessel equation is homogeneous.) Note that this solution was obtained for *integer, nonnegative values of p only.*

Next, we generalize the above case for p as a positive integer to the case of *arbitrary real p*. To do this it is necessary to replace the integer-valued function $p!$ by the Gamma function $\Gamma(p + 1)$, which is defined for arbitrary real values of p. The definition and main properties of the Gamma function are listed in Section 8.6. Using the Gamma function, the Bessel function of order p, where p is real, can be defined by a series which is constructed analogously to the series in equation (8.9):

$$a_2 = -\frac{a_0}{2(2+2p)} = -\frac{a_0}{2^2(1+p)} = -\frac{a_0 \Gamma(p+1)}{2^2 \Gamma(p+2)},$$

$$a_4 = -\frac{a_2}{2^3(p+2)} = \frac{a_0}{2!2^4(p+1)(p+2)} = \frac{a_0 \Gamma(p+1)}{2!2^4 \Gamma(p+3)},$$

$$a_{2k} = \frac{(-1)^k a_0 \Gamma(p+1)}{2^{2k} k! \Gamma(p+k+1)!},$$

$$J_p(x) = \sum_{k=0}^{\infty} \frac{(-1)^k}{\Gamma(k+1)\Gamma(k+p+1)} \left(\frac{x}{2}\right)^{p+2k}$$

$$= \left(\frac{x}{2}\right)^p \sum_{k=0}^{\infty} \frac{(-1)^k}{\Gamma(k+1)\Gamma(k+p+1)} \left(\frac{x}{2}\right)^{2k}, \tag{8.11}$$

which *converges for any p*. In particular, *replacing p by −p* we obtain

$$J_{-p}(x) = \left(\frac{x}{2}\right)^{-p} \sum_{k=0}^{\infty} \frac{(-1)^k}{\Gamma(k+1)\Gamma(k-p+1)} \left(\frac{x}{2}\right)^{2k}. \tag{8.12}$$

Since Bessel equation (8.6) contains p^2, functions $J_p(x)$ and $J_{-p}(x)$ are solutions of the equation for the same p. If p *is non-integer*, these solutions are linearly independent, since the first terms in equations (8.11) and (8.12) contain different powers of x: x^p and x^{-p}, respectively; then the general solution of equation (8.6) can be written in the form

$$y = C_1 J_p(x) + C_2 J_{-p}(x). \tag{8.13}$$

Point $x = 0$ must be excluded from the domain of definition of the function (8.12), since x^{-p} for $p > 0$ diverges at this point.

The functions, $J_{-p}(x)$ are bounded as $x \to 0$. In fact the functions $J_p(x)$ are continuous for all x since they are the sum of a converging power series. For non-integer values of p this follows from the properties of the Gamma function and the series (8.11).

Reading Exercise

1) Show that $J_{-n}(x) = (-1)^n J_n(x)$ for $n = 1, 2, 3 \ldots$

2) Show that $\lim_{x \to 0} \dfrac{J_1(x)}{x} = \dfrac{1}{2}$.

The solutions $J_p(x)$ are not the only solution of equation (8.6). All solutions of equation (8.6) have a common name and are referred to as *the cylindrical functions*. We will see in Chapters 11, 13, and 14 that these functions naturally appear in many mathematical physics problems in cylindrical and polar coordinates. They are composed of the cylindrical functions of the first kind $J_p(x)$ (*Bessel functions*), which we have just obtained;

the cylindrical functions of the second kind, $N_p(x)$ called *Neumann functions* (an alternative notation is $Y_p(x)$; and the cylindrical functions of the third kind, $H_p^{(1)}(x)$ and $H_p^{(2)}(x)$, called *Hankel functions*. The functions $N_p(x)$ are singular at $x = 0$. In Section 8.5 we shall also discuss *spherical Bessel functions* related to the solutions of some boundary value problems in spherical coordinates (these problems are discussed in book [1]).

In the case that $J_p(x)$ and $J_{-p}(x)$ are not linearly independent, the general solution can be formed from the linear combination of the functions $J_p(x)$ and $N_p(x)$ as

$$y = C_1 J_p(x) + C_2 N_p(x). \tag{8.14}$$

The functions $N_p(x)$ are singular at $x = 0$; thus if the physical formulation of the problem requires regularity of the solution at zero, the coefficient C_2 in the solution (equation (8.13)) must be zero.

We conclude this section by presenting several first Bessel functions $J_n(x)$ of integer order. The first few terms of the expansion in equation (8.6) near zero for the first three functions are

$$J_0(x) = 1 - \frac{x^2}{2^2} + \frac{x^4}{2^4 \cdot 2! \cdot 2!} - \cdots,$$

$$J_1(x) = \frac{x}{2} - \frac{x^3}{2^3 \cdot 2!} + \frac{x^5}{2^5 \cdot 2! \cdot 3!} - \cdots, \tag{8.15}$$

$$J_2(x) = \frac{x^2}{2^2 \cdot 2!} - \frac{x^4}{2^4 \cdot 3!} + \frac{x^6}{2^6 \cdot 2! \cdot 4!} - \cdots$$

Note that the functions $J_n(x)$ are even if n is an integer and odd if n is a non-integer (although in general, problems with physical boundary values have $x \geq 0$, and we do not need to consider the behavior of the function for $x < 0$). For future reference we present the useful fact that at $x = 0$ we have $J_0(0) = 1$ and $J_n(0) = 0$ for $n \geq 1$. Figure 8.1 shows graphs of functions $J_0(x)$, $J_1(x)$, and $J_2(x)$.

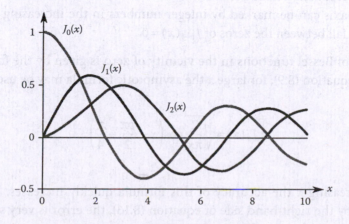

FIGURE 8.1   Graphs of functions $J_0(x)$, $J_1(x)$, and $J_2(x)$.

TABLE 8.1   Positive Roots of $J_0(x)$, $J_1(x)$, $J_2(x)$.

Function \ Roots	μ_1	μ_2	μ_3	μ_4	μ_5
$J_0(x)$	2.4048	5.5201	8.6537	11.7915	14.9309
$J_1(x)$	3.8317	7.0156	10.1735	13.3237	16.4706
$J_2(x)$	5.136	8.417	11.620	14.796	17.960

Table 8.1 lists a few first roots of Bessel functions of orders 0, 1, and 2.

A convenient notation for roots of equation $J_n(x)=0$ is $\mu_k^{(n)}$, where n stands for the order of the Bessel function and k stands for the root number. Often we will write equation $J_n(x)=0$ in the form $J_n(\mu)=0$. In cases when the value of n is clear we will omit the upper index in μ_k (like in Table 8.1).

8.3 PROPERTIES OF BESSEL FUNCTIONS

The properties of functions $J_p(x)$ listed below follow from the expansion in equation (8.9).

1. All Bessel functions are defined and continuous on the real axis and have derivatives of all orders. This is because any Bessel function can be expanded in a power series that converges for all x, and the sum of the power series is a continuous function that has derivatives of all orders.

2. For integer $p=n$ Bessel functions of even orders are even functions (since their expansion contains only even powers of the argument). Bessel functions of odd orders are odd functions.

3. Each Bessel function has an infinite number of real roots. Roots located on the positive semi-axis can be marked by integer numbers in the increasing order. Zeros of $J_p(x)=0$ fall between the zeros of $J_{p+1}(x)=0$.

4. Behavior of Bessel functions in the vicinity of zero is given by the first terms of the series in equation (8.9); for large x the asymptotic formula may be used

$$J_p(x) \approx \sqrt{\frac{2}{\pi x}} \cos\left(x - \frac{p\pi}{2} - \frac{\pi}{4}\right) \tag{8.16}$$

With increasing x the accuracy of this formula quickly increases. When $J_p(x)$ is replaced by the right-hand side of equation (8.16), the error is very small for large x and has same order as $x^{-3/2}$.

From equation (8.4) it follows, in particular, that the function $J_p(x)$ has roots that are close (for large x) to the roots of the equation

$$\cos\left(x - \frac{p\pi}{2} - \frac{\pi}{4}\right) = 0;$$

thus the difference between two adjacent roots of the function $J_p(x)$ tends to π when roots tend to infinity. A graph of $J_p(x)$ has the shape of a curve that depicts decaying oscillation; the "wavelength" is almost constant (close to π), and the amplitude decays inversely proportional to square root of x. In fact we have $\lim_{x\to\infty} J_p(x) = 0$.

5. Integration formulas are

$$\int x^{-p} J_{p+1}(x)\, dx = -x^{-p} J_p(x), \quad \int x^p J_{p-1}(x)\, dx = x^p J_p(x). \tag{8.17}$$

6. *Recurrence formulas* are

$$J_{p+1}(x) = \frac{2p}{x} J_p(x) - J_{p-1}(x),$$

$$J_{p+1}(x) = \frac{p}{x} J_p(x) - J_p'(x), \tag{8.18}$$

$$J_p'(x) = -\frac{p}{x} J_p(x) + J_{p-1}(x).$$

These identities are easily established by operating on the series that defines the function.

In many physical problems with the spherical symmetry one encounters Bessel functions of half-integer orders, where $p = (2n+1)/2$ for $n = 0, 1, 2, \ldots$ For instance, solving equation (8.6) with $p = 1/2$ and $p = -1/2$ by using the series expansion

$$y(x) = x^{1/2} \sum_{k=0}^{\infty} a_k x^k, \quad (a_0 \neq 0) \tag{8.19}$$

we obtain

$$J_{1/2}(x) = \left(\frac{2}{\pi x}\right)^{1/2} \sum_{k=0}^{\infty} (-1)^k \frac{x^{2k+1}}{(2k+1)!}, \tag{8.20}$$

and

$$J_{-1/2}(x) = \left(\frac{2}{\pi x}\right)^{1/2} \sum_{k=0}^{\infty} (-1)^k \frac{x^{2k}}{(2k)!}. \tag{8.21}$$

By comparing expansions in equations (8.20) and (8.21) to the MacLaurin series expansions in $\sin x$ and $\cos x$, we obtain

$$J_{1/2} = \left(\frac{2}{\pi x}\right)^{1/2} \sin x, \tag{8.22}$$

$$J_{-1/2} = \left(\frac{2}{\pi x}\right)^{1/2} \cos x. \tag{8.23}$$

Note that $J_{1/2}(x)$ is bounded for all x and function $J_{-1/2}(x)$ diverges at $x = 0$. Recall that equation (8.11) gives an expansion of $J_p(x)$, which is valid for any value of p. Figure 8.2 shows graphs of functions $J_{1/2}(x)$, $J_{3/2}(x)$, and $J_{5/2}(x)$; Figure 8.3 shows graphs for $J_{-1/2}(x)$, $J_{-3/2}(x)$, and $J_{-5/2}(x)$.

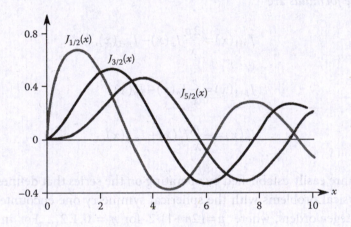

FIGURE 8.2   Graphs of functions $J_{1/2}(x)$, $J_{3/2}(x)$ and $J_{5/2}(x)$.

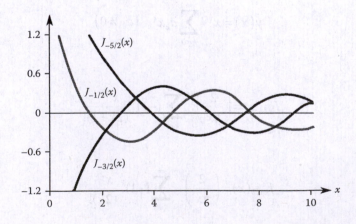

FIGURE 8.3   Graphs of $J_{-1/2}(x)$, $J_{-3/2}(x)$ and $J_{-5/2}(x)$.

Reading Exercise

Using the recurrence equations (8.18) and the expression for $J_{1/2}(x)$, obtain the functions $J_{3/2}(x)$, $J_{-3/2}(x)$, $J_{5/2}(x)$, and $J_{-5/2}(x)$. For instance, the answer for $J_{3/2}(x)$ is

$$J_{3/2} = \left(\frac{2}{\pi x}\right)^{1/2}\left(\frac{\sin x}{x} - \cos x\right). \tag{8.24}$$

Let us now briefly consider Bessel functions of the second kind, or Neumann functions $N_p(x)$. Neumann functions for non-integer p can be obtained as

$$N_p(x) = \frac{J_p(x)\cos p\pi - J_{-p}(x)}{\sin p\pi}. \tag{8.25}$$

It is easy to check that $N_p(x)$ satisfies the Bessel equation (8.6). In case of integer order n, this function is defined as $N_n(x) = \lim_{p\to n} N_p(x)$.

For integer n

$$N_{-n}(x) = (-1)^n N_n(x). \tag{8.26}$$

Other useful properties are that as $x \to 0$ the functions $N_n(x)$ diverge logarithmically; as $x \to \infty$, $N_n(x) \to 0$, oscillating with decaying amplitude. At large x we have the asymptotic form

$$N_n(x) \approx \sqrt{\frac{2}{\pi x}}\,\sin\left(x - \frac{n\pi}{2} - \frac{\pi}{4}\right). \tag{8.27}$$

8.4 BOUNDARY VALUE PROBLEMS AND FOURIER-BESSEL SERIES

In applications it is often necessary to solve Bessel equation (8.4) accompanied by boundary condition(s). For instance, function $y(x)$ defined on the interval $[0, l]$ is finite at $x = 0$ and at $x = l$ obeys a homogeneous boundary condition $\alpha y'(l) + \beta y(l) = 0$ (notice that from now on we will use letter x instead of r in equation (8.4)). When

TABLE 8.2   Positive Roots of the Functions $N_0(x)$, $N_1(x)$, and $N_2(x)$.

Roots / Function	μ_1	μ_2	μ_3	μ_4	μ_5
$N_0(x)$	0.8936	3.9577	7.0861	10.2223	13.3611
$N_1(x)$	2.1971	5.4297	8.5960	11.7492	14.8974
$N_2(x)$	3.3842	6.7938	10.0235	13.2199	16.3789

$\alpha = 0$ this mixed boundary condition becomes the Dirichlet type, when $\beta = 0$ it is the Neumann type. The Bessel equation and boundary condition(s) form the Sturm-Liouville problem; thus it has the nontrivial solutions only for nonnegative discrete eigenvalues λ_k:

$$0 \le \lambda_1 < \lambda_2 < \dots \lambda_k < \dots \tag{8.28}$$

The corresponding eigenfunctions are

$$J_p(\sqrt{\lambda_1}x),\ J_p(\sqrt{\lambda_2}x),\ \dots,\ J_p(\sqrt{\lambda_k}x),\ \dots \tag{8.29}$$

(if function $y(x)$ does not need to be finite at $x = 0$, we can also consider the set of eigenfunctions $N_p(\sqrt{\lambda_k}x)$). The boundary condition at $x = l$ gives $\alpha\sqrt{\lambda}\ J_p'(\sqrt{\lambda}l)+\beta J_p(\sqrt{\lambda}l)=0$. Setting $\sqrt{\lambda}l \equiv \mu$ we obtain a transcendental equation for μ:

$$\alpha\mu J_p'(\mu)+\beta l J_p(\mu)=0, \tag{8.30}$$

which has an infinite number of roots which we label as $\mu_k^{(p)}$. From there the eigenvalues are $\lambda_k = (\mu_k/l)^2$; we see that we need only *positive roots*, $\mu_m^{(n)}$, because negative roots do not give new values of λ_k. For example, for Dirichlet problem $\mu_k^{(p)}$ are the positive roots of the Bessel function $J_p(x)$.

Thus for fixed p we have the set of eigenfunctions (index p in $\mu_k^{(p)}$ is omitted)

$$J_p\!\left(\frac{\mu_1}{l}x\right),\ J_p\!\left(\frac{\mu_2}{l}x\right),\dots,J_p\!\left(\frac{\mu_k}{l}x\right),\dots \tag{8.31}$$

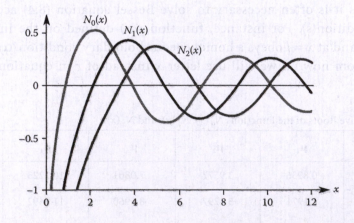

FIGURE 8.4 Graphs of $N_0\,(x)$, $N_1\,(x)$, and $N_2\,(x)$.

As follows from the Sturm-Liouville theory, these functions form a *complete set* and are pair-wise orthogonal (with weight x) on the interval $[0, l]$:

$$\int_0^l J_p\left(\frac{\mu_k}{l}x\right)J_p\left(\frac{\mu_j}{l}x\right)x\,dx = 0, \quad i \neq j. \tag{8.32}$$

A Fourier series expansion (or generalized Fourier series) of an arbitrary function $f(x)$ using the set of functions (8.31) is called a *Fourier-Bessel series* and is given by the expression

$$f(x) = \sum_{k=0}^{\infty} c_k J_p\left(\frac{\mu_k}{l}x\right). \tag{8.33}$$

The orthogonality property allows us to find the coefficients of this series. We multiply equation (8.33) by $J_p\left(\frac{\mu_k}{l}x\right)$ and integrate term by term with weight x. This gives an expression for the coefficient as

$$c_k = \frac{\int_0^l f(x)J_p\left(\frac{\mu_k}{l}x\right)x\,dx}{\int_0^l x\left[J_p\left(\frac{\mu_k}{l}x\right)\right]^2 dx}. \tag{8.34}$$

The squared norm $\|R_{pk}\|^2 = \int_0^l x\left[J_p\left(\frac{\mu_k}{l}x\right)\right]^2 dx$ is (see [1]):

1. For the *Dirichlet* boundary condition $\alpha = 0$ and $\beta = 1$, in which case eigenvalues are obtained from the equation

$$J_p(\mu) = 0,$$

and we have

$$\|R_{pk}\|^2 = \frac{l^2}{2}\left[J_p'(\mu_k^{(p)})\right]^2. \tag{8.35}$$

2. For the *Neumann* boundary condition $\alpha = 1$ and $\beta = 0$, in which case eigenvalues are obtained from the equation

$$J_p'(\mu) = 0,$$

and we have

$$\|R_{pk}\|^2 = \frac{l^2}{2}\left[1 - \frac{p^2}{\left(\mu_k^{(p)}\right)^2}\right]J_p^2(\mu_k^{(p)}). \tag{8.36}$$

3. For the *mixed* boundary condition $\alpha \equiv 1$ and $\beta \equiv h$, in which case eigenvalues are obtained from the equation

$$\mu J_p'(\mu) + h l J_p(\mu) = 0,$$

and we have

$$\|R_{pk}\|^2 = \frac{l^2}{2}\left[1 + \frac{l^2 h^2 - p^2}{\left(\mu_k^{(p)}\right)^2}\right] J_p^2\left(\mu_k^{(p)}\right). \tag{8.37}$$

The completeness of the set of functions $J_p\left(\frac{\mu_k}{l}x\right)$ on the interval $(0,l)$ means that for any square integrable on $[0,l]$ function $f(x)$ the following is true:

$$\int_0^l x f^2(x)\,dx = \sum_k \left\|J_p\left(\frac{\mu_k}{l}x\right)\right\|^2 c_k^2. \tag{8.38}$$

This is Parseval's equality for the Fourier-Bessel series. It has the same property of completeness as in the case of the trigonometric Fourier series with sines and cosines as the basis functions where the weight function equals one instead of x as in the Bessel series.

Regarding the convergence of the series (8.33), we note that the sequence of the partial sums of the series, $S_n(x)$, converges on the interval $(0, 1)$ on average (i.e., in the mean) to $f(x)$ (with weight x), which may be written as

$$\int_0^l [f(x) - S_n(x)]^2 x\,dx \to 0, \quad \text{if } n \to \infty.$$

This property is true for any function $f(x)$ from the class of piecewise-continuous functions because the orthogonal set of functions (8.31) is complete on the interval $[0,l]$. For such functions, $f(x)$, the series (8.33) converges absolutely and uniformly. We present the following theorem without the proof which states a somewhat stronger result about the convergence of the series in equation (8.33) than convergence in the mean:

Theorem

If the function $f(x)$ is piecewise-continuous on the interval $(0,l)$, then the Fourier-Bessel series converges to $f(x)$ at the points where the function $f(x)$ is continuous, and to

$$\frac{1}{2}[f(x_0 + 0) + f(x_0 - 0)],$$

if x_0 is a point of finite discontinuity of the function $f(x)$.

Because the other Bessel functions we have seen—for example, $N_p(x)$—also constitute complete sets of orthogonal functions, an arbitrary function (satisfying "reasonable" restrictions) can be resolved in a series of these sets. Neumann functions can be used to expand a function defined within a ring $a \le x \le b$ or in an infinite interval $a \le x < \infty$.

Below we consider several examples of the expansion of functions into the Fourier-Bessel series using the functions $J_p(x)$. In some cases the coefficients can be found analytically; otherwise we may calculate them numerically—for example, using the program **FourierSeries** included with this book. This program also allows the user to change the number of terms in a partial sum and investigate individual terms. All the calculations and the figures below are generated with this program, which does the calculations using the methods described in this chapter. Instructions for the use of the program **FourierSeries** are in Appendix 5.

Example 1

Let us expand the function $f(x) = A$, $A = \text{const.}$ in a series using the Bessel functions $X_k(x) = J_0\left(\mu_k^{(0)} x/l\right)$ on the interval $[0, l]$, where $\mu_k^{(0)}$ are the positive roots of the equation $J_0(\mu) = 0$.

Solution. First we calculate the norm, $\|X_k\|^2 = \left\|J_0\left(\mu_k^{(0)} x/l\right)\right\|^2$ using the relation $J_0'(x) = -J_1(x)$ to obtain

$$\left\|J_0\left(\mu_k^{(0)} x/l\right)\right\|^2 = \frac{l^2}{2}\left[J_0'\left(\mu_k^{(0)}\right)\right]^2 = \frac{l^2}{2} J_1^2\left(\mu_k^{(0)}\right).$$

Using the substitution $z = \mu_k^{(0)} x/l$ and the second of the relations (8.17) we may calculate the integral

$$\int_0^l J_0\left(\frac{\mu_k^{(0)}}{l} x\right) x\,dx = \frac{l^2}{\left(\mu_k^{(0)}\right)^2}\int_0^{\mu_k^{(0)}} J_0(z)z\,dz = \frac{l^2}{\left(\mu_k^{(0)}\right)^2}\left[zJ_1(z)\right]_0^{\mu_k^{(0)}} = \frac{l^2}{\mu_k^{(0)}} J_1\left(\mu_k^{(0)}\right).$$

For the coefficients c_k of the expansion (8.33) we have

$$c_k = \frac{1}{\left\|J_0\left(\mu_k^{(0)} x/l\right)\right\|^2}\int_0^l A\,J_0\left(\mu_k^{(0)} x/l\right) x\,dx$$

$$= \frac{2A}{l^2\left[J_1\left(\mu_k^{(0)}\right)\right]^2}\frac{l^2}{\mu_k^{(0)}} J_1\left(\mu_k^{(0)}\right) = \frac{2A}{\mu_k^{(0)} J_1\left(\mu_k^{(0)}\right)}.$$

Thus the expansion is

$$f(x) = 2A\sum_{k=0}^{\infty}\frac{1}{\mu_k^{(0)} J_1\left(\mu_k^{(0)}\right)} J_0\left(\frac{\mu_k^{(0)}}{l} x\right).$$

Figure 8.5 shows the function $f(x) = 1$ and the partial sum of its Fourier-Bessel series when $l = 1$. From this figure it is seen that the series converges very slowly (see Figure 8.5d) and even when 50 terms are kept in the expansion (Figure 8.5c) the difference from $f(x) = 1$ can easily be seen. This is because at the endpoints of the interval the value of the function $f(x) = 1$ and the functions $J_0\left(\mu_k^{(0)} x\right)$ are different. The obtained expansion does not converge well near the endpoints.

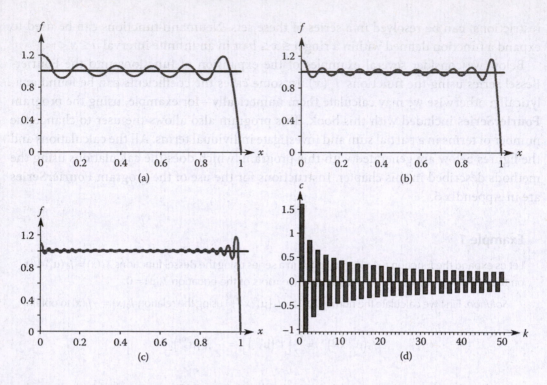

FIGURE 8.5 The function $f(x) = 1$ and the partial sum of its Fourier-Bessel series. (a) 11 terms are kept in the series ($N = 10$); (b) $N = 20$; (c) $N = 50$; (d) values of the coefficients c_k of the series.

Example 2

Let us modify the boundary condition in the previous problem. We expand the function $f(x) = A$, $A = $ const., given on the interval $[0, l]$, in a Fourier series in Bessel functions $X_k(x) = J_0\left(\mu_k^{(0)}x/l\right)$, where $\mu_k^{(0)}$ are now the positive roots of the equation $J_0'(\mu) = 0$.

Solution. For *Neumann boundary condition*, $J_0'(\mu) = 0$, for $k = 0$ we have

$$\mu_0^{(0)} = 0, \quad X_0(x) = J_0(0) = 1, \quad \|X_0\|^2 = \|J_0(0)\|^2 = \frac{l^2}{2},$$

The first coefficient of the expansion (8.33) is

$$c_0 = \frac{A}{\|J_0(0)\|^2} \int_0^l J_0(0)x\,dx = \frac{2A}{l^2}\int_0^l x\,dx = A.$$

Other coefficients c_k can be evaluated by using the substitution $z = \mu_k^{(0)}x/l$ and using the integration formula

$$\int_0^l J_0\left(\frac{\mu_k^{(0)}}{l}x\right)x\,dx = \frac{l^2}{\left(\mu_k^{(0)}\right)^2}\left[zJ_1(z)\right]_0^{\mu_k^{(0)}} = \frac{l^2}{\mu_k^{(0)}}J_1\left(\mu_k^{(0)}\right).$$

Applying the relation $J_0'(x) = -J_1(x)$ and then recalling that $J_0'\left(\mu_k^{(0)}\right) = 0$ we find

$$c_k = \frac{2A}{\mu_k^{(0)}} J_1(\mu_k^{(0)}) = 0, \text{ when } k > 0.$$

Thus, we obtain simple expansion, $f(x) = c_0 J_0\left(\mu_0^{(0)} x/l\right) = A$. In fact, this means that the given function is actually one of the functions from the set of eigenfunctions used for eigenfunction expansion.

Several other examples are presented in the library problems section of the program **FourierSeries.**

8.5 SPHERICAL BESSEL FUNCTIONS

In this section we briefly consider *spherical Bessel functions*, which are related to the solutions of certain boundary value problems in spherical coordinates.

Consider the following equation:

$$\frac{d^2 R(x)}{dx^2} + \frac{2}{x} \frac{dR(x)}{dx} + \left[1 - \frac{l(l+1)}{x^2}\right] R(x) = 0. \tag{8.39}$$

Parameter l takes discrete non-negative integer values: $l = 0,1,2,...$ Equation (8.39) is called the *spherical Bessel equation*. It differs from the cylindrical Bessel equation, equation (8.6), in that there is the coefficient 2 in the second term. Equation (8.39) can be transformed to a Bessel cylindrical equation by the substitution $R(x) = y(x)/\sqrt{x}$.

Reading Exercise

Check that equation for $y(x)$ is

$$\frac{d^2 y(x)}{dx^2} + \frac{1}{x} \frac{dy(x)}{dx} + \left[1 - \frac{(l+1/2)^2}{x^2}\right] y(x) = 0. \tag{8.40}$$

If we introduce $s = l + 1/2$ in equation (8.40) we recognize this equation as the Bessel equation which has the general solution

$$y(x) = C_1 J_s(x) + C_2 N_s(x) \tag{8.41}$$

where $J_s(x)$ and $N_s(x)$ are (cylindrical) Bessel and Neumann functions. Because $s = l + 1/2$ these functions are of half-integer order. Inverting the transformation we have that the solution, $R(x)$, to equation (8.39) is

$$R(x) = C_1 \frac{J_{l+1/2}(x)}{\sqrt{x}} + C_2 \frac{N_{l+1/2}(x)}{\sqrt{x}}. \tag{8.42}$$

If we consider a regular at $x = 0$ solution, the coefficient $C_2 \equiv 0$.

The *spherical Bessel function* $j_l(x)$ is defined to be a solution finite at $x = 0$; thus it is a multiple of $J_{l+1/2}(x)/\sqrt{x}$. The coefficient of proportionality is usually chosen to be $\sqrt{\pi/2}$ so that

$$j_l(x) = \sqrt{\frac{\pi}{2x}} J_{l+1/2}(x). \tag{8.43}$$

For $l = 0$, $J_{1/2}(x) = \sqrt{\frac{2}{\pi x}} \sin x$; thus

$$j_0(x) = \frac{\sin x}{x}. \tag{8.44}$$

Analogously we may define the *spherical Neumann functions* as

$$n_l(x) = \sqrt{\frac{\pi}{2x}} N_{l+1/2}(x), \tag{8.45}$$

from where (using equation (8.25))

$$n_0(x) = \sqrt{\frac{\pi}{2x}} J_{-1/2}(x) = -\frac{\cos x}{x}. \tag{8.46}$$

Expressions for the first few terms of the functions $j_l(x)$ and $n_l(x)$ are

$$j_1(x) = \frac{\sin x}{x^2} - \frac{\cos x}{x} + \ldots, \quad j_2(x) = \left(\frac{3}{x^3} - \frac{1}{x}\right)\sin x - \frac{3}{x^2}\cos x + \ldots, \tag{8.47}$$

$$n_1(x) = -\frac{\cos x}{x^2} - \frac{\sin x}{x} + \ldots, \quad n_2(x) = -\left(\frac{3}{x^3} - \frac{1}{x}\right)\cos x - \frac{3}{x^2}\sin x + \ldots. \tag{8.48}$$

The spherical Bessel functions with $l = 0,1,2$ are sketched in Figures 8.6 and 8.7.

The following *recurrence relations* are valid (here the symbol f is written to stand for j or n):

$$f_{l-1}(x) + f_{l+1}(x) = (2l+1)x^{-1}f_l(x), \tag{8.49}$$

$$lf_{l-1}(x) - (l+1)f_{l+1}(x) = (2l+1)\frac{d}{dx}f_l(x). \tag{8.50}$$

Differentiation formulas:

$$\frac{d}{dx}\left[x^{l+1}j_l(x)\right] = x^{l+1}j_{l-1}(x), \quad \frac{d}{dx}\left[x^{-l}j_l(x)\right] = -x^{-l}j_{l+1}(x). \tag{8.51}$$

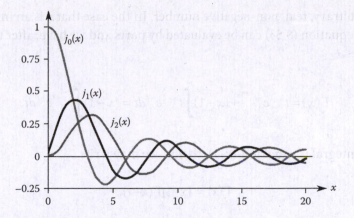

FIGURE 8.6   Graphs of functions $j_0(x)$, $j_1(x)$, and $j_2(x)$.

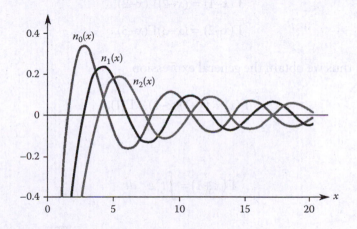

FIGURE 8.7   Graphs of functions $n_0(x)$, $n_1(x)$, and $n_2(x)$.

Asymptotic values:

$$j_l(x) \sim \frac{1}{x}\cos\left[x - \frac{\pi}{2}(l+1)\right], \quad n_l(x) \sim \frac{1}{x}\sin\left[x - \frac{\pi}{2}(l+1)\right] \text{ as } x \to \infty \qquad (8.52)$$

(the last expression has good precision for $x \gg l(l+1)$).

8.6 THE GAMMA FUNCTION

In this section we develop the essential properties of the Gamma function. One of the most important applications of the Gamma function is that it allows us to find factorials of positive numbers that are not integers. The gamma function is defined by the integral

$$\Gamma(x) = \int_0^\infty t^{x-1} e^{-t}\, dt, \; x > 0. \qquad (8.53)$$

Here x is an arbitrary, real, non-negative number. In the case that x is an integer and $x \geq 2$, the integral in equation (8.53) can be evaluated by parts and we have, after the substitution $t^{x-1} = u$, $e^{-t} dt = dv$:

$$\Gamma(x) = t^{x-1}e^{-t}\Big|_0^\infty + (x-1)\int_0^\infty t^{x-2}e^{-t}dt = (x-1)\int_0^\infty t^{x-2}e^{-t}dt. \tag{8.54}$$

The obtained integral is equal to $\Gamma(x-1)$, that is, we may write

$$\Gamma(x) = (x-1)\Gamma(x-1). \tag{8.55}$$

Substituting into equation (8.55) we have

$$\Gamma(x-1) = (x-2)\Gamma(x-2),$$

$$\Gamma(x-2) = (x-3)\Gamma(x-3),$$

and so on, and thus we obtain the general expression

$$\Gamma(x) = (x-1)(x-2)\dots\Gamma(1), \tag{8.56}$$

where

$$\Gamma(x+1) = \int_0^\infty t^x e^{-t}\, dt. \tag{8.57}$$

Substituting equation (8.57) into equation (8.56) we have

$$\Gamma(x) = (x-1)! \qquad \text{for} \qquad x = 2, 3, \dots \tag{8.58}$$

We derived equation (8.58) for integer values $x \geq 2$, but it is possible to generalize it to define *factorials of any numbers*. First we verify that equation (8.58) is valid for $x = 1$. Let $x = 1$ in equation (8.58) in which case we have $\Gamma(1) = 0! = 1$, which does agree with equation (8.57). Thus for integer values of the argument, $n = 1, 2, 3, \dots$

$$\Gamma(1) = 1,\ \Gamma(2) = 1,\ \Gamma(n) = (n-1)!. \tag{8.59}$$

Now consider non-integer values of x. For $x = 1/2$, taking into account definition (8.53) and using the substitution $t = z^2$, we obtain

$$\Gamma(1/2) = \int_0^\infty t^{-1/2}e^{-t}dt = 2\int_0^\infty e^{-z^2}dz = \sqrt{\pi}. \tag{8.60}$$

Now using equation (8.55), we find

$$\Gamma(3/2) = (1/2)\Gamma(1/2) = \sqrt{\pi}/2 \qquad (8.61)$$

Reading Exercise

Show that for any integer $n \geq 1$,

$$\Gamma\left(n + \frac{1}{2}\right) = \frac{1 \cdot 3 \cdot 5 \cdots (2n-1)}{2^n} \Gamma\left(\frac{1}{2}\right). \qquad (8.62)$$

We can also generalize definition (8.53) to negative values of x using equation (8.54). First replace x by $x + 1$, which gives

$$\Gamma(x) = \frac{\Gamma(x+1)}{x}. \qquad (8.63)$$

We may use this equation to find, for example, $\Gamma(-1/2)$ in the following way:

$$\Gamma\left(-\frac{1}{2}\right) = \frac{\Gamma(1/2)}{-1/2} = -2\sqrt{\pi}. \qquad (8.64)$$

It is clear from this that using equations (8.53) and (8.64) we can find a value of $\Gamma(x)$ for all values of x except 0 and negative integers.

The function $\Gamma(x)$ diverges at $x = 0$ as is seen from equation (8.53). Then, from equation (8.63) we see that $\Gamma(-1)$ is not defined because it involves $\Gamma(0)$. Thus, $\Gamma(x)$ does not exist for negative integer values of x. From equation (8.63) it is obvious (taking into account that $\Gamma(1) = 1$) that at all these values of x the function $\Gamma(x)$ has simple poles. A graph of $\Gamma(x)$ is plotted in Figure 8.8.

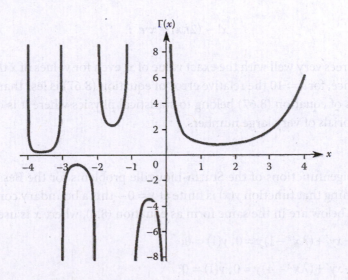

FIGURE 8.8 Graph of the Gamma function, $\Gamma(x)$.

TABLE 8.3   Values of $\Gamma(x)$ for $x \in [1, 2]$.

x	$\Gamma(x)$	x	$\Gamma(x)$	x	$\Gamma(x)$
1	1	1.35	0.8911514420	1.7	0.9086387329
1.05	0.9735042656	1.4	0.8872638175	1.75	0.9190625268
1.1	0.9513507699	1.45	0.8856613803	1.8	0.9313837710
1.15	0.9330409311	1.5	0.8862269255	1.85	0.9456111764
1.2	0.9181687424	1.55	0.8888683478	1.9	0.9617658319
1.25	0.9064024771	1.6	0.8935153493	1.95	0.9798806513
1.3	0.8974706963	1.65	0.9001168163	2	1

Equation (8.63) allows us to find the value of $\Gamma(x)$ for any non-negative real x using the value of $\Gamma(x)$ on the interval $1 \le x \le 2$. For example, $\Gamma(3.4) = 3.4 \cdot \Gamma(2.4) = 3.4 \cdot 2.4 \cdot \Gamma(1.4)$. Based on this fact, a table of values for $\Gamma(x)$ must include only the x-interval [1,2]. The minimum value of $\Gamma(x)$ is reached at $x = 1.46116321\ldots$

Equation (8.59), which is now valid for all non-negative real values of x, can be written as

$$\Gamma(x-1) = x! \tag{8.65}$$

On the other hand, from equation (8.53) we have

$$\Gamma(x+1) = \int_0^\infty t^x e^{-t} dt. \tag{8.66}$$

The integrand $t^x e^{-t}$ has a sharp peak at $t = x$ that allows us to obtain a famous approximation formula for $x!$, known as Stirling's approximation, which works very well for large x:

$$x! \sim (2\pi x)^{1/2} x^x e^{-x}. \tag{8.67}$$

This formula agrees very well with the exact value of $x!$ even for values of x that are not very large. For instance, for $x = 10$ the relative error of equation (8.67) is less than 0.8%. Most of the applications of equation (8.67) belong to statistical physics where it is often necessary to evaluate factorials of very large numbers.

Problems

1. Find the eigenfunctions of the Sturm-Liouville problems for the Bessel equation on $[0,l]$ assuming that function $y(x)$ is finite at $x = 0$—this a boundary condition at $x = 0$. Equations below are in the same form as equation (8.4), where x is used for r.

a) $x^2 y'' + xy' + (\lambda x^2 - 1) y = 0$, $y(1) = 0$;

b) $x^2 y'' + xy' + (\lambda x^2 - 4) y = 0$, $y(1) = 0$;

c) $x^2y'' + xy' + \lambda x^2 y = 0$, $y'(2) = 0$;

d) $x^2y'' + xy' + (\lambda x^2 - 9)y = 0$, $y(3) + 2y'(3) = 0$.

2. Expand the function $f(x)$, given on the interval $[0, 1]$, in a Fourier series in Bessel functions of the first kind, $X_k(x) = J_0(\mu_k^{(0)}x)$, where $\mu_k^{(0)}$ are positive roots of the equation $J_0(\mu) = 0$ (in problems 2 through 8 you can find coefficients of expansion with the help of the program **FourierSeries**), if:

a) $f(x) = \sin\pi x$;

b) $f(x) = x^2$;

c) $f(x) = \sin^2 \pi x$;

d) $f(x) = 1 - x^2$;

e) $f(x) = \cos\dfrac{\pi x}{2}$.

3. Expand the function $f(x)$, given on the interval $[0, 1]$, in a Fourier series in Bessel functions $X_k(x) = J_1(\mu_k^{(1)}x)$, where $\mu_k^{(1)}$ are positive roots of the equation $J_1(\mu) = 0$, if:

a) $f(x) = x$;

b) $f(x) = \sin\pi x$;

c) $f(x) = \sin^2 \pi x$;

d) $f(x) = x(1-x)$;

e) $f(x) = x(1-x^2)$.

4. Expand the function $f(x)$, given on the interval $[0, 1]$, in a Fourier series in Bessel functions $X_k(x) = J_0(\mu_k^{(0)}x)$, where $\mu_k^{(0)}$ are positive roots of the equation $J_0'(\mu) = 0$, if:

a) $f(x) = x(1-x)$;

b) $f(x) = x(1-x^3)$;

c) $f(x) = x(1-x^2)$;

d) $f(x) = x^3$.

5. Expand the function

$$f(x) = A\left(1 - \frac{x^2}{l^2}\right), A = \text{const},$$

given on the interval $[0,l]$, in Fourier series in Bessel functions $X_k(x) = J_0(\mu_k^{(0)}x/l)$, where $\mu_k^{(0)}$ are positive roots of the equation $J_0(\mu) = 0$.

6. Expand the function

$$f(x) = Ax, A = \text{const},$$

given on the interval $[0,l]$, in a Fourier series in Bessel functions $X_k(x) = J_1(\mu_k^{(1)}x/l)$, where $\mu_k^{(1)}$ are positive roots of the equation $J_1'(\mu) = 0$.

7. Expand the function

$$f(x) = Ax^2, A = \text{const},$$

given on the interval $[0,l]$, in Fourier series in Bessel functions of the first kind $X_k(x) = J_0\left(\mu_k^{(0)} x/l\right)$, where $\mu_k^{(0)}$ are positive roots of the equation $J_0'(\mu) = 0$.

8. Expand the function

$$f(x) = \begin{cases} x^2, & 0 \le x < 1, \\ x, & 1 \le x < 2, \end{cases}$$

given on the interval $[0,2]$, in Fourier series in Bessel functions of the first kind $X_k(x) = J_2\left(\mu_k^{(2)} x/l\right)$ $(l=2)$, where $\mu_k^{(2)}$ are positive roots of the equation $\mu J_2'(\mu) + hlJ_2(\mu) = 0$.

8.7 LEGENDRE EQUATION AND LEGENDRE POLYNOMIALS

In applications one often encounters an eigenvalue problem that contains a second-order linear homogeneous differential equation

$$(1 - x^2) y'' - 2xy' + \lambda y = 0, \quad -1 \le x \le 1, \tag{8.68}$$

where λ is a real parameter. Equation (8.68) can be rewritten in *Sturm-Liouville form* given by

$$\frac{d}{dx}\left[(1 - x^2)\frac{dy}{dx}\right] + \lambda y = 0, \quad -1 \le x \le 1 \tag{8.69}$$

and is called *the Legendre equation*. This equation frequently arises after the separation of variables procedure in spherical coordinates (in those problems the variable x is $x = \cos\theta$, where θ is a meridian angle) in many problems of mathematical physics. Prominent examples include heat conduction in a spherical domain, vibrations of spherical solids and shells, as well as boundary value problems for the electric potential in spherical coordinates (see [1]).

Because this problem is a particular example of a Sturm-Liouville problem we can expect that the eigenvalues are *nonnegative real discrete* λ_m, and the eigenfunctions corresponding to different eigenvalues are *orthogonal* on the interval $[-1, 1]$ with the weight $r(x) = 1$:

$$\int_{-1}^{1} y_n(x) y_m(x) \, dx = 0, \quad n \ne m. \tag{8.70}$$

Next, we solve equation (8.69) on the interval $x \in [-1,1]$ assuming that the function $y(x)$ is finite at the points $x = -1$ and $x = 1$. We search for a solution in the form of a power series in x:

$$y(x) = \sum_{n=0}^{\infty} a_n x^n. \tag{8.71}$$

Substitution of equation (8.71) into equation (8.69) results in the following equality:

$$\sum_{n=2}^{\infty} n(n-1)a_n x^{n-2} - \sum_{n=2}^{\infty} n(n-1)a_n x^n - \sum_{n=1}^{\infty} 2na_n x^n + \lambda \sum_{n=0}^{\infty} a_n x^n = 0.$$

Changing the index of summation in the first term from n to $n + 2$ to yield

$$\sum_{n=0}^{\infty} (n+2)(n+1)a_{n+2} x^n$$

allows us to group all the terms with $n \geq 2$ leaving the terms with $n = 0$ and $n = 1$, which we write separately, and we have

$$\left(6a_3 - 2a_1 + \lambda a_1\right)x + 2a_2 + \lambda a_0$$

$$+ \sum_{n=2}^{\infty} \left[(n+2)(n+1)a_{n+2} - \left(n^2 + n - \lambda\right)a_n\right]x^n = 0.$$

By setting coefficients of each power of x to zero we obtain an infinite system of equations for the coefficients a_n:

$$n = 0 \qquad\qquad 2a_2 + \lambda a_0 = 0 \tag{8.72}$$

$$n = 1 \qquad\qquad 6a_3 - 2a_1 + \lambda a_1 = 0 \tag{8.73}$$

$$n \geq 2 \qquad\qquad (n+2)(n+1)a_{n+2} - (n^2 + n - \lambda)a_n = 0. \tag{8.74}$$

Equation (8.74) is *the recurrence formula* for coefficients.

From equation (8.72) we have

$$a_2 = -\frac{\lambda}{2}a_0. \tag{8.75}$$

Using equations (8.75) and (8.74) we obtain

$$a_4 = \frac{6-\lambda}{3 \cdot 4}a_2 = \frac{-\lambda(6-\lambda)}{4!}a_0,$$

$$a_6 = \frac{20-\lambda}{5 \cdot 6}a_4 = \frac{-\lambda(6-\lambda)(20-\lambda)}{6!}a_0,$$

and so on. Each coefficient, a_{2n}, with even index is multiplied by a_0 and depends on n and the parameter λ.

We proceed similarly with the odd terms, a_{2k+1}. From equation (8.73) we have

$$a_3 = \frac{2-\lambda}{6}a_1 = \frac{2-\lambda}{3!}a_1.$$

Then from the recurrence formula (8.74) we obtain

$$a_5 = \frac{12-\lambda}{4\cdot5}a_3 = \frac{(2-\lambda)(12-\lambda)}{5!}a_1,$$

$$a_7 = \frac{30-\lambda}{6\cdot7}a_5 = \frac{(2-\lambda)(12-\lambda)(30-\lambda)}{7!}a_1,$$

and so on. Each coefficient, a_{2k+1}, with odd index is multiplied by a_1 and depends on n and λ. Substituting the obtained coefficients in equation (8.71) we have

$$y(x) = \sum_{n=0}^{\infty} a_n x^n = a_0\left[1 - \frac{\lambda}{2}x^2 - \frac{\lambda(6-\lambda)}{4!}x^4 - \cdots\right]$$

$$+ a_1\left[x + \frac{2-\lambda}{3!}x^3 + \frac{(2-\lambda)(12-\lambda)}{5!}x^5 + \cdots\right].$$

(8.76)

In equation (8.76), the first sum contains coefficients with even indices and the second sum coefficients with odd indices. As a result we obtain two *linearly independent* solutions of equation (8.69); one contains even powers of x, the other odd:

$$y^{(1)}(x) = \sum_{n=0}^{\infty} a_{2n} x^{2n} = a_0\left[1 - \frac{\lambda}{2}x^2 - \frac{\lambda(6-\lambda)}{4!}x^4 - \cdots\right],$$

(8.77)

$$y^{(2)}(x) = \sum_{n=0}^{\infty} a_{2n+1} x^{2n+1} = a_1\left[x + \frac{2-\lambda}{3!}x^3 + \frac{(2-\lambda)(12-\lambda)}{5!}x^5 + \cdots\right].$$

(8.78)

Now from the recurrence relation (8.74) for the coefficients,

$$a_{n+2} = \frac{(n^2+n-\lambda)}{(n+2)(n+1)}a_n,$$

(8.79)

we can state an important fact. The series in equation (8.71) converges on an open interval $-1 < x < 1$, as it can be seen from a ratio test, $\lim_{n\to\infty}\left|\frac{a_{n+2}x^{n+2}}{a_n x^n}\right| = x^2$, but diverges at the points

$x = \pm 1$. Therefore this series cannot be used as an acceptable solution of the differential equation on the entire interval $-1 \le x \le 1$ unless it *terminates as a polynomial* with a finite number of terms. This can occur if the numerator in equation (8.79) is zero for some index value, n_{max}, such that

$$\lambda = n_{max}(n_{max} + 1).$$

This gives $a_{n_{max}+2} = 0$ and consequently $a_{n_{max}+4} = 0$, $a_{n_{max}+6} = 0$, ... thus $y(x)$ will contain a finite number of terms and thus turn out to be a polynomial of degree n_{max}. In order not to overcomplicate the notation, from here on we will denote n_{max} as n. We may conclude from this discussion that λ can take only *nonnegative integer* values:

$$\lambda = n(n + 1). \tag{8.80}$$

Let us consider several particular cases. If $n = 0$ (which means that the highest degree of the polynomial is 0), $a_0 \ne 0$ and $a_2 = 0$, $a_4 = 0$, and so on. The value of λ for this case is $\lambda = 0$, and we have $y^{(1)}(x) = a_0$. If $n = 1$, $a_1 \ne 0$, and $a_3 = a_5 = \cdots = 0$, then $\lambda = 2$ and $y^{(2)}(x) = a_1 x$. If $n = 2$, the highest degree of the polynomial is 2, we have $a_2 \ne 0$, $a_4 = a_6 = \cdots = 0$, $\lambda = 6$, and from the recurrence relation we obtain $a_2 = -3a_0$. This results in $y^{(1)}(x) = a_0(1 - 3x^2)$. If $n = 3$, $a_3 \ne 0$ and $a_5 = a_7 = \ldots = 0$, $\lambda = 12$, and from the recurrence relation we obtain $a_3 = -5/3 a_1$ and as the result, $y^{(2)}(x) = a_1(1 - 5x^3/3)$. Constants a_0 and a_1 remain arbitrary unless we impose some additional requirement. A convenient requirement is that the solutions (the polynomials) obtained in this way should have the value 1 when $x = 1$.

The polynomials obtained above are denoted as $P_n(x)$ and are called *the Legendre polynomials*. The first few, which we derived above, are

$$P_0(x) = 1, \quad P_1(x) = x, \quad P_2(x) = \frac{1}{2}(3x^2 - 1), \quad P_3(x) = \frac{1}{2}(5x^3 - 3x). \tag{8.81}$$

Let us list two more (which the reader may derive as a Reading Exercise using the above relationships):

$$P_4(x) = \frac{1}{8}(35x^4 - 30x^2 + 3), \quad P_5(x) = \frac{1}{8}(63x^5 - 70x^3 + 15x). \tag{8.82}$$

Reading Exercise

Obtain $P_n(x)$ for $n = 6,7$.

Reading Exercise

Show by direct substitution that $P_2(x)$ and $P_3(x)$ satisfy equation (8.69).

As we see, $P_n(x)$ are even functions for even values of n, and odd functions for odd n.

The functions $y^{(1)}(x)$ and $y^{(2)}(x)$, bounded on the closed interval $-1 \le x \le 1$, are

$$y^{(1)}(x) = P_{2n}(x), \quad y^{(2)}(x) = P_{2n+1}(x). \tag{8.83}$$

The formula for calculating the Legendre polynomials is called Rodrigues' formula:

$$P_n(x) = \frac{1}{2^n n!} \frac{d^n}{dx^n} (x^2 - 1)^n \tag{8.84}$$

(here a zero-order derivative means the function itself).

Reading Exercise

Using Rodrigues' formula show that

$$P_n(-x) = (-1)^n P_n(x), \tag{8.85}$$

$$P_n(-1) = (-1)^n. \tag{8.86}$$

Other useful properties of Legendre polynomials are

$$P_n(1) = 1, \quad P_{2n+1}(0) = 0, \quad P_{2n}(0) = (-1)^n \frac{1 \cdot 3 \cdot \ldots \cdot (2n-1)}{2 \cdot 4 \cdot \ldots \cdot 2n}. \tag{8.87}$$

The Legendre polynomial $P_n(x)$ has n real and simple (i.e., not repeated) roots, all lying in the interval $-1 < x < 1$. Zeroes of the polynomials $P_n(x)$ and $P_{n+1}(x)$ alternate as x increases. In Figure 8.9 the first four polynomials, $P_n(x)$, are shown and their properties, as listed in equations (8.84) through (8.86), are reflected in these graphs.

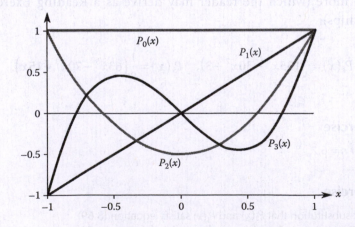

FIGURE 8.9   First four polynomials, $P_n(x)$.

The following *recurrence formula* relates three polynomials

$$(n+1)P_{n+1}(x)-(2n+1)xP_n(x)+nP_{n-1}(x)=0. \tag{8.88}$$

This formula gives a simple (and the most practical) way to obtain the Legendre polynomials of any order, one by one, starting with $P_0(x)=1$ and $P_1(x)=x$.

The following recurrence relations for Legendre polynomials are often useful:

$$P'_{n-1}(x)-xP'_n(x)+nP_n(x)=0, \tag{8.89}$$

$$P'_n(x)-xP'_{n-1}(x)-nP_{n-1}(x)=0. \tag{8.90}$$

To summarize, the solution of the Sturm-Liouville problem for equation (8.69) with boundary conditions stating that the solution is bounded on the closed interval $-1 \leq x \leq 1$ is a set of Legendre polynomials, $P_n(x)$, which are the eigenfunctions of the Sturm-Liouville operator. The eigenvalues are $\lambda = n(n + 1)$, $n = 0,1,2,...$ As a solution of the Sturm-Liouville problem, the Legendre polynomials, $P_n(x)$, form a *complete orthogonal* set of functions on the closed interval $[-1,1]$, a property we will find very useful in the applications considered below.

If the points $x = \pm 1$ are excluded from a domain, the solution in the form of an infinite series is also acceptable. In this case, functions $Q_n(x)$, which logarithmically diverge at $x = \pm 1$, are also the solutions of the Legendre equation (for details, see book [1]).

8.8 FOURIER-LEGENDRE SERIES IN LEGENDRE POLYNOMIALS

The Legendre polynomials are orthogonal on the interval $[-1,1]$:

$$\int_{-1}^{1} P_n(x)P_m(x)dx = 0, \ m \neq n. \tag{8.91}$$

The norm squared of Legendre polynomials is (see [1])

$$\|P_n\|^2 = \int_{-1}^{1} P_n^2(x)dx = \frac{2}{2n+1}. \tag{8.92}$$

Equations (8.91) and (8.92) can be combined and written as

$$\int_{-1}^{1} P_n(x)P_m(x)dx = \begin{cases} 0, & m \neq n, \\ \dfrac{2}{2n+1}, & m = n. \end{cases} \tag{8.93}$$

The Legendre polynomials form a complete set of functions on the interval [−1,1]; thus $\{P_n(x)\}$, $n = 0,1,2,...$ provide a basis for an eigenfunction expansion for functions $f(x)$, which are bounded on the interval [−1,1]:

$$f(x) = \sum_{n=0}^{\infty} c_n P_n(x). \tag{8.94}$$

Due to the orthogonality of the functions $P_n(x)$ with different indexes, the coefficients c_n are

$$c_n = \frac{1}{\|P_n\|^2} \int_{-1}^{1} f(x) P_n(x) dx = \frac{2n+1}{2} \int_{-1}^{1} f(x) P_n(x) dx. \tag{8.95}$$

As we know from the general theory discussed in Chapter 4, the sequence of the partial sums of this series, $S_N(x)$, converges on the interval (−1,1) on average (i.e., in the mean) to $f(x)$, which may be written as

$$\int_{-1}^{1} [f(x) - S_N(x)]^2 \, dx \to 0 \text{ as } N \to \infty. \tag{8.96}$$

The function $f(x)$ should be square integrable; that is, we require that the integral $\int_{-1}^{1} f^2(x) \, dx$ exists.

For an important class of piecewise-continuous functions the series in equation (8.94) converges absolutely and uniformly. The following theorem states a stronger result about the convergence of the series (8.94) than convergence in the mean.

Theorem

If the function $f(x)$ is piecewise-continuous on the interval (−1,1), then the Fourier-Legendre series converges to $f(x)$ at the points where the function $f(x)$ is continuous and to

$$\frac{1}{2}[f(x_0 + 0) + f(x_0 - 0)], \tag{8.97}$$

if x_0 is a point of finite discontinuity of the function $f(x)$.

Because the Legendre polynomials form a complete set, for any square integrable function, $f(x)$, we have

$$\int_{-1}^{1} f^2(x) dx = \sum_{n=0}^{\infty} \|P_n\|^2 c_n^2 = \sum_{n=0}^{\infty} \frac{2}{2n+1} c_n^2. \tag{8.98}$$

This is Parseval's equality (the completeness equation) for the Fourier-Legendre series. Clearly for a partial sum on the right we have Bessel's inequality

$$\int_{-1}^{1} f^2(x)dx \geq \sum_{n=0}^{N} \frac{2}{2n+1}c_n^2. \tag{8.99}$$

Below we consider several examples of the expansion of functions into the Fourier-Legendre series. In some cases the coefficients can be found analytically; otherwise we may calculate them numerically using the program **FourierSeries**. A description of how to use this program can be found in Appendix 5.

Example 1

Expand function $f(x) = A$, $A = $ const. in a Fourier-Legendre series in $P_n(x)$ on the interval $-1 \leq x \leq 1$.
   Solution. The series is

$$A = c_0 P_0(x) + c_1 P_1(x) + ...,$$

where coefficients c_n are

$$c_n = \frac{1}{\|P_n^2(x)\|} \int_{-1}^{1} A P_n(x)dx = \frac{(2n+1)A}{2} \int_{-1}^{1} P_n(x)dx.$$

From this formula it is clear that the only nonzero coefficient is $c_0 = A$.

Example 2

Expand the function $f(x) = x$ in a Fourier-Legendre series in $P_n(x)$ on the interval $-1 \leq x \leq 1$.
   Solution. The series is

$$x = c_0 P_0(x) + c_1 P_1(x) + ...,$$

where c_n are

$$c_n = \frac{1}{\|P_n^2(x)\|} \int_{-1}^{1} x P_n(x)dx = \frac{2n+1}{2} \int_{-1}^{1} x P_n(x)dx.$$

Clearly the only nonzero coefficient is

$$c_1 = \frac{3}{2} \int_{-1}^{1} x P_1(x)dx = \frac{3}{2} \int_{-1}^{1} x^2 dx = 1.$$

As in the previous example this result is apparent because one of the polynomials, $P_1(x)$ in this example, coincides with the given function, $f(x) = x$.

Example 3

Expand the function $f(x)$ given by

$$f(x) = \begin{cases} 0, & -1 < x < 0, \\ 1, & 0 < x < 1 \end{cases}$$

in a Fourier-Legendre series.

Solution. The expansion $f(x) = \sum_{n=0}^{\infty} c_n P_n(x)$ has coefficients

$$c_n = \frac{1}{\|P_n\|^2} \int_{-1}^{1} f(x) P_n(x) dx = \frac{2n+1}{2} \int_{0}^{1} P_n(x) dx.$$

The first few coefficients are

$$c_0 = \frac{1}{2} \int_{0}^{1} dx = \frac{1}{2}, \quad c_1 = \frac{3}{2} \int_{0}^{1} x dx = \frac{3}{4}, \quad c_2 = \frac{5}{2} \int_{0}^{1} \frac{1}{2}(3x^2 - 1) dx = 0.$$

Continuing, we find for the given function $f(x)$,

$$f(x) = \frac{1}{2} P_0(x) + \frac{3}{4} P_1(x) - \frac{7}{16} P_3(x) + \frac{11}{32} P_5(x) + \dots.$$

The series converges slowly because of a discontinuity of a given function $f(x)$ at the point $x = 0$.

8.9 ASSOCIATE LEGENDRE FUNCTIONS $P_n^m(x)$

In this section we consider a generalization of equation (8.68):

$$\left(1 - x^2\right) y'' - 2xy' + \left(\lambda - \frac{m^2}{1 - x^2}\right) y = 0, \quad -1 \le x \le 1, \tag{8.100}$$

where m is a specified number. Like equation (8.69), equation (8.100) has nontrivial solutions bounded at $x = \pm 1$ only for the values of $\lambda = n(n + 1)$. In mathematical physics problems the values of m are *integer*; also the values of m and n are related by inequality $|m| \le n$. Equation (8.100) is called the *associated Legendre equation of order m*. In Sturm-Liouville form this equation can be written as

$$\frac{d}{dx}\left[\left(1 - x^2\right)\frac{dy}{dx}\right] + \left(\lambda - \frac{m^2}{1 - x^2}\right) y = 0, \quad -1 \le x \le 1. \tag{8.101}$$

To solve equation (8.101) we can use a solution of equation (8.69). First we will discuss positive values of m. Let us introduce a new function, $z(x)$, to replace $y(x)$ in equation (8.101):

$$y(x) = \left(1 - x^2\right)^{\frac{m}{2}} z(x). \tag{8.102}$$

Substituting equation (8.102) into equation (8.101) we obtain

$$\left(1 - x^2\right)z'' - 2(m+1)xz' + [\lambda - m(m+1)]z = 0. \tag{8.103}$$

If $m = 0$, equation (8.103) reduces to equation (8.68); thus its solutions are Legendre polynomials $P_n(x)$.

Next, we solve equation (8.103) by expanding $z(x)$ in a power series:

$$z = \sum_{k=0}^{\infty} a_k x^k. \tag{8.104}$$

With this we have

$$z' = \sum_{k=1}^{\infty} k a_k x^{k-1} = \sum_{k=0}^{\infty} k a_k x^{k-1},$$

$$z'' = \sum_{k=2}^{\infty} k(k-1) a_k x^{k-2} = \sum_{k=0}^{\infty} (k+2)(k+1) a_{k+2} x^k,$$

$$x^2 z'' = \sum_{k=2}^{\infty} k(k-1) a_k x^k = \sum_{k=0}^{\infty} k(k-1) a_k x^k.$$

Substituting these series into equation (8.103) we obtain

$$\sum_{k=0}^{\infty} \left\{ (k+2)(k+1) a_{k+2} + [\lambda - (k+m)(k+m+1)] a_k \right\} x^k = 0.$$

Functions x^k are linearly independent; thus the coefficients of each power of x^k must be zero, which leads to a recurrence relation for coefficients a_k:

$$a_{k+2} = -\frac{\lambda - (k+m)(k+m+1)}{(k+2)(k+1)} a_k. \tag{8.105}$$

Reading Exercise

Using this recurrence relation check that the series in equation (8.104) converges for $-1 < x < 1$ and diverges at the endpoints of the interval $x = \pm 1$.

Below we will discuss only solutions that are regular on the closed interval $-1 \le x \le 1$. This means that the series (8.104) should *terminate as a polynomial* of some maximum degree. Denoting this degree as q we obtain $a_q \ne 0$ and $a_{q+2} = 0$ so that if $\lambda = (q + m)(q + m + 1)$, $q = 0, 1, \ldots$, then $a_{q+2} = a_{q+4} = \cdots = 0$. Introducing $n = q + m$, because q and m are nonnegative integers, we have $n = 0, 1, 2, \ldots$ and $n \ge m$. Thus we see that $\lambda = n(n + 1)$ as in the case of Legendre polynomials. Clearly if $n = 0$, the value of $m = 0$; thus $\lambda = 0$ and the function $z(x) = a_0$ and $y(x) = P_0(x)$.

From the above discussion we obtain that $z(x)$ is an even or odd polynomial of degree $(n-m)$:

$$z(x) = a_{n-m}x^{n-m} + a_{n-m-2}x^{n-m-2} + \cdots + \begin{cases} a_0 \\ a_1 x. \end{cases} \tag{8.106}$$

Let us present several examples for $m = 1$. If $n = 1$, then $q = 0$; thus $z(x) = a_0$. If $n = 2$, then $q = 1$; thus $z(x) = a_1 x$. If $n = 3$, then $q = 2$; thus $z(x) = a_0 + a_2 x^2$ and from the recurrence formula we have $a_2 = -5a_0$.

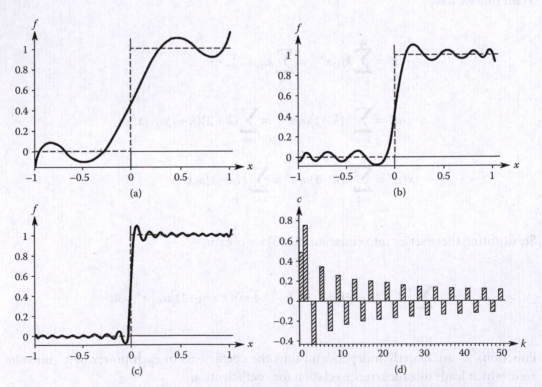

FIGURE 8.10 The function $f(x)$ and the partial sum of its Fourier-Legendre series. The graph of $f(x)$ is shown by the dashed line, and the graph of the series is shown by the solid line. $(N + 1)$ terms are kept in the series. (a) $(N = 5)$. (b) $N = 15$. (c) $N = 50$. (d) Values of the coefficients C_n of the series.

Reading Exercises

1. Find $z(x)$ for $m = 1$ and $n = 4$.
2. Find $z(x)$ for $m = 2$ and $n = 4$.
3. For all above examples check that $z(x) = \frac{d^m}{dx^m} P_n(x)$ (keeping the lowest coefficients arbitrary).

Given that $\lambda = n(n+1)$ we can obtain a solution of equation (8.103) using the solution of the Legendre equation, equation (8.69). Let us differentiate equation (8.103) with respect to the variable x:

$$(1-x^2)(z')'' - 2[(m+1)+1]x(z')' + [n(n+1)-(m+1)(m+2)]z' = 0. \tag{8.107}$$

It is seen that if in this equation we replace z' by z and $(m+1)$ by m, the obtained equation becomes equation (8.103). In other words, if $P_n(x)$ is a solution of equation (8.103) for $m = 0$, then $P_n'(x)$ is a solution of equation (8.107) for $m = 1$. Repeating this we obtain that $P_n''(x)$ is a solution for $m = 2$, $P_n'''(x)$ is a solution for $m = 3$, and so on. For arbitrary integer m, where $0 \le m \le n$, a solution of equation (8.103) is the function $\frac{d^m}{dx^m} P_n(x)$; thus

$$z(x) = \frac{d^m}{dx^m} P_n(x), \ 0 \le m \le n. \tag{8.108}$$

With equations (8.108) and (8.102) we have a solution of equation (8.101) given by

$$y(x) = (1-x^2)^{\frac{m}{2}} \frac{d^m}{dx^m} P_n(x), \ 0 \le m \le n. \tag{8.109}$$

The functions defined in equation (8.109) are called the *associated Legendre functions* and denoted as $P_n^m(x)$:

$$P_n^m(x) = (1-x^2)^{\frac{m}{2}} \frac{d^m}{dx^m} P_n(x). \tag{8.110}$$

Notice that $\frac{d^m}{dx^m} P_n(x)$ is a polynomial of degree $n-m$; thus

$$P_n^m(-x) = (-1)^{n-m} P_n^m(x) \tag{8.111}$$

which is referred to as the parity property. From equation (8.111) it is directly seen that $P_n^m(x) = 0$ for $|m| > n$ because in this case mth-order derivatives of a polynomial $P_n(x)$ of degree n are equal to zero. The graphs of several $P_n^m(x)$ are plotted in Figure 8.11.

Thus, from the above discussion, we see that equation (8.100) has eigenvalues

$$m(m+1), (m+1)(m+2), (m+2)(m+3),\ldots \tag{8.112}$$

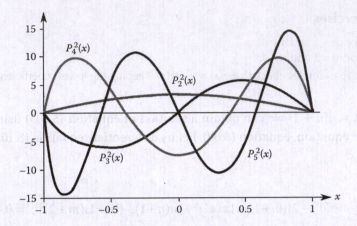

FIGURE 8.11   Graphs of $P_2^2(x)$, $P_3^2(x)$, $P_4^2(x)$, and $P_5^2(x)$.

with the corresponding eigenfunctions, bounded on $[-1,1]$:

$$P_m^m(x), \ P_{m+1}^m(x), \ P_{m+2}^m(x),\ldots \tag{8.113}$$

Equation (8.100) (or (8.101)) does not change when the sign of m changes. Therefore, a solution of equation (8.53) for positive m is also a solution for negative values $-|m|$. Thus, we can define $P_n^m(x)$ to be equal to $P_n^{|m|}(x)$ for $-n \leq m \leq n$.

$$P_n^{-|m|}(x) = P_n^{|m|}(x). \tag{8.114}$$

The first several associated Legendre functions $P_n^1(x)$ for $m = 1$ are

$$P_1^1(x) = \sqrt{1-x^2} \cdot [P_1(x)]' = \sqrt{1-x^2},$$

$$P_2^1(x) = \sqrt{1-x^2} \cdot [P_2(x)]' = \sqrt{1-x^2} \cdot 3x,$$

$$P_3^1(x) = \sqrt{1-x^2} \cdot [P_3(x)]' = \sqrt{1-x^2} \cdot \frac{3}{2}(5x^2 - 1);$$

and the first several associated Legendre functions $P_n^2(x)$ for $m = 2$ are

$$P_2^2(x) = (1-x^2) \cdot [P_2(x)]'' = (1-x^2) \cdot 3,$$

$$P_3^2(x) = (1-x^2) \cdot [P_3(x)]'' = (1-x^2) \cdot 15x,$$

$$P_4^2(x) = (1-x^2) \cdot [P_4(x)]'' = (1-x^2) \cdot \frac{15}{2}(7x^2 - 1).$$

Associated Legendre function $P_n^m(x)$ has $(n-m)$ simple (not repeating) real roots on the interval $-1 < x < 1$.

The following *recurrence formula* is often useful:

$$(2n+1)xP_n^m(x)-(n-m+1)P_{n+1}^m(x)-(n+m)P_{n-1}^m(x)=0. \qquad (8.115)$$

8.10 FOURIER-LEGENDRE SERIES IN ASSOCIATED LEGENDRE FUNCTIONS

Functions $P_n^m(x)$ *for fixed value of* $|m|$ (the upper index of the associated Legendre functions) and all possible values of the lower index,

$$P_m^m(x), \quad P_{m+1}^m(x), \quad P_{m+2}^m(x),\dots \qquad (8.116)$$

form an *orthogonal* (with respect to the weight function $r(x) = 1$) and *complete* set of functions on the interval $[-1,1]$. In other words for each value of m there is an orthogonal and complete set of equations (8.116). This follows from the fact that these functions are also the solutions of a Sturm-Liouville problem. Thus the set of equations (8.116) for any given m is a basis for an eigenfunction expansion for functions bounded on $[-1,1]$ and we may write

$$f(x)=\sum_{k=0}^{\infty}c_k P_{m+k}^m(x). \qquad (8.117)$$

The formula for the coefficients c_k ($k = 0,1,2,\dots$) follows from the orthogonality of the functions in equation (8.116):

$$c_k=\frac{1}{\left\|P_{m+k}^m\right\|^2}\int_{-1}^{1}f(x)P_{m+k}^m(x)dx=\frac{2(m+k)+1}{2}\frac{k!}{(2m+k)!}\int_{-1}^{1}f(x)P_{m+k}^m(x)dx. \qquad (8.118)$$

As previously, the sequence of the partial sums $S_N(x)$ of series (8.117) converges on the interval $(-1,1)$ in the mean to the square integrable function $f(x)$, that is,

$$\int_{-1}^{1}[f(x)-S_N(x)]^2\,dx\to 0 \text{ as } N\to\infty.$$

For piecewise-continuous functions the same theorem as in the previous section is valid.

Example 4

Expand the function

$$f(x)=\begin{cases}1+x, & -1\le x<0,\\ 1-x, & 0\le x\le 1\end{cases}$$

in terms of associated Legendre functions $P_n^m(x)$ of order $m=2$.

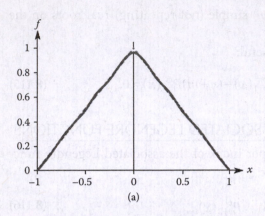
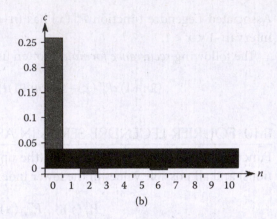

(a)
(b)

FIGURE 8.12   The function $f(x)$ and the partial sum of its Fourier-Legendre series in terms of associated Legendre functions $P_n^2(x)$. (*a*) The graph of $f(x)$ is shown by the dashed line, the graph of the partial sum with $N = 10$ terms of the series by the solid line; (*b*) values of the coefficients c_k of the series.

Solution. The series is

$$f(x) = c_0 P_2^2(x) + c_1 P_3^2(x) + c_2 P_4^2(x) + c_3 P_5^2(x)...,$$

where coefficients c_k are

$$c_k = \frac{2k+5}{2} \frac{k!}{(k+4)!} \left[\int\limits_{-1}^{0} (1+x) P_{k+2}^2(x) dx + \int\limits_{0}^{1} (1-x) P_{k+2}^2(x) dx \right].$$

Because $f(x)$ is an even function of x,

$$c_1 = c_3 = c_5 = \cdots = 0.$$

The first two coefficients with even index are

$$c_0 = \frac{5}{48} \left[\int\limits_{-1}^{0} (1+x)3(1-x^2) dx + \int\limits_{0}^{1} (1-x)3(1-x^2) dx \right] = \frac{25}{96},$$

$$c_2 = \frac{1}{80} \left[\int\limits_{-1}^{0} (1+x)\frac{15}{2}(1-x^2)(7x^2-1) dx + \int\limits_{0}^{1} (1-x)\frac{15}{2}(1-x^2)(7x^2-1) dx \right] = -\frac{1}{80}.$$

Problems

In problems 1 through 8 expand the function $f(x)$ in the Fourier-Legendre functions $P_n^m(x)$ on $[-1,1]$. Construct the expansion for (a) $m = 0$—in this case the functions $P_n^m(x)$ are the Legendre polynomials $P_n(x)$; (b) for $m = 1$; (c) for $m = 2$.

Write the formulas for the coefficients of the series expansion and the expression for the Fourier-Legendre series. If the integrals in the coefficients are not easy to evaluate, leave the evaluation to the program (they will be evaluated numerically by the program).

Using the program **FourierSeries**, obtain the pictures of several orthonormal functions $\hat{P}^m_{m+n}(x)$, plot the graphs of the given function $f(x)$ and of the partial sums $S_N(x)$, and build the histograms of coefficients c_k of the series.

Note. The program **FourierSeries** expands the function $f(x)$ in terms of *orthonormal* Legendre functions $\hat{P}^m_{m+k}(x)$, which differ from the Legendre functions $P^m_{m+k}(x)$ by a coefficient, which makes the set of functions $P^m_{m+k}(x)$ orthonormal:

$$\hat{P}^m_{m+k}(x) = \sqrt{\left(m+k+\frac{1}{2}\right)\frac{k!}{(2m+k)!}}\, P^m_{m+k}(x).$$

So, the coefficients of expansion in the program **FourierSeries** are

$$\hat{c}_k = \sqrt{\left(m+k+\frac{1}{2}\right)\frac{k!}{(2m+k)!}}\int_{-1}^{1} f(x)P^m_{m+k}(x)dx.$$

1. $f(x) = 2x-1$

2. $f(x)=1-x^2$

3. $f(x)=\begin{cases} -1, & -1<x<0 \\ 1, & 0<x<1 \end{cases}$

4. $f(x)=\begin{cases} 0, & -1<x<0 \\ x, & 0<x<1 \end{cases}$

5. $f(x)=\cos\dfrac{\pi x}{2}$ for $-1 \le x \le 1$

6. $f(x)=\begin{cases} 0, & -1<x<0 \\ \sqrt{1-x}, & 0<x<1 \end{cases}$

7. $f(x) = \sin\pi x$

8. $f(x)=e^x$

CHAPTER 9

Fourier Series

9.1 PERIODIC PROCESSES AND PERIODIC FUNCTIONS

In the sciences and in technology very often we encounter periodic phenomena. This means that some process repeats after some time interval T, called the period. Alternating electric current, an object in circular motion, and wave phenomena are examples of physical phenomena that are periodic. Such processes can be associated with periodic mathematical functions in time, t, which have the property

$$\varphi(t + T) = \varphi(t).$$

The simplest periodic function is the sine (or cosine) function, $A\sin(\omega t + \alpha)$ (or $A\cos(\omega t + \alpha)$), where ω is the *angular frequency* related to the period by the relationship

$$\omega = \frac{2\pi}{T} \tag{9.1}$$

(quantity $f = 1/T$ is called frequency, constant α is called phase).

With these simple periodic functions more complex periodic functions can be constructed, as was noted by the French mathematician Joseph Fourier. For example, if we add the functions

$$y_0 = A_0, \quad y_1 = A_1 \sin(\omega t + \alpha_1), \quad y_2 = A_2 \sin(2\omega t + \alpha_2),$$
$$y_3 = A_3 \sin(3\omega t + \alpha_3), \dots \tag{9.2}$$

with multiple frequencies ω, 2ω, 3ω, … (i.e., with the periods T, $T/2$, $T/3$,…), we obtain a periodic function (with period T), which, when graphed, has an appearance very distinct from the graphs of any of the functions in equation (9.2). Almost any periodic function can be constructed in this fashion using a combination of sine and cosine functions.

It is natural to also investigate the reverse problem. Is it possible to resolve a given arbitrary periodic function, $\varphi(t)$, with period T into a sum of simple functions such as those in equation (9.2)? As we shall see, for a very wide class of functions the answer to this question is positive, but to do so may require an infinite sequence of the functions in equation (9.2). In these cases the periodic function $\varphi(t)$ can be resolved into the infinite *trigonometric series*

$$\varphi(t) = A_0 + A_1 \sin(\omega t + \alpha_1) + A_2 \sin(2\omega t + \alpha_2) + \cdots = A_0 + \sum_{n=1}^{\infty} A_n \sin(n\omega t + \alpha_n), \quad (9.3)$$

where A_n and α_n are constants, and $\omega = 2\pi/T$. Each term in equation (9.3) is called a *harmonic* and the decomposition of periodic functions into harmonics is called *harmonic analysis*.

In many cases it is useful to introduce the variable

$$x = \omega t = \frac{2\pi t}{T}$$

and to work with the functions

$$f(x) = \varphi\left(\frac{x}{\omega}\right)$$

which are also periodic but with the *standard period* 2π: $f(x + 2\pi) = f(x)$. Using this shorthand, equation (9.3) becomes

$$f(x) = A_0 + A_1 \sin(x + \alpha_1) + A_2 \sin(2x + \alpha_2) + \cdots = A_0 + \sum_{n=1}^{\infty} A_n \sin(nx + \alpha_n). \quad (9.4)$$

With the trigonometric identity $\sin(\alpha + \beta) = \sin\alpha\cos\beta + \cos\alpha\sin\beta$ and the notation

$$A_0 = 2a_0, \quad A_n \sin\alpha_n = a_n, \quad A_n \cos\alpha_n = b_n, \quad (n = 1, 2, 3, \ldots)$$

we obtain a standardized form for the harmonic analysis of a periodic function $f(x)$ as

$$f(x) = \frac{a_0}{2} + (a_1 \cos x + b_1 \sin x) + (a_2 \cos 2x + b_2 \sin 2x) + \cdots$$

$$= \frac{a_0}{2} + \sum_{n=1}^{\infty} (a_n \cos nx + b_n \sin nx) \quad (9.5)$$

which is referred to as the trigonometric Fourier expansion.

9.2 FOURIER FORMULAS

To determine the limits of validity for the representation in equation (9.5) of a given function, $f(x)$, with period 2π and to find the coefficients a_n and b_n we follow the approach that was originally elaborated by Fourier. We first assume that the function $f(x)$ can be integrated over the interval $[-\pi,\pi]$. If $f(x)$ is discontinuous at any point, we assume that the integral of $f(x)$ converges and in this case we also assume that the integral of the absolute value of the function, $|f(x)|$, converges. A function with these properties is said to be *absolutely integrable*. Integrating the expression (9.5) term by term we obtain

$$\int_{-\pi}^{\pi} f(x)dx = \pi a_0 + \sum_{n=1}^{\infty} \left[a_n \int_{-\pi}^{\pi} \cos nx dx + b_n \int_{-\pi}^{\pi} \sin nx dx \right].$$

Since

$$\int_{-\pi}^{\pi} \cos nx dx = \frac{\sin nx}{n} \Big|_{-\pi}^{\pi} = 0 \quad \text{and} \quad \int_{-\pi}^{\pi} \sin nx dx = -\frac{\cos nx}{n} \Big|_{-\pi}^{\pi} = 0, \tag{9.6}$$

all the terms in the sum are zero and we obtain

$$a_0 = \frac{1}{\pi} \int_{-\pi}^{\pi} f(x)dx. \tag{9.7}$$

To find coefficients a_n we multiply equation (9.5) by cos mx and then integrate term by term over the interval $[-\pi,\pi]$:

$$\int_{-\pi}^{\pi} f(x)\cos mx dx = a_0 \int_{-\pi}^{\pi} \cos mx dx + \sum_{n=1}^{\infty} \left[a_n \int_{-\pi}^{\pi} \cos nx \cos mx dx + b_n \int_{-\pi}^{\pi} \sin nx \cos mx dx \right].$$

The first term is zero as was noted in equation (9.6). For any n and m we also have

$$\int_{-\pi}^{\pi} \sin nx \cos mx dx = \frac{1}{2} \int_{-\pi}^{\pi} \left[\sin(n+m)x + \sin(n-m)x \right] dx = 0, \tag{9.8}$$

and if $n \neq m$, we obtain

$$\int_{-\pi}^{\pi} \cos nx \cos mx dx = \frac{1}{2} \int_{-\pi}^{\pi} \left[\cos(n+m)x + \cos(n-m)x \right] dx = 0. \tag{9.9}$$

Using these formulas along with the identity

$$\int_{-\pi}^{\pi} \cos^2 mx\, dx = \int_{-\pi}^{\pi} \frac{1+\cos 2mx}{2}\, dx = \pi \tag{9.10}$$

we see that all the integrals in the sum are zero except the one with the coefficient a_m. We thus have

$$a_m = \frac{1}{\pi}\int_{-\pi}^{\pi} f(x)\cos mx\, dx \quad (m=1,2,3,\ldots). \tag{9.11}$$

The usefulness of introducing the factor 1/2 in the first term in equation (9.5) is now apparent since it allows the same formulas to be used for all a_n, including $n = 0$.

Similarly, multiplying equation (9.5) by sin mx and using, along with equations (9.6) and (9.8), two other simple integrals

$$\int_{-\pi}^{\pi} \sin nx \sin mx\, dx = 0, \tag{9.12}$$

if $n \neq m$, and

$$\int_{-\pi}^{\pi} \sin^2 mx\, dx = \pi, \tag{9.13}$$

we obtain the second coefficient,

$$b_m = \frac{1}{\pi}\int_{-\pi}^{\pi} f(x)\sin mx\, dx \quad (m=1,2,3,\ldots). \tag{9.14}$$

Reading Exercise:

Obtain the same result as in equations (9.8), (9.9), and (9.12) using the Euler expression

$$e^{imx} = \cos mx + i\sin mx.$$

Equations (9.8), (9.9), and (9.12) also indicate that the system of functions

$$1, \cos x, \sin x, \cos 2x, \sin 2x, \ldots, \cos nx, \sin nx. \ldots \tag{9.15}$$

is *orthogonal* on $[-\pi,\pi]$.

It is important to notice that the above system is not orthogonal on the reduced interval $[0, \pi]$ because for n and m with different parity (one odd and the other even) we have

$$\int_0^\pi \sin nx \cos mx\, dx \neq 0.$$

However, the system consisting of cosine functions only

$$1, \cos x, \cos 2x, \ldots, \cos nx, \ldots \tag{9.16}$$

is orthogonal on $[0, \pi]$ and the same is true for

$$\sin x, \sin 2x, \ldots, \sin nx. \ldots \tag{9.17}$$

A second observation, which we will need later, is that on an interval $[0, l]$ of arbitrary length l, both systems of functions

$$1, \cos \frac{\pi x}{l}, \cos \frac{2\pi x}{l}, \ldots, \cos \frac{n\pi x}{l}, \ldots \tag{9.18}$$

and

$$\sin \frac{\pi x}{l}, \sin \frac{2\pi x}{l}, \ldots, \sin \frac{n\pi x}{l}. \ldots \tag{9.19}$$

are orthogonal.

Reading Exercise

Prove the above three statements.

Equations (9.7), (9.11), and (9.14) are known as the *Fourier coefficients* and the series (9.5) with these definitions is called the *Fourier series*. Equation (9.5) is also referred to as the *Fourier expansion* of the function $f(x)$.

Notice that for the function $f(x)$ having period 2π, the integral

$$\int_\alpha^{\alpha+2\pi} f(x)\, dx$$

does not depend on the value of α. As a result we may also use the following expressions for the Fourier coefficients:

$$a_m = \frac{1}{\pi} \int_0^{2\pi} f(x) \cos mx\, dx \quad \text{and} \quad b_m = \frac{1}{\pi} \int_0^{2\pi} f(x) \sin mx\, dx. \tag{9.20}$$

It is important to realize that to obtain the results above we used a term-by-term integration of the series, which is justified only if the series converges uniformly. Until we know for sure that the series converges we can only say that the series (9.5) *corresponds* to the function $f(x)$, which usually is denoted as

$$f(x) \sim \frac{a_0}{2} + \sum_{n=1}^{\infty} (a_n \cos nx + b_n \sin nx).$$

At this point we should remind the reader what is meant by *uniform convergence*. The series $\sum_{n=1}^{\infty} f_n(x)$ converges to the sum $S(x)$ uniformly on the interval $[a,b]$ if, for any arbitrarily small $\varepsilon > 0$, we can find a number N such that for all $n \geq N$ the remainder of the series $|\sum_{n=N}^{\infty} f_n(x)| \leq \varepsilon$ for all $x \subset [a,b]$. This indicates that the series approaches its sum uniformly with respect to x. The most important features of a uniformly converging series are as follows:

1. If $f_n(x)$ for any n is a continuous function, then $S(x)$ is also a continuous function.

2. The equality $\sum_{n=1}^{\infty} f_n(x) = S(x)$ can be integrated term by term along any interval within the interval $[a,b]$.

3. If the series $\sum_{n=1}^{\infty} f_n'(x)$ converges uniformly, then its sum is equal to $S'(x)$; that is, the formula $\sum_{n=1}^{\infty} f_n(x) = S(x)$ can be differentiated term by term.

There is a simple and very practical criteria for convergence established by Karl Weierstrass that says that if $|f_n(x)| < c_n$ for each term $f_n(x)$ in the series defined on the interval $x \subset [a,b]$ (i.e., $f_n(x)$ is limited by c_n), where $\sum_{n=1}^{\infty} c_n$ is a converging numeric series, then the series

$$\sum_{n=1}^{\infty} f_n(x)$$

converges uniformly on $[a,b]$. For example, the numeric series $\sum_{n=1}^{\infty} 1/n^2$ is known to converge, so any trigonometric series with terms such as $\sin nx / n^2$ or similar will converge uniformly for all x because $|\sin nx / n^2| \leq 1/n^2$.

9.3 CONVERGENCE OF FOURIER SERIES

In this section we study the range of validity of equation (9.5) with Fourier coefficients given by equations (9.11) and (9.14). To start, it is clear that if the function $f(x)$ is finite on $[-\pi,\pi]$, then the Fourier coefficients are bounded. This is easily verified, for instance for a_n, since

$$|a_n| = \frac{1}{\pi} |\int_{-\pi}^{\pi} f(x) \cos nx \, dx| \leq \frac{1}{\pi} \int_{-\pi}^{\pi} |f(x)| \cdot |\cos nx| \, dx \leq \frac{1}{\pi} \int_{-\pi}^{\pi} |f(x)| \, dx. \tag{9.21}$$

The same result is valid in cases where $f(x)$ is not finite but is absolutely integrable, that is, the integral of its absolute value converges:

$$\int_{-\pi}^{\pi} |f(x)|\, dx < \infty. \tag{9.22}$$

The *necessary condition* that any series converges is that its terms tend to zero as $n \to \infty$. Because the absolute values of sine and cosine functions are bounded by plus and minus one, the necessary condition that the trigonometric series in equation (9.5) converges is that coefficients of expansion a_n and b_n tend to zero as $n \to \infty$. This condition is valid for functions that are integrable (or absolutely integrable in the case of functions that are not finite), which is clear from the following lemma.

Riemann's Lemma

If the function $f(t)$ is absolutely integrable on $[a,b]$, then

$$\lim_{\alpha \to \infty} \int_{a}^{b} f(t)\sin\alpha t\, dt = 0 \quad \text{and} \quad \lim_{\alpha \to \infty} \int_{a}^{b} f(t)\cos\alpha t\, dt = 0. \tag{9.23}$$

We will not prove this rigorously, but its sense should be obvious. In the case of very fast oscillations the sine and cosine functions change their sign very quickly as $\alpha \to \infty$. Thus these integrals vanish for "reasonable" (i.e., absolutely integrable) functions, $f(t)$, because they do not change substantially as the sine and cosine alternate with opposite signs in their semi-periods.

Thus, for absolutely integrating functions the necessary condition of convergence of Fourier series is satisfied. Before we discuss the problem of convergence of Fourier series in more detail, let us notice that practically any interesting for applications function can be expanded in a converging Fourier series.

It is important to know how quickly the terms in (9.5) decrease as $n \to \infty$. If they decrease rapidly, the series converges rapidly. In this case, using very few terms we have a good trigonometric approximation for $f(x)$ and the partial sum of the series, $S_n(x)$, is a good approximation to the sum $S(x) = f(x)$. If the series converges more slowly, a larger number of terms are needed to have a sufficiently accurate approximation.

Assuming that the series (9.5) converges, the speed of its convergence to $f(x)$ depends on the behavior of $f(x)$ over its period, or, in the case of nonperiodic functions, on the way it is extended from the interval $[a,b]$ to the entire axis x, as we will discuss below. Convergence is most rapid for very smooth functions (functions that have continuous derivatives of higher order). Discontinuities in the derivative of the function, $f'(x)$, substantially reduce the rate of convergence whereas discontinuities in $f(x)$ reduce the convergence rate even more with the result that many terms in the Fourier series must be used to approximate the function $f(x)$ with the necessary precision. This should be fairly obvious since the "smoothness" of $f(x)$ determines the rate of decreasing of the coefficients a_n and b_n. It can be shown [1] that the coefficients decrease (1) faster than $1/n^2$ (for example $1/n^3$) when $f(x)$ and $f'(x)$ are continuous but $f''(x)$ has a discontinuity; (2) at about the same rate as $1/n^2$ when $f(x)$ is continuous but $f'(x)$ has discontinuities; and (3) at a rate similar to the rate of convergence of $1/n$ if $f(x)$ is not continuous. It is important to note that in the first two cases the series converges uniformly, which follows from the Weierstrass criterion, because each term of equation (9.5) is bounded by the corresponding term in the converging numeric series $\sum_{n=1}^{\infty} 1/n^2 < \infty$.

The following very important theorem describes the convergence of the Fourier series given in equation (9.5) for a function $f(x)$ at a points x_0 where $f(x)$ is continuous or where it may have a discontinuity (the proof can be found in book [1]).

The Dirichlet Theorem

If the function $f(x)$ with period 2π is piecewise continuous in $[-\pi, \pi]$ and has a finite number of points of discontinuity in this interval, then its Fourier series converges to $f(x_0)$ when x_0 is a continuity point, and to

$$S(x_0) = \frac{f(x_0 + 0) + f(x_0 - 0)}{2}$$

if x_0 is a point of discontinuity.

At the ends of the interval $[-\pi, \pi]$ the Fourier series converges to

$$\frac{f(-\pi + 0) + f(\pi - 0)}{2}.$$

A function $f(x)$ defined on $[a,b]$ is called *piecewise continuous* if the following are true:

1. It is continuous on $[a,b]$ except perhaps at a finite number of points.

2. If x_0 is one such point, then the left and right limits of $f(x)$ at x_0 exist and are finite.

3. Both the limit from the right of $f(x)$ at a and the limit from the left at b exist and are finite.

Stated more briefly, for the Fourier series of a function $f(x)$ to converge, this function should be piecewise continuous with a finite number of discontinuities.

9.4 FOURIER SERIES FOR NONPERIODIC FUNCTIONS

We assumed above that the function $f(x)$ is defined on the entire x-axis and has period 2π. But very often we need to deal with nonperiodic functions defined only on the interval $[-\pi,\pi]$. The theory elaborated above can still be used if we extend $f(x)$ periodically from $(-\pi,\pi)$ to all x. In other words we assign the same values of $f(x)$ to all the intervals, $(x, 3x)$ $(3\pi,5\pi)$, ..., $(-3\pi,\pi)$, $(-5\pi,-3\pi)$, ... and then use equations (9.11) and (9.14) for the Fourier coefficients of this new function, which is periodic. Many examples of such extensions will be given below. If $f(-\pi) = f(\pi)$, we can include the end points, $x = \pm\pi$, and the Fourier series converges to $f(x)$ everywhere on $[-\pi,\pi]$. Over the entire axis the expansion gives a periodic extension of the given function $f(x)$ given originally on $[-\pi,\pi]$. In many cases $f(-\pi) \neq f(\pi)$ and the Fourier series at the ends of the interval $[-\pi,\pi]$ converges to

$$\frac{f(-\pi)+f(\pi)}{2}$$

which differs from both $f(-\pi)$ and $f(\pi)$.

The rate of convergence of the Fourier series depends on the discontinuities of the function and derivatives of the function after its extension to the entire axis. Some extensions do not increase the number of discontinuities of the original function whereas others do increase this number. In the latter case the rate of convergence is reduced. Among the examples given later in this chapter Examples 2 and 3 are of the Fourier series of $f(x)=x$ on the interval $[0,\pi]$. In the first expansion the function is extended to the entire axis as an even function and remains continuous so that the coefficients of the Fourier series decrease as $1/n^2$. In the second example this function is extended as an odd function and has discontinuities at $x = k\pi$ (integer k), in which case the coefficients decrease slower, as $1/n$.

9.5 FOURIER EXPANSIONS ON INTERVALS OF ARBITRARY LENGTH

Suppose that a function $f(x)$ is defined on some interval $[-l,l]$ of arbitrary length $2l$ (where $l > 0$). Using the substitution

$$x = \frac{ly}{\pi}(-\pi \leq y \leq \pi),$$

we obtain the function $f(yl/\pi)$ of the variable y on the interval $[-\pi,\pi]$, which can be expanded using the standard equations (9.5), (9.11), and (9.14) as

$$f\left(\frac{yl}{\pi}\right) = \frac{a_0}{2} + \sum_{n=1}^{\infty}(a_n \cos ny + b_n \sin ny),$$

with

$$a_n = \frac{1}{\pi}\int_{-\pi}^{\pi} f\left(\frac{yl}{\pi}\right)\cos nydy \quad \text{and} \quad b_n = \frac{1}{\pi}\int_{-\pi}^{\pi} f\left(\frac{yl}{\pi}\right)\sin nydy.$$

Returning to the variable x we obtain

$$f(x) = \frac{a_0}{2} + \sum_{n=1}^{\infty} \left(a_n \cos\frac{n\pi x}{l} + b_n \sin\frac{n\pi x}{l} \right), \tag{9.24}$$

with

$$a_n = \frac{1}{l} \int_{-l}^{l} f(x)\cos\frac{n\pi x}{l}dx, \qquad n = 0,1,2,...,$$

$$b_n = \frac{1}{l} \int_{-l}^{l} f(x)\sin\frac{n\pi x}{l}dx, \qquad n = 1,2,... . \tag{9.25}$$

If the function is given, but not on the interval $[-l,l]$, and instead on an arbitrary interval of length $2l$, for instance $[0,2l]$, the formulas for the coefficients of the Fourier series (9.24) become

$$a_n = \frac{1}{l} \int_{0}^{2l} f(x)\cos\frac{n\pi x}{l}dx \quad \text{and} \quad b_n = \frac{1}{l} \int_{0}^{2l} f(x)\sin\frac{n\pi x}{l}dx. \tag{9.26}$$

In both cases, the series in equation (9.24) gives a periodic function with the period $T = 2l$.

If the function $f(x)$ is given on an interval $[a,b]$ (where a and b may have the same or opposite sign, that is, the interval $[a,b]$ can include or exclude the point $x = 0$), different periodic continuations onto the entire x-axis may be made (see Figure 9.1). As an example, consider the periodic continuation $F(x)$ of the function $f(x)$, defined by the condition

$$F(x + n(b-a)) = f(x), n = 0, \pm 1, \pm 2,... \text{ for all } x.$$

In this case the Fourier series is given by equation (9.24) where $2l = b-a$. Clearly, instead of equations (9.25) the following formulas for the Fourier coefficients should be used:

$$a_n = \frac{2}{b-a} \int_{a}^{b} f(x)\cos\frac{2n\pi x}{b-a}dx, \quad b_n = \frac{2}{b-a} \int_{a}^{b} f(x)\sin\frac{2n\pi x}{b-a}dx. \tag{9.27}$$

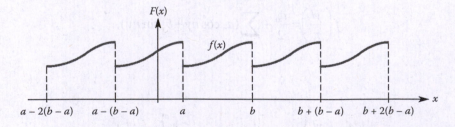

FIGURE 9.1 Arbitrary function $f(x)$ defined on the interval $[a,b]$ extended to the x-axis as the function $F(x)$.

The series in equation (9.24) gives a periodic function with the period $T = 2l = b - a$; however, the original function was defined only on the interval $[a,b]$ and is not periodic in general.

9.6 FOURIER SERIES IN COSINE OR IN SINE FUNCTIONS

Suppose that $f(x)$ is an *even* function on $[-\pi, \pi]$ so that $f(x)\sin nx$ is odd. For this case

$$b_n = \frac{1}{\pi} \int_{-\pi}^{\pi} f(x)\sin nx\, dx = 0$$

since the integral of an odd function over a symmetric interval equals zero. Coefficients a_n can be written as

$$a_n = \frac{1}{\pi} \int_{-\pi}^{\pi} f(x)\cos nx\, dx = \frac{2}{\pi} \int_{0}^{\pi} f(x)\cos nx\, dx \tag{9.28}$$

since the integrand is even. Thus, for even functions, $f(x)$ we may write

$$f(x) = \frac{a_0}{2} + \sum_{n=1}^{\infty} a_n \cos nx. \tag{9.29}$$

Similarly, if $f(x)$ is an *odd* function, we have

$$a_n = \frac{1}{\pi} \int_{-\pi}^{\pi} f(x)\cos nx\, dx = 0 \quad \text{and} \quad b_n = \frac{2}{\pi} \int_{0}^{\pi} f(x)\sin nx\, dx, \tag{9.30}$$

in which case we have

$$f(x) = \sum_{n=1}^{\infty} b_n \sin nx. \tag{9.31}$$

Thus an even on $[-\pi, \pi]$ function is expanded in the set (9.16)

$$1, \cos x, \cos 2x, \ldots, \cos nx, \ldots$$

The odd on $[-\pi,\pi]$ function is expanded in the set (9.17)

$$\sin x, \sin 2x, \ldots, \sin nx. \ldots$$

Any function can be presented as a sum of even and odd functions with set (9.15),

$$f(x) = f_1(x) + f_2(x),$$

where

$$f_1(x) = \frac{f(x) + f(-x)}{2} \quad \text{and} \quad f_2(x) = \frac{f(x) - f(-x)}{2},$$

in which case $f_1(x)$ can be expanded into a cosine Fourier series and $f_2(x)$ into a sine series.

If the function $f(x)$ is defined only on the interval $[0,\pi]$ we can extend it to the interval $[-\pi,0)$. This extension may be made in different ways corresponding to different Fourier series. In particular, such an extension can make $f(x)$ even or odd on $[-\pi,\pi]$, which leads to cosine or sine series with period 2π. In the first case on the interval $[-\pi,0)$ we have

$$f(-x) = f(x), \tag{9.32}$$

and in the second case

$$f(-x) = -f(x). \tag{9.33}$$

The points $x = 0$ and $x = \pi$ need special consideration because the sine and cosine series behave differently at these points. If $f(x)$ is continuous at these points, because of equations (9.32) and (9.33) the cosine series converges to $f(0)$ at $x = 0$ and to $f(\pi)$ at $x = \pi$. The situation is different for the sine series, however. At $x = 0$ and $x = \pi$ the sum of the sine series in equation (9.31) is zero; thus the series is equal to the functions $f(0)$ and $f(\pi)$, respectively, only when these values are zero.

If $f(x)$ is given on the interval $[0,l]$ (where $l > 0$), the cosine and sine series are

$$\frac{a_0}{2} + \sum_{n=1}^{\infty} a_n \cos\frac{n\pi x}{l} \tag{9.34}$$

and

$$\sum_{n=1}^{\infty} b_n \sin\frac{n\pi x}{l}, \tag{9.35}$$

with the coefficients

$$a_n = \frac{2}{l} \int_0^l f(x) \cos\frac{n\pi x}{l} dx, \tag{9.36}$$

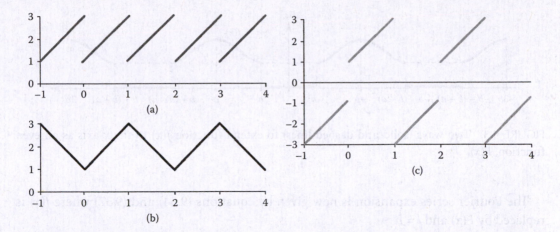

FIGURE 9.2 Three different ways of extending the function $f(x) = 2x + 1$ defined on the interval $[0,1]$ to the function $F(x)$ on the x-axis: (a) general method; (b) even terms method (cosines only); (c) odd terms method (sines only).

or

$$b_n = \frac{2}{l} \int_0^l f(x) \sin \frac{n\pi x}{l} dx. \tag{9.37}$$

If we then wish to extend these expansions to the entire axis, $-\infty < x < +\infty$, it is necessary in both cases to consider the interval $(-l, l)$, which can be expanded to the entire x-axis with the period $T = 2l$. For example, suppose we have the function $f(x) = 2x + 1$ on the interval $[0,1]$. Figure 9.2 presents different schemes of extension of this function to the x-axis.

To perform an even or odd continuation of the function $f(x)$ defined on the interval $[a,b]$ where a and b have the same sign, we can extend $f(x)$ on an interval containing the point $x = 0$. This continuation can be done in an arbitrary way; for example, we can choose $f(x) = 0$ at $x = 0$. Then the function $F(x)$ defined by

$$F(x) = \begin{cases} 0, & -a \leq x < a, \\ f(x), & a \leq x \leq b, \\ f(-x), & -b \leq x < -a \end{cases}$$

will be even and can be continued onto the entire axis with period $2l = 2b$.

To avoid the discontinuities at the points $x = \pm a$ we can set the function $F(x)$ to be equal to the value $f(a)$ for $-a \leq x \leq a$ and in this case the Fourier series will converge faster than the previous case. These two ways of even continuation are labeled as solid and dashed lines in Figure 9.3.

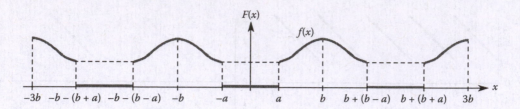

FIGURE 9.3   Two ways (solid and dashed lines) to extend function $f(x)$ to the x-axis as an even function, $F(x)$.

The Fourier series expansion is now given by equations (9.34) and (9.37) where $f(x)$ is replaced by $F(x)$ and $l = b$.

Similarly, an odd continuation can be performed. If we choose $f(x) = 0$ at $x = 0$, then the function $F(x)$ defined by

$$F(x) = \begin{cases} 0, & -a \le x < a, \\ f(x), & a \le x \le b, \\ -f(-x), & -b \le x < -a \end{cases}$$

will be odd and can be continued onto the entire axis with period $2l = 2b$. This variant is shown with a solid line in Figure 9.4. One of the other possibilities is marked by a dashed line. The Fourier series expansion is now given by equations (9.35) and (9.37) where $f(x)$ is replaced by $F(x)$ and $l = b$.

Reading Exercises

Prove the following simple properties of periodic functions:

1. Let $f(x)$ and $g(x)$ both have period T. Prove that $af(x) + bg(x)$ has the same period for any constants a and b.
2. Show that $f(ax)$ has period T/a.
3. Show that $f'(x)$ has period T.

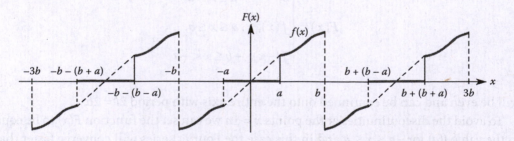

FIGURE 9.4   Two ways (solid and dashed curves) to extend the function $f(x)$ onto the x-axis as an odd function, $F(x)$.

To summarize the above discussion we see that the Fourier series provides a way to obtain an *analytic formula* for functions defined by different formulas on different intervals by combining these intervals into a larger one. Such analytic formulas replace a discontinuous function by a continuous Fourier series expansion, which is often more convenient in a given application. As we have seen above there are often many different choices of how to extend the original function, defined initially on an interval, to the entire axis. The specific choice of extension depends on the application to which the expansion is to be used. Many examples and problems demonstrating these points will be presented in the examples at the end of each of the following sections and the problems at the end of this chapter.

9.7 EXAMPLES

All the functions given below are differentiable or piecewise differentiable and can be represented by Fourier series. The expansions are given but details of the calculation are left to the reader as an exercise. An explanation of how to use the program **TrigSeries** to solve the examples is given in Appendix 5. Notice that the program uses the same formulas that the reader is directed to obtain and use while solving a problem analytically. The only numeric calculation the program performs is the evaluation of the coefficients of Fourier series (with Gauss' method and its modifications) and partial sums.

Example 1

Find the cosine series for $f(x) = x^2$ on the interval $[-\pi, \pi]$.
    Solution. The coefficients are

$$\frac{1}{2}a_0 = \frac{1}{\pi}\int_0^\pi x^2 dx = \frac{\pi^2}{3},$$

$$a_n = \frac{2}{\pi}\int_0^\pi x^2 \cos nx dx = \frac{2}{\pi}x^2 \frac{\sin nx}{n}\Big|_0^\pi - \frac{4}{n\pi}\int_0^\pi x \sin nx dx$$

$$= \frac{4}{n\pi}x\frac{\cos nx}{n}\Big|_0^\pi - \frac{4}{n^2\pi}\int_0^\pi \cos nx dx = (-1)^n \frac{4}{n^2}$$

Thus

$$x^2 = \frac{\pi^2}{3} + 4\sum_{n=1}^\infty (-1)^n \frac{\cos nx}{n^2} \quad (-\pi \le x \le \pi). \tag{9.38}$$

Reading Exercise

Using the program **TrigSeries**, plot graphs of partial sums, build the histogram of the squares of amplitudes A_n^2, and study the rate of convergence.
    In the case where $x = \pi$ we obtain a famous expansion,

$$\frac{\pi^2}{6} = \sum_{n=1}^\infty \frac{1}{n^2}. \tag{9.39}$$

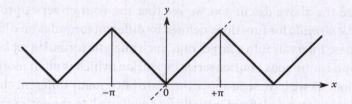

FIGURE 9.5   The function $f(x) = x$ extended to the x-axis.

Example 2

Let the function $f(x)=x$ on the interval $[0,\pi]$. Find the Fourier cosine series.

Solution. Figure 9.5 gives an even periodic continuation of $f(x) = x$ from $[0,\pi]$ onto the entire axis and it also represents the sum of the series, equation (9.43), obtained below. For coefficients we have

$$\frac{1}{2}a_0 = \frac{1}{\pi}\int_0^\pi x\,dx = \frac{\pi}{2}$$

$$a_n = \frac{2}{\pi}\int_0^\pi x\cos nx\,dx = \frac{2}{\pi}x\frac{\sin nx}{n}\Big|_0^\pi - \frac{2}{n\pi}\int_0^\pi \sin nx\,dx$$

$$= 2\frac{\cos n\pi - 1}{n^2\pi} = 2\frac{(-1)^n - 1}{n^2\pi} \qquad (n > 0);$$

that is,

$$a_{2k} = 0, \quad a_{2k-1} = -\frac{4}{(2k-1)^2\pi}, \quad (k = 1,2,3,\ldots),$$

and thus,

$$x = \frac{\pi}{2} - \frac{4}{\pi}\sum_{k=1}^\infty \frac{\cos(2k-1)x}{(2k-1)^2} \quad (0 \le x \le \pi). \tag{9.40}$$

Figure 9.6 shows the graph of the partial sum

$$y = S_5(x) = \frac{\pi}{2} - \frac{4}{\pi}\left(\cos x + \frac{1}{3^2}\cos 3x + \frac{1}{5^2}\cos 5x\right)$$

together with the graph of the extended function.

Reading Exercise

Find the sine series of the same function on the interval $(-\pi,\pi)$. The answer is

$$x = 2\sum_{n=1}^\infty (-1)^{n-1}\frac{\sin nx}{n} \tag{9.41}$$

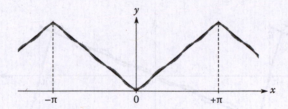

FIGURE 9.6   Original function extended to the x-axis plotted together with the partial sum of the first five terms.

which is different than the cosine series for the same function. The graphs for the series sum and the partial sum

$$y = S_5(x) = 2\left(\sin x - \frac{1}{2}\sin 2x + \frac{1}{3}\sin 3x - \frac{1}{4}\sin 4x + \frac{1}{5}\sin 5x \right)$$

are presented in Figures 9.7 and 9.8, respectively.

Example 3

Find the Fourier series for

$$f(x) = \begin{cases} 0, & \text{if } -\pi < x < 0, \\ x, & \text{if } \quad 0 \le x \le \pi. \end{cases}$$

Solution. The coefficients are

$$\frac{1}{2}a_0 = \frac{1}{2\pi}\int_0^\pi x\,dx = \frac{\pi}{4},$$

$$a_n = \frac{1}{\pi}\int_0^\pi x\cos nx\,dx = \frac{1}{\pi} x \left.\frac{\sin nx}{n}\right|_0^\pi - \frac{1}{n\pi}\int_0^\pi \sin nx\,dx = \frac{\cos n\pi - 1}{n^2\pi},$$

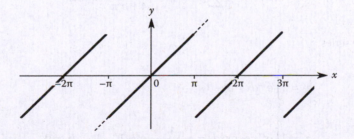

FIGURE 9.7   The function $f(x) = x$ with an alternate to Figure 9.5 extension to the x-axis.

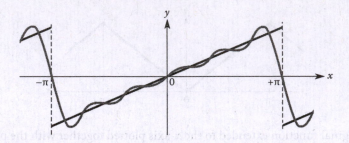

FIGURE 9.8 Original function with alternate extension to the x-axis plotted with the partial sum of the first five terms.

that is,

$$a_{2k} = 0, \quad a_{2k-1} = -\frac{2}{(2k-1)^2\pi}.$$

Similarly

$$b_n = -\frac{\cos n\pi}{n} = (-1)^{n-1}\frac{1}{n}.$$

The resulting expansion is

$$f(x) = \frac{\pi}{4} - \frac{2}{\pi}\cos x + \sin x - \frac{1}{2}\sin 2x - \frac{2}{9\pi}\cos 3x + \frac{1}{3}\sin 3x - \frac{1}{4}\sin 4x + \ldots$$

Example 4

Expand $f(x) = e^{2x}$ as a sine series and a cosine series on the interval $[0,1]$.

Solution. The coefficients can be found with formulas (9.36), (9.37).

For the sine series

$$a_n = 0, \quad b_n = 2\int_0^1 e^{2x}\sin n\pi x\, dx = \frac{2n\pi}{4+n^2\pi^2}[1-(-1)^n e^2],$$

thus

$$f(x) = e^{2x} = 2\sum_{n=1}^{\infty}[1-(-1)^n e^2]\frac{n\pi}{4+n^2\pi^2}\sin n\pi x.$$

For the cosine series

$$a_0 = 2\int_0^1 e^{2x}dx = e^2 - 1$$

$$a_n = 2\int_0^1 e^{2x}\cos n\pi x\, dx = \frac{4}{4+n^2\pi^2}[e^2(-1)^n - 1], \quad b_n = 0,$$

thus

$$f(x) = e^{2x} = \frac{e^2 - 1}{2} + 4 \sum_{n=1}^{\infty} [e^2(-1)^n - 1] \frac{1}{4 + n^2 \pi^2} \cos n\pi x.$$

Notice that both series converge to e^{2x} for $0 < x < 1$ and that at $x = 0$ and $x = 1$ the sine series converges to zero whereas the cosine series converges to 1 for $x = 0$ and to e^2 for $x = 1$.

Example 5

Find the Fourier series for $f(x) = e^{ax}$ on the interval $(-\pi, \pi)$ ($a = $ const, $a \neq 0$).
  Solution. The coefficients can be found with formulas (9.36), (9.37).

$$a_0 = \frac{1}{\pi} \int_{-\pi}^{\pi} e^{ax} dx = \frac{e^{a\pi} - e^{-a\pi}}{a\pi} = 2 \frac{\sinh a\pi}{a\pi},$$

$$a_n = \frac{1}{\pi} \int_{-\pi}^{\pi} e^{ax} \cos nx dx = \frac{1}{\pi} \frac{a\cos nx + n\sin nx}{a^2 + n^2} e^{ax} \Big|_{-\pi}^{\pi} = (-1)^n \frac{1}{\pi} \frac{2a}{a^2 + n^2} \sinh a\pi,$$

$$b_n = \frac{1}{\pi} \int_{-\pi}^{\pi} e^{ax} \sin nx dx = \frac{1}{\pi} \frac{a\sin nx - n\cos nx}{a^2 + n^2} e^{ax} \Big|_{-\pi}^{\pi} = (-1)^{n-1} \frac{1}{\pi} \frac{2n}{a^2 + n^2} \sinh a\pi.$$

Thus, for $-\pi < x < \pi$ we have

$$e^{ax} = \frac{2}{\pi} \left\{ \frac{1}{2a} + \sum_{n=1}^{\infty} \frac{(-1)^n}{a^2 + n^2} [a\cos nx - n\sin nx] \right\} \sinh a\pi. \tag{9.42}$$

Reading Exercise:

Find the series for the same function on the interval $(0, 2\pi)$. The Fourier coefficients will differ from the ones obtained above. Use the program **TrigSeries** to plot graphs of partial sums for both cases.

Example 6

Find the Fourier series for $f(x) = (\pi - x)/2$ on the interval $(0, 2\pi)$.
  Solution. The coefficients are

$$a_0 = \frac{1}{\pi} \int_0^{2\pi} \frac{\pi - x}{2} dx = \frac{1}{2\pi} \left(\pi x - \frac{1}{2} x^2 \right) \Big|_0^{2\pi} = 0,$$

$$a_n = \frac{1}{\pi} \int_0^{2\pi} \frac{\pi - x}{2} \cos nx dx = \frac{1}{2\pi} (\pi - x) \frac{\sin nx}{n} \Big|_0^{2\pi} - \frac{1}{2n\pi} \int_0^{2\pi} \sin nx dx = 0,$$

$$b_n = \frac{1}{\pi} \int_0^{2\pi} \frac{\pi - x}{2} \sin nx dx = -\frac{1}{2\pi} (\pi - x) \frac{\cos nx}{n} \Big|_0^{2\pi} - \frac{1}{2n\pi} \int_0^{2\pi} \cos nx dx = \frac{1}{n}.$$

This contains the interesting result

$$\frac{\pi - x}{2} = \sum_{n=1}^{\infty} \frac{\sin nx}{n} \quad (0 < x < 2\pi). \tag{9.43}$$

This equation is not valid at $x = 0$ and $x = 2\pi$ because the sum of the series equals zero. The equality is also violated beyond $(0, 2\pi)$ as is obvious from the plot for the partial sums that you may obtain with the program **TrigSeries**.

Notice that this series converges more slowly than Example 5; thus we need more terms to obtain the same deviation from the original function. Also this series does not converge uniformly (to understand why, attempt to differentiate it term by term and note what happens).

For $x = \pi/2$ we have another interesting result that was obtained by Leibnitz by other means:

$$\frac{\pi}{4} = 1 - \frac{1}{3} + \frac{1}{5} - \frac{1}{7} + \cdots \tag{9.44}$$

9.8 THE COMPLEX FORM OF THE TRIGONOMETRIC SERIES

For a real function, $f(x)$, with period 2π the Fourier expansion

$$f(x) = \frac{a_0}{2} + \sum_{n=1}^{\infty} \left(a_n \cos nx + b_n \sin nx \right) \tag{9.45}$$

with

$$a_n = \frac{1}{\pi} \int_{-\pi}^{\pi} f(x) \cos nx\, dx, \quad (n = 0, 1, 2, \ldots),$$

$$b_n = \frac{1}{\pi} \int_{-\pi}^{\pi} f(x) \sin nx\, dx, \quad (n = 1, 2, 3, \ldots) \tag{9.46}$$

can be rewritten in complex form. From Euler's formula

$$e^{iax} = \cos ax + i \sin ax \tag{9.47}$$

we have

$$\cos nx = \frac{1}{2}\left(e^{inx} + e^{-inx}\right),$$

$$\sin nx = \frac{1}{2i}\left(e^{inx} - e^{-inx}\right) = \frac{i}{2}\left(e^{-inx} - e^{inx}\right)$$

from which we obtain

$$f(x) = \frac{a_0}{2} + \sum_{n=1}^{\infty}\left[\frac{1}{2}(a_n - b_n i)e^{inx} + \frac{1}{2}(a_n + b_n i)e^{-inx}\right].$$

Using the notations

$$c_0 = \frac{1}{2}a_0, \quad c_n = \frac{1}{2}(a_n - b_n i), \quad c_{-n} = \frac{1}{2}(a_n + b_n i),$$

we have

$$f(x) = \sum_{n=-\infty}^{\infty} c_n e^{inx}. \tag{9.48}$$

With the Fourier equations for a_n and b_n (9.46), it is easy to see that the coefficients c_n can be written as

$$c_n = \frac{1}{2\pi}\int_{-\pi}^{\pi} f(x)e^{-inx}\,dx \quad (n = 0, \pm 1, \pm 2, \ldots). \tag{9.49}$$

It is clear that for functions with period $2l$, the equations (9.48) and (9.49) have the form

$$f(x) = \sum_{n=-\infty}^{\infty} c_n e^{\frac{in\pi x}{l}} \tag{9.50}$$

and

$$c_n = \frac{1}{2l}\int_{-l}^{l} f(x)e^{-\frac{in\pi x}{l}}\,dx \quad (n = 0, \pm 1, \pm 2, \ldots). \tag{9.51}$$

For periodic functions in time t and processes with a period T, the same formulas can be written as

$$f(t) = \sum_{n=-\infty}^{\infty} c_n e^{\frac{2in\pi t}{T}} \tag{9.52}$$

and

$$c_n = \frac{1}{T} \int_{-T/2}^{T/2} f(t) e^{-\frac{2in\pi x}{T}} dt \quad (n = 0, \pm 1, \pm 2, ...). \tag{9.53}$$

Several useful properties of these results can be easily verified:

1. Because $f(x)$ is real, c_n and c_{-n} are complex conjugate and we have $c_{-n} = c_n^*$.

2. If $f(x)$ is even, all c_n are real.

3. If $f(x)$ is odd, $c_0 = 0$ and all c_n are pure imaginary.

If the function, $f(x)$, of the real variable x is complex we have

$$f(x) = f_1(x) + if_2(x),$$

where $f_1(x)$, $f_2(x)$ are real functions, in which case the Fourier series for $f(x)$ is the sum of Fourier series for $f_1(x)$ and $f_2(x)$ where the second series is multiplied by the imaginary number i. Equations (9.48) and (9.49) remain unchanged, but the above three properties of the coefficients are not valid (in particular coefficients c_n and c_{-n} are not complex conjugate). Instead of the above properties in this case we have the following:

1. If $f(x)$ is even, then $c_{-n} = c_n$.

2. If $f(x)$ is odd, then $c_{-n} = -c_n$.

Example 7

Represent the function

$$f(x) = \begin{cases} 0, & -\pi < x \le 0, \\ 1, & 0 < x \le \pi, \end{cases}$$

by a complex Fourier series.

Solution. The coefficients are

$$c_0 = \frac{1}{2\pi} \int_0^\pi dx = \frac{1}{2},$$

and

$$c_n = \frac{1}{2\pi} \int_0^\pi e^{-inx} dx = \frac{1-e^{-in\pi}}{2\pi ni} = \begin{cases} 0, & n = \text{even}, \\ \dfrac{1}{\pi ni}, & n = \text{odd}. \end{cases}$$

Thus

$$f(x) = \frac{1}{2} + \frac{1}{\pi i} \sum_{\substack{n=-\infty \\ n=\text{odd}}}^{+\infty} \frac{1}{n} e^{inx}.$$

Reading Exercise

Using Euler's formula check that from this expression it follows that
Im $f(x) = 0$ (as should be) and

$$\text{Re } f(x) = \frac{1}{2} + \frac{2}{\pi} \sum_{n=1,3,\dots}^{\infty} \frac{\sin nx}{n}.$$

The same result can be obtained if we apply the real form of the Fourier series from the beginning.

Example 8

Find the Fourier series of the function $f(x) = e^{-x}$ on the interval $(-\pi, \pi)$.
   Solution. First use the complex Fourier series with coefficients

$$c_n = \frac{1}{2\pi} \int_{-\pi}^\pi e^{-x} e^{-inx} dx = \frac{1}{2\pi} \int_{-\pi}^\pi e^{-(1+in)x} dx = \frac{e^\pi e^{in\pi} - e^{-\pi} e^{-in\pi}}{2\pi(1+in)}.$$

Then with $e^{\pm in\pi} = \cos n\pi \pm i\sin n\pi = (-1)^n$ we have $c_n = (-1)^n (e^\pi - e^{-\pi}) / [2\pi(1+in)]$; thus

$$e^{-x} = \sum_{n=-\infty}^\infty c_n e^{\frac{in\pi x}{l}} = \frac{e^\pi - e^{-\pi}}{2\pi} \sum_{n=-\infty}^\infty \frac{(-1)^n e^{inx}}{1+in}.$$

In the interval $(-\pi, \pi)$ this series converges to e^{-x} and at points $x = \pm\pi$ its sum is $(e^\pi + e^{-\pi})/2$.

Reading Exercise

Apply Euler's formula and check that this series in real form becomes

$$e^{-x} = \frac{e^\pi - e^{-\pi}}{\pi} \left[\frac{1}{2} + \sum_{n=1}^\infty \frac{(-1)^n}{1+n^2} (\cos nx + n\sin nx) \right].$$

The same result is obtained if we apply the real form of the Fourier series from the beginning.

9.9 FOURIER SERIES FOR FUNCTIONS OF SEVERAL VARIABLES

In this section we extend the previous ideas to generate the Fourier series for functions of two variables, $f(x,y)$, which have period 2π in both the variables x and y. Analogous to the development of equation (9.48) we write a double Fourier series for the function $f(x,y)$ as

$$f(x,y)= \sum_{n,m=-\infty}^{+\infty} \alpha_{nm} e^{i(nx+my)} \tag{9.54}$$

in the domain $(D) = (-\pi \le x \le \pi, -\pi \le y \le \pi)$.

The coefficients α_{nm} can be obtained by multiplying equation (9.54) by $e^{-i(nx+my)}$ and integrating over the domain (D), performing this integration for the series term by term. Because the e^{inx} form a complete set of orthogonal functions on $[-\pi,\pi]$ (and the same for e^{imy}), we obtain

$$\alpha_{nm} = \frac{1}{4\pi^2} \iint\limits_{(D)} f(x,y)e^{-i(nx+my)}\, dx dy \quad (n,m=0,\pm1,\pm2,...). \tag{9.55}$$

The previous two formulas give the Fourier series for $f(x,y)$ in complex form. For the real Fourier series instead of equation (9.54) we have

$$f(x,y)= \sum_{n,m=0}^{+\infty} \big[a_{nm} \cos nx \cos my + b_{nm} \cos nx \sin my$$

$$+ c_{nm} \sin nx \cos my + d_{nm} \sin nx \sin my \big] \tag{9.56}$$

where

$$a_{00} = \frac{1}{4\pi^2} \iint\limits_{(D)} f(x,y)dx\,dy, \quad a_{n0} = \frac{1}{2\pi^2} \iint\limits_{(D)} f(x,y)\cos nx\, dx\,dy,$$

$$a_{0m} = \frac{1}{2\pi^2} \iint\limits_{(D)} f(x,y)\cos my\, dx\,dy, \quad a_{nm} = \frac{1}{\pi^2} \iint\limits_{(D)} f(x,y)\cos nx \cos my\, dx\,dy,$$

$$b_{0m} = \frac{1}{2\pi^2} \iint\limits_{(D)} f(x,y)\sin my\, dx\,dy, \quad b_{nm} = \frac{1}{\pi^2} \iint\limits_{(D)} f(x,y)\cos nx \sin my\, dx\,dy, \tag{9.57}$$

$$c_{n0} = \frac{1}{2\pi^2} \iint\limits_{(D)} f(x,y)\sin nx\, dx\,dy, \quad c_{nm} = \frac{1}{\pi^2} \iint\limits_{(D)} f(x,y)\sin nx \cos my\, dx\,dy,$$

and

$$d_{nm} = \frac{1}{\pi^2} \iint\limits_{(D)} f(x,y)\sin nx \sin my \, dx \, dy \quad \text{for } n,m = 1,2,3,\dots$$

9.10 GENERALIZED FOURIER SERIES

Consider expansions similar to trigonometric Fourier series using a set of orthogonal functions as a basis for the expansion. Recall that two complex functions, $\varphi(x)$ and $\psi(x)$, of a real variable x are said to be orthogonal on the interval $[a,b]$ (which can be an infinite interval) if

$$\int_a^b \varphi(x)\psi^*(x)\,dx = 0, \tag{9.58}$$

where $\psi^*(x)$ is the complex conjugate of $\psi(x)$ (when $\psi(x)$ is real $\psi^* = \psi$).

Let us expand some function $f(x)$ into a set of orthogonal functions $\{\varphi_n(x)\}$:

$$f(x) = c_1\varphi_1(x) + c_2\varphi_2(x) + \cdots + c_n\varphi_n(x) + \cdots = \sum_{n=1}^{\infty} c_n\varphi_n(x). \tag{9.59}$$

Multiplying by $\varphi_n(x)$, integrating, and using the orthogonality condition, we obtain the coefficients

$$c_n = \frac{\int_a^b f(x)\varphi_n^*(x)\,dx}{\int_a^b \varphi_n(x)\varphi_n^*(x)\,dx} = \frac{1}{\lambda_n} \int_a^b f(x)\varphi_n^*(x)\,dx, \tag{9.60}$$

where

$$\lambda_n = \int_a^b |\varphi_n(x)|^2 (x)\,dx$$

are real numbers—squared *norms* of functions $\varphi_n(x)$.

Series (9.59) with coefficients (9.60) is called a *generalized Fourier series*.

If the set $\{\varphi_n(x)\}$ is normalized, $\lambda_n = 1$, and the previous formula becomes

$$c_n = \int_a^b f(x)\varphi_n^*(x)\,dx. \tag{9.61}$$

In the case of trigonometric Fourier series (9.5), the orthogonal functions $\varphi_n(x)$ are

$$1, \cos x, \sin x, \cos 2x, \sin 2x, \ldots, \cos nx, \sin nx, \ldots \tag{9.62}$$

This set is *complete* on the interval $[-\pi, \pi]$. Rather than the standard interval $[-\pi, \pi]$, we may also wish to consider any interval of length 2π, or the interval $[-l, l]$ where instead of the argument x in equation (9.62) we have $n\pi x/l$, and so on.

Other sets of orthogonal functions are a system of sines (9.17) or a system of cosines (9.16) on the interval $[0, \pi]$.

Set of exponential functions

$$\ldots, e^{-i2x},\ e^{-ix},\ 1,\ e^{ix},\ e^{i2x}, \ldots \tag{9.63}$$

which are orthogonal on $[-\pi, \pi]$, we use in the expansion (9.48). Another complex set of functions that we use for expansion (9.50)

$$\ldots, e^{-\frac{i2\pi x}{l}},\ e^{-\frac{i\pi x}{l}},\ 1,\ e^{\frac{i\pi x}{l}},\ e^{\frac{i2\pi x}{l}}, \ldots \tag{9.64}$$

is complete and orthogonal on $[-l, l]$. Notice that for this set $\lambda_n = b - a$ and the functions $\{\varphi_n(x)/\lambda_n\}$ are normalized, that is, have norms equal to one.

Let us multiply equation (9.59) by its complex conjugated, , and integrate over the interval $[a, b]$ (or the entire axis). This gives, due to the orthogonality of the functions $\{\varphi_n(x)\}$,

$$\int_a^b |f|^2(x)\,dx = \sum_{n=1}^\infty |c_n|^2 \int_a^b |\varphi_n(x)|^2\,dx = \sum_{n=1}^\infty |c_n|^2 \lambda_n. \tag{9.65}$$

Equation (9.65) is known as the *completeness equation* or *Parsevale's equality*. If this equation is satisfied, the set of functions $\{\varphi_n(x)\}$ is *complete*. Equation (9.65) is an extension of the Pythagorean theorem to a space with an infinite number of dimensions; the square of the diagonal of an (infinite dimensional) parallelepiped is equal to the sum of the squares of all its sides.

The completeness of set $\{\varphi_n(x)\}$ means that any function $f(x)$ (for which $\int_a^b |f(x)|^2\,dx < \infty$) can be expanded in this set (formula (9.59)) and no other functions except $\{\varphi_n(x)\}$ need to be included.

When the norms of the functions equal unity, that is, $\lambda_n = 1$, equation (9.65) has its simplest form:

$$\int_a^b |f(x)|^2\,dx = \sum_{n=1}^\infty |c_n|^2. \tag{9.66}$$

For real functions $f(x)$ and $\{\varphi_n(x)\}$

$$\int_a^b f^2(x)\,dx = \sum_{n=1}^{\infty} c_n^2. \tag{9.67}$$

Note that from formulas (9.66) and (9.67) it follows that $c_n \to 0$ as $n \to \infty$.

For the trigonometric Fourier series (9.5) on $[-\pi,\pi]$, equation (9.66) becomes

$$\int_{-\pi}^{\pi} f^2(x)\,dx = \frac{1}{2}\pi a_0^2 + \pi \sum_{n=1}^{\infty} \left(a_n^2 + b_n^2\right), \tag{9.68}$$

and for a series on the interval we have

$$\int_{-l}^{l} f^2(x)\,dx = \frac{1}{2}l a_0^2 + l \sum_{n=1}^{\infty} \left(a_n^2 + b_n^2\right). \tag{9.69}$$

9.11 SOLUTION OF DIFFERENTIAL EQUATIONS USING FOURIER SERIES

Fourier series expansions provide the means for solving many types of differential equations. Here we consider periodic solutions of linear differential equation for a function $y(x)$

$$y^{(n)} + p_{n-1}y^{(n-1)} + p_{n-2}y^{(n-2)} + \cdots + p_1 y' + p_0 y = f(x) \tag{9.70}$$

with constant coefficients $p_i = \text{const}$, and $f(x + 2\pi) = f(x)$ a given function with the period 2π (we take this period just to simplify the formulas below).

We may search for the solution to this equation in the form of the Fourier series given by

$$y(x) = \sum_{k=-\infty}^{\infty} y_k e^{ikx} \tag{9.71}$$

with coefficients y_k needed to be found.

The periodic function $f(x)$ can also be represented by its Fourier series,

$$f(x) = \sum_{k=-\infty}^{\infty} f_k e^{ikx} \tag{9.72}$$

where

$$f_k = \frac{1}{2\pi} \int_{-\pi}^{\pi} f(x) e^{-ikx}\,dx.$$

Let substitute equations (9.71) and (9.72) into equation (9.70) to obtain

$$\sum_{k=-\infty}^{\infty} \left[p_n(ik)^n + p_{n-1}(ik)^{n-1} + \cdots + p_0 ik \right] y_k e^{ikx} = \sum_{k=-\infty}^{\infty} f_k e^{ikx}, \quad k = 0, \pm 1, \pm 2, \ldots$$

From here we obtain for the coefficients y_k

$$y_k = \frac{g_k}{a_0(ik)^n + a_1(ik)^{n-1} + \cdots + a_{n-1}ik},$$

thus for $y(x)$ we have the Fourier series

$$y(x) = \sum_{k=-\infty}^{\infty} \frac{f_k}{p_n(ik)^n + p_{n-1}(ik)^{n-1} + \cdots + p_1 ik} e^{ikx}. \tag{9.73}$$

From this we see the solution to the differential equation (9.70) is represented as a sum of harmonics with periods $2\pi/k$. A real part of formula (9.73) gives the solution of equation (9.70) in real form.

Similarly we can obtain the solution using a real Fourier series for $y(x)$,

$$y(x) = \frac{a_0}{2} + \sum_{k=1}^{\infty} a_k \cos kx + b_k \sin kx \tag{9.74}$$

using the expansion for $f(x)$:

$$f(x) = \frac{A_0}{2} + \sum_{k=1}^{\infty} A_k \cos kx + B_k \sin kx. \tag{9.75}$$

Reading Exercise

Following a line of reasoning similar to the one presented above, develop the periodic solution of equation (9.70) when the function $f(x)$ (a) has an arbitrary period $2l$; (b) is defined on $[0,l]$ and extended to the entire axis as an even function; (c) is defined on $[0,l]$ and extended to the entire axis as an odd function.

Example 9

Solve the equation $y'' + 4y' = f(x)$ with $f(x) = \text{sgn}\, x$, $x \in (-\pi, \pi)$.

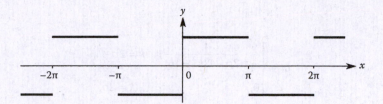

FIGURE 9.9 Periodic function $f(x) = \operatorname{sgn} x$, $x \in (-\pi, \pi)$.

Solution. The Fourier series of the function $\operatorname{sgn} x$ is

$$\operatorname{sgn} x = \frac{4}{\pi} \sum_{k=1}^{\infty} \frac{\sin(2k-1)x}{2k-1}.$$

This formula gives the periodic extension to the x-axis with period $T=2\pi$, of the function given initially on $(-\pi, \pi)$. The graph of this periodic function is shown in Figure 9.9.

Using this result the differential equation becomes

$$y'' + 4y' = \frac{4}{\pi} \sum_{k=1}^{\infty} \frac{\sin(2k-1)x}{2k-1}.$$

Let us search for the solution, $y(x)$, as a real Fourier series (9.74). Substituting this series for $y(x)$ into the expression $y'' + 4y'$, we have

$$y'' + 4y' = \sum_{k=1}^{\infty} \left[-k^2 a_k \cos kx - k^2 b_k \sin kx + 4b_k \cos kx - 4a_k k \sin kx \right]$$

$$= \sum_{k=1}^{\infty} \left[\left(-k^2 a_k + 4k b_k \right) \cos kx + \left(-k^2 b_k + 4k a_k \right) \sin kx \right].$$

Because the expansion is unique we can equate the coefficients of sines and cosines in this equation with the coefficients for the expansion of $\operatorname{sgn} x$, which yields

$$-k^2 a_k + 4k b_k = 0,$$

from which we have $a_k = 4b_k/k$, and

$$-k^2 b_k + 4k a_k = \begin{cases} 0, & k = 2q, \\ \dfrac{4}{\pi(2p-1)}, & k = 2p-1, \ p = 1, 2 \ldots \end{cases}$$

Using the expression $a_k = 4b_k/k$, we have for $k = 2p-1$

$$-k^2 b_k - 16 b_k = \frac{4}{\pi(2p-1)}.$$

Thus

$$
b_k = \begin{cases} -\dfrac{4}{\pi(2p-1)(k^2+16)}, & k = 2p-1, \\[3mm] 0, & k = 2p, \end{cases}
$$

$$
a_{2k-1} = -\frac{16}{\pi(2k-1)^2[(2k-1)^2+16]}.
$$

Using these results the solution to the differential equation becomes

$$
y = -\frac{4}{\pi}\sum_{k=1}^{\infty}\frac{4\cos(2k-1)x+(2k-1)\sin(2k-1)x}{(2k-1)^2\left[(2k-1)^2+16\right]}
$$

$$
= -\frac{4}{\pi}\left[\frac{4\cos x}{17}+\frac{\sin x}{17}+\frac{4\cos 3x}{9\cdot 25}+\frac{\sin 3x}{3\cdot 25}+\cdots\right].
$$

The solution found is the particular solution of a nonhomogeneous equation. A general solution of $y'' + 4y' = 0$, which is $C_1 + C_2 e^{-4x}$, should be added to the result above.

Example 10

As another useful example, let us consider the application of the Fourier series method to the common problem of forced oscillations (physical examples might include mechanical systems such as a body attached to a string, electrical circuits driven by an oscillating voltage, etc.). The linear differential equation

$$
L(y) \equiv y'' + 2\lambda y' + \omega_0^2 y = f(t) \tag{9.76}
$$

describes such forced oscillations, damped due to some form of dissipation (for instance by viscous friction in a mechanical system). In this equation $f(t)$ is a periodic external force with the period T, ω_0 is the frequency of free oscillations obeying the equation $y'' + \omega_0^2 y = 0$, and λ is the damping coefficient. The values $y(0)$ and $y'(0)$ define the initial position and velocity (in case of mechanical oscillations).

The solution to this problem can be expressed as the sum of functions

$$
y(t) = \bar{y}(t) + C_1 y_1(t) + C_2 y_2(t), \tag{9.77}
$$

where $\bar{y}(t)$ is a particular solution to the nonhomogeneous problem and $y_1(t)$, $y_2(t)$ are the fundamental solutions (two linearly independent solutions) for the respective homogeneous equation, $L(y) \equiv 0$. It is easy to check that:

a) If $\lambda < \omega_0$, then $y_1(t) = e^{-\lambda t}\cos\tilde{\omega}t$, $y_2(t) = e^{-\lambda t}\sin\tilde{\omega}t$.

b) If $\lambda = \omega_0$, then $y_1(t) = e^{-\lambda t}$, $y_2(t) = t\cdot e^{-\lambda t}$. $\qquad\qquad$ (9.78)

c) If $\lambda > \omega_0$, then $y_1(t) = e^{k_1 t}$, $y_2(t) = e^{k_2 t}$.

Here

$$\tilde{\omega} = \sqrt{\omega_0^2 - \lambda^2}, \quad k_1 = -\lambda - \sqrt{\lambda^2 - \omega_0^2}, \quad k_2 = -\lambda + \sqrt{\lambda^2 - \omega_0^2} \quad (\text{clearly } k_{1,2} < 0).$$

Now expand the external force, $f(t)$, in a Fourier series with the period T

$$f(t) = \frac{a_0}{2} + \sum_{n=1}^{\infty} [a_n \cos n\omega t + b_k \sin n\omega t], \tag{9.79}$$

where $\omega = 2\pi / T$.

Let us seek the particular solution of the nonhomogeneous equation as Fourier series

$$\bar{y}(t) = \sum_{n=0}^{\infty} [A_n \cos \omega_n t + B_n \sin \omega_n t]. \tag{9.80}$$

Comparing the coefficients of $\cos n\omega t$ in both sides of equation $\bar{y}'' + 2\lambda \bar{y}' + \omega_0^2 \bar{y} = f(t)$ with from the expansions (9.79), find that

$$A_n = \frac{a_n \left(\omega_0^2 - \omega_n^2\right) - b_n \left(2\lambda\omega_n\right)}{\left(\omega_0^2 - \omega_n^2\right)^2 + \left(2\lambda\omega_n\right)^2}, \quad B_n = \frac{b_n \left(\omega_0^2 - \omega_n^2\right) + a_n \left(2\lambda\omega_n\right)}{\left(\omega_0^2 - \omega_n^2\right)^2 + \left(2\lambda\omega_n\right)^2}.$$

These coefficients give the particular solution of equation (9.76) with the right side given by expansion (9.79).

The initial conditions $y(0)$ and $y'(0)$ determine the constants C_1 and C_2. It is obvious that for big t a general solution $C_1 y_1(t) + C_2 y_2(t)$ vanishes because of dissipation and only a particular solution describing the forced oscillations survives. For a harmonic ω_n close to ω_0 we have resonance with the amplitude, bigger than in other harmonics.

9.12 FOURIER TRANSFORMS

A Fourier series is a representation of a function that uses a *discrete* system of orthogonal functions. This idea may be expanded to a *continuous* set of orthogonal functions. The corresponding expansion in this case is referred to as a *Fourier transform*.

Let us start with the complex form of the Fourier series for function $f(x)$ on the interval $[-l,l]$

$$f(x) = \sum_{n=-\infty}^{\infty} c_n e^{\frac{in\pi x}{l}} \tag{9.81}$$

with coefficients

$$c_n = \frac{1}{2l} \int_{-l}^{l} f(x) e^{-\frac{in\pi x}{l}} \, dx \quad (n = 0, \pm 1, \pm 2, \dots). \tag{9.82}$$

In physics terminology, equation (9.81) gives a discrete spectrum of function $f(x)$ with *wave numbers* $k_n = n\pi/l$. Here $c_n e^{ik_n x}$ is a harmonic with complex amplitude c_n defined by equation (9.82) or

$$c_n = \frac{1}{2l} \int_{-l}^{l} f(x) e^{-ik_n x} \, dx. \tag{9.83}$$

Suppose now that l is very large; thus the distance between two neighboring wave numbers, $\Delta k = \pi/l$, is very small. Using the notation

$$\hat{f}(k) = \int\limits_{-\infty}^{\infty} f(x)e^{-ikx}\, dx \tag{9.84}$$

we may write equation (9.83) in the form

$$c_n = \frac{1}{2\pi} \int\limits_{-\infty}^{\infty} f(x)e^{-ik_n x}\, dx \cdot \frac{\pi}{l} = \frac{1}{2\pi} \hat{f}(k_n)\Delta k. \tag{9.85}$$

Using this definition equation (9.81) can be written as

$$f(x) = \sum_{n} c_n e^{ik_n x} = \frac{1}{2\pi} \sum_{n} \hat{f}(k_n) e^{ik_n x} \Delta k \quad (-l < x < l). \tag{9.86}$$

In the limit $l \to \infty$ this becomes the integral

$$f(x) = \frac{1}{2\pi} \int\limits_{-\infty}^{\infty} \hat{f}(k) e^{ikx}\, dk \quad (-\infty < x < \infty). \tag{9.87}$$

In this limit the wave number takes all the values from $-\infty$ to ∞, that is, when $l \to \infty$ the spectrum is continuous. The amplitudes are distributed continuously and for each infinitesimal interval from k to $k + dk$ there is an infinitesimal amplitude

$$dc = \frac{1}{2\pi} \hat{f}(k)dk. \tag{9.88}$$

With this equation as a definition, $\hat{f}(k)$ is called the *spectral density* of $f(x)$.

Equations (9.84) and (9.87) define the *Fourier transform*. Equation (9.84) is called the direct Fourier transform and equation (9.87) is referred to as the inverse Fourier transform. These formulas are valid if the function $f(x)$ is absolutely integrable on $(-\infty,\infty)$:

$$\int\limits_{-\infty}^{\infty} |f(x)|\, dx < \infty. \tag{9.89}$$

It should be noted that there are different ways to deal with the factor $1/2\pi$ in the formulas for direct and inverse transforms. Often, this factor is placed in the direct transform formula while other authors split this factor into two identical factors, $1/\sqrt{2\pi}$, one in each equation. Using the definition given by equations (9.84) and (9.87) has the advantage that the Fourier transform of the Dirac delta function

$$\delta(x) = \frac{1}{2\pi} \int_{-\infty}^{\infty} e^{ikx}\, dk \tag{9.90}$$

equals one, as can be seen by comparing equations (9.87) and (9.90). Here we remind the reader that the most useful property of the delta function is

$$\int_{-\infty}^{\infty} f(x')\delta(x-x')\, dx' = f(x). \tag{9.91}$$

The delta function defined with the coefficient in equation (9.90) obeys the normalization condition

$$\int_{-\infty}^{\infty} \delta(x-x')\, dx' = 1. \tag{9.92}$$

The step or Heaviside function is defined as

$$H(x) = \begin{cases} 1, & x \geq 0, \\ 0, & x < 0, \end{cases}$$

and is related to the delta function by the relation

$$\frac{d}{dx} H(x) = \delta(x). \tag{9.93}$$

Reading Exercise

Prove the following two properties of delta function:

$$\delta(-x) = \delta(x) \quad \text{and} \quad \delta(ax) = \frac{1}{|a|}\delta(x). \tag{9.94}$$

For many practical applications it is useful to present Fourier transform formulas for another pair of physical variables, time and frequency. Using equations (9.84) and (9.87) we may write the direct and inverse transforms as

$$\hat{f}(\omega) = \int_{-\infty}^{\infty} f(t)e^{-i\omega t}\, dt \quad \text{and} \quad f(t) = \frac{1}{2\pi} \int_{-\infty}^{\infty} \hat{f}(\omega)e^{i\omega t}\, d\omega. \tag{9.95}$$

Fourier transform equations are easy to generalize to cases of higher dimensions. For instance, for an application with spatial variables represented as vectors, equations (9.84) and (9.87) become

$$\hat{f}(\vec{k}) = \int_{-\infty}^{\infty} f(\vec{x}) e^{-i\vec{k}\vec{x}}\, d\vec{x} \quad \text{and} \quad f(\vec{x}) = \frac{1}{2\pi} \int_{-\infty}^{\infty} \hat{f}(\vec{k}) e^{i\vec{k}\vec{x}}\, d\vec{k}. \tag{9.96}$$

Next we briefly discuss *Fourier transforms of even or odd functions*. If the function $f(x)$ is even, we have

$$\hat{f}(k) = \int_{-\infty}^{\infty} f(x)\cos kx\, dx - i \int_{-\infty}^{\infty} f(x)\sin kx\, dx = 2\int_{0}^{\infty} f(x)\cos kx\, dx.$$

From here we see that $\hat{f}(k)$ is also even and with equation (9.87) we obtain

$$f(x) = \frac{1}{\pi} \int_{0}^{\infty} \hat{f}(k)\cos kx\, dk. \tag{9.97}$$

These formulas give what is known as the *Fourier cosine transform*. Similarly if $f(x)$ is odd we obtain the *Fourier sine transform*

$$i\hat{f}(k) = 2\int_{0}^{\infty} f(x)\sin kx\, dx, \quad f(x) = \frac{1}{\pi} \int_{0}^{\infty} i\hat{f}(k)\sin kx\, dk. \tag{9.98}$$

In this case usually $i\hat{f}(k)$ (rather than $\hat{f}(k)$) is called the Fourier transform. We leave it to the reader to obtain equations (9.98) as a Reading Exercise.

If the function $f(x)$ is given on the interval $0 < x < \infty$, it can be extended to $-\infty < x < 0$ in either an even or odd way and we may use either sine or cosine transforms.

Example 11

Let $f(x) = 1$ on $-1 < x < 1$ and zero outside this interval. This is an even function; thus with the cosine Fourier transform we have

$$\hat{f}(k) = 2\left(\int_{0}^{1} 1 \cdot \cos kx\, dx + \int_{1}^{\infty} 0 \cdot \cos kx\, dx \right) = \frac{2\sin k}{k}.$$

The inverse transform gives

$$f(x) = 2\int_{0}^{\infty} \frac{\sin k}{\pi k} \cos kx\, dk. \tag{9.99}$$

As in the case for the regular Fourier series, if we substitute some value, x_0, into the formula for the inverse transform, we obtain the value $f(x_0)$ at the point where this function is continuous. The equation

$$[f(x_0 + 0) + f(x_0 - 0)]/2$$

gives the value at a point where it has a finite discontinuity. For instance, substituting $x = 0$ in (9.99) gives

$$1 = 2\int_0^\infty \frac{\sin k}{\pi k}\, dk,$$

from which we obtain the interesting result

$$\int_0^\infty \frac{\sin k}{k}\, dk = \frac{\pi}{2}. \qquad (9.100)$$

Reading Exercises

Here are several important properties of Fourier transform, proofs of which we leave to the reader.

1. Prove that the Fourier transform of $f(-x)$ is equal to $\hat{f}(-k)$.
2. Prove that the Fourier transform of $f'(x)$ is equal to $ik\hat{f}(k)$. *Hint*: The Fourier transform of $f'(x)$ is $\int_{-\infty}^\infty f'(x)e^{-ikx}dx$; differentiate it by parts and take into account that $\int_{-\infty}^\infty |f(x)|\, dx < \infty$.
3. Prove that the Fourier transform of $f(x - x_0)$ (a shift of origin) is equal to $e^{-kx_0}\hat{f}(k)$.
4. Prove that the Fourier transform of $f(\alpha x)$ (where α is constant) is equal to $\hat{f}(k/\alpha)/\alpha$. This property shows that if we stretch the size of an "object" along the x-axis, then the size of Fourier "image" compresses by the same factor. This means that it is not possible to localize a function in both "x- and k-spaces," which is a mathematical expression representing the *uncertainty principle* in quantum mechanics.

Problems

In problems 1 through 5 expand the function $f(x)$ defined on the interval $[a,b]$ into a trigonometric Fourier series using the following:

1. The *general* method of expansion (cosine and sine series)

2. The *even terms only* method of expansion (cosine series)

3. The *odd terms only* method of expansion (sine series)

Using the program **TrigSeries**, plot the graphs of the periodic continuations on different intervals and build the histograms of the squared amplitudes A_n^2. Study the behavior of individual harmonics of the received series and their partial sums with the help option *Choose terms to be included.*

1. Expand the function $f(x) = x$ into a trigonometric Fourier series in the interval $[a,b]$.

2. Expand the function $f(x) = x^2$ into a trigonometric Fourier series in the interval $[0,\pi]$.

3. Expand the function $f(x) = e^{ax}$ into a trigonometric Fourier series in the interval $[0,\pi]$.

4. Expand the function $f(x) = \cos ax$ into a trigonometric Fourier series in the interval $[-\pi,\pi]$.

5. Expand the function $f(x) = \sin ax$ into a trigonometric Fourier series in the interval $[-\pi,\pi]$.

In problems 6 through 11 expand the function $f(x)$ defined on the interval $[a,b]$ into a trigonometric Fourier series using the *general* method of expansion (cosine and sine series).

Using the program **TrigSeries**, plot the graphs the periodic continuations with different number of terms ($N = 10, 20, 30$). Make sure you notice that the *Gibbs's phenomenon* may be observed in the neighborhood of the points of discontinuity.

6. Expand the function

$$f(x) = \frac{\pi - x}{2}$$

defined on the interval $[0,2\pi]$ into a trigonometric Fourier series using the *general* method of expansion.

Draw the graph of the partial sum in the interval $[0,0.5]$, and try to evaluate the height of the first peak of $S_N(x)$ with the help of the screen cursor (use option "*Graph of the Partial Sum*"). In theory $|S_N(x)| \to 1.85...$ as $|x| \to 0$ and $N \to \infty$.

7. Expand the function $f(x) = \pi - 2x$ defined on the interval $[0,\pi]$ into a trigonometric Fourier series using the *general* method of expansion.

8. Expand the function

$$f(x) = \begin{cases} -h, & -\pi \le x \le 0, \\ h, & 0 \le x \le \pi, \end{cases}$$

defined on the interval $[-\pi,\pi]$ into a trigonometric Fourier series using *general* method of expansion.

9. Expand the function

$$f(x) = \begin{cases} 1 & \text{if } 0 \le x \le h, \\ 0 & \text{if } h < x \le \pi \end{cases}$$

defined on the interval $[0,\pi]$ into a trigonometric Fourier series using *general* method of expansion.

10. Expand the function

$$f(x) = \begin{cases} -x, & -\pi \le x \le 0, \\ 0, & 0 \le x \le \pi. \end{cases}$$

defined on the interval $[-\pi,\pi]$ into a trigonometric Fourier series using *general* method of expansion.

11. Expand the function

$$f(x) = \begin{cases} -\pi/2, & -\pi \le x \le 0, \\ \pi/2, & 0 \le x \le \pi. \end{cases}$$

defined on the interval $[-\pi,\pi]$ into a trigonometric Fourier series using *general* method of expansion ($N = 2m$, $m = 5,10,15$).

Draw the graph of the partial sum in the interval $[0,\pi]$, and try to evaluate the coordinates of its extremums with the help of the screen cursor (use option "*Graph of the Partial Sum*"). Ensure that the points of maximums and minimums are

$$x_k = k\frac{\pi}{2m}$$

(maximums for the odd k and minimums for the even k).

In problems 12 through 15 the function $g(x)$ is defined on the interval $[a,b]$. Extend the function $h(x)$ in the interval $[b,c]$, and expand the function

$$f(x) = \begin{cases} h(x), & a \le x < b, \\ g(x), & b \le x \le c \end{cases}$$

into a trigonometric Fourier series using the general method of expansion (cosine and sine series). Define the function $g(x)$ in such a way that it satisfies given conditions at the ends of the interval $[b,c]$.

12. Let the function $h(x)=\cos x$ be given in the interval $[0,\pi]$. Extend the function $h(x)$ in the interval $[\pi,2\pi]$, and expand the function

$$f(x) = \begin{cases} h(x) = \cos x, & 0 \le x < \pi, \\ g(x), & \pi \le x \le 2\pi \end{cases}$$

into a trigonometric Fourier series using the general method of expansion. Define the function $g(x)$ in such a way that it satisfies one of the following sets of conditions at the ends of the interval $[0,2\pi]$:

1) $g(\pi) = h(\pi) = -1$; $g(2\pi) \neq h(0) = 1$;

2) $g(\pi) = h(\pi) = -1$; $g(2\pi) = h(0) = 1$.

13. Let the function $h(x) = e^{-x}$ be given in the interval $[0,1]$. Extend the function $h(x)$ in the interval $[1,2]$, and expand the function

$$f(x) = \begin{cases} h(x) = e^{-x}, & 0 \leq x < 1, \\ g(x), & 1 \leq x \leq 2 \end{cases}$$

into a trigonometric Fourier series using the general method of expansion. Define the function $g(x)$ in such a way that it satisfies one of the following sets of conditions at the ends of the interval $[1,2]$:

1) $g(1) = h(1) = e^{-1}$; $g(2) \neq h(0)$;

2) $g(1) = h(1) = e^{-1}$; $g(2) = h(0) = 1$.

14. Let the function $h(x) = x(1-x)$ be given in the interval $[0,1]$. Extend the function $h(x)$ in the interval $[1,2]$, and expand the function

$$f(x) = \begin{cases} h(x) = x(1-x), & 0 \leq x < 1, \\ g(x), & 1 \leq x \leq 2 \end{cases}$$

into a trigonometric Fourier series using the general method of expansion. Define the function $g(x)$ in such a way that it satisfies one of the following sets of conditions at the ends of the interval $[1,2]$:

1) $g(1) = h(1) = 0$, $g(2) = h(0) = 0$;

2) $g(1) = h(1) = 0, g(2) = h(0) = 0$; $g'(1) = h'(1) = -1$, and $g'(2) = h'(0) = 1$.

15. Let the function $h(x) = x$ be given in the interval $[0,1]$. Extend the function $h(x)$ in the interval $[1,2]$, and expand the function

$$f(x) = \begin{cases} h(x) = x, & 0 \leq x < 1, \\ g(x), & 1 \leq x \leq 2 \end{cases}$$

into a trigonometric Fourier series using the general method of expansion. Define the function $g(x)$ in such a way that it satisfies one of the following sets of conditions at the ends of the interval [1,2]:

1) $g(1) = h(1) = 1, g(2) = h(0) = 0$;

2) $g(1) = h(1) = 1, g(2) = h(0) = 0; \; g'(1) = h'(1) = 1, \quad$ and $\quad g'(2) = h'(0) = 1$.

In problems 16 through 21 using the method of Fourier series expansion find *particular solutions* on interval [a,b] of the following *differential equations* with functions $f(x) = f_k(x)$:

$$f_1(x) = x, \; f_2(x) = x^2, \; f_3(x) = \cos x.$$

Resolve function $f(x)$ in a general Fourier series and find a particular solution of differential equation as a series: to do this resolve the unknown function $y(x)$ in a general Fourier series and express the coefficients of this series through the coefficients of $f(x)$ Fourier series.

16. $y'' + 6y' + 5y = f(x)$, $x \in [-2,2]$

17. $y'' + 4y' + 4y = f(x)$, $x \in [-\pi,\pi]$

18. $4y'' + 4y' + y = f(x)$, $x \in [-1,1]$

19. $y'' + 6y' + 9y = f(x)$, $x \in [-2,2]$

20. $y'' + 2y' + 2y = f(x)$, $x \in [-1,1]$

21. $y'' + 4y' + 5y = f(x)$, $x \in [-2,2]$

II

Partial Differential Equations

Partial differential equations (PDEs) are differential equations for unknown functions of several variables, let's say for a function of a position and the time $u(x,t)$. The unknown function can be dependent on more than two variables—for instance, in the three-dimensional case $u = u(\vec{r},t)$, where $\vec{r} = (x_1, x_2, x_3)$ is the position vector of a particle or object.

A general form of a second order PDE for the function $u(x_1, x_2, ... x_n)$ is

$$F\left(x_1, ... x_n, u, \frac{\partial u}{\partial x_1}, ..., \frac{\partial u}{\partial x_n}, \frac{\partial^2 u}{\partial x_1^2}, \frac{\partial^2 u}{\partial x_1 \partial x_2}, ..., \frac{\partial^2 u}{\partial x_n^2} \right) = 0, \tag{1}$$

where F is a given function.

The PDE itself is not sufficient for the determination of its solution describing a particular natural phenomenon; additional information, which takes the form of the initial and boundary conditions, is required. This gives rise to the existence and necessity of the initial-boundary value problems (or boundary value problems) for PDEs.

Many natural phenomena—such as sound, electromagnetic, and other types of waves, fluid flow, diffusion, propagation of heat, and electrons' motion in an atom—are described by PDEs. Also, countless models and applications in many areas of pure and applied science and technology are largely based on PDEs. Several important PDEs are presented below.

Electromagnetic phenomena, from radio and computer communication to medical imaging, probably have the largest impact on modern life. Maxwell's equations for the electric, $\vec{E}(\vec{r},t)$, and magnetic, $\vec{H}(\vec{r},t)$, fields, which describe electromagnetic phenomena, can be reduced to *D'Alembert's* equation

$$\nabla^2 \vec{A}(\vec{r},t) - \frac{1}{c^2} \frac{\partial^2 \vec{A}(\vec{r},t)}{\partial t^2} = 0 \tag{2}$$

for a vector potential $\vec{A}(\vec{r},t)$. From vector potential the electric and magnetic fields can be obtained. Equation (2) is written for the electromagnetic wave in a vacuum, where

$$\nabla^2 = \frac{\partial^2}{\partial x^2} + \frac{\partial^2}{\partial y^2} + \frac{\partial^2}{\partial z^2}$$

is the three-dimensional Laplace operator and c is speed of light.

Equation (2) is the particular case of the *wave equation*, which describes wave propagation. Similar equations describe other periodic or near-periodic behavior such as sound waves, waves propagating in strings, membranes, and many other types of waves. The simplest one-dimensional wave equation is

$$\frac{\partial u^2(x,t)}{\partial t^2} = \frac{1}{c^2} \frac{\partial^2 u(x,t)}{\partial x^2}. \tag{3}$$

The equation for the electrostatic potential in an empty space, $\varphi(\vec{r})$, is

$$\nabla^2 \varphi(\vec{r}) = 0. \tag{4}$$

This is the *Laplace equation*; it too can be derived from Maxwell's equations. Unlike equations (2) and (3), equation (4) is time-independent (static); thus its complete formulation requires boundary conditions only. If electric charges are present, the equation for $\varphi(\vec{r})$ becomes

$$\nabla^2 \varphi(\vec{r}) = -4\pi\rho(\vec{r}), \tag{5}$$

where $\rho(\vec{r})$ is the charge density distribution. Equation (5) is called the *Poisson equation*.

The one-dimensional *heat equation*

$$\frac{\partial u(x,t)}{\partial t} = a^2 \frac{\partial^2 u(x,t)}{\partial x^2} \tag{6}$$

(in a simple case a^2 is a constant) describes a process with a loss (dissipation) of energy, such as heat flow and diffusion. The solution of equation (6) decays exponentially in time, which distinguishes it from equation (3).

The Helmholtz equation

$$\nabla^2 u(\vec{r}) + k^2 u(\vec{r}) = 0, \tag{7}$$

where k^2 is a constant or some function of coordinates, represents the spatial (time independent) part of either the wave or heat equations. The stationary Schrödinger equation of quantum mechanics also has a form of the Helmholtz equation.

The equations shown above are classified as either *hyperbolic (wave equation, Schrödinger equation)*, *parabolic (heat equation)*, or *elliptic (Laplace, Poisson, Helmholtz equations)*.

This terminology comes from the classification based on conic sections. The nontrivial solutions of a second-order algebraic equation

$$ax^2 + bxy + cy^2 + dx + ey + f = 0 \tag{8}$$

describe hyperbolas, parabolas, or ellipses depending on the sign of the discriminant $b^2 - 4ac$. Similarly, considering for simplicity functions of two variables, $u(x,y)$, the *second-order general linear PDE*

$$au''_{xx}(x,y) + bu''_{xy}(x,y) + cu''_{yy}(x,y) + du'_x(x,y) + eu'_y(x,y) + fu = g(x,y) \tag{9}$$

is called hyperbolic if $b^2 - 4ac > 0$, elliptic if $b^2 - 4ac < 0$, and parabolic if $b^2 - 4ac = 0$. The sign of the discriminant for the equations (2)–(7) confirms their classification as stated previously. This classification is useful and important because, as we mentioned earlier, the solutions of these different types of equations exhibit very different physical behavior, but within a given class the solutions have many similar features. For instance, solutions of all hyperbolic equations describing very different natural phenomena are oscillatory.

CHAPTER 10

One-Dimensional Hyperbolic Equations

10.1 WAVE EQUATION

We start with the problem of small transverse oscillations of a thin, stretched string. We assume that the movement of each point of the string is perpendicular to the x-axis with no displacements or velocities along this axis. Let $u(x,t)$ represent displacements of the points of the string from the equilibrium so that a graph of the function $u(x,t)$ gives the string's amplitude at the location x and time t (u plays the role of the y coordinate; see Figure 10.1). Small oscillations mean small relative to the length of the string displacements, $u(x,t)$, and what is important, we assume the partial derivative $u_x(x,t)$ to be small for all values of x and t (i.e., the slope is small everywhere during the string's motion), and its second power can be neglected: $(u_x)^2 \ll 1$ (u_x has no dimension). With these assumptions, which are justified in many applications, the equations we will derive will be *linear* partial differential equations.

Consider an interval $(x, x + \Delta x)$ in Figure 10.1. Points on the string move perpendicular to the x direction; thus the sum of the x components of the tension forces at points x and $x + \Delta x$ equals zero. Because the tension forces are directed along tangent lines, we have

$$- T(x)\cos\alpha(x) + T(x + \Delta x)\cos\alpha(x + \Delta x) = 0.$$

Clearly, $\tan\alpha = u_x$ and for small oscillations $\cos\alpha = 1/\sqrt{1 + \tan^2\alpha} = 1/\sqrt{1 + u_x^2} \approx 1$; thus

$$T(x) \approx T(x + \Delta x),$$

that is, the value of tension T does not depend on x, and for all x and t it approximately equals its value in the equilibrium state.

287

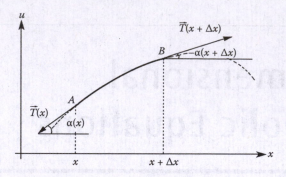

FIGURE 10.1 Small oscillations of a string.

In the situation shown in Figure 10.1, at point x the vertical component of force of tension is

$$T_{vert} = -T \sin \alpha(x).$$

The same expression with a positive sign holds at the point $x + \Delta x$: $T_{vert} = T \sin \alpha(x + \Delta x)$. The signs of T_{vert} at x and $x + \Delta x$ depend on the orientation of the segment of the string and are opposite for the two ends of the segment.

For small oscillations $\sin \alpha = \tan \alpha / \sqrt{1 + \tan^2 \alpha} = u_x / \sqrt{1 + u_x^2} \approx u_x$, so that

$$T_{vert} = -T u_x(x,t) \tag{10.1}$$

and the sum of vertical components of forces of tension at points x and $x + \Delta x$ is

$$T_{vert}^{net} = T \left[\sin \alpha(x + \Delta x) - \sin \alpha(x) \right] = T \left[u_x(x + \Delta x, t) - u_x(x,t) \right].$$

As $\Delta x \to 0$ we arrive to

$$T_{vert}^{net} = T \frac{\partial^2 u}{\partial x^2} dx.$$

From the other side the force T_{vert}^{net} on segment Δx is equal to mass of this segment, $\rho(x)dx$ (where $\rho(x)$ is a linear mass density of the string) times acceleration:

$$T_{vert}^{net} = \rho(x) \frac{\partial^2 u}{\partial t^2} dx.$$

If there is also an additional external force $F(x,t)$ per unit length acting on the string perpendicular to the x-axis (for small oscillations the force should be small with respect to tension, T), we obtain the equation for *forced transverse oscillations of a string*:

$$\rho(x) \frac{\partial^2 u}{\partial t^2} = T \frac{\partial^2 u}{\partial x^2} + F(x,t). \tag{10.2}$$

For the case of a constant mass density, $\rho =$ const, that is, for a uniform string, this equation can be written as

$$\frac{\partial^2 u}{\partial t^2} = a^2 \frac{\partial^2 u}{\partial x^2} + f(x,t), \tag{10.3}$$

where $a = \sqrt{T/\rho} =$ const, $f(x,t) = F(x,t)/\rho$. For instance, if the weight of the string cannot be neglected and the gravity force is directed down perpendicular to the x-axis, we have $f(x,t) = -mg/l\rho = -g$. If there is no external force, $F(x,t) \equiv 0$, we have the equation for free oscillations of a string

$$\frac{\partial^2 u}{\partial t^2} = a^2 \frac{\partial^2 u}{\partial x^2}, \tag{10.4}$$

which is referred to as the *homogeneous wave equation*; equation (10.3) is called the *nonhomogeneous wave equation*. With subscripts for derivatives, equation (10.4) is

$$u_{tt}(x,t) = a^2 u_{xx}(x,t).$$

It may be immediately verified that

$$u(x,t) = f_1(x - at) \quad \text{and} \quad u(x,t) = f_2(x + at)$$

are solutions of equation (10.4), where f_1 and f_2 are arbitrary, twice differentiable functions. Each of these solutions has a simple physical interpretation. In the first case, the displacement $u = f_1$ at point x and time t is the same as that at point $x + a\Delta t$ at time $t + \Delta t$. Thus, the disturbance moves in the direction of increasing x with velocity a. The quantity a is therefore the "velocity of propagation" of the disturbance, or the *wave speed*. In the second case, the displacement $u = f_2$ at point x at time t is found at the point with coordinate $x - a\Delta t$ at a later time $t + \Delta t$. This disturbance therefore travels in the direction of decreasing x with velocity a. The general solution of equation (10.4) can be written as the sum of f_1 and f_2:

$$u(x,t) = f_1(x - at) + f_2(x + at). \tag{10.5}$$

This solution is described with more details in the section devoted to the method of D'Alembert.

Reading Exercise*: Select two arbitrary, twice differentiable functions $f_1(x - at)$ and $f_2(x + at)$ (for example, sine or exponential functions) and show that they are solutions to equation (10.4). Show that their sum is a solution. As an example show that function $\sin x \sin at$ can be presented as a sum of $f_1(x - at)$ and $f_2(x + at)$.

Although a general solution in the form (10.5) is known, physical problems impose certain conditions on the solution $u(x,t)$. For instance, for a finite string with the ends $x = 0$

---

* The book's Reading Exercises can be also considered as homework problems.

and $x = l$ fixed, function (10.5) does not satisfy the conditions $u(0,t) = u(l,t) = 0$ and is not a proper particular solution of equation (10.4).

Equation (10.4) describes the simplest situation with no external forces (including string's weight) and no dissipation. For a string vibrating in an elastic medium when the force on the string from the medium is proportional to string's deflection, $F = -\alpha u$ (that is, Hooke's law; α is a coefficient with the dimension force per length squared, F is a force per unit length) we have the wave equation in the form

$$\rho\,\frac{\partial^2 u}{\partial t^2} = T\,\frac{\partial^2 u}{\partial x^2} - \alpha u. \tag{10.6}$$

When a string oscillates in a medium with force of friction proportional to the speed, the force per unit length, F, is given by $F = -ku_t$, where k is a coefficient of friction. For this case, the equation contains the time derivative $u_t(x,t)$:

$$\frac{\partial^2 u}{\partial t^2} = a^2\,\frac{\partial^2 u}{\partial x^2} - 2\kappa\,\frac{\partial u}{\partial t}, \tag{10.7}$$

where $2\kappa = k/\rho$ (κ has dimension of inversed time).

All equations (10.4), (10.6), and (10.7) are hyperbolic type (as we discussed in the Introduction).

10.2 BOUNDARY AND INITIAL CONDITIONS

For a problem described by a differential equation, additional conditions are needed to find a particular solution describing the behavior of the system. These additional conditions are determined by the nature of the system and should obey the following demands:

i) They should guarantee the *uniqueness* of the solution; that is, there should not be two different functions satisfying the equation and additional conditions;

ii) They should guarantee the *stability* of the solution; that is, any small variation of these additional conditions or the coefficients of the differential equation result in only insignificant variations in the solution. In other words, the solution should depend *continuously* on these additional conditions and the coefficients of the equation.

These additional conditions may be classified as two distinct types: initial conditions and boundary conditions.

Initial conditions characterize the function satisfying the equation at the initial moment $t = 0$. Equations that are second order in time have two initial conditions. For example, in the problem of transverse oscillations of a string, the initial conditions define the string's shape and speed distribution at zero time:

$$u(x,0) = \varphi(x), \quad \text{and} \quad \frac{\partial u}{\partial t}(x,0) = \psi(x), \tag{10.8}$$

where $\varphi(x)$ and $\psi(x)$ are given functions of x.

Boundary conditions characterize the behavior of the function satisfying the equation at the boundary of the physical region of interest for all moments of time t. In most cases, the boundary conditions for partial differential equations give the function $u(x,t)$ and/or $u_x(x,t)$ at the boundary.

Let us consider various boundary conditions for transverse oscillations of a string over the finite interval $0 \leq x \leq l$ from a physical point of view.

1. If the left end of the string, located at $x = 0$, is rigidly fixed, the boundary condition at $x = 0$ is

$$u(0,t) = 0. \tag{10.9}$$

A similar condition exists for the right end, located at $x = l$, if it is fixed. These are called *fixed end* boundary conditions.

2. If the motion of left end of the string is driven with the function $g(t)$, then

$$u(0,t) = g(t), \tag{10.10}$$

in which case we have *driven end* boundary conditions.

3. If the end at $x = 0$ can move and experiences a perpendicular to the x-axis force, $f(t)$ (e.g., a string attached to a ring which is driven up and down on a vertical rod), then from equation (10.1) we have

$$-Tu_x(0,t) = f(t). \tag{10.11}$$

If this boundary condition is applied, instead, to the right end of the string at $x = l$, the left-hand side of this formula will have a positive sign. These are called *forced end* boundary conditions.

4. If the end at $x = 0$ moves freely, but is still attached (e.g., a string attached to a ring that can slide up and down on a vertical rod with no friction), then the slope at the end will be zero. In this case, the last equation gives

$$u_x(0,t) = 0 \tag{10.12}$$

with similar equations for the right end. The conditions in this case are called *free end* boundary conditions.

5. If the left end is attached to a surface that can stretch, we have an *elastic boundary*, in which case a vertical component of elastic force is $T_{vert} = -\alpha u(0,t)$. Together with equation (10.1) this results in the boundary condition

$$u_x(0,t) - hu(0,t) = 0, \quad h = \alpha/T > 0. \tag{10.13}$$

For the right end we have

$$u_x(l,t) + hu(l,t) = 0. \tag{10.14}$$

If the point to which the string is elastically attached is also moving and its deviation from the initial position is described by the function $g(t)$, replacing $u(0,t)$ in equation (10.14) by $u(0,t) - g(t)$ leads to the boundary condition

$$u_x(0,t) - h[u(0,t) - g(t)] = 0. \tag{10.15}$$

It can be seen that for stiff attachment (large h) when even a small shift of the end causes strong tension ($h \to \infty$), the boundary condition (10.15) becomes $u(0,t) = g(t)$. For weak attachment ($h \to 0$, weak tensions), this condition (10.15) becomes the condition for a free end, $u_x(0,t) = 0$.

In general, for one-dimensional problems, the boundary conditions at the ends $x = 0$ and $x = l$ can be summarized in the form

$$\alpha_1 u_x + \beta_1 u\big|_{x=0} = g_1(t), \quad \alpha_2 u_x + \beta_2 u\big|_{x=l} = g_2(t), \tag{10.16}$$

where $g_1(t)$ and $g_2(t)$ are given functions, and α_1, β_1, α_2, β_2 are real constants. As is discussed in Chapter 4, due to physical constraints the normal restrictions on these constants are $\beta_1 / \alpha_1 < 0$ and $\beta_2 / \alpha_2 > 0$.

When functions on the right-hand sides of equation (10.16) are zero (i.e., $g_{1,2}(t) \equiv 0$), the boundary conditions are said to be *homogeneous*. In this case, if $u_1(x,t), u_2(x,t),\ldots, u_n(x,t)$ satisfy these boundary conditions, then any linear combination of these functions

$$C_1 u_1(x,t) + C_2 u_2(x,t) + \cdots + C_n u_n(x,t)$$

(where $C_1,\ldots,C_n$ are constants) also satisfies these conditions. This property will be used frequently in the following discussion.

We may classify the above physical notions of boundary conditions as formally belonging to one of three main types:

1. *Boundary conditions of the first kind (Dirichlet boundary condition)*. For this case we are given $u\big|_{x=A} = g(t)$, where here and below $A = 0$ or l. This describes a given boundary regime; for example, if $g(t) \equiv 0$, we have fixed ends.

2. *Boundary conditions of the second kind (Neumann boundary condition)*. In this case, we are given $u_x\big|_{x=A} = g(t)$, which describes a given force acting at the ends of the string; for example, if $g(t) \equiv 0$, we have free ends;

3. *Boundary conditions of the third kind (mixed boundary condition)*. Here we have $u_x \pm hu\big|_{x=A} = g(t)$ (minus sign for $A = 0$, plus sign for $A = l$); for example, an elastic attachment for the case $h = $ const.

Applying these three conditions alternately to the two ends of the string result in nine types of boundary problems. A list classifying all possible combinations of boundary conditions can be found in Appendix 1.

As mentioned previously, the initial and boundary conditions completely determine the solution of the wave equation. It can be proved that under certain conditions of smoothness of the functions $\varphi(x)$, $\psi(x)$, $g_1(t)$, and $g_2(t)$ defined in equations (10.8) and (10.16), a unique solution always exists; therefore these conditions are, in general, necessary. The following sections investigate many examples of the dependence of the solutions on the boundary conditions.

In some physical situations, either the initial conditions or the boundary conditions may be ignored, leaving only one condition to determine the solution. For instance, suppose the point M_0 is rather distant from the boundary and the boundary conditions are given such that the influence of these conditions at M_0 is exposed after a rather long time interval. In such cases, if we investigate the situation for a relatively short time interval, one can ignore the boundaries and study the *initial value problem* (or *the Cauchy problem*). These solutions for an infinite region can be used for a finite one for times short enough that the boundary conditions have not had time to have an effect. For instance, in the one-dimensional case for short time periods, we may ignore the boundary conditions and search for the solution of the equation

$$u_{tt} = a^2 u_{xx} + f(x,t) \text{ for } -\infty < x < \infty,$$

with the initial conditions

$$\left. \begin{array}{l} u(x,0) = \varphi(x), \\ u_t(x,0) = \psi(x) \end{array} \right\} \text{ for } -\infty < x < \infty.$$

Similarly, if we study a process close enough to one boundary (at one end for the one-dimensional case) and rather far from the other boundary for some characteristic time of that process, the boundary condition at the distant end may be insignificant. For the one-dimensional case, we arrive at a boundary value problem for a semi-infinite region, $0 \le x < \infty$, where in addition to the differential equation we have initial conditions

$$u(0,t) = g(t), \, t > 0,$$

$$\left. \begin{array}{l} u(x,0) = \varphi(x), \\ u_t(x,0) = \psi(x) \end{array} \right\} \, 0 \le x < \infty.$$

On the other hand, it is clear physically that for substantially large time, the behavior of the system mostly depends on the boundary conditions. These are problems *with no initial conditions* (the *steady-state regime*). For the one-dimensional case, they are formulated in the following way: Find the solution of the equation at hand (e.g., the homogeneous wave equation) for the region $0 \le x \le l$ for times $t \gg 0$ with the boundary conditions

$$u(0,t) = g_1(t), \, u(l,t) = g_2(t).$$

Here, as well as in the previous situation, other kinds of boundary conditions described above can be applied.

10.3 LONGITUDINAL VIBRATIONS OF A ROD AND ELECTRICAL OSCILLATIONS

In this section we consider other boundary value problems that are similar to the vibrating string problem. The intent here is to show the similarity in the approach to solving physical problems, which on the surface appear quite different but in fact have a similar mathematical structure.

10.3.1 Rod Oscillations: Equations and Boundary Conditions

Consider a thin elastic rod of cylindrical, rectangular, or other uniform cross section. In this case forces applied along the axis, perpendicular to the (rigid) cross section, will cause changes in the length of the rod. We will assume that the forces act along the rod axis and each cross-sectional area can move only in the x direction. Such assumptions can be justified if the transverse dimensions are substantially smaller compared to the length of the rod and the forces acting on the rod are comparatively weak.

If a force compresses the rod along its axis and is then released, the rod will begin to vibrate along this axis—contrary to transverse string oscillations, such considered rod oscillations are *longitudinal*. Let the ends of the rod be located at the points $x = 0$ and $x = l$ when it is at rest. The location of some cross section at rest will be given by x. Let the function $u(x,t)$ be the longitudinal shift of this cross section from equilibrium at time t. The force of tension is proportional to $u_x(x,t)$, the relative length change at location x, and the cross-sectional area of the rod (Hooke's law); thus $T(x,t) = ESu_x(x,t)$, where E is the elasticity modulus.

Consider the element of the rod between two cross sections S and S_1 (Figure 10.2), with coordinates at rest of x and $x + dx$. The forces of tension at these cross sections, T_x and T_{x+dx}, act along the x-axis. For small deflections, the resultant of these two forces is

$$T_{x+dx} - T_x = ES\frac{\partial u}{\partial x}\bigg|_{x+dx} - ES\frac{\partial u}{\partial x}\bigg|_x \approx ES\frac{\partial^2 u}{\partial x^2}dx.$$

The acceleration of this element is $\partial^2 u / \partial t^2$ in the direction of the resultant force. Together, these two equations give the equation of longitudinal motion of a cross-sectional element as

$$\rho S dx \frac{\partial^2 u}{\partial t^2} = ES\frac{\partial^2 u}{\partial x^2}dx, \tag{10.17}$$

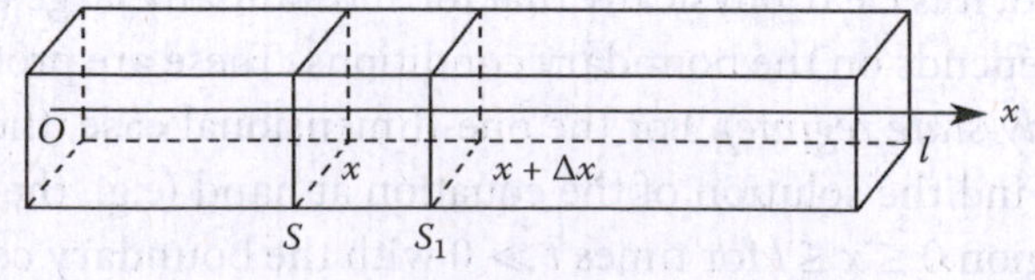

FIGURE 10.2 Arbitrary segment of a rod of length Δx.

where ρ is the linear rod density. Using the notation

$$a = \sqrt{E/\rho}, \tag{10.18}$$

we obtain the differential equation for longitudinal free oscillations of a uniform rod as

$$\frac{\partial^2 u}{\partial t^2} = a^2 \frac{\partial^2 u}{\partial x^2}. \tag{10.19}$$

As discussed earlier, solutions of such hyperbolic equations have a wave character with the speed of wave propagation, a, given by (10.18).

If there is also an external force per unit volume, $F(x,t)$ we obtain instead the equation

$$\rho S dx \frac{\partial^2 u}{\partial t^2} = ES \frac{\partial^2 u}{\partial x^2} dx + F(x,t)S dx,$$

or

$$\frac{\partial^2 u}{\partial t^2} = a^2 \frac{\partial^2 u}{\partial x^2} + \frac{1}{\rho} F(x,t). \tag{10.20}$$

This is the equation for *forced oscillations of a uniform rod*. Note the similarity of equation (10.20) to equation (10.3) for a string under forced oscillations: the two equations are equivalent.

The initial conditions are similar to those for a string

$$u(x,0) = \varphi(x), \quad \frac{\partial u}{\partial t}(x,0) = \psi(x),$$

which are initial deflection and initial speed of points of a rod, respectively.

Now consider boundary conditions for a rod.

1. For a rod with *rigid fixed ends* with the left end located at $x = 0$, the boundary condition is

$$u(0,t) = 0.$$

2. If the left end at $x = 0$ is *driven* by the function $g_1(t)$, then

$$u(0,t) = g_1(t),$$

where $g_1(t)$ is a given function of t.

3. If the end at $x = 0$ can move and experiences a force $f(t)$ along the x-axis, then from equation $T(x,t) = ESu_x(x,t)$ we have

$$-u_x(0,t) = f(t)/ES.$$

If a force $f(t)$ is applied, instead, to the right end at $x = l$, the left-hand side of this formula will have a positive sign. These are *forced end* boundary conditions.

4. If the left end is *free*, the tension at that location is zero, $T(a,t) = 0$. Then, from $T(x,t) = ESu_x(x,t)$, the condition follows that

$$u_x(0,t) = 0.$$

5. If the either end of the rod is attached to an elastic material (a wall that "gives" in the horizontal direction) obeying Hooke's law, $F = -\alpha u$, it is an *elastic end* boundary condition. For the left end of the rod at $x = 0$, similarly to (10.13) we have

$$u_x(0,t) - hu(0,t) = 0.$$

For the right end of the rod, located at $x = l$, similarly to (10.14) we have

$$u_x(l,t) + hu(l,t) = 0,$$

where $h = \alpha/ES > 0$.

10.3.2 Electrical Oscillations in a Circuit

Let us briefly set up the boundary value problem for a current and a voltage in a circuit that contains resistance, capacitance, and inductance R, C, and L as well as the possibility of leakage, G. For simplicity, these quantities are considered uniformly distributed in a wire placed along the x-axis, and defined to be per unit length. The functions $i(x,t)$ and $V(x,t)$ represent current and voltage at a location x along the wire and at time t. Applying Ohm's law for a circuit with nozero self-inductance and using charge conservation, the so-called *telegraph equations* can be obtained (for more details see book [1]):

$$i_{xx}(x,t) = LCi_{tt}(x,t) + (RC + LG)i_t(x,t) + RGi(x,t) \tag{10.21}$$

and

$$V_{xx}(x,t) = LCV_{tt}(x,t) + (RC + LG)V_t(x,t) + RGV(x,t). \tag{10.22}$$

These equations are similar to the equations of string oscillations; therefore they describe electrical oscillations in an *RCL* circuit, which can be considered as a longitudinal wave along a conductor. When $R = 0$ and $G = 0$, the equations have the simplest form. For instance, for current

$$\frac{\partial^2 i}{\partial t^2} = a^2 \frac{\partial^2 i}{\partial x^2},$$

where $a^2 = 1/LC$, and once again we have the wave equation (10.4). If $G = 0$, the equation is similar to equation (10.7) describing oscillations in a medium with the force of resistance proportional to speed:

$$\frac{\partial^2 i}{\partial t^2} = a^2 \frac{\partial^2 i}{\partial x^2} - 2\kappa \frac{\partial i}{\partial t},$$

where $a^2 = 1/LC$, $2\kappa = (R/L + G/C)$.

Consider initial and boundary conditions for a current and voltage. Let initially, at $t = 0$, the current be $\varphi(x)$, and the voltage be $\psi(x)$ along the wire.

The equation for current contains the second-order time derivative; thus we need two initial conditions. One initial condition is the initial current in the wire

$$i(x,0) = \varphi(x). \tag{10.23}$$

The second initial condition for the current is

$$i_t(x,0) = -\frac{1}{L}[\psi'(x) - R\varphi(x)]. \tag{10.24}$$

If we know the voltage at various points in the circuit at some instant of time, we have an initial condition for $V(x,t)$, which is

$$V(x,0) = \psi(x). \tag{10.25}$$

The second initial condition for the voltage is

$$V_t(x,0) = -\frac{1}{C}[\varphi'(x) - G\psi(x)]. \tag{10.26}$$

For the details how to obtain conditions (10.24) and (10.26) see [1].

Let us give two examples of the boundary conditions.

1. One end of the wire of length l is attached to a source of electromotive force (*emf*) $E(t)$, and the other end is grounded. The boundary conditions are

$$V(0,t) = E(t), \quad V(l,t) = 0.$$

2. A sinusoidal voltage with frequency ω is applied to the left end of the wire, and the other end is insulated. The boundary conditions are

$$V(0,t) = V_0 \sin \omega t, \quad i(l,t) = 0.$$

One-dimensional wave equations describe many other phenomena with periodic nature, among others the acoustic longitudinal waves propagating in different materials, and transverse waves in fluids in shallow channels—for details see [1]. For longitudinal waves there is an important physical restriction: the derivative $u_t(x,t)$ must be small in comparison to the speed of wave propagation: $u_t \ll a$.

10.4 TRAVELING WAVES: D'ALEMBERT'S METHOD

In this section, we study waves propagating along infinite $(-\infty < x < \infty)$ interval. For this case, a physically intuitive method to solve the wave equation, D'Alembert's method, avoids using the more mathematically sophisticated Fourier method. Our physical model will be waves on a string; however, as shown in the previous sections, the same wave equation describes many different physical phenomena and the results derived here apply to those cases as well. The material related with semi-infinite intervals, $0 \le x < \infty$ and $-\infty < x \le 0$, is discussed in more detail in book [1].

It is clear that oscillations in the parts of a very long string very distant from its ends do not depend on the behavior of the ends. For oscillations we consider, it should be the case that $(u_x)^2 \ll 1$; also $u(x,t)$ is supposed to be small relative to some characteristic scale (instead of string's length), which always appears in scientific problems. Also, in most physical applications u_t should be small comparing to the speed of propagation of a wave, $u_t \ll a$.

When the ends of the string are not considered, the solution of the equation

$$\frac{\partial^2 u}{\partial t^2} - a^2 \frac{\partial^2 u}{\partial x^2} = 0 \tag{10.27}$$

can be presented as a sum of two twice differentiable functions f_1 and f_2:

$$u = f_1(x - at) + f_2(x + at). \tag{10.28}$$

As we have discussed earlier, the functions f_1 and f_2 represent the waves moving with constant speed, a, to the right (along the x-axis), and to the left, respectively.

The solution in equation (10.28) of equation (10.27) is called *D'Alembert's solution*. In order to describe the solution for free oscillations in a particular physical case we should find the functions f_1 and f_2 for that particular situation. Since we are considering boundaries at infinity, these functions will be determined only by the initial conditions of the string.

The Cauchy problem for the infinite string is defined by the case for which the solutions of equation (10.27) satisfy the conditions

$$u\big|_{t=0} = \varphi(x) \quad \text{and} \quad \frac{\partial u}{\partial t}\bigg|_{t=0} = \psi(x). \tag{10.29}$$

Substituting equation (10.28) into equation (10.29) gives

$$f_1(x) + f_2(x) = \varphi(x),$$

$$-af_1'(x) + af_2'(x) = \psi(x).$$

From the second relation

$$-af_1(x) + af_2(x) = \int_0^x \psi(x)dx + aC,$$

(where C is an arbitrary constant); thus

$$f_1(x) = \frac{1}{2}\left[\varphi(x) - \frac{1}{a}\int_0^x \psi(x)dx - C \right],$$

$$f_2(x) = \frac{1}{2}\left[\varphi(x) + \frac{1}{a}\int_0^x \psi(x)dx + C \right].$$

Therefore

$$u(x,t) = \frac{1}{2}\left[\varphi(x-at) - \frac{1}{a}\int_0^{x-at} \psi(x)dx - C + \varphi(x+at) + \frac{1}{a}\int_0^{x+at} \psi(x)dx + C \right],$$

which finally gives

$$u(x,t) = \frac{\varphi(x-at) + \varphi(x+at)}{2} + \frac{1}{2a}\int_{x-at}^{x+at} \psi(x)dx. \tag{10.30}$$

If the function $\psi(x)$ is differentiable and $\varphi(x)$ twice differentiable, this solution satisfies equation (10.27) and initial conditions (10.29). (The reader may check this as a Reading Exercise.) The method of construction of the solution in equation (10.30) proves its uniqueness. The solutions will be stable solutions if functions $\varphi(x)$ and $\psi(x)$ depend continuously on the variables x and t.

Another way to write the solution of equation (10.27) is to use *the characteristic triangle* shown in Figure 10.3. Suppose we want to find the solution at some point (x_0, t_0). The vertex of this triangle is the point (x_0, t_0) and the two sides are given by the equations $x - at = x_0 - at_0$ and $x + at = x_0 + at_0$. The base of this triangle, that is, the line between the points $P(x_0 - at_0, 0)$ and $Q(x_0 + at_0, 0)$, determines the wave amplitude at point (x_0, t_0); all other points beyond this base do not contribute to the solution at this point, as follows from equation (10.30):

$$u(x_0, t_0) = \frac{\varphi(P) + \varphi(Q)}{2} + \frac{1}{2a}\int_P^Q \psi(x)dx. \tag{10.31}$$

We now discuss, in more detail, two physical situations that are typically encountered: waves created by a displacement and waves created by a pulse.

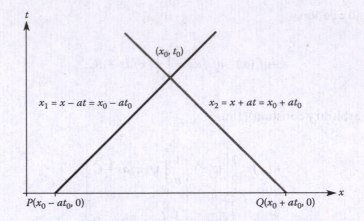

FIGURE 10.3 The characteristic triangle with vertex at point (x_0, t_0).

a) *Waves created by a displacement.* Let the initial speeds of points on the string be zero, but the initial displacements are not zero. The solution given by equation (10.30) in this case is

$$u(x,t) = \frac{1}{2}\left[\varphi(x-at) + \varphi(x+at)\right]. \tag{10.32}$$

The first term, $\varphi(x-at)/2$, is a constant shape disturbance propagating with speed a in the positive x direction, and the term $\varphi(x+at)/2$ is the same shaped disturbance moving in the opposite direction. Suppose that the initial disturbance exists only on a limited interval $-l \le x \le l$. This kind of disturbance is shown schematically in Figure 10.4, where the function $\varphi(x)$ is plotted with a solid line in the upper part of the figure. The dashed line shows the function $\varphi(x)/2$. We can consider $u(x, t = 0)$ given by equation (10.32) as two independent disturbances, $\varphi(x)/2$, each propagating in opposite directions with unchanged amplitude. Initially, at $t = 0$, the profiles

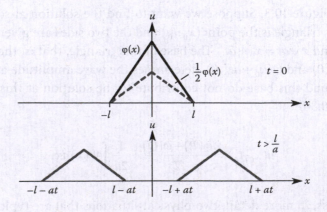

FIGURE 10.4 Propagation of an initial displacement.

of both waves coincide, after which they separate and the distance between them increases. The bottom part of Figure 10.4 shows these waves after some time $t > l/a$. As they pass a given section of the string, this part returns to rest. Only two intervals of the string of length $2l$ each are deflected at any instant t.

Reading Exercise

Let the initial displacement be $\varphi(x) = \exp(-x^2 / b^2)$; b is a constant. Use equation (10.32) to obtain

$$u(x,t) = \frac{1}{2}\left[\exp\left(-(x-at)^2 / b^2\right) + \exp\left(-(x+at)^2 / b^2\right)\right].$$

Notice that the introduction of constant b (as well as constants to keep proper dimension in the examples and problems everywhere in the book) is not necessary if x is dimensionless—it is very common that problems parameters are used in dimensionless form.

b) *Waves created by a pulse.* For this case we let the initial displacement be zero, $\varphi(x) = 0$, but the initial velocities are given as a function of position, $\psi(x)$. In other words, points on the string are given some initial velocity by an external agent. An example of a distribution of velocities is schematically shown in Figure 10.5, where, to simplify the plot, a constant function $\psi(x)$ on $-l \le x \le l$ was chosen. The function $\psi(x)$ is plotted with a solid line in the upper part of Figure 10.5. Consider the integral

$$\Psi(x) = \frac{1}{2a} \int_{-\infty}^{x} \psi(x)\,dx,$$

which is zero on the interval $-\infty < x \le -l$; for $x \ge l$ the function $\Psi(x)$ is constant and equal to

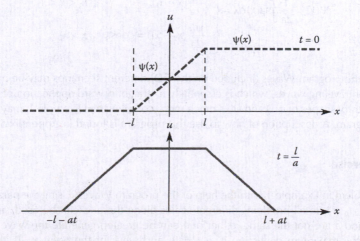

FIGURE 10.5   Propagation of an initial pulse.

$$\frac{1}{2a}\int_{-l}^{l}\psi(x)dx.$$

The graph for $\Psi(x)$ is shown with a dashed line in the Figure 10.5. From equation (10.30) we have

$$u(x,t)=\frac{1}{2a}\int_{x-at}^{x+at}\psi(x)dx=\Psi(x+at)-\Psi(x-at). \tag{10.33}$$

Equation (10.33) means that we again have two waves moving in opposite directions, but now the waves have opposite signs.

At some location x for large enough time, we have $x + at > l$ and the wave $\Psi(x + at)$ becomes a constant and at the same instant of time $x - at < - l$ and the wave $\Psi(x - at)$ is equal to zero. As a result, the perturbation propagates in both directions but, contrary to the case of a wave created by a displacement, none of the elements of the string return to the initial position existing at $t = 0$. The bottom part of Figure 10.5 illustrates this situation for an instant of time $t = l/a$. When $t > l/a$ we have a similar trapezoid for $u(x,t)$ (with the same height as at $t = l/a$), which expands in both directions uniformly with time.

Example 1

Let $\varphi(x) = 0$, $\psi(x) = \psi_0 = $ const for $x_1 < x < x_2$ and $\psi(x) = 0$ outside of this interval.

Solution. In this case we have

$$\Psi(x)=\frac{1}{2a}\int_{x_1}^{x}\psi(x)dx=\begin{cases} 0, & x < x_1, \\ \dfrac{\psi_0}{2a}(x-x_1), & x_1 < x < x_2, \\ \dfrac{\psi_0}{2a}(x_2-x_1)=\text{const}, & x > x_2. \end{cases}$$

With help of the program Waves (included on the CD) different scenarios may be modeled using the method of traveling waves, which is essentially a straightforward application of D'Alembert's formula. Arbitrary shapes of $\varphi(x)$ and $\psi(x)$, much more complex than Example 1, may be examined using the program. A description of how to use this program is found in Appendix 5.

Reading Exercise

Solve the problem in Example 1 with the help of the program Waves. Using the parameter values $a = 0.5$, $l = 1$, $\psi_0 = 1$ (in all problems different scales along the axis can be used, and unless the units for ψ_0 and a are not the same, either one can be greater) generate the wave propagation. Explain why the maximum deflection is $u_{max} = 2$, and why, for the point $x = 0$, this maximum deflection value is reached at time $t = l/a = 2$.

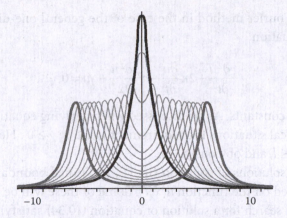

FIGURE 10.6   Graph of the solution to Example 2.

Example 2

An infinite stretched string is excited by the initial deflection

$$\varphi(x) = \frac{A}{x^2 + B}$$

with no initial velocities. Find the vibrations of the string. Write an analytical solution representing the motion of the string, and simulate the vibrations using the program Waves.

Solution. In the given problem, the initial speeds of points on the string are zero ($\psi(x) \equiv 0$), so the solution given by equation (10.32) in this case is

$$u(x,t) = \frac{\varphi(x-at) + \varphi(x+at)}{2} = \frac{A}{2}\left[\frac{1}{(x-at)^2 + B} + \frac{1}{(x+at)^2 + B}\right].$$

Figure 10.6 shows the solution for the case when $a^2 = 0.25$, $A = B = 1$. The bold black line represents the initial deflection and the bold gray line is the string profile at time $t = 12$. The gray lines show the evolution of the string profile within the period of time from $t = 0$ to $t = 12$. The graph was obtained with the program Waves.

Notice that for proper dimension of $\varphi(x)$, the values of A and B should have dimensions of length cubed and squared, correspondingly (if we consider string oscillations). In the problems following this and other chapters, in the most cases we use dimensionless parameters and leave it to the reader to discuss particular physical situations.

10.5 FINITE INTERVALS: THE FOURIER METHOD FOR HOMOGENEOUS EQUATIONS

In this and the following sections, we introduce a powerful Fourier method for solving partial differential equations for *finite intervals*. The Fourier method, or the method of separation of variables, is one of the most widely used methods for analytical solution of boundary value problems in mathematical physics. The method gives a solution in terms of a series of eigenfunctions of the corresponding Sturm-Liouville problem (discussed previously).

Let us apply the Fourier method in the case of the general one-dimensional homogeneous hyperbolic equation

$$\frac{\partial^2 u}{\partial t^2} + 2\kappa \frac{\partial u}{\partial t} - a^2 \frac{\partial^2 u}{\partial x^2} + \gamma u = 0, \tag{10.34}$$

where a, κ, and γ are constants. As we discussed when deriving equations (3.6) and (3.7) in Section 3.1, for physical situations the requirement is $\kappa \geq 0$, $\gamma \geq 0^{\cdot}$. Here we will work on a finite interval, $0 \leq x \leq l$, and obviously $t \geq 0$.

To obtain unique solutions of equation (10.34), additional boundary and initial conditions must be imposed on the function $u(x,t)$. Some of these will be homogeneous, some not. Initially, we will search for a solution of equation (10.34) satisfying the *homogeneous* boundary conditions

$$\alpha_1 u_x + \beta_1 u\big|_{x=0} = 0, \quad \alpha_2 u_x + \beta_2 u\big|_{x=l} = 0, \tag{10.35}$$

with constants α_1, β_1, α_2, and β_2, and initial conditions

$$u(x,t)\big|_{t=0} = \varphi(x), \quad u_t(x,t)\big|_{t=0} = \psi(x), \tag{10.36}$$

where $\varphi(x)$ and $\psi(x)$ are given functions. As we discussed in Chapter 2, normally there are physical restrictions on the signs of the coefficients in equation (10.35) so that we have $\alpha_1 / \beta_1 < 0$ and $\alpha_2 / \beta_2 > 0$. Obviously, only these ratios are significant in equations (10.35), but to formulate a solution of the Sturm-Liouville problem for functions $X(x)$ it is more convenient to keep all the constants α_1, β_1, α_2, and β_2.

We begin by assuming that a nontrivial (nonzero) solution of equation (10.34) can be found that is a product of two functions, one depending only on x, another depending only on t:

$$u(x,t) = X(x)T(t). \tag{10.37}$$

This assumption that the variables x and t can be separated will be justified if it leads to a unique solution of the boundary value problem consisting of equation (10.34) and the conditions in equations (10.35) and (10.36). Substituting equation (10.37) into equation (10.34), we obtain

$$X(x)T''(t) + 2\kappa X(x)T'(t) - a^2 X''(x)T(t) + \gamma X(x)T(t) = 0$$

or, by rearranging terms,

$$\frac{T''(t) + 2\kappa T'(t) + \gamma T(t)}{a^2 T(t)} = \frac{X''(x)}{X(x)}, \tag{10.38}$$

$^{\cdot}$ Typical wave problems do not contain terms like bu_x, but if included in equation (10.34), the substitution $u(x,t) = e^{bx/2a^2} v(x,t)$ leads to the equation for function $v(x,t)$ without the v_x term.

where primes indicate derivatives with t or x The left side of this equation depends only on t, and the right side only on x, which is possible only if each side equals a constant. By using $-\lambda$ for this constant, we obtain

$$\frac{T''(t)+2\kappa T'(t)+\gamma T(t)}{a^2 T(t)} \equiv \frac{X''(x)}{X(x)} = -\lambda$$

(it is seen from this relation that λ has dimension of inversed length squared).

Thus, equation (10.38) gives two ordinary second-order linear homogeneous differential equations:

$$T''(t)+2\kappa T'(t)+(a^2\lambda+\gamma)T(t)=0, \tag{10.39}$$

and

$$X''(x)+\lambda X(x)=0. \tag{10.40}$$

Thus, we see that we have successfully separated the variables, resulting in separate equations for functions $X(x)$ and $T(t)$. These equations share the common parameter λ.

To find λ, we apply the boundary conditions. The homogenous boundary conditions of equation (10.35), imposed on $u(x,y)$, give the homogeneous boundary conditions on the function $X(x)$:

$$\alpha_1 X' + \beta_1 X\big|_{x=0} = 0, \quad \alpha_2 X' + \beta_2 X\big|_{x=l} = 0 \tag{10.41}$$

with restrictions $\alpha_1/\beta_1 < 0$ and $\alpha_2/\beta_2 > 0$.

This result therefore leads to the *Sturm-Liouville boundary value problem*, which may be stated in the present case as the following:

Find values of the parameter λ (eigenvalues) for which nontrivial (not identically equal to zero) solutions to equation (10.40), $X(x)$ (eigenfunctions), satisfying boundary conditions (10.41) exist.

Let us briefly remind the reader of the main properties of eigenvalues and eigenfunctions of the Sturm-Liouville problem given in equations (10.40) and (10.41).

1. There exists an *infinite set* of real nonnegative discrete eigenvalues $\{\lambda_n\}$ and corresponding eigenfunctions $\{X_n(x)\}$. The eigenvalues increase as the number n increases:

$$0 \leq \lambda_1 < \lambda_2 < \lambda_3 < \ldots < \lambda_n < \ldots \quad (\lim \lambda_n = +\infty).$$

2. Eigenfunctions corresponding to different eigenvalues are *linearly independent and orthogonal*:

$$\int_0^l X_i(x)X_j(x)dx = 0, \quad i \neq j. \tag{10.42}$$

3. *The completeness property* states that any function $f(x)$ that is twice differentiable on $(0,l)$ and satisfies the homogeneous boundary conditions in equations (10.41) can be resolved in an absolutely and uniformly converging series with eigenfunctions of the boundary value problem given in equations (10.40) and (10.41):

$$f(x) = \sum_{n=1}^{\infty} f_n X_n(x), \quad f_n = \frac{1}{\|X_n\|^2} \int_0^l f(x) X_n(x) dx. \tag{10.43}$$

Eigenvalues and eigenfunctions of a boundary value problem depend on the type of the boundary conditions: Dirichlet, Neumann, or mixed. The values of the constants α_i and β_i in equation (10.41) determine one of these three possible types. All possible variants are presented in Appendix 1. For all of these variants, the solution for the problem given in equations (10.40) and (10.41) is

$$X(x) = C_1 \cos\sqrt{\lambda} x + C_2 \sin\sqrt{\lambda} x. \tag{10.44}$$

The coefficients C_1 and C_2 are determined from the system of equations (10.41):

$$\begin{cases} C_1 \beta_1 + C_2 \alpha_1 \sqrt{\lambda} = 0, \\ C_1 \left[-\alpha_2 \sqrt{\lambda} \sin\sqrt{\lambda}l + \beta_2 \cos\sqrt{\lambda}l \right] + C_2 \left[\alpha_2 \sqrt{\lambda} \cos\sqrt{\lambda}l + \beta_2 \sin\sqrt{\lambda}l \right] = 0. \end{cases} \tag{10.45}$$

This system of linear homogeneous algebraic equations has nontrivial solution only when its determinant equals zero:

$$(\alpha_1 \alpha_2 \lambda + \beta_1 \beta_2) \tan\sqrt{\lambda}l - \sqrt{\lambda}(\alpha_1 \beta_2 - \beta_1 \alpha_2) = 0. \tag{10.46}$$

It is easy to determine (for instance, by using graphical methods) that equation (10.46) has an infinite number of roots $\{\lambda_n\}$, which conforms to general Sturm-Liouville theory. For each root λ_n, we obtain a nonzero solution of equation (10.45).

It is often convenient to present the solution in a form that allows us to do in a unified way a mixed boundary condition and two other kinds of boundary conditions as well, when some of the constants α_i or β_i may be equal to zero. We will do this in the following way. Using the first expression in equations (10.45), we represent C_1 and C_2 as

$$C_1 = C\alpha_1 \sqrt{\lambda_n}, \quad C_2 = -C\beta_1,$$

where $C \neq 0$ is an arbitrary constant (because the determinant is equal to zero, the same C_1 and C_2 satisfy the second equation of the system of equations (10.45)). For these constraints the choice

$$C = 1/\sqrt{\lambda_n \alpha_1^2 + \beta_1^2}$$

(with positive square root and $\alpha_1^2 + \beta_1^2 \neq 0$) allows us to obtain a simple set of coefficients C_1 and C_2. For Dirichlet boundary conditions, we assign $\alpha_1 = 0, \beta_1 = -1$ (we may also use $\beta_1 = 1$ because the overall sign of $X_n(x)$ is not important) so that $C_1 = 0$ and $C_2 = 1$. For Neumann boundary conditions $\beta_1 = 0, \alpha_1 = 1$, and we have $C_1 = 1, C_2 = 0$.

With this choice of C, the functions $X_n(x)$ are *bounded by the values* ±1. The alternative often used for the coefficients C_1 and C_2 corresponds to the normalizations $\|X_n\|^2 = 1$. Here and elsewhere in the book we will use the first choice because in this case the graphs for $X_n(x)$ are easier to plot.

From the above discussion, the eigenfunctions of the Sturm-Liouville problem given by the equations

$$X'' + \lambda X = 0,$$

$$\alpha_1 X' + \beta_1 X\big|_{x=0} = 0, \quad \alpha_2 X' + \beta_2 X\big|_{x=l} = 0$$

can be written as

$$X_n(x) = \frac{1}{\sqrt{\alpha_1^2 \lambda_n + \beta_1^2}} \left[\alpha_1 \sqrt{\lambda_n} \cos \sqrt{\lambda_n} x - \beta_1 \sin \sqrt{\lambda_n} x \right]. \tag{10.47}$$

The orthogonality property in equation (10.42) can be easily verified by the reader as a Reading Exercise. The square norms of eigenfunctions are

$$\|X_n\|^2 = \int_0^l X_n^2(x)\,dx = \frac{1}{2}\left[l + \frac{(\beta_2\alpha_1 - \beta_1\alpha_2)(\lambda_n\alpha_1\alpha_2 - \beta_1\beta_2)}{(\lambda_n\alpha_1^2 + \beta_1^2)(\lambda_n\alpha_2^2 + \beta_2^2)} \right]. \tag{10.48}$$

The eigenvalues are $\lambda_n = \left(\dfrac{\mu_n}{l}\right)^2$, where μ_n is the nth root of the equation

$$\tan\mu = \frac{(\alpha_1\beta_2 - \alpha_2\beta_1)l\mu}{\mu^2\alpha_1\alpha_2 + l^2\beta_1\beta_2}. \tag{10.49}$$

This equation remains unchanged when the sign of μ changes, which indicates that positive and negative roots are placed symmetrically on the μ-axis. Because the eigenvalues λ_n do not depend on the sign of μ, it is *enough to find only positive roots* μ_n of equation (10.49) since negative roots do not give new values of λ_n. Clearly, μ has no dimension.

It is not difficult to demonstrate (for instance, using a graphical solution of this equation) that $\mu_{n+1} > \mu_n$, again in accordance with Sturm-Liouville theory. The cases when either both boundary conditions are of Dirichlet type, or Neumann type, or when one is Dirichlet type and the other is Neumann type, a graphical solution of equation (10.49) is not needed and we leave to the reader as the Reading Exercises to obtain analytical expressions for $X_n(x)$. The obtained results can be compared with those collected in Appendix 1.

Let us find the roots of equation (10.49) for mixed boundary conditions

$$u_x(0,t) - hu(0,t) = 0, \quad \text{and} \quad u_x(l,t) + hu(l,t) = 0. \tag{10.50}$$

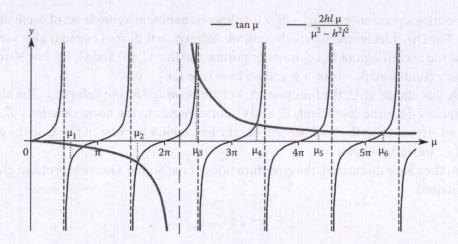

FIGURE 10.7   A graphical solutions of equation (10.51) for $l = 100$ and $h = 0.07$.

Such a particular case with equal values of h in both equations (10.50) is rather common.

The eigenvalues are determined from the equation for $\tan\mu$, equation (10.49), where $\alpha_1 = 1, \beta_1 = -h, \alpha_2 = 1, \beta_2 = h$:

$$\tan\mu = \frac{2hl\mu}{\mu^2 - h^2 l^2}. \tag{10.51}$$

This equation has an infinite number of roots, μ_n. Figure 10.7, obtained with the program Waves, shows curves of the two functions $y = \tan\mu$ and $y = 2hl\mu/(\mu^2 - h^2 l^2)$, plotted on the same set of axes. The values of μ at the intersection points of these curves are the roots of the eigenvalue equation (10.51).

For the values $l = 100$ and $h = 0.07$ the first six positive roots of this equation are

$$\mu_1 = 2.464 \in \left(\frac{\pi}{2}, \frac{3\pi}{2}\right), \mu_2 = 5.036 \in \left(\frac{3\pi}{2}, \frac{5\pi}{2}\right), \mu_3 = 7.752 \in \left(\frac{3\pi}{2}, \frac{5\pi}{2}\right),$$

$$\mu_4 = 10.59 \in \left(\frac{5\pi}{2}, \frac{7\pi}{2}\right), \mu_5 = 13.25 \in \left(\frac{7\pi}{2}, \frac{9\pi}{2}\right), \mu_6 = 16.51 \in \left(\frac{9\pi}{2}, \frac{11\pi}{2}\right).$$

The line $\mu = hl$ (the dashed line in Figure 10.7) is the asymptote of the graph $y = 2hl\mu/(\mu^2 - h^2 l^2)$. In our example $\mu = 7 \in (3\pi/2, 5\pi/2)$, and we notice that this interval includes *two roots*, μ_2 and μ_3, whereas each of the other intervals, $((2k-1)\pi/2, (2k+1)\pi/2)$, contains *one root* of equation (10.51). This is a commonly met situation.

Each eigenvalue

$$\lambda_n = \left(\frac{\mu_n}{l}\right)^2, \quad n = 1, 2, 3, \dots \tag{10.52}$$

corresponds to an eigenfunction

$$X_n(x) = \frac{1}{\sqrt{\lambda_n + h^2}}\left[\sqrt{\lambda_n}\cos\sqrt{\lambda_n}x + h\sin\sqrt{\lambda_n}x\right]. \tag{10.53}$$

The norms of these eigenfunctions are

$$\|X_n\|^2 = \frac{l}{2} + \frac{h}{\lambda_n + h^2}. \tag{10.54}$$

The next step of the solution is to find the function $T(t)$. Equation (10.39) is an ordinary linear second-order homogeneous differential equation. For $\lambda = \lambda_n$

$$T_n''(t) + 2\kappa T_n''(t) + (a^2\lambda_n + \gamma)T_n(t) = 0, \tag{10.55}$$

and a general solution of this equation is

$$T_n(t) = a_n y_n^{(1)}(t) + b_n y_n^{(2)}(t), \tag{10.56}$$

where a_n and b_n are arbitrary constants. Two particular solutions, $y_n^{(1)}(t)$ and $y_n^{(2)}(t)$, are

$$y_n^{(1)}(t) = \begin{cases} e^{-\kappa t}\cos\omega_n t, & \text{if } \kappa^2 < a^2\lambda_n + \gamma, \\ e^{-\kappa t}\cosh\omega_n t, & \text{if } \kappa^2 > a^2\lambda_n + \gamma, \\ e^{-\kappa t}, & \text{if } \kappa^2 = a^2\lambda_n + \gamma, \end{cases} \tag{10.57}$$

and

$$y_n^{(2)}(t) = \begin{cases} e^{-\kappa t}\sin\omega_n t, & \text{if } \kappa^2 < a^2\lambda_n + \gamma, \\ e^{-\kappa t}\sinh\omega_n t, & \text{if } \kappa^2 > a^2\lambda_n + \gamma, \\ te^{-\kappa t}, & \text{if } \kappa^2 = a^2\lambda_n + \gamma, \end{cases} \tag{10.58}$$

where

$$\omega_n = \sqrt{|a^2\lambda_n + \gamma - \kappa^2|} \tag{10.59}$$

(obviously, ω_n has dimension of inversed time).

Reading Exercise

Verify the above expressions.

It is clear that each function

$$u_n(x,t) = T_n(t)X_n(x) = \left[a_n y_n^{(1)}(t) + b_n y_n^{(2)}(t)\right]X_n(x) \tag{10.60}$$

is a solution of equation (10.34) and satisfies boundary conditions in equation (10.35).
The next step is to compose the series

$$u(x,t) = \sum_{n=1}^{\infty} \left[a_n y_n^{(1)}(t) + b_n y_n^{(2)}(t)\right]X_n(x), \tag{10.61}$$

which will allow us to satisfy the initial conditions in equation (10.36). If this series and the series obtained from differentiating this one by x and t converges uniformly, its sum gives a solution of equation (10.34) and satisfies the boundary conditions in equation (10.35). The initial conditions in equation (10.36) give

$$u\big|_{t=0} = \varphi(x) = \sum_{n=1}^{\infty} a_n X_n(x) \tag{10.62}$$

and

$$\frac{\partial u}{\partial t}\bigg|_{t=0} = \psi(x) = \sum_{n=1}^{\infty} \left[\omega_n b_n - \kappa a_n\right] X_n(x) \tag{10.63}$$

(here we have to set $\omega_n \equiv 1$ if $\kappa^2 = a^2 \lambda_n + \gamma$ to consider all three cases for the value of κ^2 with this formula).

These two equations represent the expansions of the functions $\varphi(x)$ and $\psi(x)$ in a series using the eigenfunctions of the boundary value problem given in equations (10.40) and (10.41).

Assuming uniform convergence of the series (10.62) and (10.63), we can find coefficients a_n and b_n. By multiplying both sides of these equations by $X_n(x)$, integrating from 0 to l, and using the orthogonality condition defined in equation (10.42), we obtain

$$a_n = \frac{1}{\|X_n\|^2} \int_0^l \varphi(x) X_n(x)\, dx \tag{10.64}$$

and

$$b_n = \frac{1}{\omega_n}\left[\frac{1}{\|X_n\|^2} \int_0^l \psi(x) X_n(x)\, dx + \kappa a_n \right]. \tag{10.65}$$

Note that to simplify the formulas and the graphs for $X_n(x)$, we take them in dimensionless form; in particular $X_n(x)$ are bounded by the values ± 1. From that it follows that the coefficients a_n and b_n have dimension of function $u(x,t)$, which, for a case of string oscillations, is a length.

The series (10.61) with these coefficients gives the solution of the original problem given in equations (10.34) through (10.36) describing *free oscillations*. This series can be considered as the expansion of the unknown function $u(x,t)$ into a Fourier series using an orthogonal system of functions $\{X_n(x)\}$. The series converges for "reasonably smooth" boundary and initial conditions. In physical problems $\varphi(x)$ is continuous, and $\psi(x)$ is sometimes assumed to be discontinuous, but it is invariable piecewise continuous. Therefore, as we discussed in the chapter on Fourier series, the coefficients a_n and b_n can be determined.

Recall that the success of this method is based on the following details: the functions $\{X_n(x)\}$ are orthogonal to each other and form a complete set (i.e., a basis for an expansion of $u(x,t)$); the functions $\{X_n(x)\}$ satisfy the same boundary conditions as the solutions,

$u(x,t)$; and solutions to linear equations obey the superposition principle (i.e., sums of solutions are also solutions).

Processes with not very big damping, that is, $\kappa^2 < a^2\lambda_n + \gamma$, are periodic (or quasi-periodic) and have a physical interest. For pure periodic motion, $\kappa \approx \gamma \approx 0$; thus $\omega_n = a\sqrt{\lambda_n}$ are the frequencies. The partial solutions $u_n(x,t) = T_n(t)X_n(x)$ are called *normal modes*. The first term, $u_1(x,t)$, called the *first (or fundamental) harmonic,* has time dependence with frequency ω_1 and period $2\pi/\omega_1$. The *second harmonic (or first overtone),* $u_2(x,t)$, oscillates with greater frequency ω_2 (for Dirichlet- or Neumann-type boundary conditions and $\kappa = \gamma = 0$, it is twice ω_1); the *third harmonic* is called *second overtone,* and so on. The points where $X_n(x) = 0$ are not moving are called *nodes* of the harmonic $u_n(x,t)$. Between the nodes the string oscillates up and down. The waves $u_n(x,t)$ are also called *standing waves* because the positions of the nodes are fixed in time. The general solution, $u(x,t)$, is a *superposition of standing waves;* thus any oscillation can be presented in this way.

Clearly, $\omega_n = a\sqrt{\lambda_n}$ increase with the tension and decrease with the length and density: tuning any stringed instrument is based on changing the tension, and the bass strings are longer and heavier. The loudness of a sound is characterized by the energy or amplitude of the oscillations; tone by the period of oscillations; and timbre by the ratio of energies of the main mode and overtones. The presence of high overtones destroys the harmony of a sound, producing dissonance. Low overtones, in contrast, give a sense of completeness to a sound.

We give two examples for homogeneous wave equations with homogeneous boundary conditions.

Example 3

The ends of a uniform string of length *l* are fixed and all external forces including the gravitational force can be neglected. Displace the string from equilibrium by shifting the point $x = x_0$ by distance A at time $t = 0$ and then release it with zero initial speed. Find the displacements $u(x,t)$ of the string for times $t > 0$.

Solution. The boundary value problem is

$$u_{tt} - a^2 u_{xx} = 0,\ 0 < x < l,\ t > 0,$$

$$u(x,0) = \begin{cases} \dfrac{A}{x_0}\,x, & 0 < x \le x_0 \\[2ex] \dfrac{A(l-x)}{l-x_0}, & x_0 < x < l, \end{cases}$$

$$u_t(x,0) = 0,$$

$$u(0,t) = 0,\ u(l,t) = 0.$$

The eigenvalues and eigenfunctions are those of the Dirichlet problem on $0 < x < l$:

$$\lambda_n = \left(\frac{n\pi}{l}\right)^2, \quad X_n(x) = \sin\frac{n\pi x}{l}, \quad \|X_n\|^2 = \frac{l}{2}, \quad n = 1,2,3,\dots.$$

Figure 10.8 shows the first four eigenfunctions obtained with the program Waves for $l = 100$.

Using equations (10.64) and (10.65), we obtain

$$a_n = \frac{2Al^2}{\pi^2 x_0(l-x_0)n^2}\sin\frac{n\pi x_0}{l}, \quad b_n = 0.$$

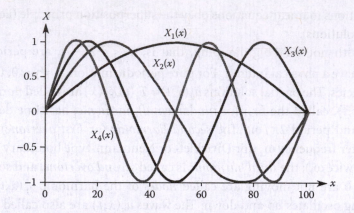

FIGURE 10.8   Eigenfunctions $X_1(x)$ through $X_4(x)$ for Example 3.

Therefore, vibrations on the string are given by the series

$$u(x,t) = \frac{2Al^2}{\pi^2 x_0(l - x_0)} \sum_{n=1}^{\infty} \frac{1}{n^2} \sin\frac{n\pi x_0}{l} \sin\frac{n\pi x}{l} \cos\frac{n\pi a t}{l}.$$

Figure 10.9 shows the spatial-time-dependent solution $u(x,t)$ for Example 3. This solution was obtained with the program Waves for the case when $a^2 = 1$, $l = 100$, $A = 6$, and $x_0 = 25$.

Here and in the following examples the bold black line represents the initial deflection of the string; the bold gray line is the deflection of the string at some final time. The gray lines show the deflection at some time instants. Because there is no dissipation, it is sufficient to run the simulation until the time is equal to the one period of the main harmonic (until $2l/a = 200$ in this example). In this and other examples and problems, the values of parameters are chosen in some arbitrary units.

In the solution $u(x,t)$ the terms for which $\sin(n\pi x_0 / l) = 0$ vanish from the series, that is, the solution does not contain overtones for which $x = x_0$ is a node. For instance, if x_0 is at the middle of the string, the solution does not contain harmonics with even numbers.

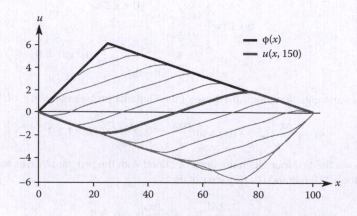

FIGURE 10.9   Solutions, $u(x,t)$, at various times (gray curves) for Example 3.

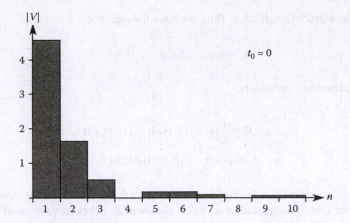

FIGURE 10.10   Bar charts of $|V_n(t)|$ for Example 3.

The formula for $u(x,t)$ can be presented in a more physically intuitive form. Let us denote $\sqrt{\lambda_n}$ as k_n, where $k_n = n\pi/l$ which are called *wave numbers*. The frequencies $\omega_n = n\pi a/l$ give *the frequency spectrum* of the problem. The solution to the problem now can be written as

$$u(x,t) = \sum_{n=1}^{\infty} A_n \cos\omega_n t \cdot \sin k_n x,$$

where the amplitude $A_n = (D/n^2)\sin(n\pi x_0/l)$ with $D \equiv 2Al^2/[\pi^2 x_0(l-x_0)]$.

The first harmonic (or mode), $u_1(x,t)$, is zero for all x when $\omega_1 t = \pi/2$, that is, when $t = l/2a$. It becomes zero again for $t = 3l/2a$, $t = 5l/2a$, and so on. The second harmonic $u_2(x,t)$ is zero for all x for the moments of time $t = l/4a$, $t = 3l/4a$, and so on.

The bar chart in Figure 10.10, obtained with the program Waves, represents $V_n(t) = |A_n|\cos\omega_n t$ (in units of D) for the case $x_0 = l/4 = 25$ and $t = 0$ (i.e., $V_n(t) = |A_n|$). The terms with numbers $n = 4k$ vanish from the series. Note that the amplitudes A_n decrease as $1/n^2$.

Analytical solution of this and other problems allows one to find (as well as simulated with the program allows one to estimate) the derivatives u_x and u_t—if necessary, the last one can be compared with a, and the smallness of u_x can be discussed.

Example 4

A homogeneous rod of length l elastically fixed at the end $x = l$ is stretched by a longitudinal force $F_0 = $ const, applied to the end at $x = 0$. At time $t = 0$ the force F_0 stops acting. Find the longitudinal oscillations of the rod if initial velocities are zero; the resistance of a medium as well as external forces are absent.

Solution. Let us first find initial displacements of locations along the rod, $u|_{t=0} = \varphi(x)$. Because in each cross section the force of tension T is constant and equals F_0 we have $\varphi'(x) = -F_0/ES$, where E is Young's modulus and S is the cross-sectional area of the rod. Negative $\varphi'(x)$ corresponds to decrease of longitudinal shift from the left end to the right. If F_0 is a compressing force, the expression for $\varphi'(x)$ will have the opposite sign. Integrating and using $\varphi'(l) + h\varphi(l) = 0$ (since the rod is elastically fixed at $x = l$) we obtain

$$\varphi(x) = \frac{F_0}{ES}(l + 1/h - x).$$

Here, h is the elasticity coefficient. Thus, we have the equation

$$u_{tt} = a^2 u_{xx}, \quad 0 < x < l, \quad t > 0,$$

with initial and boundary conditions

$$u(x,0) = \frac{F_0}{ES}(l + 1/h - x), \quad u_t(x,0) = 0,$$

$$u_x(0,t) = 0, \quad u_x(l,t) + hu(l,t) = 0.$$

The boundary conditions are the Neumann type at $x = 0$ (the free end) and mixed type at $x = l$ (an elastic connection). The eigenvalues are easily obtained (also, they may be found in Appendix 1) and are

$$\lambda_n = \left(\frac{\mu_n}{l}\right)^2, \quad n = 1,2,3,\ldots,$$

where μ_n is the nth root of the equation $\tan\mu = hl/\mu$.

Figure 10.11, obtained with the program Waves, shows curves of the two functions $\tan\mu$ and hl/μ, plotted on the same set of axes. The roots μ_n are at the intersection points of these curves. Each eigenvalue corresponds to an eigenfunction $X_n(x) = \cos\sqrt{\lambda_n}\,x$ with the norm

$$\| X_n \|^2 = \frac{1}{2}\left(l + \frac{h}{\lambda_n + h^2}\right).$$

Figure 10.12 shows first four eigenfunctions obtained with the program Waves for $l = 100$, $h = 0.1$. Because the initial speeds are zero, all coefficients, b_n, are zero. (See Figure 10.13.) Coefficients a_n are found using equation (10.63):

$$a_n = \frac{1}{\|X_n\|^2}\int_0^l \varphi(x)\cos(\sqrt{\lambda_n}\,x)\,dx = \frac{2F_0}{ES}\frac{1 - \cos\sqrt{\lambda_n}\,l + (\sqrt{\lambda_n}\,/\,h)\sin\sqrt{\lambda_n}\,l}{\lambda_n\left[l + h/(\lambda_n + h^2)\right]}.$$

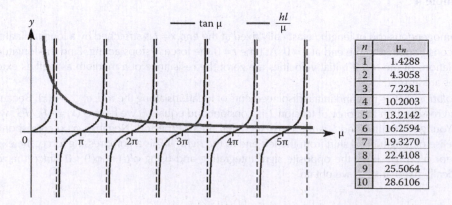

n	μ_n
1	1.4288
2	4.3058
3	7.2281
4	10.2003
5	13.2142
6	16.2594
7	19.3270
8	22.4108
9	25.5064
10	28.6106

FIGURE 10.11   Graphical solution of the eigenvalue equation for $l = 100$, $h = 0.1$.

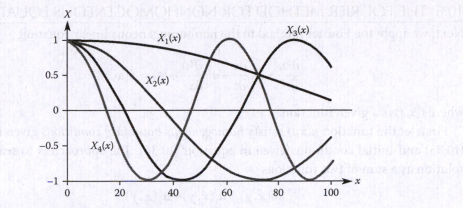

FIGURE 10.12   Eigenfunctions $X_1(x)$ through $X_4(x)$ for Example 4.

Finally, we have the series that describes the rod's oscillations

$$u(x,t) = \frac{2F_0}{ES} \sum_{n=1}^{\infty} \frac{1 - \cos\sqrt{\lambda_n}\, l + (\sqrt{\lambda_n}\,/\,h)\sin\sqrt{\lambda_n}\, l}{\lambda_n\left[l + h\,/\,(\lambda_n + h^2)\right]} \cos\sqrt{\lambda_n}\, at \cdot \cos\sqrt{\lambda_n}\, x.$$

With the program Waves (you can start with the values of parameters $a^2 = 1, l = 100, h = 0.1, E = 1,$ $S = 1, F_0 = 0.05$) it is interesting to see the time behavior of individual harmonics (with Bar Chart and Energy of the System options). Clearly, because of the elastic fixing, the rod's energy is not conserved. Factor $\cos\sqrt{\lambda_n}\, at$, describing the time dependence of the solution, can be written as $\cos\omega_n t$; thus the harmonic's periods are $2\pi\,/\,\omega_n = 2\pi\,/\,\sqrt{\lambda_n}\, a$. It makes sense to simulate the solution for the duration of a period of the main harmonic (which is the biggest period); for periodic motion with dissipation it is interesting to study the system's behavior for longer time.

Notice that the program Waves uses the same formulas as those used to solve a problem analytically. The only numeric calculations the program performs are numeric evaluations of the coefficients of Fourier series and numeric solutions of transcendental equations for eigenvalues in the cases of mixed boundary conditions. Thus, the reader can compare each step of an analytical solution with analogous, obtained with the program. A description of how to use the program is found in Appendix 5.

n	λ_n	$\|x_n\|^2$	a_n	b_n
1	0.0002042	54.9000	4.460800	0
2	0.0018540	54.2180	0.497415	0
3	0.0052246	53.2842	0.179607	0
4	0.0104045	52.4504	0.091622	0
5	0.0174615	51.8207	0.055257	0
6	0.0264367	51.3722	0.036816	0
7	0.0373534	51.0559	0.026218	0
8	0.0502246	50.8302	0.019585	0
9	0.0650576	50.6662	0.015169	0
10	0.0818565	50.5443	0.012085	0

FIGURE 10.13   Eigenvalues, norms $\|X_n\|^2$, coefficients a_n and b_n for Example 4.

10.6 THE FOURIER METHOD FOR NONHOMOGENEOUS EQUATIONS

Next, we apply the Fourier method to the nonhomogeneous linear equation

$$\frac{\partial^2 u}{\partial t^2} + 2\kappa \frac{\partial u}{\partial t} - a^2 \frac{\partial^2 u}{\partial x^2} + \gamma u = f(x,t), \tag{10.66}$$

where $f(x,t)$ is a given function.

First, let the function $u(x,t)$ satisfy *homogeneous boundary conditions* given in equation (10.35) and initial conditions given in equation (10.36). The approach is to search for the solution as a sum of two functions

$$u(x,y) = u_1(x,y) + u_2(x,y), \tag{10.67}$$

where $u_1(x,t)$ is the solution of the homogeneous equation satisfying given boundary and initial conditions

$$\frac{\partial^2 u_1}{\partial t^2} + 2\kappa \frac{\partial u_1}{\partial t} - a^2 \frac{\partial^2 u_1}{\partial x^2} + \gamma u_1 = 0, \tag{10.68}$$

$$u_1(x,t)\big|_{t=0} = \varphi(x), \quad \frac{\partial u_1}{\partial t}(x,t)\bigg|_{t=0} = \psi(x), \tag{10.69}$$

$$\alpha_1 u_{1x} + \beta_1 u_1\big|_{x=0} = 0, \quad \alpha_2 u_{1x} + \beta_2 u_1\big|_{x=l} = 0, \tag{10.70}$$

and $u_2(x,t)$ is the solution of nonhomogeneous equation satisfying the same boundary conditions and *zero initial conditions*

$$\frac{\partial^2 u_2}{\partial t^2} + 2\kappa \frac{\partial u_2}{\partial t} - a^2 \frac{\partial^2 u_2}{\partial x^2} + \gamma u_2 = f(x,t), \tag{10.71}$$

$$u_2(x,t)\big|_{t=0} = 0, \quad \frac{\partial u_2}{\partial t}(x,t)\bigg|_{t=0} = 0, \tag{10.72}$$

$$\alpha_1 u_{2x} + \beta_1 u_2\big|_{x=0} = 0, \quad \alpha_2 u_{2x} + \beta_2 u_2\big|_{x=l} = 0. \tag{10.73}$$

Clearly, the function $u_1(x,t)$ represents *free oscillations* (i.e., the oscillations due to initial perturbation only) and the function $u_2(x,t)$ represents *forced oscillations* (i.e., the oscillations due to an external force when initial disturbances are zero).

The methods for finding the solution in the case of free oscillations, $u_1(x,t)$, have been discussed in the previous section; therefore we turn our attention in this section to finding the forced oscillation solutions, $u_2(x,t)$. Similar to the case of free oscillations, let us search the solution for the function $u_2(x,t)$ as the series

$$u_2(x,t) = \sum_{n=1}^{\infty} T_n(t) X_n(x), \tag{10.74}$$

where $X_n(x)$ are eigenfunctions of the corresponding boundary value problem for $u_1(x,t)$, and $T_n(t)$ are unknown functions of t (they are different than the functions $T_n(t)$ in equation (10.56) obtained for the homogeneous wave equation).

We choose the functions $T_n(t)$ in such a way that the series (10.74) satisfies the nonhomogeneous equation (10.71) and *zero initial conditions* given in equation (10.72). Substituting the series (10.74) into equation (10.71), we obtain (assuming that this series can be differentiated necessary number of times)

$$\sum_{n=1}^{\infty}\left\{\left[T_n'''(t)+2\kappa T_n'(t)+\gamma T_n(t)\right]X_n(x)-a^2 T_n(t)X_n''(x)\right\}=f(x,t).$$

Because $X_n(x)$ are the eigenfunctions of the corresponding homogeneous boundary value problem (10.68)–(10.70), they satisfy equation

$$X_n''(x)+\lambda_n X_n(x)=0.$$

Using this, we obtain

$$\sum_{n=1}^{\infty}\left[T_n''(t)+2\kappa T_n'(t)+(a^2\lambda_n+\gamma)T_n(t)\right]X_n(x)=f(x,t). \tag{10.75}$$

Because of the completeness of the set of functions $\{X_n(x)\}$, we can expand the function $f(x,t)$ on the interval $(0,l)$ into a Fourier series of the functions $X_n(x)$ so that

$$f(x,t)=\sum_{n=1}^{\infty}f_n(t)X_n(x), \tag{10.76}$$

where

$$f_n(t)=\frac{1}{\|X_n\|^2}\int_0^l f(x,t)X_n(x)\,dx. \tag{10.77}$$

Comparing the series in equation (10.75) and that in equation (10.76), we obtain an ordinary second-order linear differential equation with constant coefficients for the functions $T_n(t)$:

$$T_n''(t)+2\kappa T_n'(t)+(a^2\lambda_n+\gamma)T_n(t)=f_n(t), \quad n=1,2,3,\dots \tag{10.78}$$

In order that the function $u_2(x,t)$ represented by the series given in equation (10.74) satisfy initial conditions in equation (10.72), it is clear that functions $T_n(t)$ should satisfy the conditions

$$T_n(0)=0,\, T_n'(0)=0, \quad n=1,2,3,\dots. \tag{10.79}$$

Solutions of linear equation (10.78) can be easily obtained in a standard way (see part of this book devoted to ODE), but here we give the solution of the Cauchy problem given in equations (10.78) and (10.79) for functions $T_n(t)$ as an integral representation which is very convenient for our further purposes:

$$\omega_n T_n(t) = \int_0^t f_n(\tau) Y_n(t-\tau) d\tau. \tag{10.80}$$

Here

$$Y_n(t) = \begin{cases} e^{-\kappa t} \sin \omega_n t, & \text{if } \kappa^2 < a^2 \lambda_n + \gamma, \\ e^{-\kappa t} \sinh \omega_n t, & \text{if } \kappa^2 > a^2 \lambda_n + \gamma, \\ t e^{-\kappa t}, & \text{if } \kappa^2 = a^2 \lambda_n + \gamma, \end{cases} \tag{10.81}$$

where $\omega_n = \sqrt{|a^2 \lambda_n + \gamma - \kappa^2|}$.

To prove that the representation in equation (10.81) yields equation (10.78), we differentiate equation (10.80) by t twice to get

$$\omega_n T_n'(t) = \int_0^t f_n(\tau) \frac{\partial}{\partial t} Y_n(t-\tau) d\tau + Y_n(0) f_n(t),$$

and

$$\omega_n T_n''(t) = \int_0^t f_n(\tau) \frac{\partial^2}{\partial t^2} Y_n(t-\tau) d\tau + Y_n'(0) f_n(t)$$

(here we have to use $\omega_n \equiv 1$ for $\kappa^2 = a^2 \lambda_n + \gamma$).

By using these formulas, we can prove that equation (10.78) and the initial conditions in equation (10.79) are satisfied.

Reading Exercise

Verify this result for $\kappa = 0$ (i.e., for $Y_n(t) = \sin \omega_n t$).

With expression (10.80) for $T_n(t)$ the series (10.74) gives, assuming that it converges uniformly as well as the series obtained by differentiating by x and t up to two times, the solution of the boundary value problem given in equations (10.71)–(10.73).

Combining the results for the functions $u_1(x,t)$ and $u_2(x,t)$, the solution of the forced oscillations problem with initial conditions (10.36) and homogeneous boundary conditions (10.35) is

$$u(x,t)=u_1(x,t)+u_2(x,t)=\sum_{n=1}^{\infty}\left\{\left[a_n y_n^{(1)}(t)+b_n y_n^{(2)}(t)\right]+T_n(t)\right\}X_n(x), \qquad (10.82)$$

where the $T_n(t)$ are defined by equation (10.82) and $y_n^{(1)}(t)$, $y_n^{(2)}(t)$, a_n, and b_n are given by the formulas (10.57), (10.58), (10.64), and (10.65).

Let us consider two examples of nonhomogeneous problems with homogeneous boundary conditions.

Example 5

Consider a homogeneous string of mass density ρ with rigidly fixed ends. Starting at time $t = 0$ a uniformly distributed harmonic force with linear density

$$F(x,t) = F_0 \sin \omega t$$

acts on the string. The initial deflection and speed are zero. Neglecting friction find the resulting oscillations and investigate the resonance behavior.

Solution. The boundary value problem modeling this process is

$$\frac{\partial^2 u}{\partial t^2} - a^2 \frac{\partial^2 u}{\partial x^2} = \frac{F_0}{\rho}\sin \omega t, \quad 0 < x < l, t > 0, \qquad (10.83)$$

$$u(x,0) = 0, \, u_t(x,0) = 0,$$

$$u(0,t) = u(l,t) = 0.$$

The boundary conditions here are homogeneous Dirichlet conditions; thus the eigenvalues and eigenfunctions are

$$\lambda_n = \left(\frac{n\pi}{l}\right)^2, \quad X_n(x) = \sin\frac{n\pi x}{l}, \quad \|X_n\|^2 = \frac{l}{2}, \quad n = 1,2,3,....$$

Because the initial conditions are $\varphi(x) = \psi(x) = 0$, the solution of homogeneous wave equation, $u_1(x,t) = 0$, and the only nonzero contribution is the solution of nonhomogeneous equation, $u_2(x,t)$, which is given by the series

$$u(x,t) = \sum_{n=1}^{\infty} T_n(t)\sin\frac{n\pi x}{l}. \qquad (10.84)$$

We can find $T_n(t)$ by using equation (10.80):

$$T_n(t) = \frac{1}{\omega_n}\int_0^t f_n(\tau)Y_n(t-\tau)d\tau,$$

where $\omega_n = a\sqrt{\lambda_n} = n\pi a/l$ are the natural frequencies and $Y_n(t) = \sin(n\pi a t/l)$.

We then have

$$f_n(t) = \frac{2}{l} \int_0^l f(x,t)\sin\frac{n\pi x}{l}\,dx = \frac{2}{l}\frac{F_0}{\rho}\sin\omega t \int_0^l \sin\frac{n\pi x}{l}\,dx,$$

which gives

$$f_{2n-1}(t) = \frac{4F_0}{(2n-1)\pi\rho}\sin\omega t, \quad f_{2n}(t) = 0.$$

Then, for $T_n(t)$ we have

$$T_{2n-1}(t) = \frac{l}{(2n-1)\pi a}\frac{4F_0}{(2n-1)\pi\rho}\int_0^t \sin\omega\tau\,\sin[\omega_{2n-1}(t-\tau)]\,d\tau$$

$$= \frac{4F_0 l}{a\pi^2(2n-1)^2\rho}\frac{\omega_{2n-1}\sin\omega t - \omega\sin\omega_{2n-1}t}{\omega_{2n-1}^2 - \omega^2},$$

$$T_{2n}(t) = 0.$$

These expressions apply when, for any n, the frequency of the external force, ω, is not equal to any of the natural frequencies of the string. Substituting $T_n(t)$ into equation (10.84) we obtain the solution

$$u(x,t) = \frac{4F_0 l}{\rho a\pi^2}\sum_{n=1}^{\infty}\frac{1}{(2n-1)^2}\frac{\omega_{2n-1}\sin\omega t - \omega\sin\omega_{2n-1}t}{\omega_{2n-1}^2 - \omega^2}\sin\frac{n\pi x}{l}. \tag{10.85}$$

Because ω_n is proportional to n, terms in (10.85) with $\sin\omega t$, which are purely periodic with frequency ω, decrease as $1/n$, whereas terms with $\sin\omega_{2n-1}t$, which represent oscillations (eigenmodes) with different frequencies, decrease as $1/n^2$.

Figure 10.14 shows the spatial-time-dependent solution $u(x,t)$ for Example 5. This solution was obtained with the program Waves for the case when $a^2 = 1$, $l = 100$, $\rho = 1$, $F_0 = 0.025$, $\omega = 0.3$. With these parameters the period of the main harmonic is $2l/a = 200$.

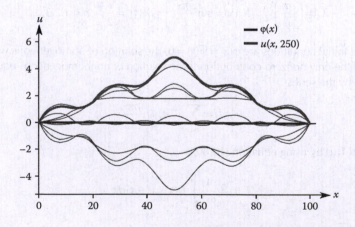

FIGURE 10.14   Graph of solutions, $u(x,t)$, at various times for Example 5.

If for some $n = k$, the frequency of the external force $\omega = \omega_{2k-1}$, we have a case of *resonance*. To proceed, we apply L'Hôpital's rule to this term, taking the derivatives of the numerator and denominator with respect to ω_{2k-1} to yield

$$-\frac{2F_0 l}{\rho a\pi^2 (2k-1)^2}\frac{\omega_{2n-1}t\cos\omega_{2k-1}t - \sin\omega_{2n-1}t}{\omega_{2k-1}}$$

$$=\frac{2F_0 l^2}{\rho a^2\pi^3 (2k-1)^3}\left[\sin\omega_{2n-1}t - \omega_{2n-1}t\cos\omega_{2k-1}t\right].$$

In this case, can rewrite the solution in the form

$$u(x,t) = \frac{4F_0 l}{\rho a\pi^2}\sum_{\substack{n=1 \\ n\neq k}}^{\infty}\frac{1}{(2n-1)^2}\frac{\omega_{2n-1}\sin\omega t - \omega\sin\omega_{2n-1}t}{\omega_{2n-1}^2 - \omega^2}\sin\frac{n\pi x}{l}$$

$$+\frac{2F_0 l^2}{\rho a^2\pi^3 (2k-1)^3}\left[\sin\omega t - \omega t\cos\omega t\right]\sin\frac{(2k-1)\pi x}{l}.$$

(10.86)

With the program Waves you can see the contribution of individual harmonics in the solution (10.85). For example, consider the case of resonance in the fifth harmonic (mode) taking $\omega = \omega_5 = 5\pi/100$. At small times few odd harmonics contribute (you can see it with the option Bar Charts), but when time increases, the fifth (resonant) harmonic dominates.

Intuitively, it is clear that under the action of a periodic external force the motion eventually becomes periodic with the frequency coinciding with force's frequency. The case in which the eigenmodes are practically gone and the motion is governed by external influence only is referred to as steady-state motion. This also means that the role of initial conditions becomes negligible after some time interval that depends on the parameters of the problem.

Let us obtain a periodic solution for Example 5. Contrary to the solution obtained above in the form of an infinite series, a steady-state solution is described by a finite formula, which is more convenient for the analysis of the properties of the solution. Let us write the equation as

$$\frac{\partial^2 u}{\partial t^2} - a^2\frac{\partial^2 u}{\partial x^2} = A\sin\omega t$$

(10.87)

and seek a steady-state solution in the form

$$u(x,t) = X(x)\sin\omega t.$$

(10.88)

For $X(x)$ we obtain an ordinary linear differential equation

$$X''(x) + \left(\frac{\omega}{a}\right)^2 X(x) = -\frac{A}{a^2}$$

(10.89)

which has a general solution

$$X(x) = c_1 \cos kx + c_2 \sin kx - \frac{A}{\omega^2}, \qquad (10.90)$$

where $k = \omega/a$. The boundary conditions $X(0) = X(l) = 0$ give $c_1 = A/\omega^2$ and $c_2 = \frac{A}{\omega^2} \frac{1 - \cos kl}{\sin kl}$. Then, using standard trigonometric formulas for double angles we obtain

$$u(x,t) = \frac{2A}{\omega^2} \frac{\sin(kx/2)\sin[k(l-x)/2]}{\cos(kl/2)} \sin \omega t. \qquad (10.91)$$

Reading Exercise

Should this solution give essentially the same results as the solution (10.85) obtained with the Fourier method? To verify, expand the solution (10.91) into a Fourier $\sin(n\pi x/l)$ series. Using the program Waves, play with the parameters to see when you can neglect the terms with $\omega \sin \omega_{2n-1} t$ in the solution (10.85).

Example 6

The upper end of an elastic homogeneous heavy rod is rigidly attached to the ceiling of a free-falling elevator. When the elevator reaches the speed v_0 it stops instantly. Set up the boundary value problem for vibrations of the rod.

Solution. The mathematical model for this problem is

$$u_{tt} - a^2 u_{xx} = -g, \quad 0 < x < l, \quad t > 0,$$
$$u(x,0) = 0, \quad u_t(x,0) = v_0,$$
$$u(0,t) = 0, \quad u_x(l,t) = 0,$$

where g is the acceleration of gravity.

The boundary conditions are Dirichlet type at $x = 0$ (the secured end) and Neumann type at $x = l$ (the free end). From our previous work, we have eigenvalues and eigenfunctions for the problem given by

$$\lambda_n = \left[\frac{(2n-1)\pi}{2l}\right]^2, \quad X_n(x) = \sin\frac{(2n-1)\pi x}{2l}, \quad n = 1,2,3,....$$

From equations (10.64), (10.65), (10.77), and (10.80), we have

$$a_n = 0, \quad b_n = \frac{8v_0 l}{a\pi^2(2n-1)^2},$$

$$f_n(t) = -\frac{4g}{\pi(2n-1)}, \quad T_n(t) = -\frac{16gl^2}{a^2\pi^3(2n-1)^3}\left[1 - \cos\frac{(2n-1)a\pi t}{2l}\right]$$

Thus, a general solution of this boundary value problem is

$$u(x,t) = \frac{8l}{a\pi^2} \sum_{n=1}^{\infty} \frac{1}{(2n-1)^2}$$

$$\times \left\{ \frac{-2gl}{a\pi(2n-1)} \left[1 - \cos\frac{(2n-1)a\pi t}{2l} \right] + v_0\sin\frac{(2n-1)\pi a t}{2l} \right\} \sin\frac{(2n-1)\pi x}{2l}.$$

10.7 EQUATIONS WITH NONHOMOGENEOUS BOUNDARY CONDITIONS

With the preceding development, we are prepared to study a general boundary value problem for nonhomogeneous equations with nonhomogeneous boundary conditions defined by

$$\frac{\partial^2 u}{\partial t^2} + 2\kappa\frac{\partial u}{\partial t} - a^2\frac{\partial^2 u}{\partial x^2} + \gamma u = f(x,t), \tag{10.92}$$

$$u(x,t)\big|_{t=0} = \varphi(x), \quad u_t(x,t)\big|_{t=0} = \psi(x), \tag{10.93}$$

$$\alpha_1 u_x + \beta_1 u\big|_{x=0} = g_1(t), \quad \alpha_2 u_x + \beta_2 u\big|_{x=l} = g_2(t). \tag{10.94}$$

The Fourier method cannot be applied to this problem directly because the boundary conditions are nonhomogeneous. However, we can reduce this problem to the previously investigated case with homogeneous boundary conditions.

To proceed, let us search for the solution as a sum of two functions

$$u(x,t) = v(x,t) + w(x,t), \tag{10.95}$$

where $v(x,t)$ is a new, unknown function and the function $w(x,t)$ is chosen in a way that satisfies the given nonhomogeneous boundary conditions

$$\alpha_1 w_x + \beta_1 w\big|_{x=0} = g_1(t), \quad \alpha_2 w_x + \beta_2 w\big|_{x=l} = g_2(t). \tag{10.96}$$

The function $w(x,t)$ should also have the necessary number of continuous derivatives in x and t.

For the function $v(x,t)$, we obtain the boundary value problem with *homogeneous boundary conditions*

$$\frac{\partial^2 v}{\partial t^2} + 2\kappa\frac{\partial v}{\partial t} - a^2\frac{\partial^2 v}{\partial x^2} + \gamma v = f^*(x,t), \tag{10.97}$$

$$v(x,t)\big|_{t=0} = \varphi^*(x), \quad v_t(x,t)\big|_{t=0} = \psi^*(x), \tag{10.98}$$

$$\alpha_1 v_x + \beta_1 v\big|_{x=0} = 0, \quad \alpha_2 v_x + \beta_2 v\big|_{x=l} = 0, \tag{10.99}$$

where

$$f^*(x,t) = f(x,t) - \frac{\partial^2 w}{\partial t^2} - 2\kappa \frac{\partial w}{\partial t} + a^2 \frac{\partial^2 w}{\partial x^2}, \tag{10.100}$$

$$\varphi^*(x) = \varphi(x) - w(x,0), \tag{10.101}$$

$$\psi^*(x) = \psi(x) - w_t(x,0). \tag{10.102}$$

The solution to this problem has been described above.

Reading Exercise

Verify equations (10.97)–(10.102).

For the auxiliary function, $w(x,t)$, a number of choices are possible. One criterion is to simplify the form of the equation for the function, $v(x,t)$. To this end, we search for $w(x,t)$ in with the form

$$w(x,t) = P_1(x)g_1(t) + P_2(x)g_2(t), \tag{10.103}$$

where $P_1(x)$ and $P_2(x)$ are polynomials; we will show that in some cases we need polynomials of the first order, and in some, of the second order. Coefficients of these polynomials should be chosen in such a way that the function $w(x,t)$ satisfies the boundary conditions (10.96); therefore the function $v(x,t)$ satisfies homogeneous boundary conditions (10.99).

First, consider the situation when parameters β_1 and β_2 are not zero simultaneously.

Let us take $P_{1,2}(x)$ as polynomials of the first order and search for the function $w(x,t)$ in the form

$$w(x,t) = (\gamma_1 + \delta_1 x)g_1(t) + (\gamma_2 + \delta_2 x)g_2(t). \tag{10.104}$$

Substituting this into boundary conditions (10.96), we obtain two equations that may be written as

$$\begin{cases} \alpha_1[\delta_1 g_1(t) + \delta_2 g_2(t)] + \beta_1[\gamma_1 g_1(t) + \gamma_2 g_2(t)] = g_1(t), \\ \alpha_2[\delta_1 g_1(t) + \delta_2 g_2(t)] + \beta_2[(\gamma_1 + \delta_1 l)g_1(t) + (\gamma_2 + \delta_2 l)g_2(t)] = g_2(t), \end{cases}$$

or as

$$\begin{cases} (\alpha_1 \delta_1 + \beta_1 \gamma_1 - 1)g_1(t) + (\alpha_1 \delta_2 + \beta_1 \gamma_2)g_2(t) = 0, \\ (\alpha_2 \delta_1 + \beta_2 \gamma_1 + \beta_2 \delta_1 l)g_1(t) + (\alpha_2 \delta_2 + \beta_2 \gamma_2 + \beta_2 \delta_2 l - 1)g_2(t) = 0. \end{cases}$$

To be true, the coefficients of $g_1(t)$ and $g_2(t)$ should be zero, in which case the following system of equations is valid for arbitrary t:

$$\begin{cases} \alpha_1\delta_1 + \beta_1\gamma_1 - 1 = 0, \\ \alpha_1\delta_2 + \beta_1\gamma_2 = 0, \\ \alpha_2\delta_1 + \beta_2\gamma_1 + \beta_2\delta_1 l = 0, \\ \alpha_2\delta_2 + \beta_2\gamma_2 + \beta_2\delta_2 l - 1 = 0. \end{cases} \tag{10.105}$$

From this system of equations we obtain coefficients γ_1, δ_1, γ_2, and δ_2 as

$$\gamma_1 = \frac{\alpha_2 + \beta_2 l}{\beta_1\beta_2 l + \beta_1\alpha_2 - \beta_2\alpha_1}, \qquad \delta_1 = \frac{-\beta_2}{\beta_1\beta_2 l + \beta_1\alpha_2 - \beta_2\alpha_1}, \tag{10.106}$$

$$\gamma_2 = \frac{-\alpha_1}{\beta_1\beta_2 l + \beta_1\alpha_2 - \beta_2\alpha_1}, \qquad \delta_2 = \frac{\beta_1}{\beta_1\beta_2 l + \beta_1\alpha_2 - \beta_2\alpha_1}. \tag{10.107}$$

Therefore, the choice of function $w(x,t)$ in the form (10.104) is consistent with the boundary conditions (10.106).

Reading Exercise

Obtain the results (10.106) and (10.107).

Now, consider the case when $\beta_1 = \beta_2 = 0$, that is, $u_x\big|_{x=0} = g_1(t)$, $u_x\big|_{x=l} = g_2(t)$.

In this situation, the system of equations (10.105) is inconsistent, which is why the polynomials $P_{1,2}(x)$ in the expression for $w(x,t)$ must be of higher order. Let us take the second-order polynomials:

$$P_1(x) = \gamma_1 + \delta_1 x + \xi_1 x^2, \qquad P_2(x) = \gamma_2 + \delta_2 x + \xi_2 x^2. \tag{10.108}$$

Equations (10.96) yield a system of simultaneous equations that, to be valid for arbitrary t, leads to $\gamma_1 = \gamma_2 = \delta_2 = 0$ and the expression for $w(x,t)$ given by

$$w(x,t) = \left[x - \frac{x^2}{2l} \right] g_1(t) + \frac{x^2}{2l} g_2(t). \tag{10.109}$$

Reading Exercise

Check that when $\beta_1 = \beta_2 = 0$ such a choice of $w(x,t)$ satisfies the boundary conditions (10.96).

Combining all described types of boundary conditions, we obtain nine different auxiliary functions, which are given in Appendix 2.

The following two examples are applications of the wave equation with nonhomogeneous boundary conditions.

Example 7

The left end of a string is moving according to

$$g(t) = A\sin\omega t,$$

and the right end, $x = l$, is secured. Initially, the string is at rest. Describe oscillations when there are no external forces and the resistance of the surrounding medium is zero. Find the solution in the case of resonance.

Solution. The boundary value problem is

$$u_{tt} = a^2 u_{xx}, \quad 0 < x < l, \quad t > 0,$$

$$u(x,0) = 0, \quad u_t(x,0) = 0,$$

$$u(0,t) = A\sin\omega t, \quad u(l,t) = 0.$$

Let us search for the solution as the sum

$$u(x,t) = v(x,t) + w(x,t),$$

where we use an auxiliary function

$$w(x,t) = \left(1 - \frac{x}{l}\right) A\sin\omega t,$$

which satisfies the same as $u(x,t)$, boundary conditions, that is, $w(0,t) = A\sin\omega t$ and $w(l,t) = 0$. This obvious choice of $w(x,t)$ can also be obtained from equations (10.106) and (10.107), which give $\gamma_1 = 1$, $\delta_1 = -1/l$, $\gamma_2 = 0$, $\delta_2 = 1/l$. For $v(x,t)$ we arrive to the boundary value problem for the non-homogeneous wave equation with external force (10.100):

$$f^*(x,t) = A\omega^2\left(1 - \frac{x}{l}\right)\sin\omega t,$$

initial conditions (10.101) and (10.102),

$$\varphi^*(x,t) = 0, \quad \psi^*(x,t) = A\omega\left(1 - \frac{x}{l}\right)$$

and homogeneous boundary conditions.

The eigenvalues and eigenfunctions are

$$\lambda_n = \left(\frac{n\pi}{l}\right)^2, \quad X_n(x) = \sin\frac{n\pi x}{l}, \quad \|X_n\|^2 = \frac{l}{2}, \quad n = 1,2,3,\ldots.$$

From equations (10.64), (10.65), and (10.77), we have

$$a_n = 0, \quad b_n = \frac{1}{\omega_n\|X_n\|^2}\int_0^l \psi^*(x)\sin\frac{n\pi x}{l}x\,dx = -\frac{2A\omega l}{an^2\pi^2},$$

$$f_n(t) = \frac{2A\omega^2}{n\pi}\sin\omega t.$$

We can find $T_n(t)$ by using equation (10.80):

$$T_n(t) = \frac{2A\omega^2 l}{a\pi^2 n^2} \frac{\omega \sin\omega_n t - \omega_n \sin\omega t}{\omega^2 - \omega_n^2}.$$

This expression apply, when, for any n, the frequency of the external force $f^*(x,t)$ is not equal to any of the natural frequencies $\omega_n = n\pi a/l$ of the string (i.e., $\omega \neq \omega_n$). In this case, the solution is

$$u(x,t) = w(x,t) + \sum_{n=1}^{\infty} [T_n(t) + b_n \sin\omega_n t] \sin\frac{n\pi x}{l}$$

$$= \left(1 - \frac{x}{l}\right) A\sin\omega t + \frac{2A\omega l}{a\pi^2} \sum_{n=1}^{\infty} \frac{1}{n^2} \frac{\omega_n (\omega_n \sin\omega_n t - \omega\sin\omega t)}{\omega^2 - \omega_n^2} \sin\frac{n\pi x}{l}.$$

Figure 10.15 shows the spatial-time-dependent solution $u(x,t)$ for Example 7. This solution was obtained with program Waves for the case when $a^2 = 1$, $l = 100$, $A = 4$, $\omega = 0.3$. With these parameters the period of the main harmonic is $2l/a = 200$.

If, for some $n = k$, the frequency of the external force $\omega = \omega_k$, we have a case of resonance. For the kth term we have

$$b_k = -\frac{2A\omega l}{ak^2\pi^2} = -\frac{2A\omega}{k\pi\omega_k} = -\frac{2A}{k\pi},$$

$$T_k(t) = \frac{1}{\omega_k} \int_0^t f_k(\tau)\sin\omega_k(t-\tau)d\tau = \frac{2A\omega^2}{\omega n\pi} \int_0^t \sin\omega t \sin\omega(t-\tau)d\tau$$

$$= \frac{A\omega}{n\pi} \int_0^t [\cos\omega(2\tau-t) - \cos\omega t] d\tau = \frac{A}{\pi k}(\sin\omega t - \omega t\cos\omega t),$$

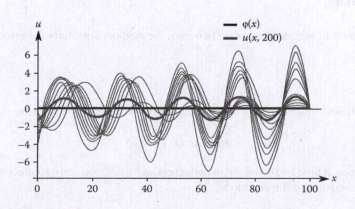

FIGURE 10.15   Graph of solutions, $u(x,t)$ at various times for Example 7.

so that

$$T_k(t) + b_k \sin \omega_k t = -\frac{A}{k\pi}(\sin \omega t + \omega t \cos \omega t).$$

Thus, we can rewrite the solution in the form

$$u(x,t) = w(x,t) + \left[T_k(t) + b_k \sin \omega_k t\right]\sin\frac{k\pi x}{l} + \sum_{\substack{n=1 \\ n \neq k}}^{\infty}\left[T_n(t) + b_n \sin \omega_n t\right]\sin\frac{n\pi x}{l}$$

$$= \left(1 - \frac{x}{l}\right)A\sin\omega t - \frac{A}{k\pi}(\sin\omega t + \omega t\cos\omega t)\sin\frac{k\pi x}{l}$$

$$+ \frac{2A\omega l}{a\pi^2}\sum_{\substack{n=1 \\ n \neq k}}^{\infty}\frac{1}{n^2}\frac{\omega_n(\omega_n\sin\omega_n t - \omega\sin\omega t)}{\omega^2 - \omega_n^2}\sin\frac{n\pi x}{l}.$$

With the program Waves you can simulate the behavior of the system in the case of resonance in, for instance, the third mode, $\omega = \omega_3 = 3\pi/100$. With the Bar Chart option it is interesting to compare how the harmonics contribution changes in time for resonant and nonresonant situations.

In the case of periodic boundary condition, a steady-state solution clearly is also periodic with the same period.

Reading Exercise

Obtain a steady-state solution for Example 7.
  Hint: Search for a solution in the form $u(x,t) = v(x,t) + w(x,t)$ and then substitute $v(x,t) = X(x)\sin\omega t$.

Reading Exercise

Modify Example 7 to include force of friction from the medium and study the resonance case.

Example 8

Find the longitudinal vibrations of a rod, $0 \leq x \leq l$, with the left end fixed. To the right end at $x = l$ the force

$$F(t) = At \quad (A = \text{Const})$$

is applied starting at time $t = 0$. The initial deflection and speed are zero. Neglect friction.
Solution. We should solve the equation

$$u_{tt} = a^2 u_{xx}, \quad 0 < x < l, \quad t > 0$$

with the boundary and initial conditions

$$u(x,0) = 0, \quad u_t(x,0) = 0,$$

$$u(0,t) = 0, \quad u_x(l,t) = At / ES.$$

Assuming

$$u(x,t) = v(x,t) + w(x,t)$$

it is clear that we may choose the function $w(x,t)$ to have the form

$$w(x,t) = \frac{A}{ES} xt$$

such that $w(x,t)$ satisfies both boundary conditions (this choice also can be formally obtained as explained above). For $v(x,t)$, we obtain the boundary value problem

$$v_{tt} = a^2 v_{xx},$$

$$v(x,t)\big|_{t=0} = 0, \quad v_t(x,t)\big|_{t=0} = -Ax / ES,$$

$$v(0,t) = 0, \quad v_x(l,t) = 0.$$

Eigenvalues and eigenfunctions of this problem are

$$\lambda_n = \frac{(2n-1)\pi}{2l}, \quad X_n(x) = \sin\frac{(2n-1)\pi x}{2l}, \quad n = 1,2,3,....$$

From equations (10.61) and (10.62) we have

$$a_n = 0, \quad b_n = (-1)^n \frac{16Al^2}{ESa\,(2n-1)^3 \pi^3}$$

in which case the solution for $v(x,t)$ is

$$v(x,t) = \frac{16Al^2}{ESa\pi^3} \sum_{n=1}^{\infty} \frac{(-1)^n}{(2n-1)^3} \sin\frac{(2n-1)a\pi t}{2l} \sin\frac{(2n-1)\pi x}{2l}.$$

Thus, the solution of to this problem is

$$u(x,t) = \frac{A}{ES} xt + \frac{16Al^2}{ESa\pi^3} \sum_{n=1}^{\infty} \frac{(-1)^n}{(2n-1)^3} \sin\frac{(2n-1)a\pi t}{2l} \sin\frac{(2n-1)\pi x}{2l}.$$

Figure 10.16 shows the solution $u(x,t)$ as a function of t ($0 \le t \le 500$) for several values of coordinate x for Example 8. This solution was obtained with program Waves for the case when $a^2 = 1$, $l = 100$, $E = 1$, $S = 1$, $A = 0.0002$.

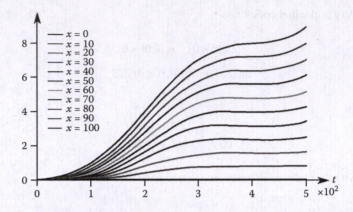

FIGURE 10.16   Time-traces of solutions $u(x,t)$ at various values of x for Example 8.

10.8 THE CONSISTENCY CONDITIONS AND GENERALIZED SOLUTIONS

Let us briefly discuss the consistency between initial and boundary conditions and situations when functions representing these conditions are not perfectly smooth. Assuming that the solution of a boundary value problem, $u(x,t)$, and its first derivatives are continuous on the closed interval $[0,l]$, the initial condition $u(x,t)\big|_{t=0} = \varphi(x)$ and $u_t(x,t)\big|_{t=0} = \psi(x)$ at $x = 0$ and $x = l$ lead to the relations

$$\alpha_1 \varphi_x + \beta_1 \varphi\big|_{x=0} = g_1(0), \quad \alpha_2 \varphi_x + \beta_2 \varphi\big|_{x=l} = g_2(0), \tag{10.110}$$

and

$$\alpha_1 \psi_x + \beta_1 \psi\big|_{x=0} = \frac{\partial g_1}{\partial t}(0), \quad \alpha_2 \psi_x + \beta_2 \psi\big|_{x=l} = \frac{\partial g_2}{\partial t}(0). \tag{10.111}$$

If not all of these *consistency relations* are satisfied, or the functions $f(x,t)$, $\varphi(x)$, $\psi(x)$, $g_1(x,t)$ $g_2(x,t)$ and their derivatives are not continuous on the entire interval $[0,l]$, we still can solve a problem by the methods described above. In such cases, as well as for real, physical problems in which these functions always have some limited precision, such obtained solutions are called *generalized solutions*.

A simple example is the problem of string oscillations in a gravitational field. The external force (nonhomogeneous term) is not zero at the ends of the string and if the ends are fixed, the boundary conditions are not consistent with the equation. But the situation is not that bad because the formal (generalized) solution converges uniformly at any point in the interval.

Also notice that in Example 4 from Section 10.5.1 the boundary $u_x(0,t) = 0$ and initial condition $u(x,0) = F_0(l+1/h-x)/ES$ are inconsistent. For long enough time this inconsistency plays an insignificant role.

Reading Exercise

Solve the problem of a string oscillating in gravity field

$$\frac{\partial^2 u}{\partial t^2} = a^2 \frac{\partial^2 u}{\partial x^2} - g,$$

if the ends are fixed and initially the string was displaced at some point from the equilibrium position and released with zero speed. Solve the problem analytically.

Discuss the properties of the analytical solution and its behavior at the end of the interval. Compare with the results generated with the program Waves at short and longer times. Suggest not consistent initial and boundary conditions and analyze the obtained solution.

10.9 ENERGY IN THE HARMONICS

Next we consider the energy associated with the motion of the string. The kinetic energy density (energy per unit length) of the string is

$$\frac{1}{2} \rho \left(\frac{\partial u}{\partial t} \right)^2$$

and the total kinetic energy then is

$$E_{kin}(t) = \frac{1}{2} \int_0^l \rho \left(\frac{\partial u}{\partial t} \right)^2 dx, \qquad (10.112)$$

the integral being taken over the whole length of the string.

Next, we determine an expression for the potential energy. If a portion of the string of initial length dx is stretched to a length ds when displaced, the increase in length is

$$ds - dx = \left[\sqrt{1 + \left(\frac{\partial u}{\partial x} \right)^2} - 1 \right] dx.$$

In the approximation of "small vibrations" ($\partial u/\partial x \ll 1$), this becomes

$$\frac{1}{2} \left(\frac{\partial u}{\partial x} \right)^2 dx.$$

Since this stretching takes place against a force of tension T, the potential energy gain on the interval dx, which is the work done against tension, is

$$\frac{1}{2} T \left(\frac{\partial u}{\partial x} \right)^2 dx.$$

Thus, the potential energy of the string is

$$E_{pot}(t) = \frac{1}{2}\int_0^l T\left(\frac{\partial u}{\partial x}\right)^2 dx. \tag{10.113}$$

The total energy is therefore

$$E_{tot} = E_{kin} + E_{pot} = \frac{\rho}{2}\int_0^l \left[\left(\frac{\partial u}{\partial t}\right)^2 + a^2\left(\frac{\partial u}{\partial x}\right)^2\right]dx, \tag{10.114}$$

where we have used the fact that $T = \rho a^2$.

It is easy to see (we leave to prove it as the Reading Exercise) that for homogeneous boundary conditions of Dirichlet or Neumann type (e.g., the ends of a string are rigidly fixed or free) the derivatives of functions $X_n(x)$ are orthogonal (like the functions $X_n(x)$ themselves):

$$\int_0^l \frac{dX_n}{dx}\frac{dX_m}{dx}dx = 0. \tag{10.115}$$

The physical significance of this equation is that in this case the individual terms in the Fourier series solution are independent and energy in a harmonic cannot be exchanged with the energies associated with the other terms.

The nth harmonic is

$$u_n(x,t) = \left\{T_n(t) + \left[a_n y_n^{(1)}(t) + b_n y_n^{(2)}(t)\right]\right\}X_n(x) \tag{10.116}$$

(for homogeneous boundary conditions the auxiliary function $w(x,t) \equiv 0$).

Its kinetic energy is (due to orthogonality of $X_n(x)$ and their derivatives)

$$E_{kin}^{(n)}(t) = \frac{\rho}{2}\int_0^l \left(\frac{\partial u_n}{\partial t}\right)^2 dx = \frac{\rho}{2}\left[\frac{dT_n}{dt} + a_n\frac{dy_n^{(1)}}{dt} + b_n\frac{dy_n^{(2)}}{dt}\right]^2 \int_0^l X_n^2(x)dx \tag{10.117}$$

and its potential energy is

$$E_{pot}^{(n)}(t) = \frac{T}{2}\int_0^l \left(\frac{\partial u_n}{\partial x}\right)^2 dx = \frac{T}{2}\left[T_n(t) + a_n y_n^{(1)}(t) + b_n y_n^{(2)}(t)\right]^2 \int_0^l \left(\frac{dX_n}{dx}\right)^2 dx. \tag{10.118}$$

It is evident that there are no terms representing interactions of the harmonics in either the kinetic or potential energy expressions. The total energy of the nth harmonic is the sum of these two energies.

Consider, as an example, free vibrations that occur in medium without damping governed by the wave equation,

$$\frac{\partial^2 u}{\partial t^2} = a^2 \frac{\partial^2 u}{\partial t^2}. \tag{10.119}$$

In this case

$$T_n(t) = 0, \; y_n^{(1)}(t) = \cos a \sqrt{\lambda_n} t, \; y_n^{(2)}(t) = \sin a \sqrt{\lambda_n} t \tag{10.120}$$

and the kinetic energy of nth harmonic is

$$E_{kin}^{(n)}(t) = \frac{\rho a^2 \lambda_n}{2} \left[-a_n \sin a \sqrt{\lambda_n} t + b_n \cos a \sqrt{\lambda_n} t \right]^2 \|X_n\|^2; \tag{10.121}$$

the potential energy of nth harmonic is

$$E_{pot}^{(n)}(t) = \frac{T}{2} \left[a_n \cos a \sqrt{\lambda_n} t + b_n \sin a \sqrt{\lambda_n} t \right]^2 \left\| \frac{dX_n}{dx} \right\|^2. \tag{10.122}$$

For boundary conditions of Dirichlet or Neumann type we have

$$\|X_n\|^2 = \int_0^l X_n^2(x)\,dx = \frac{l}{2}, \; \left\| \frac{dX_n}{dx} \right\|^2 = \int_0^l \left(\frac{dX_n}{dx} \right)^2 dx = \lambda_n \|X_n\|^2 = \frac{\lambda_n l}{2}. \tag{10.123}$$

The potential and kinetic energies of a harmonic taken together give a constant, the total energy, which is (taking into account $T = \rho a^2$)

$$E_{tot}^{(n)}(t) = E_{kin}^{(n)}(t) + E_{pot}^{(n)}(t) = \frac{\rho a^2 \lambda_n l}{4} (a_n^2 + b_n^2). \tag{10.124}$$

In each harmonic, the energy oscillates between kinetic and potential forms as the string itself oscillates. The periods of the oscillations of the string are

$$\tau_n = \frac{2\pi}{a \sqrt{\lambda_n}} \tag{10.125}$$

and those of the energy are half that value.

Reading Exercises

We leave it to the reader to work out few more results:
1. If both ends of the string are rigidly fixed or both ends are free, then $\lambda_n = \pi^2 n^2 / l^2$, and the total energy is

$$E_{tot}^{(n)}(t) = \frac{\rho a^2 \pi^2 n^2}{4l} (a_n^2 + b_n^2). \tag{10.126}$$

2. If one end of the string is rigidly fixed and the other is free, then $\lambda_n = \pi^2(2n+1)^2/(4l^2)$, and the total energy is

$$E_{tot}^{(n)}(t) = \frac{\rho a^2 \pi^2 (2n+1)^2}{16l}\left(a_n^2 + b_n^2\right). \tag{10.127}$$

3. As an example of another physical system, consider energy oscillations in the *RLC* circuit. Magnetic and electric energies are

$$E_L = \frac{Li^2}{2} \text{ and } E_C = \frac{CV^2}{2}.$$

With $i = \partial q/\partial t$ this gives (for a wire from 0 to l)

$$E_L = \frac{1}{2}\int_0^l L\left(\frac{\partial q}{\partial t}\right)^2 dx \text{ and } E_C = \frac{1}{2}\int_0^l R^2 C\left(\frac{\partial q}{\partial t}\right)^2 dx. \tag{10.128}$$

Reading Exercises

Using the material described in Section 10.3, obtain the expressions for these energies for boundary conditions of Dirichlet and Neumann type.

Let us use Examples 3 and 5 from Sections 10.5 and 10.6 to find the energy of a string under various conditions. In both examples, as the Reading Exercises, find analytical expressions for kinetic, potential, and total energies of the string as functions of time, and compare these solution with those simulated with the program Waves.

In Example 3 (fixed ends, no external forces), the string was displaced from equilibrium by shifting the point $x = x_0$ by a distance h at time $t = 0$ and then releasing with zero initial speed. Figure 10.17 shows the energies on the string obtained with the program

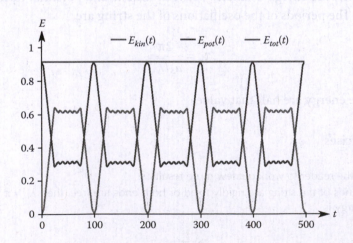

FIGURE 10.17   Energies of the string for Example 3.

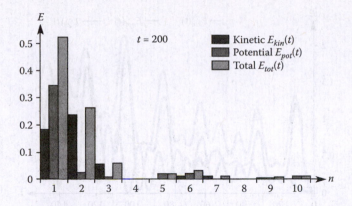

FIGURE 10.18   Distribution of kinetic, potential, and total energies in harmonics at $t = 200$ for Example 3.

Waves. The following values of the parameters were chosen: $a^2 = 1$, $\kappa = 0$, $l = 100$, $h = 6$, and $x_0 = 25$.

Figures 10.18 demonstrate energy in harmonics at time $t = 200$.

We may solve the same problem when the force of friction is not zero. Here we again use the program Waves with the same values of the parameters and with the friction coefficient, $\kappa = 0.001$.

In Example 5 from Section 10.6, we considered a string with rigidly fixed ends and no friction. Starting at time $t = 0$ a harmonic force with linear density $F(x,t) = F_0 \sin \omega t$ acts on the string. In Figure 10.20 we present the energy graphs obtained with the program Waves for the following values of the parameters: $a^2 = 1$, $\kappa = 0$, $l = 100$, $\rho = 1$, $F_0 = 0.025$, $\omega = 0.3$. Obviously, the energy is not conserved.

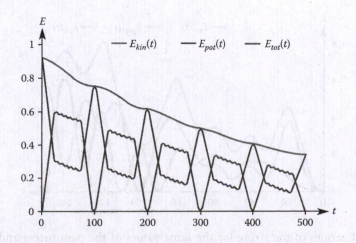

FIGURE 10.19   Energies of the string for Example 3 with dissipation included.

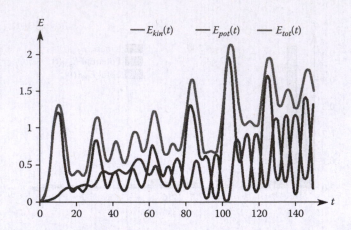

FIGURE 10.20 Energies of the string for Example 5.

Figure 10.21 represents a resonance case. As Reading Exercises, describe and explain the difference between the cases presented in Figures 10.20 and 10.21.

Problems

The following selection of problems has been composed to be solved using the software **Waves** included with the book as well as analytically. Solving these problems analytically means the following: formulate the equation and initial and boundary conditions, obtain the eigenvalues and eigenfunctions, and write the formulas for coefficients of the series expansion and the expression for the solution of the problem. If the integrals in the coefficients are not easy to evaluate, leave it for the program, which evaluates them numerically. Then study the problem in detail with the program **Waves**. Obtain the pictures of several eigenfunctions and screenshots of solution in different moments of time. Directions on how to use the program are found in Appendix 5.

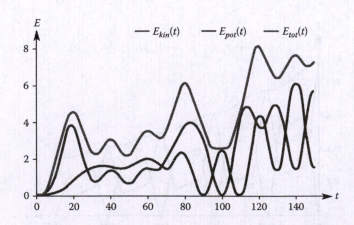

FIGURE 10.21 Energies of the string for the same values of the parameters and with a driving frequency $\omega = \omega_5 = 5\pi/100$.

Dimensionless parameters may be used, but for the physicist or engineer it is also useful, if possible, to model a realistic situation. Solving problems analytically allows one to find the derivatives u_x and u_t —if necessary, the latter derivative can be compared with a and the smallness of u_x can be discussed. It is also often useful to estimate the contribution of different terms in the wave equation after the solution has been obtained: for instance, in equation (10.2) the term ρu_{tt} may be large due to fast oscillations; usually the term Tu_{xx} is large because T is large.

The text of the software is written for the case of a vibrating string, but it is easy to model, for instance, electrical oscillations by inputting the corresponding "electrical" parameters.

One of the advantages of using the software is that a range of parameters may be explored very quickly. For example, it is very simple to find a medium resistance so that the main mode is reduced by a factor two during some number of periods, or to study the role of a parameter's effect in the behavior of different modes. This opens the way to exploring a wide variety of different variants of basically the same problem. It is always useful to solve a problem analytically, and if this solution is correct, it should give the same results that the software provides.

Infinite strings

In problems 1 through 5 an infinite stretched string is excited by the initial deflection $u(x,0) = \varphi(x)$ with no initial velocities.

Find the vibrations of the string. Write analytical solution, representing the motion of the string for $t > 0$, and simulate the vibrations using the program **Waves**.

1. $\varphi(x) = \dfrac{A}{1 + Bx^2}$

2. $\varphi(x) = Ae^{-Bx^2}$

3. $\varphi(x) = \begin{cases} A(l^2 - x^2), & x \in [-l,l], \\ 0, & x \notin [-l,l] ; \end{cases}$

4. $\varphi(x) = \begin{cases} A\sin\dfrac{\pi x}{l}, & x \in [-l,l], \\ 0, & x \notin [-l,l] ; \end{cases}$

In problems 5 through 8 an infinite stretched string is initially at rest. Assume at time $t = 0$ the initial distribution of velocities is given by $u_t(x,0) = \psi(x)$.

Find the vibrations of the string. Write analytical solution, representing the motion of the string for $t > 0$, and simulate the vibrations using the program **Waves**.

5. $\psi(x) = \dfrac{A}{1 + x^2}$

6. $\psi(x) = Axe^{-Bx^2}$

7. $\psi(x) = \begin{cases} v_0 = \text{const}, & x \in [-l,l], \\ 0, & x \notin [-l,l]; \end{cases}$

8. $\psi(x) = \begin{cases} x, & x \in [-l,l], \\ 0, & x \notin [-l,l]. \end{cases}$

9. Let an infinite string be at rest prior to $t = 0$. At time $t = 0$ it is excited by a sharp blow from a hammer that transmits an impulse I at point $x = x_0$ to the string.

   Find the vibrations of the string. Write formulas representing the motion of the string and simulate the vibrations using the program **Waves**.

10. Consider an infinite thin wire with resistance R, capacitance C, inductance L, and leakage G distributed along its length. Find the electrical current oscillations in a circuit if $GL = CR$. For the following initial voltage and current in the wire

$$V(x,0) = f_1(x) = e^{-x^2}, \quad i(x,0) = f_2(x) = e^{-x^2}.$$

Find electrical voltage oscillations in the wire. Write an analytical solution and simulate the oscillations using the program **Waves**.

Transverse oscillations in finite strings

Problems 11 through 34 refer to uniform finite strings with the ends at $x = 0$ and $x = l$.

Note. When you choose the parameters of the problem and coefficients of the functions (initial and boundary conditions, and external forces) do not forget that the amplitudes of oscillations should remain small.

In problems 11 through 15 the initial shape of a string with fixed ends is; $u(x,0) = \varphi(x)$, the initial speed is $u_t(x,0) = \psi(x)$. External forces and dissipation are absent. Find the vibrations of the string.

Choose tension T (or the wave speed a) in order the period of vibrations decrease (1) twice, (2) four times, (3) five times. Obtain the same results changing the length l of the string.

11. $\varphi(x) = Ax\left(1 - \dfrac{x}{l}\right), \psi(x) = 0$

12. $\varphi(x) = A\sin\dfrac{\pi x}{l}, \psi(x) = 0$

13. $\varphi(x) = 0, \psi(x) = v_0 = \text{const}$

14. $\varphi(x) = 0, \psi(x) = \dfrac{B}{l}x(l-x)$

15. $\varphi(x) = Ax\left(1 - \dfrac{x}{l}\right), \psi(x) = \dfrac{B}{l}x(l-x)$

16. A string with fixed ends is displaced at point $x = x_0$ by a small distance h from equilibrium and released at $t = 0$ without initial speed. No external forces or dissipation act. Describe the string oscillations. Find the location of x_0 so that the following overtones are absent: (a) 3rd; (b) 5th; (c) 7th.

17. A string with fixed ends is displaced at point $x = x_0$ by a small distance h from the x-axis and released at $t = 0$ without initial speed. Find the vibrations of the string if it vibrates in the constant gravitational field g, and the resistance of a medium is proportional to speed. (Take the value of gravitational force $g = -0.001$).

    Choose damping coefficient κ such that oscillations decay (with precision of about 5%) during time (*a*) $t = 2000$; (*b*) $t = 1500$; (*c*) $t = 1000$. Read the coordinates of the bottom point of the string in the equilibrium state under the influence of the gravitational force.

In problems 18 through 22 the end at $x = 0$ of a string is fixed while the end at $x = l$ is attached to a massless ring that can slide along a frictionless rod perpendicular to the x-axis such that the tangent line to the string is always horizontal. The initial shape and speed of the string are $u(x,0) = \varphi(x)$ and $u_t(x,0) = \psi(x)$. No external forces act except that the surrounding medium has a resistance with coefficient κ. Find string oscillations.

Find the coefficient κ such that oscillations decay (with precision of about 5%) during (a) two periods; (b) three periods; (c) four periods of the main mode.

18. $\varphi(x) = A\dfrac{x}{l}$, $\psi(x) = 0$

19. $\varphi(x) = A\sin\dfrac{\pi x}{2l}$, $\psi(x) = 0$

20. $\varphi(x) = 0$, $\psi(x) = v_0 = \text{const}$

21. $\varphi(x) = 0$, $\psi(x) = Ax\left(1 - \dfrac{x}{l}\right)$

22. $\varphi(x) = Ax\left(1 - \dfrac{x}{l}\right)$, $\psi(x) = v_0\cos\dfrac{\pi x}{l}$

23. The ends of a string are rigidly fixed. The string is excited by the sharp blow of a hammer, supplying an impulse I at point x_0. No external forces act, but the surrounding medium supplies a resistance with coefficient κ. Model the string oscillations.

    Find x_0 such that the energies of the seventh and eighth overtones are reduced. Find the value of the coefficient, such that oscillations decay (with precision of about 5%) during (a) two periods; (b) three periods; (c) four periods of the main mode.

In problems 24 through 27 the ends of a string are rigidly fixed. An external force $F(t)$ begins to acts at point x_0 at time $t = 0$. Model and solve analytically string oscillations for zero initial conditions where the string is in a resisting medium with resistance coefficient κ.

24. $F(t) = F_0 \sin \omega t F$

25. $F(t) = F_0 \cos \omega t$

26. $F(t) = F_0 e^{-At} \sin \omega t$

27. $F(t) = F_0 e^{-At} \cos \omega t$

In problems 28 through 31 the ends of a string are rigidly fixed. The initial conditions are zero with no resistance. Starting at time $t = 0$ a uniformly distributed force with linear density $f(x,t)$ acts on the string. Find oscillations of the string.

28. $f(x) = Ax(x/l - 1)$

29. $f(x,t) = Axe^{-Bt}$

30. $f(x,t) = Ax(x/l - 1)\sin \omega t$

31. $f(x,t) = Ax(x/l - 1)\cos \omega t$

In problems 32 through 34 the left end $(x = 0)$ of a string is driven according to $u(0,t) = g_1(t)$; the right end $(x = l)$ is fixed. Initially the string is at rest.

Find oscillations of the string when there are no external forces and the resistance of a medium is zero. Find the solution in the case of resonance.

32. $g_1(t) = A \sin wt$.

33. $g_1(t) = A(1 - \cos \omega t)$

34. $g_1(t) = A(\sin \omega t + \cos \omega t)$

Longitudinal oscillations in rods

The following two problems are formulated in terms of waves in a thin elastic straight rod of density ρ, length l (the ends at $x = 0$ and $x = l$), the elasticity modulus E, and the cross section S.

In problems 35 and 36 a rod elastically fixed at one end is stretched by longitudinal force $F_0 = \text{const}$ applied to the other end. At time $t = 0$ the force F_0 ceases to act. Find the longitudinal oscillations of the rod if initial velocities and the resistance of a medium are zero, and there are no external forces. Describe oscillations for small h ("soft" attachment) and large h ("stiff" attachment). Compare with a similar case with free ends.

35. A rod is elastically fixed at the end $x = l$ and stretched by longitudinal force $F_0 = \text{const}$ applied to the end $x = 0$.

36. A rod is elastically fixed at the end $x = 0$ and stretched by longitudinal force $F_0 = \text{Const}$ applied to the end $x = l$.

In problems 37 through 39 starting at $t = 0$ the end of a rod at $x = 0$ is moving horizontally according to $u(0,t) = g(t)$ and an external force $F = F(t)$ is applied to the end at $x = l$ along the axis. Assume zero initial conditions and an embedding medium that has a resistance proportional to speed. Describe the oscillations $u(x,t)$ of the rod.

Hint: The equation and the boundary conditions are

$$u_{tt} + 2\kappa u_t - a^2 u_{xx} = 0,$$

$$u(x,0) = 0, \quad u_t(x,0) = 0,$$

$$u(0,t) = g(t), \quad u_x(l,t) = F(t)/ES.$$

37. $F(t) = F_0(1 - \cos \omega t),\ g(t) = g_0$

38. $F(t) = F_0,\ g(t) = g_0 \sin \omega t$

39. $F(t) = F_0 e^{-At} \sin \omega t,\ g(t) = g_0 \sin \omega t$

Electrical oscillations in circuits

In problems 40 through 44 a conductor of length l is perfectly insulated ($G = 0$) and R, L, and C are known. Current at $t = 0$ is absent and the voltage is $V(x,t)\big|_{t=0} = V(x)$. The end at $x = 0$ is insulated and the end at $x = l$ is grounded. Find the current $i(x,t)$ in the wire for $t > 0$.

40. $V(x) = V_0 \sin \dfrac{\pi x}{l}$

41. $V(x) = V_0 \left(1 - \cos \dfrac{\pi x}{l}\right)$

42. $V(x) = V_0 x(l - x)$

43. $V(x) = \dfrac{V_0}{(x - l/2)^2 + A}$

44. $V(x) = V_0 \exp\left[-A(x - l/2)^2\right]$

In problems 45 through 47 initially the current and voltage in an insulated conductor are zero. The left end at $x = 0$ is insulated; the right end at $x = l$ is attached to a source of *emf* $E(t)$ at $t = 0$. The parameters R, L, G, and C are known. (The leakage current G is zero by the assumption that the wire is insulated.) Find the voltage $V(x,t)$ in the wire.

45. $E(t) = E_0 \sin \omega t$

46. $E(t) = E_0(1 - \cos \omega t)$

47. $E(t) = E_0 e^{-At}(\sin \omega t + \cos \omega t)$

CHAPTER 11

Two-Dimensional Hyperbolic Equations

In this chapter we consider physical problems related to two-dimensional flexible surfaces called membranes. A membrane may be defined as a thin film that bends but, in the present development, does not stretch. The boundary of the membrane may be fixed or free or have forces applied to it. We will also consider cases where the membrane interacts with the material in which it is embedded and is thus subject to external forces such as driving forces or friction. Examples of membranes include drum heads, flags, trampolines, soap films, and biological barriers such as cellular membranes. Surfaces of liquids may be treated as membranes under the appropriate circumstances.

Our development of the behavior of a membrane will parallel our previous discussion of a vibrating string, but in this case the motion is of a two-dimensional object oscillating in a third direction. First let us consider a membrane in equilibrium in the x-y plane limited by a smooth, closed boundary, L, under tension, T, which acts tangent to the surface of the membrane. In the following we will treat external forces acting on the membrane in a direction perpendicular to the x-y plane only, except at the boundary of the membrane. Under the action of such a force or in the case of an initial perturbation from equilibrium, points on the membrane move to a new position, which we will describe by the distance from equilibrium, $u = u(x, y)$ at location (x, y). The distance of the membrane surface from equilibrium may also vary in time so that the displacement $u(x, y, t)$ is a function of time as well as location.

We consider only cases where the curvature of the membrane is small, in which case we can neglect powers of u squared (and higher orders) and derivatives: $u^2 \approx 0$, $u_x^2 \approx 0$, and so on.

In Figure 11.1 a small section, σ, of the membrane whose equilibrium position is limited by the closed curve l is shown. When the membrane is displaced from the equilibrium position this section is deformed to the area σ', limited by the closed curve l' as shown in Figure 11.1. The new area σ' at some instant of time is given by

$$\sigma' = \iint_\sigma \sqrt{1 + u_x^2 + u_y^2}\, dx\, dy \approx \iint_\sigma dx\, dy = \sigma.$$

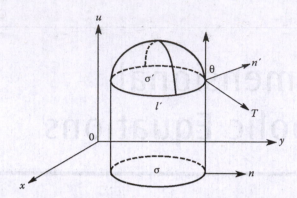

FIGURE 11.1 Small surface element of a membrane, σ, displaced from equilibrium into stretched element σ′. The angle θ is between the force of tension, T, which is tangent to the curved surface element and the direction of the displacement, u. The vectors n and n' are the normal vectors to the surfaces σ and σ′, respectively.

From this result we see that, for small oscillations with low curvature, we may neglect changes of area of the membrane. As was the case for small vibrations of a string, this allows us to assume that the tension, T, in the membrane does not vary with x or y.

11.1 DERIVATION OF THE EQUATIONS OF MOTION

To derive an equation of motion for the membrane let us consider its fragment, the deformed area σ′ limited by the curve l'. The tension, T, acting on this area is evenly distributed on the contour, l', and is perpendicular to the contour and tangent to the surface of the deformed area. For a segment ds' of the curve l' the tension will be Tds' acting on the segment. Since motions of the membrane are constrained to be perpendicular to the x-y plane, we consider the component of the tension in the direction u (perpendicular to the x-y plane), which is $Tds'\cos\theta$ where θ is the angle between T and the direction of the displacement u. For small oscillations of the membrane, $cos\theta$ is approximately equal to $\partial u/\partial n$ where n is the normal perpendicular to the curve l, the boundary of the original equilibrium area σ. From this we have that the component of tension acting on element ds' of contour l' in the direction of displacement u is

$$T\frac{\partial u}{\partial n}ds'.$$

We now integrate over the contour l' to find the component of tension acting on area element σ′ and perpendicular to the equilibrium surface as

$$T\int_{l'}\frac{\partial u}{\partial n}ds'.$$

For small oscillations of the membrane $ds \approx ds'$ (i.e., the boundary l does not deform much as the element σ is stretched). Using Green's formula (see Appendix 4) we have, in rectangular coordinates,

$$T\int_l \frac{\partial u}{\partial n} ds = \iint_\sigma T\left(\frac{\partial^2 u}{\partial x^2} + \frac{\partial^2 u}{\partial y^2}\right) dx\,dy. \tag{11.1}$$

The above only includes forces due to the original tension on the membrane. If an additional external force per unit area $F(x, y, t)$ (which may vary in time) acts parallel to the direction $u(x, y, t)$, then the component in the u direction of this force acting on area σ' of the membrane is given by

$$\iint_\sigma F(x, y, t) dx\,dy. \tag{11.2}$$

The two forces in equations (11.1) and (11.2) cause an acceleration of the area element σ'. If the mass per area of the membrane is given by the surface density, $\rho(x, y)$, the right-hand side of Newton's second law for the motion of this area element becomes

$$\iint_\sigma \rho(x, y) \frac{\partial^2 u}{\partial t^2} dx\,dy.$$

Setting the forces acting on this element equal to the mass times acceleration of the area element we have

$$\iint_\sigma \left[\rho(x, y)\frac{\partial^2 u}{\partial t^2} - T\left(\frac{\partial^2 u}{\partial x^2} + \frac{\partial^2 u}{\partial y^2}\right) + F(x, y, t)\right] dx\,dy = 0.$$

We began with an arbitrary surface element, σ, from which it follows that

$$\rho(x, y)\frac{\partial^2 u}{\partial t^2} - T\left(\frac{\partial^2 u}{\partial x^2} + \frac{\partial^2 u}{\partial y^2}\right) = F(x, y, t). \tag{11.3}$$

Equation (11.3) is the linear partial differential equation, which describes *small, transverse, forced oscillations of a membrane*.

In the case of a membrane of uniform mass density ($\rho = $ const) we may write this equation as

$$\frac{\partial^2 u}{\partial t^2} - a^2\left(\frac{\partial^2 u}{\partial x^2} + \frac{\partial^2 u}{\partial y^2}\right) = f(x, y, t), \tag{11.4}$$

where $a = \sqrt{T/\rho}$, $f(x, y, t) = F(x, y, t)/\rho$. In cases where the external force is absent, that is, $F(x, y, t) = 0$, then from equation (11.4) we obtain the *homogeneous* equation for *free oscillations of a uniform membrane* given by

$$\frac{\partial^2 u}{\partial t^2} = a^2\left(\frac{\partial^2 u}{\partial x^2} + \frac{\partial^2 u}{\partial y^2}\right). \tag{11.5}$$

If, in addition to the internal tension, the membrane is subject to an external linear restoring force proportional to displacement, we may add a force $F = -\alpha u$ per unit of area of the membrane, where α is the spring coefficient of the ambient material. For such a membrane embedded in a springy or spongy environment equation (11.4) becomes

$$\frac{\partial^2 u}{\partial t^2} - a^2 \left(\frac{\partial^2 u}{\partial x^2} + \frac{\partial^2 u}{\partial y^2} \right) + \gamma u = f(x, y, t), \tag{11.6}$$

where $\gamma = \alpha/\rho$.

If the membrane is embedded in a material that produces a drag on the motion of the membrane such as the case for biological membranes, which are normally immersed in a liquid environment, a friction term must be added to equation (11.4). Friction forces are generally proportional to velocity, and we have $F = -ku_t$ as the force per unit area of membrane where k is the coefficient of friction. The equation of oscillation in this case includes the time derivative of displacement, $u_t(x, y, t)$ and we have

$$\frac{\partial^2 u}{\partial t^2} + 2\kappa \frac{\partial u}{\partial t} - a^2 \left(\frac{\partial^2 u}{\partial x^2} + \frac{\partial^2 u}{\partial y^2} \right) = f(x, y, t), \tag{11.7}$$

where $2\kappa = k/\rho$.

All the equations from (11.4) through (11.7) are linear partial differential equations of hyperbolic type. In the following we solve the above equations for various cases and give examples. First we consider the physical limitations presented by requirements at the boundaries of the membrane.

11.1.1 Boundary and Initial Conditions

The equations of motion (11.4), (11.5), (11.6), and (11.7) are not by themselves sufficient to entirely specify the motion of a membrane. Additional conditions need be specified: initial conditions and boundary conditions.

If the position and velocity of points on the membrane are known at some initial time, $t = 0$, and are given by the functions $\phi(x, y)$ and $\psi(x, y)$, respectively, we have the *initial conditions*

$$u|_{t=0} = \varphi(x, y), \quad \frac{\partial u}{\partial t}\bigg|_{t=0} = \psi(x, y). \tag{11.8}$$

As in the case of the vibrating string we may be given, along with initial conditions, information about the behavior of the membrane at its edges at all times, t, in which case we have *boundary conditions*. In the following we outline several variants of conditions on the boundary, L, of the membrane.

1. If edge of the membrane is rigidly fixed, then motion of membrane on the border L does not occur and we have as the boundary condition

$$u|_L = 0$$

which is referred to as a *fixed edge* boundary condition.

2. If the behavior over time of the displacement, $u(x, y, t)$, of the boundary is given by some function $g(t)$, then we have

$$u|_L = g(t)$$

which is called a *driven edge* boundary condition.

3. In the case of a boundary that is free (for example, the edges of a flag under small oscillations) so that the displacement is only in a direction perpendicular to the x-y plane, we have *free edge* boundary conditions given by

$$\frac{\partial u}{\partial n}\bigg|_L = 0.$$

4. The edge may also be subject to a force with linear density, f_1 in the x-y plane, which affects the tension at the boundary. In this case we have the *stretched edge* boundary condition,

$$\left(-T\frac{\partial u}{\partial n} + f_1\right)\bigg|_L = 0. \tag{11.9}$$

5. If the force density, f_1, in the stretched edge condition, equation (11.10) is a linear spring-like force (for example, the boundary of a trampoline fixed to its support with springs), we may write $-ku$ for f_1 and we have

$$\left(\frac{\partial u}{\partial n} + hu\right)\bigg|_L = 0, \text{ where } h = k/T. \tag{11.10}$$

6. If the edges to which a membrane is elastically attached are moving in some pre-scribed way, the right sides of equations (11.9) and (11.10) will contain some function of time, $g(t)$, describing the motion of the edges. In this case we have nonhomogeneous boundary conditions.

We may combine all of these conditions in a generic form given by

$$\alpha\frac{\partial u}{\partial n} + \beta u\bigg|_L = g(t). \tag{11.11}$$

Dirichlet boundary condition correspond to $\alpha = 0$, $\beta = 1$ in which case we have the driven edge situation. Neumann boundary condition correspond to $\alpha = 1$, $\beta = 0$ and we have the stretched edge condition or the free edge if $g(t) = 0$. Mixed boundary conditions, when both $\alpha \neq 0$ and $\beta \neq 0$, correspond to two last cases; if $g(t) = 0$ we have homogeneous ($g(t) \neq 0$ nonhomogeneous) mixed boundary conditions. Clearly, the types of boundary conditions can vary along the boundary, and we will consider such a situation in the following section.

11.2 OSCILLATIONS OF A RECTANGULAR MEMBRANE

In this section we consider the method of Fourier series for a rectangular membrane limited by the straight lines $x = 0$, $x = l_x$, $y = 0$, and $y = l_y$ (Figure 11.2).

We begin with the most general case of a membrane subject to friction forces, a linear restoring force and external forcing, $f(x, y, t)$. From the previous discussion we see that the equation of motion for such a problem is given by

$$\frac{\partial^2 u}{\partial t^2} + 2\kappa \frac{\partial u}{\partial t} - a^2 \left(\frac{\partial^2 u}{\partial x^2} + \frac{\partial^2 u}{\partial y^2} \right) + \gamma u = f(x, y, t) \tag{11.12}$$

with generic boundary conditions given on the boundary of the rectangle as

$$P_1[u] \equiv \alpha_1 \frac{\partial u}{\partial x} + \beta_1 u \bigg|_{x=0} = g_1(y, t), \quad P_2[u] \equiv \alpha_2 \frac{\partial u}{\partial x} + \beta_2 u \bigg|_{x=l_2} = g_2(y, t),$$

$$P_3[u] \equiv \alpha_3 \frac{\partial u}{\partial y} + \beta_3 u \bigg|_{y=0} = g_3(x, t), \quad P_4[u] \equiv \alpha_4 \frac{\partial u}{\partial y} + \beta_4 u \bigg|_{y=l_y} = g_4(x, t), \tag{11.13}$$

where $g_1(y, t), \dots, g_4(x, t)$ are the given functions of time and respective variable and α_1, β_1, $\alpha_2, \beta_2, \alpha_3, \beta_3, \alpha_4$, and β_4 are constants subject to the same restrictions from physical arguments which we saw in Chapter 4. We also consider initial conditions

$$u|_{t=0} = \varphi(x, y), \quad \frac{\partial u}{\partial t} \bigg|_{t=0} = \psi(x, y), \tag{11.14}$$

where $\varphi(x,y)$ and $\psi(x,y)$ are given functions.

As in the case for the movement of a string, we will use the Fourier method and separation of variables to solve this equation. In a manner exactly parallel to the solution of a vibrating string, but instead for an object described initially by two spatial dimensions, we will obtain solutions in the form of a series of eigenfunctions of the corresponding Sturm-Liouville problem.

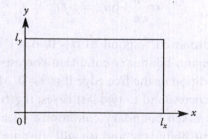

FIGURE 11.2 Rectangular membrane in its equilibrium position.

11.2.1 The Fourier Method for Homogeneous Equations with Homogeneous Boundary Conditions

We start with the *homogeneous equation* (i.e., no external forcing)

$$\frac{\partial^2 u}{\partial t^2} + 2\kappa \frac{\partial u}{\partial t} - a^2 \left(\frac{\partial^2 u}{\partial x^2} + \frac{\partial^2 u}{\partial y^2} \right) + \gamma u = 0 \qquad (11.15)$$

with homogeneous boundary conditions

$$P_1[u] \equiv \alpha_1 \frac{\partial u}{\partial x} + \beta_1 u \Big|_{x=0} = 0 \quad P_2[u] \equiv \alpha_2 \frac{\partial u}{\partial x} + \beta_2 u \Big|_{x=l_x} = 0,$$

$$P_3[u] \equiv \alpha_3 \frac{\partial u}{\partial y} + \beta_3 u \Big|_{y=0} = 0, \quad P_4[u] \equiv \alpha_4 \frac{\partial u}{\partial y} + \beta_4 u \Big|_{y=l_y} = 0, \qquad (11.16)$$

and initial conditions in equation (11.14) given by

$$u|_{t=0} = \varphi(x, y), \quad \frac{\partial u}{\partial t} \Big|_{t=0} = \psi(x, y).$$

To begin, we first assume that nontrivial (nonzero) solutions can be written as the product of two functions, one a function of time the second a function of x and y:

$$u(x, y, t) = V(x, y)T(t). \qquad (11.17)$$

Substituting equation (11.17) into equation (11.15), we get

$$V(x, y)T''(t) + 2\kappa V(x, y)T'(t) - a^2[V_{xx}(x, y) + V_{yy}(x, y)]T(t) + \gamma V(x, y)T(t) = 0$$

or, upon rearranging terms,

$$\frac{T''(t) + 2\kappa T'(t) + \gamma T(t)}{a^2 T(t)} = \frac{V_{xx}(x, y) + V_{yy}(x, y)}{V(x, y)},$$

where we have used the shorthand notation for the derivatives in x and y

$$V_{xx}(x, y) \equiv \frac{\partial^2 V}{\partial x^2} \quad \text{and} \quad V_{yy}(x, y) \equiv \frac{\partial^2 V}{\partial y^2}.$$

The left side of the previous equality is a function t only and the right side only of x and y, which is possible only if both sides are equal to some constant value. Denoting this constant as $-\lambda$ we have

$$\frac{T''(t) + 2\kappa T'(t) + \gamma T(t)}{a^2 T(t)} \equiv \frac{V_{xx}(x, y) + V_{yy}(x, y)}{V(x, y)} = -\lambda.$$

Using the left side of this equation for the function $T(t)$ we get the homogeneous linear differential equation of second order

$$T''(t) + 2\kappa T'(t) + (a^2\lambda + \gamma)T(t) = 0, \tag{11.18}$$

where primes denote derivatives with respect to time. For the function $V(x, y)$ we have the equation

$$V_{xx}(x, y) + V_{yy}(x, y) + \lambda V(x, y) = 0 \tag{11.19}$$

with boundary conditions

$$P_1[V] \equiv \alpha_1 V_x + \beta_1 V\big|_{x=0} = 0, \quad P_2[V] \equiv \alpha_2 V_x + \beta_2 V\big|_{x=l_x} = 0,$$
$$P_3[V] \equiv \alpha_3 V_y + \beta_3 V\big|_{y=0} = 0, \quad P_4[V] \equiv \alpha_4 V_y + \beta_4 V\big|_{y=l_y} = 0.$$

To solve equation (11.19) for $V(x, y)$ we again make the assumption that the variables are independent and attempt to separate them using the substitution

$$V(x, y) = X(x)Y(y).$$

Reading Exercise

Show that making the above substitution into equation (11.19) and applying the boundary conditions yields

$$X''(x) + \lambda_x X(x) = 0, \tag{11.20}$$

with boundary conditions

$$\alpha_1 X'(0) + \beta_1 X(0) = 0, \quad \alpha_2 X'(l_x) + \beta_2 X(l_x) = 0,$$

and

$$Y''(y) + \lambda_y Y(y) = 0, \tag{11.21}$$

with boundary conditions

$$\alpha_3 Y'(0) + \beta_3 Y(0) = 0, \quad \alpha_4 Y'(l_y) + \beta_4 Y(l_y) = 0,$$

where λ_x and λ_y are constants from the division of variables linked by the correlation

$$\lambda_x + \lambda_y = \lambda.$$

Hint: The boundary conditions for $X(x)$ and $Y(y)$ follow from the corresponding conditions for the function $V(x,y)$. For example, from the condition

$$\alpha_1 V_x(0, y) + \beta_1 V(0, y) = \alpha_1 X'(0)Y(y) + \beta_1 X(0)Y(y)$$

$$= [\alpha_1 X'(0) + \beta_1 X(0)]Y(y) = 0$$

it follows that

$$\alpha_1 X'(0) + \beta_1 X(0) = 0,$$

since $Y(y) \neq 0$ (i.e., we are searching for nontrivial solutions).

Solutions to equations (11.20) and (11.21) (given in Appendix 1) depend on the boundary conditions and have the generic form

$$X(x) = C_1 \cos\sqrt{\lambda_x}\, x + C_2 \sin\sqrt{\lambda_x}\, x,$$

and

$$Y(y) = D_1 \cos\sqrt{\lambda_y}\, y + D_2 \sin\sqrt{\lambda_y}\, y.$$

Applying the respective boundary conditions for equation (11.20) and (11.21) allows us to determine the coefficients C_1, C_2 and D_1, D_2 (as it was done in Section 10.5):

$$C_1 = \frac{\alpha_1\sqrt{\lambda_{xn}}}{\sqrt{\lambda_{xn}\alpha_1^2 + \beta_1^2}}, \quad C_2 = -\frac{\beta_1}{\sqrt{\lambda_{xn}\alpha_1^2 + \beta_1^2}},$$

$$D_1 = \frac{\alpha_3\sqrt{\lambda_{ym}}}{\sqrt{\lambda_{ym}\alpha_3^2 + \beta_3^2}}, \quad D_2 = -\frac{\beta_3}{\sqrt{\lambda_{ym}\alpha_3^2 + \beta_3^2}}.$$

Collecting the above results we have eigenfunctions for the Sturm-Liouville problem defined by equations (11.20) and (11.21) given by

$$X_n(x) = \frac{1}{\sqrt{\alpha_1^2\lambda_{xn} + \beta_1^2}}\left[\alpha_1\sqrt{\lambda_{xn}}\cos\sqrt{\lambda_{xn}}\, l_x - \beta_1\sin\sqrt{\lambda_{xn}}\, l_x\right],$$

$$Y_m(y) = \frac{1}{\sqrt{\alpha_3^2\lambda_{ym} + \beta_3^2}}\left[\alpha_3\sqrt{\lambda_{ym}}\cos\sqrt{\lambda_{ym}}\, l_y - \beta_3\sin\sqrt{\lambda_{ym}}\, l_y\right] \tag{11.22}$$

(square roots should be taken with positive signs).

The eigenvalues of the problem are

$$\lambda_{xn} = \left[\frac{\mu_{xn}}{l_x}\right]^2, \quad \text{and} \quad \lambda_{ym} = \left[\frac{\mu_{ym}}{l_y}\right]^2, \tag{11.23}$$

where μ_{xn} is the nth root of the equation

$$\tan\mu_x = \frac{(\alpha_1\beta_2 - \alpha_2\beta_1)\, l_x\mu_x}{\mu_x^2\alpha_1\alpha_2 + l_x^2\beta_1\beta_2}, \tag{11.24}$$

and μ_{ym} is the mth root of the equation

$$\tan\mu_y = \frac{(\alpha_3\beta_4 - \alpha_4\beta_3)\, l_y\mu_y}{\mu_y^2\alpha_3\alpha_4 + l_y^2\beta_3\beta_4}. \tag{11.25}$$

The norms of the eigenfunctions are given by

$$\|X_n\|^2 = \int_0^{l_x} X_n^2(x)\,dx = \frac{1}{2}\left[l_x + \frac{(\alpha_1\beta_2 - \alpha_2\beta_1)(\lambda_{xn}\alpha_1\alpha_2 - \beta_1\beta_2)}{(\lambda_{xn}\alpha_1^2 + \beta_1^2)(\lambda_{xn}\alpha_2^2 + \beta_2^2)} \right],$$

$$\|Y_m\|^2 = \int_0^{l_y} Y_m^2(y)\,dy = \frac{1}{2}\left[l_y + \frac{(\alpha_3\beta_4 - \alpha_4\beta_3)(\lambda_{ym}\alpha_3\alpha_4 - \beta_3\beta_4)}{(\lambda_{ym}\alpha_3^2 + \beta_3^2)(\lambda_{ym}\alpha_4^2 + \beta_4^2)} \right].$$

(11.26)

in which case functions $X_n(x)$ and $Y_m(y)$ are bounded by the values ± 1.

Returning to the two-dimensional problem defined by equation (11.19) and subsequent boundary conditions we make the following observations based on the linearity of the original differential equation (11.15). If λ_{xn} and $X_n(x)$ are eigenvalues and eigenfunctions of equation (11.20), and λ_{ym} and $Y_m(y)$ are eigenvalues and eigenfunctions of equation (11.21), then

$$\lambda_{nm} = \lambda_{xn} + \lambda_{ym} \tag{11.27}$$

and

$$V_{nm}(x,y) = X_n(x)Y_m(y) \tag{11.28}$$

are eigenvalues and eigenvectors, respectively, of the problem in equation (11.19). We remind the reader that the success of the method of separation of variables is based on the fact that the sets of orthogonal functions $X_n(x)$ and $Y_n(x)$ satisfy the same boundary conditions as $u(x, y, t)$. The functions $V_{nm}(x, y)$ are orthogonal and the norms are given by

$$\|V_{nm}\|^2 = \|X_n\|^2 \|Y_m\|^2. \tag{11.29}$$

The system of eigenfunctions, V_{nm}, given in equation (11.28) form a complete set of basis functions for a two-dimensional rectangular membrane. By this we mean that any smooth (i.e., twice differentiable) shape of the deformed rectangular membrane with the generic boundary conditions given above can be expanded in a converging series of the functions V_{nm}.

We now return to the equation describing the time evolution of the membrane, equation (11.18). This is an ordinary linear differential equation of second order, which we have seen previously for one-dimensional oscillations. It should clear in this case, however, that $T(t)$ now depends on two indexes corresponding to the eigenfunctions $X(x)$ and $Y(y)$. Specifically we may write $\lambda = \lambda_{nm}$ and denote $T(t)$ as $T_{nm}(t)$, which is defined by

$$T_{nm}(t) = a_{nm} y_{nm}^{(1)}(t) + b_{nm} y_{nm}^{(2)}(t), \tag{11.30}$$

where a_{nm} and b_{nm} are arbitrary constants. Similar to the case for the one-dimensional problem, we have

$$y_{nm}^{(1)}(t) = \begin{cases} e^{-\kappa t}\cos\omega_{nm}t, & \omega_{nm} = \sqrt{a^2\lambda_{nm} + \gamma - \kappa^2}, & \kappa^2 < a^2\lambda_{nm} + \gamma, \\ e^{-\kappa t}\cosh\omega_{nm}t, & \omega_{nm} = \sqrt{\kappa^2 - a^2\lambda_{nm} - \gamma}, & \kappa^2 > a^2\lambda_{nm} + \gamma, \\ e^{-\kappa t}, & \omega_{nm} = 1, & \kappa^2 = a^2\lambda_{nm} + \gamma, \end{cases}$$

$$y_{nm}^{(2)}(t) = \begin{cases} e^{-\kappa t}\sin\omega_{nm}t, & \omega_{nm} = \sqrt{a^2\lambda_{nm} + \gamma - \kappa^2}, & \kappa^2 < a^2\lambda_{nm} + \gamma, \\ e^{-\kappa t}\sinh\omega_{nm}t, & \omega_{nm} = \sqrt{\kappa^2 - a^2\lambda_{nm} - \gamma}, & \kappa^2 > a^2\lambda_{nm} + \gamma, \\ te^{-\kappa t}, & \omega_{nm} = 1, & \kappa^2 = a^2\lambda_{nm} + \gamma. \end{cases} \quad (11.31)$$

Reading Exercise

Following the arguments used in the previous chapter for the one-dimensional case, verify the above formulas.

Thus particular solutions for the free oscillations of a rectangular membrane may be written as

$$u_{nm}(x,y,t) = T_{nm}(t)V_{nm}(x,y) = [a_{nm}y_{nm}^{(1)}(t) + b_{nm}y_{nm}^{(2)}(t)]V_{nm}(x,y) \quad (11.32)$$

which will form a complete set of solutions to equation (11.15) satisfying boundary conditions in equation (11.16).

To satisfy the initial conditions of equation (11.14) we first expand the displacement $u(x,y,t)$ in a Fourier series as

$$u(x,y,t) = \sum_n \sum_m [a_{nm}y_{nm}^{(1)}(t) + b_{nm}y_{nm}^{(2)}(t)]V_{nm}(x,y). \quad (11.33)$$

If this series converges uniformly as well as the series obtained by twice differentiation term by term with respect the variables x, y, and t, then its sum will be the solution of equation (11.15) satisfying boundary conditions (11.16). Under the same assumptions we may also expand the boundary conditions given by equation (11.14) using the same set of basis functions:

$$u|_{t=0} = \varphi(x,y) = \sum_n \sum_m a_{nm}V_{nm}(x,y), \quad (11.34)$$

$$\frac{\partial u}{\partial t}\bigg|_{t=0} = \psi(x,y) = \sum_n \sum_m [\omega_{nm}b_{nm} - \kappa a_{nm}]V_{nm}(x,y). \quad (11.35)$$

Formulas (11.34) and (11.35) show that functions $\varphi(x,y)$ and $\psi(x,y)$ can be expanded in a complete set of functions, $V_{nm}(x,y)$, which form the solution of the Sturm-Liouville boundary value problem of equation (11.19).

Again supposing the series in equations (11.34) and (11.35) converge uniformly we may determine the coefficients a_{nm} and b_{nm} by making use of the orthogonality of the eigenfunctions $V_{nm}(x,y)$. First we multiply equations (11.34) and (11.35) by $V_{nm}(x,y)$ and integrate over x from 0 to l_x and over y from 0 to l_y. This yields the Fourier coefficients

$$a_{nm} = \frac{1}{\|V_{nm}\|^2} \int_0^{l_x} \int_0^{l_y} \varphi(x,y) V_{nm}(xy) \, dx \, dy, \tag{11.36}$$

$$b_{nm} = \frac{1}{\omega_{nm}} \left[\frac{1}{\|V_{nm}\|^2} \int_0^{l_x} \int_0^{l_y} \psi(x,y) V_{nm}(xy) \, dx \, dy + \kappa a_{nm} \right]. \tag{11.37}$$

The coefficients a_{nm} and b_{nm} substituted into the series (11.33) yield a complete solution to equation (11.15) with boundary conditions (11.16), and initial conditions (11.14), under the assumption that the series in equation (11.33) converges uniformly and can be differentiated term by term in x, y, and t. Therefore we may say that equation (11.33) completely describes the *free oscillations* of a membrane. This solution thus has the form of a Fourier series on the orthogonal system of functions $\{V_{nm}(x,y)\}$, each function of which represents a mode characterized by two numbers, n and m. This series converges under "reasonable" assumptions about initial and boundary conditions.

Particular solutions $u_{nm}(x,y,t) = T_{nm}(t) V_{nm}(x,y)$ where the time and space components are separate are called *standing wave* solutions and are analogous to standing waves on a one-dimensional string. The profile of the standing wave is defined by the function $V_{nm}(x,y)$, which varies as a function of time $T_{nm}(t)$. Lines, along which $V_{nm}(x,y) = 0$ do not change with time, are called *node lines* of the standing wave. Loose sand placed on a vibrating membrane will collect along node lines because there is no motion at those locations. Locations where $V_{nm}(x,y)$ has a relative maximum or minimum at some instant of time are called antinodes of the standing wave. The general solution, $u(x,y,t)$, is an infinite sum of these standing waves as was the case for the vibrating string. This property of being able to construct arbitrary shapes from a sum of component waves (or modes) is referred to as the *superposition of standing waves* and is a general property of linear systems of all dimensions.

Consider a simple case of a rectangular membrane with sides clamped at the boundary. The vibrations are caused only by initial conditions; thus we want to solve the equation

$$\frac{\partial^2 u}{\partial t^2} = a^2 \left(\frac{\partial^2 u}{\partial x^2} + \frac{\partial^2 u}{\partial y^2} \right)$$

satisfying boundary conditions

$$u(0,y,t) = u(l_x,y,t) = u(x,0,t) = u(x,l_y,t) = 0,$$

and initial conditions in equation (11.14).

We leave to the reader to check as a Reading Exercise that, using the results from the generic case investigated above, eigenvalues and eigenfunctions for this problem are

$$\lambda_{xn} = \left(\frac{n\pi}{l_x}\right)^2, \quad X_n(x) = \sin\frac{n\pi x}{l_x}, \quad \|X_n\|^2 = \frac{l_x}{2}, \quad n = 1,2,3,...,$$

$$\lambda_{ym} = \left(\frac{m\pi}{l_y}\right)^2, \quad Y_m(y) = \sin\frac{m\pi y}{l_y}, \quad \|Y_m\|^2 = \frac{l_y}{2}, \quad m = 1,2,3,...,$$

with

$$\lambda_{nm} = \lambda_{xn} + \lambda_{ym} = \pi^2\left(\frac{n^2}{l_x^2} + \frac{m^2}{l_y^2}\right),$$

and we have that

$$V_{nm}(x,y) = X_n(x)Y_m(y) = \sin\frac{n\pi x}{l_x}\sin\frac{m\pi y}{l_y} \quad \text{with} \quad \|V_{nm}\|^2 = \frac{l_x l_y}{4}.$$

It is obvious that these functions form a complete set of orthogonal functions for oscillations of the rectangular membrane. The time evolution function is

$$T_{nm}(t) = a_{nm}\cos\omega_{nm}t + b_{nm}\sin\omega_{nm}t,$$

where the frequencies are $\omega_{nm} = a\lambda_{nm} = a\pi\sqrt{(n/l_x)^2 + (m/l_y)^2}$. Each pair of integers (n, m) corresponds to a particular characteristic mode (called a normal mode) of vibration of the membrane. An arbitrary membrane deflection may then be represented as a superposition of normal modes:

$$u(x,y,t) = \sum_{n=1}^{\infty}\sum_{m=1}^{\infty} T_{nm}V_{nm} = \sum_{n=1}^{\infty}\sum_{m=1}^{\infty} c_{nm}\sin\frac{n\pi x}{l_x}\sin\frac{m\pi y}{l_y}\cos(\omega_{nm}t + \delta_{nm}),$$

where we have introduced coefficients c_{nm} and phase shifts δ_{nm} via the relations $a_{nm} = c_{nm}\cos\delta_{nm}$, $b_{nm} = -c_{nm}\sin\delta_{nm}$.

If the membrane vibrates in one of its normal modes, then all points on the membrane participate in harmonic motion with frequency ω_{nm}. As an example consider the (2,1) mode (i.e., $n = 2$, $m = 1$). The eigenfunction is

$$V_{21}(x,y) = X_2(x)Y_1(y) = \sin\frac{2\pi x}{l_x}\sin\frac{\pi y}{l_y}.$$

The only nodal line is the straight line $x = l_x/2$. Similarly, the (1,2) mode ($n = 1$, $m = 2$) has the nodal line $y = l_y/2$ (see Figure 11.3). Nodal lines split the membrane into zones and all

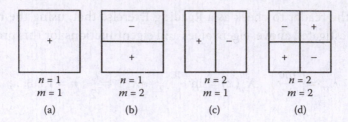

$$
\begin{array}{cccc}
n = 1 & n = 1 & n = 2 & n = 2 \\
m = 1 & m = 2 & m = 1 & m = 2 \\
\text{(a)} & \text{(b)} & \text{(c)} & \text{(d)}
\end{array}
$$

FIGURE 11.3   Modes of vibration $V_{nm}(x, y) = X_n(x)Y_m(y)$. Plus signs indicate motion out of the page; minus signs indicate simultaneous motion into the page.

points of each zone move with the same phase, that is, all up or all down (labeled with + and –) at some instant (although not necessarily with the same amplitude).

Generally speaking, each node vibrates with its own frequency, ω_{nm}. However, if l_y / l_x is a rational number, two or more modes could possess the same frequency. As an example consider a square membrane where $l_x = l_y$, in which case $\omega_{12} = \omega_{21}$. This frequency is said to be twofold degenerate, by which we mean there are two linearly independent eigenfunctions corresponding the same eigenvalue.

Below we consider two examples of physical problems for free oscillations of a membrane with homogeneous boundary conditions.

Example 1

Find the transverse oscillations of a uniform rectangular membrane ($0 \le x \le l_x$, $0 \le y \le l_y$) having fixed edges and with an initial displacement of

$$
u(x, y, 0) = Axy(l_x - x)(l_y - y),
$$

assuming interactions with the surrounding medium can be neglected and the initial velocities of points on the membrane are zero.

Solution. This is an example of the case discussed above with the specified initial conditions, thus the problem reduces to solutions of the equation

$$
\frac{\partial^2 u}{\partial t^2} - a^2 \left(\frac{\partial^2 u}{\partial x^2} + \frac{\partial^2 u}{\partial y^2} \right) = 0,
$$

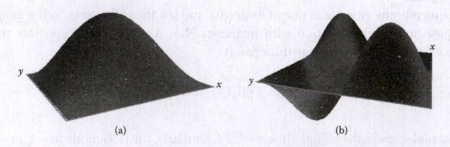

$$
\begin{array}{cc}
\text{(a)} & \text{(b)}
\end{array}
$$

FIGURE 11.4   Eigenfunctions (a) $V_{11}(x, y)$ and (b) $V_{22}(x, y)$ for a membrane with fixed edges.

with initial and boundary conditions

$$u(x,y,0) = Axy(l_x - x)(l_y - y), \quad \frac{\partial u}{\partial t}(x,y,0) = 0,$$

$$u(0,y,t) = u(l_x,y,t) = u(x,0,t) = u(x,l_y,t) = 0.$$

As we obtained above, the eigenfunctions are

$$V_{nm}(x,y) = X_n(x)Y_m(y) = \sin\frac{n\pi x}{l_x}\sin\frac{m\pi y}{l_y}, \quad \text{with} \quad \|V_{nm}\|^2 = \frac{l_x l_y}{4}.$$

The three-dimensional view shown in Figure 11.4 depicts two eigenfunctions, $V_{11}(x,y)$ and $V_{22}(x,y)$, chosen as examples, for this problem. The values of λ are $\lambda_{11} = 0.891$ and $\lambda_{22} = 3.564$.

We leave to the reader as a Reading Exercise to check, using equations (11.30) and (11.31), that the expressions for the coefficients of the series are

$$a_{nm} = \begin{cases} \dfrac{64A\,l_x^2 l_y^2}{\pi^2 n^2 m^2}, & \text{if } n \text{ and } m - \text{ odd}, \\ 0, & \text{if } n \text{ or } m - \text{ even}, \end{cases} \quad \text{and} \quad b_{nm} = 0,$$

and the time evolution is given by

$$T_{nm} = a_{nm}\cos\omega_{nm}t = \begin{cases} \dfrac{64A\,l_x^2 l_y^2}{\pi^2 n^2 m^2}\cos\omega_{nm}t, & \text{if } n \text{ and } m - \text{ odd}, \\ 0, & \text{if } n \text{ or } m - \text{ even}, \end{cases}$$

where $\omega_{nm} = a\lambda_{nm} = a\pi\sqrt{(n/l_x)^2 + (m/l_y)^2}$. Consequently the displacements of the membrane as a function of time for this problem can be expressed by the series

$$u(x,y,t) = \frac{64Al_x^2 l_y^2}{\pi^2}\sum_{n=1}^{\infty}\sum_{m=1}^{\infty}\frac{\cos\omega_{(2n-1)(2m-1)}t}{(2n-1)^2(2m-1)^2}\sin\frac{(2n-1)\pi x}{l_x}\sin\frac{(2m-1)\pi y}{l_y}.$$

Figure 11.5 shows two snapshots of the solution at the times $t = 2$ and $t = 10$. This solution was obtained with program **Waves** for the case $a^2 = 1$, $l_x = 4$, $l_y = 6$, and $A = 0.01$. A description of the program **Waves** and directions for its use are given in Appendix 5.

(a)   (b)

FIGURE 11.5   Graph of the membrane in Example 1 at (a) $t = 7$, (b) $t = 10$.

Example 2

A uniform rectangular membrane $(0 \leq x \leq l_x,\ 0 \leq y \leq l_y)$ has edges $x = l_x$ and $y = l_y$, which are free, and edges at $x = 0$ and $y = 0$, which are firmly fixed. Find the transverse oscillations of the membrane caused by an initial displacement $u(x, y, 0) = Axy$ assuming interactions with the surrounding medium can be neglected and the initial velocities of points on the membrane are zero.

Solution. This problem reduces to finding the solution of the equation

$$\frac{\partial^2 u}{\partial t^2} - a^2 \left(\frac{\partial^2 u}{\partial x^2} + \frac{\partial^2 u}{\partial y^2} \right) = 0,$$

with initial and boundary conditions

$$u(x, y, 0) = Axy, \quad \frac{\partial u}{\partial t}(x, y, 0) = 0,$$

$$u(0, y, t) = \frac{\partial u}{\partial x}(l_x, y, t) = 0, \quad u(0, y, t) = \frac{\partial u}{\partial y}(x, l_y, t) = 0.$$

Eigenvalues and eigenfunctions of the problem are

$$\lambda_{nm} = \lambda_{xn} + \lambda_{ym} = \pi^2 \left[\frac{(2n-1)^2}{4l_x^2} + \frac{(2m-1)^2}{4l_y^2} \right], \quad n, m = 1, 2, \dots.$$

and

$$V_{nm}(x, y) = X_n(x)Y_m(y) = \sin\frac{(2n-1)\pi x}{2l_x} \sin\frac{(2m-1)\pi y}{2l_y}, \quad \|V_{nm}\|^2 = \frac{l_x l_y}{4}.$$

The three-dimensional picture shown in Figure 11.6 depicts two eigenfunctions, $V_{11}(x, y)$ and $V_{22}(x, y)$, chosen as examples for this problem. The values of λ are $\lambda_{11} = 0.223$ and $\lambda_{22} = 2.005$. Using formulas (11.30) and (11.31) we obtain the coefficients

$$a_{nm} = (-1)^{n+m} \frac{64 A l_x l_y}{\pi^4 (2n-1)^2 (2m-1)^2} \quad \text{and} \quad b_{nm} = 0.$$

In this case displacements of the membrane as a function of time is expressed by the series

$$u(x, y, t) = \frac{64 A l_x l_y}{\pi^4} \sum_{n=1}^{\infty} \sum_{m=1}^{\infty} \frac{(-1)^{n+m} \cos\omega_{nm} t}{(2n-1)^2 (2m-1)^2} \sin\frac{(2n-1)\pi x}{2l_x} \sin\frac{(2m-1)\pi y}{2l_y},$$

where

$$\omega_{nm} = a\lambda_{nm} = a\pi \sqrt{\frac{(2n-1)^2}{4l_x^2} + \frac{(2m-1)^2}{4l_y^2}}.$$

11.2.2 The Fourier Method for Nonhomogeneous Equations

Building on the previous sections we now consider the problem of solutions of the *nonhomogeneous* linear equation (11.12) for a two-dimensional membrane:

$$\frac{\partial^2 u}{\partial t^2} + 2\kappa \frac{\partial u}{\partial t} - a^2 \left(\frac{\partial^2 u}{\partial x^2} + \frac{\partial^2 u}{\partial y^2} \right) + \gamma u = f(x, y, t),$$

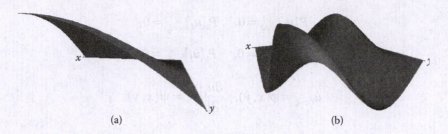

(a)   (b)

FIGURE 11.6   Sample eigenfunctions (a) $V_{11}(x,y)$ and (b) $V_{22}(x,y)$ for Example 2.

where $f(x,y,t)$ is a given function. First we search for solutions that satisfy the *homogeneous boundary conditions* in equation (11.16) given by

$$P_1[u] \equiv \alpha_1 \frac{\partial u}{\partial x} + \beta_1 u \bigg|_{x=0} = 0, \quad P_2[u] \equiv \alpha_2 \frac{\partial u}{\partial x} + \beta_2 u \bigg|_{x=l_x} = 0,$$

$$P_3[u] \equiv \alpha_3 \frac{\partial u}{\partial y} + \beta_3 u \bigg|_{y=0} = 0, \quad P_4[u] \equiv \alpha_4 \frac{\partial u}{\partial y} + \beta_4 u \bigg|_{y=l_y} = 0,$$

and nonhomogeneous initial conditions given in equation (11.14) as

$$u|_{t=0} = \varphi(x,y), \quad \frac{\partial u}{\partial t}\bigg|_{t=0} = \psi(x,y).$$

Because the equation of membrane oscillations is linear, the displacement, $u(x,y,t)$, may be written as the sum

$$u(x,y,t) = u_1(x,y,t) + u_2(x,y,t),$$

where $u_1(x,y,t)$ is the solution of the *homogeneous* equation with *homogeneous* boundaries and *nonhomogeneous* initial conditions:

$$\frac{\partial^2 u_1}{\partial t^2} + 2\kappa \frac{\partial u_1}{\partial t} - a^2 \left(\frac{\partial^2 u_1}{\partial x^2} + \frac{\partial^2 u_1}{\partial y^2} \right) + \gamma u_1 = 0, \tag{11.38}$$

(a)   (b)

FIGURE 11.7   Graph of the membrane in Example 2 at (a) $t = 7$, (b) $t = 10$.

$$P_1[u_1]\big|_{x=0} = 0, \quad P_2[u_1]\big|_{x=l_x} = 0,$$

$$P_3[u_1]\big|_{y=0} = 0, \quad P_4[u_1]\big|_{y=l_y} = 0,$$

(11.39)

$$u_1\big|_{t=0} = \varphi(x,y), \quad \frac{\partial u_1}{\partial t}\bigg|_{t=0} = \psi(x,y).$$

(11.40)

The function $u_2(x,y,t)$ is the solution of the *nonhomogeneous* equation with *homogeneous* boundary conditions and initial conditions:

$$\frac{\partial^2 u_2}{\partial t^2} + 2\kappa \frac{\partial u_2}{\partial t} - a^2 \left(\frac{\partial^2 u_2}{\partial x^2} + \frac{\partial^2 u_2}{\partial y^2} \right) + \gamma u_2 = f(x,y,t),$$

(11.41)

$$P_1[u_2]\big|_{x=0} = 0, \quad P_2[u_2]\big|_{x=l_x} = 0,$$

$$P_3[u_2]\big|_{y=0} = 0, \quad P_4[u_2]\big|_{y=l_y} = 0,$$

(11.42)

$$u_2\big|_{t=0} = 0, \quad \frac{\partial u_2}{\partial t}\bigg|_{t=0} = 0.$$

(11.43)

In other words, the solution $u_1(x,y,t)$ is for *free oscillations*, that is, such oscillations that occur only as a consequence of an initial perturbation, and the solution $u_2(x,y,t)$ is for the case of *forced oscillations*, that is, such oscillations that occur under the action of an external force $f(x,y,t)$ when initial perturbations are absent.

The problem of free oscillations was considered in the previous section for which case the solution $u_1(x,y,t)$ is known. To proceed we need only to find the solution $u_2(x,y,t)$ for forced oscillations. As in the case for free oscillations we may expand $u_2(x,y,t)$ in the series

$$u_2(x,y,t) = \sum_n \sum_m T_{nm}(t) V_{nm}(x,y),$$

(11.44)

where $V_{nm}(x,y)$ are eigenfunctions of the corresponding homogeneous boundary problem and $T_{nm}(t)$ are, at this stage, unknown functions of t. Any choice of functions $T_{nm}(t)$ satisfies the homogeneous boundary conditions (11.42) for the function $u_2(x,y,t)$ because the functions $V_{nm}(x,y)$ satisfy these conditions.

To find the functions $T_{nm}(t)$ we proceed as follows. Substituting the series in equation (11.44) into equation (11.41) we have

$$\sum_n \sum_m [T''_{nm}(t) + 2\kappa T'_{nm}(t) + (a^2 \lambda_{nm} + \gamma) T_{nm}(t)] V_{nm}(x,y) = f(x,y,t).$$

(11.45)

We may also expand the function $f(x,y,t)$ in a Fourier series using the basis functions $V_{nm}(x,y)$ on the rectangle $[0,l_x;0,l_y]$:

$$f(x,y,t) = \sum_n \sum_m f_{nm}(t) V_{nm}(x,y),$$

(11.46)

where the coefficients of expansion are given by

$$f_{nm}(t) = \frac{1}{\|V_{nm}\|^2} \int_0^{l_x} \int_0^{l_y} f(x,y,t)V_{nm}(x,y)\,dx\,dy. \tag{11.47}$$

Comparing the expansions in equations (11.45) and (11.46) for same function $f(x,y,t)$, obtain a differential equation for the functions $T_{nm}(t)$:

$$T''_{nm}(t) + 2\kappa T'_{nm}(t) + (a^2\lambda_{nm} + \gamma)T_{nm}(t) = f_{nm}(t). \tag{11.48}$$

The solution $u_2(x,y,t)$, defined by the series in equation (11.44) and satisfying initial conditions (11.43), requires that the functions $T_{nm}(t)$ in turn satisfy the conditions

$$T_{nm}(0) = 0, \quad T'_{nm}(0) = 0. \tag{11.49}$$

The solution of the Cauchy problem defined by equations (11.48) and (11.49) may be written as

$$T_{nm}(t) = \frac{1}{\omega_{nm}} \int_0^t f_{nm}(\tau)Y_{nm}(t-\tau)\,d\tau,$$

where

$$Y_{nm}(t) = \begin{cases} e^{-\kappa t}\sin\omega_{nm}t, & \omega_{nm} = \sqrt{a^2\lambda_{nm} + \gamma - \kappa^2}, & \kappa^2 < a^2\lambda_{nm} + \gamma, \\ e^{-\kappa t}\sinh\omega_{nm}t, & \omega_{nm} = \sqrt{\kappa^2 - a^2\lambda_{nm} - \gamma}, & \kappa^2 > a^2\lambda_{nm} + \gamma, \\ te^{-\kappa t}, & \omega_{nm} = 1, & \kappa^2 = a^2\lambda_{nm} + \gamma. \end{cases}$$

Reading Exercise

Verify the above formulas.

We may substitute the expression for $f_{nm}(p)$ given in equation (11.47) to yield, finally,

$$T_{nm}(t) = \frac{1}{\omega_{nm}\|V_{nm}\|^2} \int_0^t d\tau \int_0^{l_x} \int_0^{l_y} f(x,y,\tau)V_{nm}(x,y)Y_{nm}(t-\tau)\,dx\,dy. \tag{11.50}$$

Substituting the above formulas for $T_{nm}(t)$ into the series (11.48) yields the solution of the boundary value problem defined by equations (11.41) through (11.43) under the condition that the series (11.44) and the series obtained from equation (11.48) by term-by-term differentiation (up to second order with respect to x, y, and t) converge uniformly. Thus, the solution of the original problem of forced oscillations with zero initial conditions is given by

$$u(x,y,t) = u_1(x,y,t) + u_2(x,y,t)$$

$$= \sum_n \sum_m \{ T_{nm}(t) + [a_{nm} y_{nm}^{(1)}(t) + b_{nm} y_{nm}^{(2)}(t)] \} V_{nm}(x,y),$$

where coefficients $T_{nm}(t)$ are defined by equation (11.50) and a_{nm}, b_{nm} are defined in the previous section for free oscillations.

We now consider examples of solutions of physical problems involving nonhomogeneous equation of oscillations with homogeneous boundary conditions.

Example 3

Consider transverse oscillations of a rectangular membrane $[0, l_x; 0, l_y]$ with fixed edges, subjected to a transverse driving force

$$F(t) = A\sin\omega t,$$

attached at the point (x_0, y_0), $0 < x_0 < l_x$, $0 < y_0 < l_y$. Assume the reaction of the surrounding medium can be ignored.

Solution. The boundary problem may be defined from the above equations as

$$\frac{\partial^2 u}{\partial t^2} - a^2 \left(\frac{\partial^2 u}{\partial x^2} + \frac{\partial^2 u}{\partial y^2} \right) = \frac{A}{\rho} \delta(x - x_0) \delta(y - y_0) \sin\omega t$$

with initial conditions

$$u(x,y,0) = 0, \quad \frac{\partial u}{\partial t}(x,y,0) = 0,$$

and Dirichlet homogeneous boundary conditions

$$u(0,y,t) = u(l_x,y,t) = u(x,0,t) = u(x,l_y,t) = 0.$$

From the previous development, eigenvalues and eigenfunctions for these boundary conditions are

$$\lambda_{nm} = \lambda_{xn} + \lambda_{ym} = \pi^2 \left(\frac{n^2}{l_x^2} + \frac{m^2}{l_y^2} \right),$$

$$V_{nm}(x,y) = X_n(x)Y_m(y) = \sin\frac{n\pi x}{l_x} \sin\frac{m\pi y}{l_y}, \quad \|V_{nm}\|^2 = \frac{l_x l_y}{4},$$

Using the initial conditions, $\varphi(x,y) = \psi(x,y) = 0$, the solution $u(x,y,t)$ is defined by the series

$$u(x,y,t) = \sum_{n=1}^{\infty} \sum_{m=1}^{\infty} T_{nm}(t) \sin\frac{n\pi x}{l_x} \sin\frac{m\pi y}{l_y},$$

where

$$T_{nm}(t) = \frac{1}{\omega_{nm}} \int_0^t f_{nm}(\tau) \sin\omega_{nm}(t-\tau)\,d\tau,$$

$$f_{nm}(t) = \frac{4A}{\rho l_x l_y} \sin\omega t \sin\frac{n\pi x_0}{l_x} \sin\frac{m\pi y_0}{l_y},$$

and

$$\omega_{nm} = a\lambda_{nm} = a\pi\sqrt{\frac{n^2}{l_x^2} + \frac{m^2}{l_y^2}}.$$

Figure 11.8 shows the solution profile $u(x,2.5,t)$ for Example 3. This solution was obtained with the program **Waves** for the case $a^2 = 1$, $l_x = 4$, $l_y = 6$, $\rho = 1$, $A = 0.5$, $x_0 = 1.5$, $y_0 = 2.5$, and $\omega = 1.5$ (the frequency of the external force).

If the frequency of the driving force is not equal to any of the natural frequencies of membrane, that is, $\omega \neq \omega_{nm}$, $n,m = 1,2,3\ldots$, then

$$T_{nm}(t) = \frac{4A}{\rho l_x l_y (\omega_{nm}^2 - \omega^2)} \sin\frac{n\pi x_0}{l_x} \sin\frac{m\pi y_0}{l_y} \left[\sin\omega t - \frac{\omega}{\omega_{nm}}\sin\omega_{nm}t\right]$$

and

$$u(x,y,t) = \frac{4A}{\rho l_x l_y} \sum_{n=1}^{\infty} \sum_{m=1}^{\infty} \frac{1}{(\omega_{nm}^2 - \omega^2)} \left[\sin\omega t - \frac{\omega}{\omega_{nm}}\sin\omega_{nm}t\right]$$

$$\times \sin\frac{n\pi x_0}{l_x} \sin\frac{m\pi y_0}{l_y} \sin\frac{n\pi x}{l_x} \sin\frac{m\pi y}{l_y}.$$

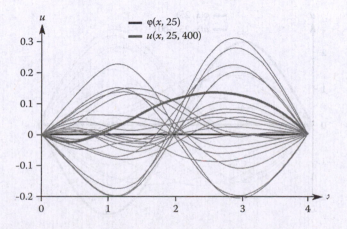

FIGURE 11.8   Solution profile $u(x,2.5,t)$ for Example 3 at a driving frequency other than resonance.

In the case of *resonance*, where the frequency of the driving force *does* coincide with one of the normal mode frequencies of the membrane, (n_0, m_0), that is, $\omega = \omega_{n_0 m_0}$, we have

$$T_{n_0 m_0}(t) = \frac{2A}{\rho l_x l_y \omega} \sin \frac{n_0 \pi x_0}{l_x} \sin \frac{m_0 \pi y_0}{l_y} \left[\sin \omega t - \omega t \frac{\omega}{\omega_{n_0 m_0}} \cos \omega t \right]$$

and

$$u(x,y,t) = \frac{4A}{\rho l_x l_y} \sum_{n \neq n_0}^{\infty} \sum_{m \neq m_0}^{\infty} \frac{1}{(\omega_{nm}^2 - \omega^2)} \left[\sin \omega t - \frac{\omega}{\omega_{nm}} \sin \omega_{nm} t \right]$$

$$\times \sin \frac{n \pi x_0}{l_x} \sin \frac{m \pi y_0}{l_y} \sin \frac{n \pi x}{l_x} \sin \frac{m \pi y}{l_y}$$

$$+ \frac{2A}{\rho l_x l_y \omega} \left[\sin \omega t - \omega t \frac{\omega}{\omega_{n_0 m_0}} \cos \omega t \right] \sin \frac{n_0 \pi x_0}{l_x} \sin \frac{m_0 \pi y_0}{l_y} \sin \frac{n_0 \pi x}{l_x} \sin \frac{m_0 \pi y}{l_y}.$$

Figure 11.9 shows the solution profile $u(x, 2.5, t)$ for the case of resonance where the frequency of the external force $\omega = \omega_{23} = \pi / \sqrt{2}$. Other parameters are the same as in Figure 11.8.

If the frequency $\omega_{n_0 m_0}$ is a multiple of a natural frequency, then instead of one of resonance term appearing in the solution there will be several.

Example 4

Consider the oscillations of the surface of a body of water in a reservoir with a rectangular surface $[0, l_x; 0, l_y]$ under the action of a variable external pressure on the surface given by

$$p_0(x, y, t) = -A\omega \cos \frac{\pi x}{l_x} \cos \frac{\pi y}{l_y} \sin \omega t,$$

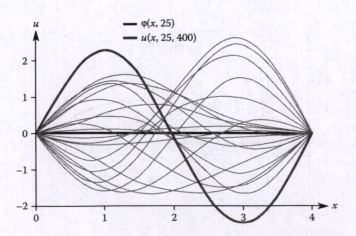

FIGURE 11.9   Solution profile $u(x, 2.5, t)$ for Example 3 at resonance.

if the depth of water in the equilibrium state is h.

Solution. The vertical displacement is described by equation (11.4):

$$\frac{\partial^2 u}{\partial t^2} - a^2 \left(\frac{\partial^2 u}{\partial x^2} + \frac{\partial^2 u}{\partial y^2} \right) = -\frac{A\omega}{\rho} \cos\frac{\pi x}{l_x} \cos\frac{\pi y}{l_y} \sin\omega t, \quad a^2 = gh$$

under zero initial conditions, $u(x,y,0) = 0$, $\partial u/\partial t(x,y,0) = 0$, and Neumann boundary conditions

$$\frac{\partial u}{\partial x}(0,y,t) = \frac{\partial u}{\partial x}(l_x,y,t) = \frac{\partial u}{\partial y}(x,0,t) = \frac{\partial u}{\partial y}(x,l_y,t) = 0.$$

Eigenvalues and eigenfunctions for these boundary conditions are

$$\lambda_{nm} = \lambda_{xn} + \lambda_{ym} = \pi^2 \left(\frac{n^2}{l_x^2} + \frac{m^2}{l_y^2} \right),$$

$$V_{nm}(x,y) = \cos\frac{n\pi x}{l_x} \cos\frac{m\pi y}{l_y}, \quad \|V_{nm}\|^2 = \begin{cases} l_x l_y, & \text{if } n=0,\ m=0, \\ l_x l_y / 2, & \text{if } n=0,\ m>0, \\ & \text{or } n>0,\ m=0, \\ l_x l_y / 4, & \text{if } n>0,\ m>0. \end{cases}$$

The three-dimensional picture shown in Figure 11.10 depicts two eigenfunctions (chosen as examples), $V_{11}(x,y)$ and $V_{22}(x,y)$. The values of λ for these modes are $\lambda_{11} = 0.891$ and $\lambda_{22} = 3.564$. As a result of initial conditions being equal to zero, the solution $u(x,y,t)$ is defined by the series

$$u(x,y,t) = \sum_{n=0}^{\infty} \sum_{m=0}^{\infty} T_{nm}(t) \cos\frac{n\pi x}{l_x} \cos\frac{m\pi y}{l_y},$$

where

$$T_{nm}(t) = \frac{1}{\omega_{nm}} \int_0^t f_{nm}(\tau) \sin\omega_{nm}(t-\tau)d\tau,$$

$$f_{nm}(t) = -\frac{A\omega\sin\omega t}{\rho} \cdot \frac{1}{\|V_{nm}\|^2} \int_0^{l_x}\int_0^{l_y} \cos\frac{\pi x}{l_x} \cos\frac{\pi y}{l_y} \cos\frac{n\pi x}{l_x} \cos\frac{m\pi y}{l_y}\ dx\,dy,$$

and $\omega_{nm} = a\lambda_{nm} = a\pi\sqrt{(n/l_x)^2 + (m/l_y)^2}$.

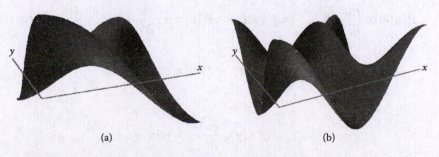

(a)　　　　　　　　　　　　　(b)

FIGURE 11.10 Eigenfunctions (a) $V_{11}(x,y)$ and (b) $V_{22}(x,y)$ for the free surface described in Example 4.

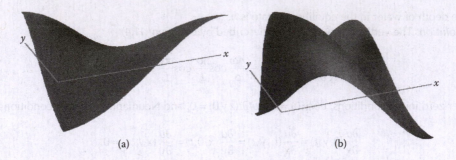

(a)   (b)

FIGURE 11.11   Graph of the membrane in Example 4.4 at (a) $t = 0.8$, (b) $t = 1.6$. The values of the parameters are $A = 0.1$, $\omega = 5$, $h = 1$, $g = 9.8$, $l_{x=4}$, $l_{y=6}$.

It is obvious that $f_{nm}(t) \neq 0$ only when $n = m = 1$ and that $f_{11}(t) = -(A\omega/\rho)\sin\omega t$. Thus we have

$$T_{11}(t) = -\frac{A\omega}{\rho(\omega_{11}^2 - \omega^2)}\left[\sin\omega t - \frac{\omega}{\omega_{11}}\sin\omega_{11}t\right]$$

and

$$u(x,y,t) = -\frac{A\omega}{\rho(\omega_{11}^2 - \omega^2)}\left[\sin\omega t - \frac{\omega}{\omega_{11}}\sin\omega_{11}t\right]\cos\frac{\pi x}{l_x}\cos\frac{\pi y}{l_y}.$$

11.2.3 The Fourier Method for Equations with Nonhomogeneous Boundary Conditions

Consider now the boundary problem of forced oscillations of a membrane given by equation (11.12) with *nonhomogeneous boundary and initial conditions* given by equations (11.13) and (11.14):

$$\frac{\partial^2 u}{\partial t^2} + 2\kappa\frac{\partial u}{\partial t} - a^2\left(\frac{\partial^2 u}{\partial x^2} + \frac{\partial^2 u}{\partial y^2}\right) + \gamma u = f(x,y,z),$$

$$P_1[u] \equiv \alpha_1\frac{\partial u}{\partial x} + \beta_1 u\bigg|_{x=0} = g_1(y,t), \quad P_2[u] \equiv \alpha_2\frac{\partial u}{\partial x} + \beta_2 u\bigg|_{x=l_x} = g_2(y,t),$$

$$P_3[u] \equiv \alpha_3\frac{\partial u}{\partial y} + \beta_3 u\bigg|_{y=0} = g_3(x,t), \quad P_4[u] \equiv \alpha_4\frac{\partial u}{\partial y} + \beta_4 u\bigg|_{y=l_y} = g_4(x,t),$$

$$u\big|_{t=0} = \varphi(x,y), \quad \frac{\partial u}{\partial t}\bigg|_{t=0} = \psi(x,y).$$

We will consider situations when the boundary conditions along the edges of membrane are consistent at the corners of a membrane (which would be required in a physical occurrence), that is, when the following conforming conditions are valid:

$$P_3[g_1] \equiv \alpha_3 \frac{\partial g_1}{\partial y} + \beta_3 g_1 \bigg|_{y=0} = P_1[g_3] \equiv \alpha_1 \frac{\partial g_3}{\partial y} + \beta_1 g_3 \bigg|_{x=0},$$

$$P_4[g_1] \equiv \alpha_4 \frac{\partial g_1}{\partial y} + \beta_4 g_1 \bigg|_{y=l_y} = P_1[g_4] \equiv \alpha_1 \frac{\partial g_4}{\partial y} + \beta_1 g_4 \bigg|_{x=0},$$

$$P_3[g_2] \equiv \alpha_3 \frac{\partial g_2}{\partial y} + \beta_3 g_2 \bigg|_{y=0} = P_2[g_3] \equiv \alpha_2 \frac{\partial g_3}{\partial y} + \beta_2 g_3 \bigg|_{x=l_x},$$

$$P_4[g_2] \equiv \alpha_4 \frac{\partial g_2}{\partial y} + \beta_4 g_2 \bigg|_{y=l_y} = P_2[g_4] \equiv \alpha_2 \frac{\partial g_4}{\partial y} + \beta_2 g_4 \bigg|_{x=l_x}.$$

If in some particular problem such consistency conditions do not hold, the way to resolve such a difficulty may require a smearing of boundary conditions at the corners using simple smooth functions to make them consistent. The solutions of an actual problem and the modified one will differ at the vicinity of the corners, but hopefully this difference will be not that substantial at points far enough away from these corners.

Let us return to consistent boundary conditions. Notice that it is not possible to use the Fourier method immediately since the boundary conditions are nonhomogeneous. However this problem is easily reduced to the problem with zero boundary conditions. To proceed, let us search for solutions of the problem in the form

$$u(x, y, t) = v(x, y, t) + w(x, y, t), \tag{11.51}$$

where $v(x, y, t)$ is an unknown function and the function $w(x, y, t)$ satisfies the given nonhomogeneous boundary conditions

$$P_1[w] \equiv \alpha_1 w_x + \beta_1 w \big|_{x=0} = g_1(y, t), \quad P_2[w] \equiv \alpha_2 w_x + \beta_2 w \big|_{x=l_x} = g_2(y, t),$$

$$P_3[w] \equiv \alpha_3 w_y + \beta_3 w \big|_{y=0} = g_3(x, t), \quad P_4[w] \equiv \alpha_4 w_y + \beta_4 w \big|_{y=l_y} = g_4(x, t) \tag{11.52}$$

and possesses the necessary number of continuous derivatives with respect to x, y and t.

For the function $v(x, y, t)$ we have following boundary value problem (check this result as a Reading Exercise):

$$\frac{\partial^2 v}{\partial t^2} + 2\kappa \frac{\partial v}{\partial t} - a^2 \left(\frac{\partial^2 v}{\partial x^2} + \frac{\partial^2 v}{\partial y^2} \right) + \gamma v = f^*(x, y, t),$$

$$P_1[v]\big|_{x=0} = 0, \quad P_2[v]\big|_{x=l_x} = 0,$$

$$P_3[v]\big|_{y=0} = 0, \quad P_4[v]\big|_{y=l_y} = 0,$$

$$v\big|_{t=0} = \varphi^*(x,y), \quad \frac{\partial v}{\partial t}\bigg|_{t=0} = \psi^*(x,y),$$

where

$$f^*(x,y,t) = f(x,y,t) - \frac{\partial^2 w}{\partial t^2} - 2\kappa \frac{\partial w}{\partial t} + a^2\left(\frac{\partial^2 w}{\partial x^2} + \frac{\partial^2 w}{\partial y^2}\right) - \gamma w, \tag{11.53}$$

$$\varphi^*(x,y) = \varphi(x,y) - w(x,y,0),$$

$$\psi^*(x,y) = \psi(x,y) - w_t(x,y,0). \tag{11.54}$$

Solutions of this problem were considered in the previous section.

Let us search for the auxiliary function $w(x,y,t)$ in the form

$$w(x,y,t) = g_1(y,t)\bar{X} + g_2(y,t)\bar{\bar{X}} + g_3(x,t)\bar{Y} + g_4(x,t)\bar{\bar{Y}}$$

$$+ A(t)\bar{X}\bar{Y} + B(t)\bar{X}\bar{\bar{Y}} + C(t)\bar{\bar{X}}\bar{Y} + D(t)\bar{\bar{X}}\bar{\bar{Y}}. \tag{11.55}$$

We will choose the functions $\{\bar{X}, \bar{\bar{X}}\}$ in such a way that the function $\bar{X}(x)$ satisfies homogeneous boundary condition at $x = l_x$ and the function $\bar{\bar{X}}(x)$ satisfies homogeneous boundary condition at $x = 0$,

$$P_2[\bar{X}(l_x)] = 0, \quad P_1[\bar{\bar{X}}(0)] = 0.$$

Also, it is convenient to normalize the functions $\bar{X}(x)$ and $\bar{\bar{X}}(x)$ so that

$$P_1[\bar{X}(0)] = 1, \quad P_2[\bar{\bar{X}}(l_x)] = 1.$$

The final choice of functions $\{\bar{X}, \bar{\bar{X}}\}$ depends on the type of boundary conditions for the function $u(x,y,t)$.

Suppose β_1 and β_2 are not both zero. In this case we can search for $\bar{X}(x)$ and $\bar{\bar{X}}(x)$ as polynomials of the first order:

$$\bar{X}(x) = \gamma_1 + \delta_1 x \quad \text{and} \quad \bar{\bar{X}}(x) = \gamma_2 + \delta_2 x. \tag{11.56}$$

This choice yields the system of equations

$$\begin{cases} P_1[\bar{X}] \equiv \alpha_1 \dfrac{\partial \bar{X}}{\partial x} + \beta_1 \bar{X}\bigg|_{x=0} = \beta_1 \gamma_1 + \alpha_1 \delta_1 = 1, \\[3mm] P_2[\bar{X}] \equiv \alpha_2 \dfrac{\partial \bar{X}}{\partial x} + \beta_2 \bar{X}\bigg|_{x=l_x} = \beta_2 \gamma_1 + (\alpha_2 + \beta_2 l_x)\delta_1 = 0, \end{cases}$$

and

$$\begin{cases} P_1[\bar{\bar{X}}] \equiv \alpha_1 \dfrac{\partial \bar{\bar{X}}}{\partial x} + \beta_1 \bar{\bar{X}}\bigg|_{x=0} = \beta_1 \gamma_2 + \alpha_1 \delta_2 = 0, \\[3mm] P_2[\bar{\bar{X}}] \equiv \alpha_2 \dfrac{\partial \bar{\bar{X}}}{\partial x} + \beta_2 \bar{\bar{X}}\bigg|_{x=l_x} = \beta_2 \gamma_2 + (\alpha_2 + \beta_2 l_x)\delta_2 = 1. \end{cases}$$

From the above conditions we may find a unique solution for coefficients γ_1, δ_1, γ_2, δ_2:

$$\gamma_1 = \frac{\alpha_2 + \beta_2 l_x}{\beta_1 \beta_2 l_x + \beta_1 \alpha_2 - \beta_2 \alpha_1}, \quad \delta_1 = \frac{-\beta_2}{\beta_1 \beta_2 l_x + \beta_1 \alpha_2 - \beta_2 \alpha_1},$$

$$\gamma_2 = \frac{-\alpha_1}{\beta_1 \beta_2 l_x + \beta_1 \alpha_2 - \beta_2 \alpha_1}, \quad \delta_2 = \frac{\beta_1}{\beta_1 \beta_2 l_x + \beta_1 \alpha_2 - \beta_2 \alpha_1}. \tag{11.57}$$

Reading Exercise

Obtain these formulas for γ_1, δ_1, γ_2, and δ_2.

If $\beta_1 = \beta_2 = 0$, we search for $\bar{X}(x)$ and $\bar{\bar{X}}(x)$ as polynomials of second order from which it is easy to see that

$$\bar{X}(x) = x - \frac{x^2}{2l_x} \quad \text{and} \quad \bar{\bar{X}}(x) = \frac{x^2}{2l_x} \tag{11.58}$$

will serve our needs.

Similarly we will choose the functions $\{\bar{Y}, \bar{\bar{Y}}\}$ in such a way that

$$P_3[\bar{Y}(0)] = 1, \quad P_3[\bar{Y}(l_y)] = 0,$$

$$P_4[\bar{\bar{Y}}(0)] = 0, \quad P_4[\bar{\bar{Y}}(l_y)] = 1. \tag{11.59}$$

If $\beta_3 \neq 0$ or $\beta_4 \neq 0$, then the functions $\bar{Y}(y)$ and $\bar{\bar{Y}}(y)$ are polynomials of the first order and we may write

$$\bar{Y}(y) = \gamma_3 + \delta_3 y, \quad \bar{\bar{Y}}(y) = \gamma_4 + \delta_4 y. \tag{11.60}$$

The coefficients $\gamma_3, \delta_3, \gamma_4, \delta_4$ of these polynomials are defined uniquely and depend on the types of boundary conditions:

$$\gamma_3 = \frac{\alpha_4 + \beta_4 l_y}{\beta_3 \beta_4 l_y + \beta_3 \alpha_4 - \beta_4 \alpha_3}, \quad \delta_3 = \frac{-\beta_4}{\beta_3 \beta_4 l_y + \beta_3 \alpha_4 - \beta_4 \alpha_3},$$

$$\gamma_4 = \frac{-\alpha_3}{\beta_3 \beta_4 l_y + \beta_3 \alpha_4 - \beta_4 \alpha_3}, \quad \delta_4 = \frac{\beta_3}{\beta_3 \beta_4 l_y + \beta_3 \alpha_4 - \beta_4 \alpha_3}. \tag{11.61}$$

If $\beta_3 = \beta_4 = 0$, then $\bar{Y}(y)$ and $\bar{\bar{Y}}(y)$ can be taken as polynomials of the second order:

$$\bar{Y}(y) = y - \frac{y^2}{2l_y}, \quad \bar{\bar{Y}}(y) = \frac{y^2}{2l_y}. \tag{11.62}$$

Using the above results we can find the coefficients $A(t)$, $B(t)$, $C(t)$, and $D(t)$ in the auxiliary function $w(x, y, t)$.

At the boundary $x = 0$ we have

$$P_1[w]_{x=0} = g_1(y, t) + (P_1[g_3(0, t)] + A)\bar{Y} + (P_1[g_4(0, t)] + B)\bar{\bar{Y}}.$$

At the boundary $x = l_x$ we have

$$P_2[w]_{x=l_x} = g_2(y, t) + (P_2[g_3(l_x, t)] + C)\bar{Y} + (P_2[g_4(l_x, t)] + D)\bar{\bar{Y}}.$$

At the boundary $y = 0$ we have

$$P_3[w]_{y=0} = g_3(x, t) + (P_3[g_1(0, t)] + A)\bar{X} + (P_3[g_2(0, t)] + C)\bar{\bar{X}}.$$

At the boundary $y = l_y$ we have

$$P_4[w]_{y=l_y} = g_4(x, t) + (P_4[g_1(l_y, t)] + B)\bar{X} + (P_4[g_2(l_y, t)] + D)\bar{\bar{X}}.$$

To simplify the above we may choose

$$A(t) = -P_3[g_1(y, t)]_{y=0}, \quad B(t) = -P_4[g_1(y, t)]_{y=l_y},$$

$$C(t) = -P_3[g_2(y, t)]_{y=0}, \quad D(t) = -P_4[g_2(y, t)]_{y=l_y}.$$

For the above choices the boundary conditions conform at the edges:

$$P_3[g_1(y,t)]_{y=0} = P_1[g_3(x,t)]_{x=0}, \quad P_3[g_2(y,t)]_{y=0} = P_2[g_3(x,t)]_{x=l_x},$$

$$P_4[g_1(y,t)]_{y=l_y} = P_1[g_4(x,t)]_{x=0}, \quad P_4[g_2(y,t)]_{y=l_y} = P_2[g_4(x,t)]_{x=l_x}.$$

It is easy to check that the above auxiliary functions, $w(x,y,t)$, satisfy the given boundary conditions. We leave to verify as a Reading Exercise that

$$P_1[w]_{x=0} = g_1(y,t), \quad P_2[w]_{x=l_x} = g_2(y,t),$$

$$P_3[w]_{y=0} = g_3(x,t), \quad P_4[w]_{y=l_y} = g_4(x,t).$$

We now consider an example of the solution of a specific problem for oscillations with nonhomogeneous boundary conditions.

Example 5

Consider oscillations of a homogeneous rectangular membrane ($0 \le x \le l_x, \ 0 \le y \le l_y$), if the boundary conditions are given by

$$u(0,y,t) = u(l_x,y,t) = 0 \quad \text{and} \quad u(x,0,t) = u(x,l_y,t) = h\sin\frac{\pi x}{l_x}.$$

Initially the membrane has shape and velocity given, respectively, by

$$\varphi(x,y,0) = h\sin\frac{\pi x}{l_x} \quad \text{and} \quad \psi(x,y,0) = v_0\sin\frac{\pi x}{l_x}.$$

Solution. To proceed we must solve the equation

$$\frac{\partial^2 u}{\partial t^2} - a^2\left(\frac{\partial^2 u}{\partial x^2} + \frac{\partial^2 u}{\partial y^2}\right) = 0$$

under the conditions

$$u(x,y,0) = h\sin\frac{\pi x}{l_x}, \quad \frac{\partial u}{\partial t}(x,y,0) = v_0\sin\frac{\pi x}{l_x},$$

$$u(0,y,t) = u(l_x,y,t) = 0, \quad u(x,0,t) = u(x,l_y,t) = h\sin\frac{\pi x}{l_x}.$$

We search for a solution of this problem as the sum

$$u(x,y,t) = v(x,y,t) + w(x,y,t),$$

where $w(x,y,t)$, chosen to be

$$w(x,y,t) = h\sin\frac{\pi x}{l_x},$$

FIGURE 11.12   Graph of the auxiliary function $w(x,y,t)$ for Example 5.

satisfies the boundary conditions of the problem and therefore obviously provides homogeneous boundary conditions for the function $v(x,y,t)$ (this result can be obtained from general formulas presented above).

For the function $v(x,y,t)$ we have the boundary problem for the nonhomogeneous equation of oscillation where

$$f^*(x,y,t) = -a^2 \frac{h\pi^2}{l_x^2} \sin\frac{\pi x}{l_x},$$

$$\varphi^*(x,y) = 0, \quad \psi^*(x,y) = v_0 \sin\frac{\pi x}{l_x}.$$

and homogeneous boundary conditions. The solution to the problem is thus

$$u(x,y,t) = h\sin\frac{\pi x}{l_x} + \sum_{m=1}^{\infty}\left\{\frac{4v_0}{(2m-1)\pi\omega_{1(2m-1)}}\sin\omega_{1(2m-1)}t\right.$$

$$\left. - \frac{4ha^2\pi}{(2m-1)\,l_x^2\omega_{1(2m-1)}^2}\left[1-\cos\omega_{1(2m-1)}t\right]\right\}\sin\frac{\pi x}{l_x}\sin\frac{(2m-1)\pi y}{l_y}.$$

Figures 11.12 and 11.13 present the graphs of the auxiliary function and profile of the membrane for the following values of the parameters: $a^2 = 1$, $l_x = 4$, $l_y = 6$, $v_0 = 0.5$, $h = 0.1$.

11.3 THE FOURIER METHOD APPLIED TO SMALL TRANSVERSE OSCILLATIONS OF A CIRCULAR MEMBRANE

Suppose a membrane in its equilibrium position has the form of a circle with radius l, is located in the x-y plane, and has its center as the origin of coordinates. As before for the case of rectangular membranes we will consider transverse oscillations of the membrane only, for which all points on the membrane move perpendicular to the x-y plane. In polar

(a)   (b)

FIGURE 11.13   Graph of the membrane in Example 5 at (a) $t = 5$, (b) $t = 8.5$.

(or circular) coordinates with variables r and φ, the displacement, u, of points on the membrane will be a function of r, φ, and time t: $u = u(r,\varphi,t)$.

The Laplace operator in polar coordinates is given by

$$\nabla^2 u = \frac{\partial^2 u}{\partial r^2} + \frac{1}{r}\frac{\partial u}{\partial r} + \frac{1}{r^2}\frac{\partial^2 u}{\partial \varphi^2}$$

with the result that the equation of oscillations of a membrane in polar coordinates has the form

$$\frac{\partial^2 u}{\partial t^2} + 2\kappa\frac{\partial u}{\partial t} - a^2\left(\frac{\partial^2 u}{\partial r^2} + \frac{1}{r}\frac{\partial u}{\partial r} + \frac{1}{r^2}\frac{\partial^2 u}{\partial \varphi^2}\right) + \gamma u = f(r,\varphi,t). \tag{11.63}$$

The domains of the independent variables are $0 \le r \le l$, $0 \le \varphi < 2\pi$, and $0 \le t < \infty$, respectively.

Boundary conditions in polar coordinate are particularly simple and in general from can be written as

$$\alpha\frac{\partial u}{\partial r} + \beta u\bigg|_{r=l} = g(\varphi,t), \tag{11.64}$$

where α and β are constants that are not zero simultaneously, that is, $|\alpha| + |\beta| \ne 0$. If $\alpha = 0$, we have Dirichlet boundary condition, and if $\beta = 0$, we have Neumann boundary condition. If $\alpha \ne 0$ and $\beta \ne 0$, then we have mixed boundary conditions. From physical arguments it will normally be the case that $\beta/\alpha > 0$.

It should be clear from our previous discussion that in order to describe the membrane oscillation for $t > 0$ we will need to know the initial displacement and initial velocity of the membrane in order to solve various physical problems. In an analogous manner to the rectangular case, the initial conditions may be stated as

$$u|_{t=0} = \phi(r,\varphi), \quad \frac{\partial u}{\partial t}\bigg|_{t=0} = \psi(r,\varphi). \tag{11.65}$$

Thus the deviation of points of membrane with coordinates (r,φ) at some arbitrary initial moment of time is $\phi(r,\varphi)$ and initial velocities of these points is given by the function $\psi(r,\varphi)$. It should be clear from physical arguments that the solution $u(r,\varphi,t)$ is to be single-valued, periodic in φ with period 2π and remains finite at all points of the membrane, including the center of membrane where $r = 0$.

11.3.1 The Fourier Method for Homogeneous Equations with Homogeneous Boundary Conditions

We begin by assuming that the membrane is set in motion by some combination of initial displacements and/or initial velocities. To solve the equation of motion for the displacement from equilibrium, $u(r, \varphi, t)$, of the membrane at all points and times we start by solving the *homogeneous equation* of oscillations

$$\frac{\partial^2 u}{\partial t^2} + 2\kappa \frac{\partial u}{\partial t} - a^2 \left(\frac{\partial^2 u}{\partial r^2} + \frac{1}{r} \frac{\partial u}{\partial r} + \frac{1}{r^2} \frac{\partial^2 u}{\partial \varphi^2} \right) + \gamma u = 0 \tag{11.66}$$

with *homogeneous* boundary conditions

$$\alpha \frac{\partial u}{\partial r} + \beta u \bigg|_{r=l} = 0 \tag{11.67}$$

and *nonhomogeneous* initial conditions from equations (11.64)

$$u|_{t=0} = \phi(r, \varphi), \quad \frac{\partial u}{\partial t} \bigg|_{t=0} = \psi(r, \varphi).$$

Let us represent the function $u(r, \varphi, t)$ as a product of two functions. The first depends only on r and φ and we denote it as $V(r, \varphi)$; the second depends only on t and is denoted as $T(t)$:

$$u(r, \varphi, t) = V(r, \varphi) T(t), \tag{11.68}$$

which will be required to satisfy the boundary condition in equation (11.66).

Substituting equation (11.67) in equation (11.65) and separating variables we obtain

$$\frac{T''(t) + 2\kappa T'(t) + \gamma T(t)}{a^2 T(t)} \equiv \frac{1}{V(r, \varphi)} \left[\frac{\partial^2 V}{\partial r^2} + \frac{1}{r} \frac{\partial V}{\partial r} + \frac{1}{r^2} \frac{\partial^2 V}{\partial \varphi^2} \right] = -\lambda$$

where λ is a separation of variables constant (we already know that a choice of minus sign before λ is convenient). Thus the function $T(t)$ satisfies the ordinary linear homogeneous differential equation of second order

$$T''(t) + 2\kappa T'(t) + (a^2 \lambda + \gamma) T(t) = 0, \tag{11.69}$$

and the function $V(r, \phi)$ satisfies the equation

$$\frac{\partial^2 V}{\partial r^2} + \frac{1}{r} \frac{\partial V}{\partial r} + \frac{1}{r^2} \frac{\partial^2 V}{\partial \varphi^2} + \lambda V = 0, \tag{11.70}$$

with

$$\alpha \frac{\partial V(l, \varphi)}{\partial r} + \beta V(l, \varphi) = 0, \tag{11.71}$$

as a boundary condition. Using physical arguments we also require that the solutions remain finite (everywhere, including point $r = 0$) so that

$$V(r, \varphi)| < \infty \qquad (11.72)$$

and require the solutions to be periodic which may be defined as

$$V(r, \varphi) = V(r, \varphi + 2\pi). \qquad (11.73)$$

For the boundary value problem defined by equations (11.69) through (11.72) we again may separate the variables, in this case r and φ, using the substitution

$$V(r, \varphi) = R(r)\Phi(\varphi). \qquad (11.74)$$

Substituting equation (11.73) into equation (11.69) and dividing by $R(r)\Phi(\varphi)$, we obtain

$$\frac{r^2 R''(r) + r R'(r)}{R(r)} + \lambda r^2 \equiv -\frac{\Phi''(\varphi)}{\Phi(\varphi)} = \nu,$$

where ν is another separation of variables constant. The above equations result in two eigenvalue problems, each of which is one-dimensional. The first is given by

$$\Phi''(\varphi) + \nu \Phi(\varphi) = 0 \qquad (11.75)$$

with boundary condition

$$\Phi(\varphi) = \Phi(\varphi + 2\pi), \qquad (11.76)$$

(with a change of φ by the value 2π we return to the same point on the membrane; consequently the displacement of the membrane $u(r, \varphi, t)$ does not change under the substitution φ by $\varphi + 2\pi$) and the second is given by

$$\frac{d^2 R}{dr^2} + \frac{1}{r}\frac{dR}{dr} + \left(\lambda - \frac{\nu}{r^2}\right)R = 0, \qquad (11.77)$$

with boundary condition

$$\alpha \frac{\partial R}{\partial r} + \beta R\bigg|_{r=l} = 0, \ |R(0)| < \infty. \qquad (11.78)$$

Boundary condition (11.75) indicates that ν cannot be negative; otherwise solutions of equation (11.74) would include exponentially increasing or decreasing functions that are not periodic.

If $v = 0$, then equation (11.74) reduces to $\Phi''(\varphi) = 0$ having the solution $\Phi(\varphi) = C_1\varphi + C_2$, which will satisfy condition (11.75) only in the case that $C_1 = 0$. In other words, $\Phi(\varphi) = C_2$ where C_2 is a constant and, in particular, we may take $\Phi(\varphi) \equiv 1$. For the above choices $v_0 = 0$ is the eigenvalue and the function $\Phi_0(\varphi) \equiv 1$ is the corresponding eigenfunction.

If $v > 0$ linear-independent solutions of equation (11.74) are

$$\Phi^{(1)}(\varphi) = \cos\sqrt{v}\varphi \quad \text{and} \quad \Phi^{(2)}(\varphi) = \sin\sqrt{v}\varphi.$$

The functions $\Phi^{(1)}(\varphi)$ and $\Phi^{(2)}(\varphi)$ each have period $2\pi/\sqrt{v}$. To maintain the periodic boundary conditions this period must equal 2π or an integer number times 2π, which will be the case only if $\sqrt{v}$ is an integer number. We require, therefore, that $v = n^2$ where $n = 1,2,3,\ldots$ We may use these integer values of v to label the eigenvalues for the equation as v_n and corresponding eigenfunctions by $\Phi_n^{(1)}(\varphi)$ and $\Phi_n^{(2)}(\varphi)$. Thus the solutions to equation (11.74) are given by

$$v = n^2, \quad \Phi_n^{(1)}(\varphi) = \cos n\varphi, \quad \Phi_n^{(2)}(\varphi) = \sin n\varphi, \quad n = 0,1,2\ldots \tag{11.79}$$

Note that the boundary value problem given by equations (11.74) and (11.75) is the Sturm-Liouville problem with periodic boundary conditions on the interval $[0,2\pi]$, which is why the two linearly independent eigenfunctions $\cos n\phi$ and $\sin n\varphi$ belong to the same eigenvalue $v_n = n^2$.

To find the solutions $R(r)$ we use equation (11.76) with homogeneous boundary conditions (11.77) where the eigenvalues $v = n^2$ found previously for $\Phi(\varphi)$ are now included in equation (11.76). Equation (11.76) is the Bessel equation—see Chapter 8. The second of the conditions given in equation (11.77) imposed on the function $R(r)$ is that it should be bounded at $r = 0$, which is a singular point of the equation. For singular points this kind of restriction may be considered as a boundary condition.

The general solution of equation (11.76) with $v = n^2$ is

$$R_n(r) = d_{1n}J_n(\sqrt{\lambda}r) + d_{2n}N_n(\sqrt{\lambda}r), \tag{11.80}$$

where $J_n(\sqrt{\lambda}r)$ is the Bessel function of order n and $N_n(\sqrt{\lambda}r)$ is the Neumann function of order n, $n = 0,1,2\ldots$ Clearly only positive λ are allowed. The functions $J_n(\sqrt{\lambda}r)$ are bounded as $r \to 0$ whereas the functions $N_n(\sqrt{\lambda}r)$ diverge as $r \to 0$. The requirement that the function $R(r)$ be bounded leads to the result that the coefficients of $N_n(\sqrt{\lambda}r)$ must equal zero, in which case we have $d_{2n} = 0$.

From the homogeneous boundary condition given in equation (11.77) we have

$$\alpha\sqrt{\lambda}\,J_n'(\sqrt{\lambda}l) + \beta J_n(\sqrt{\lambda}l) = 0.$$

Setting $\sqrt{\lambda}l \equiv \mu$ we obtain a transcendental equation defining μ

$$\alpha\mu J_n'(\mu) + \beta l J_n(\mu) = 0, \tag{11.81}$$

which has an infinite number of roots which we label as

$$\mu_0^{(n)},\ \mu_1^{(n)},\mu_2^{(n)},\ldots$$

The corresponding values of λ are thus

$$\lambda_{nm} = \left(\frac{\mu_m^{(n)}}{l}\right)^2,\ n,m = 0,1,2,\ldots. \tag{11.82}$$

From here we see that we need only *positive roots*, $\mu_m^{(n)}$, because negative roots do not give new values of λ_{nm}.

Eigenfunctions are

$$R_{nm}(r) = J_n\left(\frac{\mu_m^{(n)}}{l}r\right). \tag{11.83}$$

The index $m = 0$ corresponds to the first root of equation (11.80). (It should be noted that very often the roots are labeled with the starting value $m = 1$ in the literature.). The eigenvalues λ_{nm} and eigenfunctions $R_{nm}(r)$ are the eigenvalues and eigenfunctions of the Sturm-Liouville problem given in equations (11.76) and (11.77) with $v = n^2$.

Eigenfunctions $R_{nm}(r)$ belonging to different eigenvalues λ_{nm} for some fixed value of n are orthogonal with weight r (see Chapter 8):

$$\int_0^l r R_{nm_1}(r)R_{nm_2}(r)dr = 0 \tag{11.84}$$

or

$$\int_0^l r J_n\left(\mu_{m_1}^{(n)}r/l\right)J_n\left(\mu_{m_2}^{(n)}r/l\right)dr = 0 \quad \text{for} \quad m_1 \neq m_2. \tag{11.85}$$

For each eigenvalue

$$\lambda_{nm} = \left(\frac{\mu_m^{(n)}}{l}\right)^2$$

there are two linearly independent eigenfunctions

$$V_{nm}^{(1)}(r,\varphi) = J_n\left(\frac{\mu_m^{(n)}}{l}r\right)\cos n\varphi \quad \text{and} \quad V_{nm}^{(2)}(r,\varphi) = J_n\left(\frac{\mu_m^{(n)}}{l}r\right)\sin n\varphi. \tag{11.86}$$

Since

$$\int_0^{2\pi} d\varphi = 2\pi,\ \int_0^{2\pi}\cos^2 n\varphi\ d\varphi = \pi,\ \int_0^{2\pi}\sin^2 n\varphi\ d\varphi = \pi\ (n > 0)$$

the norms of eigenfunctions $V_{nm}^{(1)}(r,\varphi)$ and $V_{nm}^{(2)}(r,\varphi)$ are

$$\|V_{0m}^{(1)}\|^2 = 2\pi\|R_{nm}(r)\|^2, \quad \|V_{nm}^{(1)}\|^2 = \|V_{nm}^{(2)}\|^2 = \pi\|R_{nm}(r)\|^2 \quad \text{for} \ \ n>0. \tag{11.87}$$

The squared norm $\|R_{nm}\|^2 = \int\limits_0^l rJ_n^2\left(\dfrac{\mu_m^{(n)}}{l}r\right) dr$ is (see [1]):

1. For the *Dirichlet* boundary condition $\alpha = 0$ and $\beta = 1$, in which case eigenvalues are obtained from the equation

$$J_n(\mu) = 0$$

and we have

$$\|R_{nm}\|^2 = \frac{l^2}{2}[J_n'(\mu_m^{(n)})]^2. \tag{11.88}$$

2. For the *Neumann* boundary condition $\alpha = 1$ and $\beta = 0$, in which case eigenvalues are obtained from the equation

$$J_n'(\mu) = 0$$

and we have

$$\|R_{nm}\|^2 = \frac{l^2}{2(\mu_m^{(n)})^2}\left[(\mu_m^{(n)})^2 - n^2\right] J_n^2(\mu_m^{(n)}). \tag{11.89}$$

3. For the *mixed* boundary condition $\alpha = 1$ and $\beta = h$, in which case eigenvalues are obtained from the equation

$$\mu J_n'(\mu) + hlJ_n(\mu) = 0$$

and we have

$$\|R_{nm}\|^2 = \frac{l^2}{2}\left[1 + \frac{l^2h^2 - n^2}{(\mu_m^{(n)})^2}\right] J_n^2(\mu_m^{(n)}). \tag{11.90}$$

These results completely define λ_{nm}, $V_{nm}^{(1)}$, and $V_{nm}^{(2)}$, the eigenvalues and eigenfunctions of equations (11.69) through (11.72) for the problem of free oscillations of a circular membrane for the case of homogeneous boundary conditions.

To determine the time evolution of the oscillating membrane we return to equation (11.68). This is an ordinary linear homogeneous differential equation of second order.

Substituting $\lambda = \lambda_{nm}$ in the equation and denoting the corresponding solution of this equation as $T_{nm}(t)$, we have

$$T_{nm}''(t) + 2\kappa T_{nm}'(t) + (a^2\lambda_{nm} + \gamma)T_{nm}(t) = 0. \tag{11.91}$$

This linear second-order equation with constant coefficients has two linearly independent solutions

$$y_{nm}^{(1)}(t) = \begin{cases} e^{-\kappa t}\cos\omega_{nm}t, & \kappa^2 < a^2\lambda_{nm} + \gamma, \\ e^{-\kappa t}\cosh\omega_{nm}t, & \kappa^2 > a^2\lambda_{nm} + \gamma, \\ e^{-\kappa t}, & \kappa^2 = a^2\lambda_{nm} + \gamma, \end{cases}$$

$$\tag{11.92}$$

$$y_{nm}^{(2)}(t) = \begin{cases} e^{-\kappa t}\sin\omega_{nm}t, & \kappa^2 < a^2\lambda_{nm} + \gamma, \\ e^{-\kappa t}\sinh\omega_{nm}t, & \kappa^2 > a^2\lambda_{nm} + \gamma, \\ te^{-\kappa t}, & \kappa^2 = a^2\lambda_{nm} + \gamma, \end{cases}$$

with $\omega_{nm} = \sqrt{|a^2\lambda_{nm} + \gamma - \kappa^2|}$.

A general solution of equation (11.91) is a linear combination of these $y_{nm}^{(1)}(t)$ and $y_{nm}^{(2)}(t)$. Collecting the functions $\Phi(\varphi)$, $R(r)$, and $T(t)$ and substituting them into identity (11.68) gives solutions to equation (11.66) in the form of a product of functions of one variable satisfying the given boundary conditions:

$$u_{nm}^{(1)}(r,\varphi,t) = T_{nm}^{(1)}V_{nm}^{(1)}(r,\varphi) = [a_{nm}y_{nm}^{(1)} + b_{nm}y_{nm}^{(2)}]V_{nm}^{(1)}(r,\varphi),$$

$$u_{nm}^{(2)}(r,\varphi,t) = T_{nm}^{(2)}V_{nm}^{(2)}(r,\varphi) = [c_{nm}y_{nm}^{(1)} + d_{nm}y_{nm}^{(2)}]V_{nm}^{(2)}(r,\varphi).$$

To find solutions to the equation of motion for a membrane satisfying not only the boundary conditions above but also various initial conditions, let us sum these functions as a series, superposing all $u_{nm}^{(1)}(r,\varphi,t)$ and $u_{nm}^{(2)}(r,\varphi,t)$:

$$u(r,\varphi,t) = \sum_{n=0}^{\infty}\sum_{m=0}^{\infty}\left\{[a_{nm}y_{nm}^{(1)}(t) + b_{nm}y_{nm}^{(2)}(t)]V_{nm}^{(1)}(r,\varphi)\right.$$

$$\tag{11.93}$$

$$\left. + [c_{nm}y_{nm}^{(1)}(t) + d_{nm}y_{nm}^{(2)}(t)]V_{nm}^{(2)}(r,\varphi)\right\}.$$

If this series and the series obtained from it by twice differentiating term by term with respect to the variables r, φ, and t converge uniformly, then their sum will be a solution to equation (11.66), satisfying boundary condition (11.67).

To satisfy the initial conditions given in equation (11.65) we require that

$$u|_{t=0} = \phi(r,\varphi) = \sum_{n=0}^{\infty}\sum_{m=0}^{\infty}[a_{nm}V_{nm}^{(1)}(r,\varphi) + c_{nm}V_{nm}^{(2)}(r,\varphi)] \tag{11.94}$$

and

$$\frac{\partial u}{\partial t}\bigg|_{t=0} = \psi(r,\varphi) = \sum_{n=0}^{\infty} \sum_{m=0}^{\infty} \{[\omega_{mn}b_{mn} - \kappa a_{mn}]V_{nm}^{(1)}(r,\varphi) $$
$$+ [\omega_{mn}d_{mn} - \kappa c_{mn}]V_{nm}^{(2)}(r,\varphi)\}. \tag{11.95}$$

From here the coefficients a_{nm}, b_{mn}, c_{nm}, and d_{nm} can be obtained:

$$a_{nm} = \frac{1}{\|V_{nm}^{(1)}\|^2} \int_0^l \int_0^{2\pi} \phi(r,\varphi)\, V_{nm}^{(1)}(r,\varphi)\, r\,dr\,d\varphi,$$

$$b_{nm} = \frac{1}{\omega_{nm}}\left[\frac{1}{\|V_{nm}^{(1)}\|^2} \int_0^l \int_0^{2\pi} \psi(r,\varphi)\, V_{nm}^{(1)}(r,\varphi)\, r\,dr\,d\varphi + \kappa a_{nm} \right], \tag{11.96}$$

$$c_{nm} = \frac{1}{\|V_{nm}^{(2)}\|^2} \int_0^l \int_0^{2\pi} \phi(r,\varphi)\, V_{nm}^{(2)}(r,\varphi)\, r\,dr\,d\varphi,$$

$$d_{nm} = \frac{1}{\omega_{nm}}\left[\frac{1}{\|V_{nm}^{(2)}\|^2} \int_0^l \int_0^{2\pi} \psi(r,\varphi)\, V_{nm}^{(2)}(r,\varphi)\, r\,dr\,d\varphi + \kappa c_{nm} \right].$$

Equation (11.93) gives the evolution of *free oscillations* of a circular membrane when boundary conditions are homogeneous. It can be considered as the expansion of the (unknown) function $u(r,\varphi,t)$ in a Fourier series using the orthogonal system of functions $\{V_{nm}(r,\varphi)\}$. This series converges under sufficiently reasonable assumptions about initial and boundary conditions.

The particular solutions $u_{nm}(r,\varphi,t) = T_{nm}^{(1)}(t)V_{nm}^{(1)}(r,\varphi) + T_{nm}^{(2)}(t)V_{nm}^{(2)}(r,\varphi)$ are called *standing wave* solutions. From this we see that the profile of a standing wave depends on the functions $V_{nm}(r,\varphi)$; the functions $T_{nm}^{(1)}(t)$ and $T_{nm}^{(2)}(t)$ only change the amplitude of the standing wave over time, as was the case for standing waves on a string and the rectangular membrane. Lines on the membrane defined by $V_{nm}(r,\varphi) = 0$ remain at rest for all times and are called *nodal lines* of the standing wave $V_{nm}(r,\varphi)$. Points where $V_{nm}(r,\varphi)$ reaches a relative maximum or minimum for all times are called *antinodes* of this standing wave. From the above discussion of the Fourier expansion we see that an arbitrary motion of the membrane may be thought of as an infinite sum of these standing waves.

Each mode $u_{nm}(r,\varphi,t)$ possesses a characteristic pattern of nodal lines. The first few of these normal vibration modes for $V_{nm}^{(1)}(r,\varphi) = J_n\left(\frac{\mu_m^{(n)}}{l}r\right)\cos n\varphi$ are sketched in Figure 11.14 with similar pictures for the modes $V_{nm}^{(2)}(r,\varphi)$. In the fundamental mode of vibration corresponding to $\mu_0^{(0)}$, the membrane vibrates as a whole. In the mode corresponding to $\mu_1^{(0)}$ the membrane vibrates in two parts as shown with the part labeled with a plus sign initially above the equilibrium level and the part labeled with a minus sign initially below the

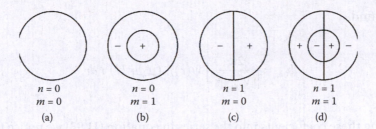

$n = 0$	$n = 0$	$n = 1$	$n = 1$
$m = 0$	$m = 1$	$m = 0$	$m = 1$
(a)	(b)	(c)	(d)

FIGURE 11.14 Drawing of the first few modes of vibration for the mode $V_{nm}^{(1)}(r,\varphi)$.

equilibrium. The nodal line in this case is a circle, which remains at rest as the two sections reverse location. The mode characterized by $\mu_0^{(1)}$ is equal to zero when $\varphi = \pm\pi/2$ and is positive and negative as shown.

11.3.2 Radial Oscillations of a Membrane

Oscillations of a circular membrane are said to be radial if they do not depend on the polar angle φ (i.e., the deviation of an arbitrary point M from its position of equilibrium at time t depends only on t and the distance between point M and the center of the membrane). Solutions for radial oscillations will have a simpler form than more general types of oscillations. Physically we see that radial oscillations will occur when initial displacements and initial velocities do not depend on φ, but rather are functions only of r:

$$u(r,\varphi,t)\big|_{t=0} = \phi(r), \quad \frac{\partial u}{\partial t}(r,\varphi,t)\bigg|_{t=0} = \psi(r). \tag{11.97}$$

In this case all coefficients, a_{nm}, b_{nm}, c_{nm}, and d_{nm} with $n \geq 1$ equal zero. We may easily verify this, for example, for a_{nm}:

$$a_{nm} = \frac{1}{\| V_{nm}^{(1)} \|^2} \int_0^l \int_0^{2\pi} \phi(r) J_n \left(\mu_m^{(n)} r/l\right) \cos n\varphi \, r \, dr \, d\varphi.$$

Because $\int_0^{2\pi} \cos n\varphi \, d\varphi = 0$ for any integer $n \geq 1$ we have $a_{nm} = 0$.

Similarly $b_{nm} = 0$ for $n \geq 1$, $c_{nm} = 0$, $d_{nm} = 0$ for all n. Thus, *the solution does not contain the functions $V_{nm}^{(2)}(r,\varphi)$*. If $n = 0$, the coefficients a_{0m} and b_{0m} are nonzero and the formulas used to calculate them can be simplified. Putting factors that are independent of φ outside the integral and using $\int_0^{2\pi} d\varphi = 2\pi$, we have, after simplification,

$$a_{0m} = \frac{2\pi}{\| V_{0m}^{(1)} \|^2} \int_0^l \varphi(r) J_0 \left(\mu_m^{(0)} r/l\right) r \, dr. \tag{11.98}$$

Similarly we find

$$b_{0m} = \frac{2\pi}{\| V_{0m}^{(1)} \|^2} \int_0^l \psi(r) J_0(\mu_m^{(0)} r / l) \, r \, dr. \tag{11.99}$$

Substituting these coefficients into the series in equation (11.93) we notice that the series reduces from a double series to a single one since all terms in the second sum of this series disappear. Only those terms in the first sum remain for which $n = 0$, making it necessary to sum only on m but not on n. The final result is

$$u(r, \varphi, t) = \sum_{m=0}^{\infty} [a_{0m} y_{0m}^{(1)}(t) + b_{0m} y_{0m}^{(2)}(t)] J_0\left(\frac{\mu_m^{(0)}}{l} r\right). \tag{11.100}$$

Thus, for radial oscillations the solution contains only Bessel functions of zero order.

Example 6

Find the transverse oscillations of a circular membrane with radius l with a fixed edge. Assume the initial displacement has the form of a paraboloid of rotation, initial velocities are zero, and the reaction of the environment is small enough to be neglected.

Solution. Drawing from the above discussion we have the following boundary value problem of a circular membrane with fixed edge:

$$\frac{\partial^2 u}{\partial t^2} - a^2 \left[\frac{\partial^2 u}{\partial r^2} + \frac{1}{r} \frac{\partial u}{\partial r} \right] = 0 \quad (0 \le r < l, \, 0 \le \varphi < 2\pi, \, t > 0),$$

$$u(r,0) = A\left(1 - \frac{r^2}{l^2}\right), \quad \frac{\partial u}{\partial t}(r, \varphi, 0) = 0, \quad u(l, \varphi, t) = 0.$$

The oscillations of the membrane are radial since the initial displacement and the initial velocities do not depend on the polar angle φ. Thus, only terms with $n = 0$ are not zero.

Boundary conditions of the problem are of *Dirichlet type*, in which case eigenvalues $\mu_m^{(0)}$ are the solutions of the equation $J_0(\mu) = 0$; the eigenfunctions are

$$V_{0m}^{(1)}(r, \varphi) = J_0\left(\frac{\mu_m^{(0)}}{l} r\right).$$

The three-dimensional picture shown in Figure 11.15 depicts the two eigenfunctions for the given problem,

$$V_{00}^{(1)}(r, \varphi) = J_0(\mu_0^{(0)} r / l) \text{ and } V_{02}^{(1)}(r, \varphi) = J_0(\mu_2^{(0)} r / l)$$

(these two eigenfunctions are chosen as examples), the corresponding values of λ are $\lambda_{00} = 1.4458$ and $\lambda_{02} = 18.7218$.

(a)   (b)

FIGURE 11.15   Two eigenfunctions, (a) $V_{00}^{(1)}(r,\varphi)$ and (b) $V_{02}^{(1)}(r,\varphi)$, for the Dirichlet boundary conditions in Example 6.

The solution, $u(r,\varphi,t)$, is given by the series

$$u(r,\varphi,t) = \sum_{m=0}^{\infty} a_{0m} \cos\frac{a\mu_m^{(0)}t}{l} J_0(\mu_m^{(0)}r\,/\,l).$$

The coefficients a_{0m} are given by equation (11.97):

$$a_{0m} = \frac{2\pi}{\|V_{0m}^{(1)}\|^2} \int_0^l A\left(1-\frac{r^2}{l^2}\right) J_0(\mu_m^{(0)}r\,/\,l)\,r\,dr.$$

The program **Waves** calculates these coefficients (for values of l assigned by the reader), thus allowing one to concentrate on the concept of solution and modeling the resulting function $u(r,\varphi,t)$. To see the way of calculating the coefficients a_{0m} we refer the reader (here and in other similar situations) to book [1]: the analytical formula for a_{0m} is

$$a_{0m} = \frac{8A}{\left(\mu_m^{(0)}\right)^3 J_1\left(\mu_m^{(0)}\right)},$$

thus

$$u(r,\varphi,t) = 8A \sum_{m=0}^{\infty} \frac{1}{\left(\mu_m^{(0)}\right)^3 J_1(\mu_m^{(0)})} \cos\frac{a\mu_m^{(0)}t}{l} J_0(\mu_m^{(0)}r\,/\,l).$$

Figure 11.16 shows two snapshots of the solution at the times $t=2.5$ and $t=5$. This solution was obtained with program **Waves** for the case $a^2=1$, $l=2$, and $A=0.1$.

(a)   (b)

FIGURE 11.16   Graph of the membrane in Example 6 at (a) $t=2.5$ and (b) $t=5$.

Example 7

Find the transverse oscillations of a homogeneous circular membrane of radius l with a rigidly fixed edge where the oscillations are initiated by a localized impact, normal to a surface of the membrane. This impact is applied at the point (r_0, φ_0) and supplies an impulse I $(0 < r_0 < l)$ to the membrane. Any initial displacement is absent and the reaction of the environment is negligible.

Solution. The boundary value problem describing the oscillations of the membrane reduces to the solution of the equation

$$\frac{\partial^2 u}{\partial t^2} = a^2 \left[\frac{\partial^2 u}{\partial r^2} + \frac{1}{r}\frac{\partial u}{\partial r} + \frac{1}{r^2}\frac{\partial^2 u}{\partial \varphi^2} \right] \quad (0 \le r < l, 0 \le \varphi < 2\pi, t > 0)$$

under the conditions

$$u(r, \varphi, 0) = 0, \quad \frac{\partial u}{\partial t}(r, \varphi, 0) = \frac{I}{\rho}\delta(r - r_0)\delta(\varphi - \varphi_0), \quad u(l, \varphi t) = 0.$$

The product $\delta(r - r_0)\delta(\varphi - \varphi_0)$ is a δ-function in two (polar) dimensions; multiplying by the area element in polar coordinates given by $r dr d\varphi$ and integrating over this area we obtain 1 or 0, depending on whether the point (r_0, φ_0) belongs to this area or not.

The boundary condition of the problem is of *Dirichlet type*, the eigenvalues are given by equation $J_n(\mu) = 0$, and the eigenfunctions are given by equations (11.86):

$$V_{nm}^{(1)}(r, \varphi) = J_n(\mu_m^{(n)} r / l)\cos n\varphi, \quad V_{nm}^{(2)}(r, \varphi) = J_n(\mu_m^{(n)} r / l)\sin n\varphi.$$

The initial displacement of the membrane is zero, in which case the solution $u(r, \varphi, t)$ is given by the series

$$u(r, \varphi, t) = \sum_{n=0}^{\infty}\sum_{m=0}^{\infty} [b_{nm}\cos n\varphi + d_{nm}\sin n\varphi] \cdot J_n(\mu_m^{(n)} r / l)\sin\frac{a\mu_m^{(n)}t}{l}.$$

Next we calculate the coefficients b_{nm} and d_{nm} in (11.96) to get (using the norms (11.87) and (11.88))

$$b_{nm} = \frac{I}{\rho\omega_{nm}\|V_{nm}^{(1)}\|^2} \int_0^l \int_0^{2\pi} \delta(r - r_0)\delta(\varphi - \varphi_0)\cos n\varphi J_n(\mu_m^{(n)} r / l)\ r\, dr\, d\varphi$$

$$= \frac{2I\cos n\varphi_0}{\varepsilon_n a\pi\rho/\mu_m^{(n)}[J_n'(\mu_m^{(n)})]^2} J_n(\mu_m^{(n)} r_0 / l),$$

and

$$d_{nm} = \frac{I}{\rho\omega_{nm}\|V_{nm}^{(2)}\|^2} \int_0^l \int_0^{2\pi} \delta(r - r_0)\delta(\varphi - \varphi_0)\sin n\varphi J_n(\mu_m^{(n)} r / l)\ r\, dr\, d\varphi$$

$$= \frac{2I\sin n\varphi_0}{\varepsilon_n a\pi\rho/\mu_m^{(n)}[J_n'(\mu_m^{(n)})]^2} J_n(\mu_m^{(n)} r_0 / l).$$

Therefore the evolution of the displacements of points on the membrane is described by series

$$u(r, \varphi, t) = \frac{2I}{a\pi\rho l}\sum_{n=0}^{\infty}\sum_{m=0}^{\infty} \frac{\cos n(\varphi - \varphi_0)J_n(\mu_m^{(n)} r_0 / l)}{\varepsilon_n \mu_m^{(n)}[J_n'(\mu_m^{(n)})]^2} J_n\left(\frac{\mu_m^{(n)}}{l}r\right)\sin\frac{a\mu_m^{(n)}t}{l}.$$

(a)   (b)

FIGURE 11.17   Graph of the membrane in Example 7 at (a) $t = 0.3$, (b) $t = 4.3$.

Figure 11.17 shows two snapshots of the solution at the times $t = 0.3$ and $t = 4.3$. This solution was obtained with program **Waves** for the case $a^2 = 1$, $l = 2$, $r_0 = 1$, $\varphi_0 = \pi$, $I = 10$, and $\rho = 1$.

Example 8

A container in the form of a vertical circular cylinder of radius l with horizontal bottom is partially filled with water and moves for a long time with velocity $v_0 = $ const. in a direction perpendicular to the axis of the container.

Find the oscillations of the surface of the water in the container for $t > 0$ if, at time $t = 0$, the container stops instantly. Assume that for times $t < 0$ the water is not moving relative to the container. The pressure on the free surface of the water may be considered to be constant and other interactions with the environment may be neglected.

Solution. The equation describing the surface before $t = 0$ when the container is moving at constant speed is given by

$$\frac{\partial^2 u}{\partial t^2} = a^2 \left[\frac{\partial^2 u}{\partial r^2} + \frac{1}{r}\frac{\partial u}{\partial r} + \frac{1}{r^2}\frac{\partial^2 u}{\partial \varphi^2} \right] \quad (0 \le r < l,\ 0 \le \varphi < 2\pi,\ t > 0).$$

The initial and boundary conditions are

$$u(r,\varphi,0) = v_0 r \, \cos\varphi, \quad \frac{\partial u}{\partial t}(r,\varphi,0) = 0, \quad \text{and} \quad \frac{\partial u}{\partial r}(l,\varphi,t) = 0.$$

The boundary condition of the problem is of *Neumann type*, in which case eigenvalues are given by the equation

$$J_n'(\mu) = 0.$$

Obviously in this case $\lambda_{00} = 0$ and $V_{00}^{(1)} = 0$. The other eigenfunctions are

$$V_{nm}^{(1)}(r,\varphi) = J_n(\mu_m^{(n)} r / l) \cos n\varphi \quad \text{and} \quad V_{nm}^{(2)}(r,\varphi) = J_n(\mu_m^{(n)} r / l) \sin n\varphi.$$

The three-dimensional picture shown in Figure 11.18 shows the two eigenfunctions $V_{11}^{(1)}(r,\varphi) = J_1(\mu_1^{(1)} r / l) \cos \varphi$ and $V_{13}^{(1)}(r,\varphi) = J_1(\mu_3^{(1)} r / l) \cos \varphi$ (chosen as examples) for Example 8; the corresponding values of λ are $\lambda_{11} = 7.1061$ and $\lambda_{13} = 34.2574$.

FIGURE 11.18   Eigenfunctions (a) $V_{11}^{(1)}(r,\varphi)$ and (b) $V_{13}^{(1)}(r,\varphi)$ for the Neumann boundary conditions of Example 8.

Let us find the coefficients of the series (11.96). Because the second initial condition is homogeneous and $\kappa = 0$, we have $b_{nm} = d_{nm} = 0$ for all n and m. For the remaining coefficients we have (with the norms (11.86), (11.88))

$$a_{nm} = \frac{1}{||V_{nm}^{(1)}||^2} \int_0^l \int_0^{2\pi} v_0 r \, \cos\varphi \, V_{nm}^{(1)}(r,\varphi) \, r \, dr \, d\varphi$$

$$= \frac{v_0}{||V_{nm}^{(1)}||^2} \int_0^{2\pi} \cos\varphi \, \cos n\varphi \, d\varphi \int_0^l r^2 J_n(\mu_m^{(n)} r / l) \, dr,$$

$$c_{nm} = \frac{v_0}{||V_{nm}^{(2)}||^2} \int_0^{2\pi} \cos\varphi \, \sin n\varphi \, d\varphi \int_0^l r^2 J_n(\mu_m^{(n)} r / l) \, dr.$$

Obviously, $a_{nm} = 0$ for $n \neq 1$ and $c_{nm} = 0$ for all n. Thus the expansion contains only the Bessel function J_1:

$$a_{1m} = \frac{2v_0 (\mu_m^{(1)})^2}{l^2 \left[(\mu_m^{(1)})^2 - 1\right] J_1^2(\mu_m^{(1)})} \int_0^l r^2 J_1\left(\frac{\mu_m^{(1)}}{l} r\right) dr = \frac{2v_0 l \mu_m^{(1)} J_2(\mu_m^{(1)})}{\left[(\mu_m^{(1)})^2 - 1\right] J_1^2(\mu_m^{(1)})}.$$

Finally we have

$$u(r,\varphi,t) = v_0 \cos\varphi \sum_{m=0}^{\infty} a_{1m} J_1\left(\frac{\mu_m^{(1)}}{l} r\right) \cos\frac{a\mu_m^{(1)} t}{l}.$$

Figure 11.19 shows two snapshots of the solution at the times $t = 0.3$ and $t = 4.3$. This solution was obtained with program **Waves** for the case $a^2 = 0.25$, $l = 2$ and $v_0 = 5$.

FIGURE 11.19   Graph of the membrane in Example 8 at (a) $t = 1.4$ and (b) $t = 5.4$.

(a)   (b)

FIGURE 11.20 Eigenfunctions (a) $V_{01}^{(1)}(r,\varphi)$ and (b) $V_{02}^{(1)}(r,\varphi)$ for the mixed boundary conditions of Example 9.

Example 9

The periphery of a flexible circular membrane of radius l is fixed elastically with coefficient h. The initial displacement is zero. Find the transversal vibrations of the membrane if the initial velocities of the membrane are described by the function

$$\psi(r) = A(l-r).$$

Solution. The boundary value problem consists of the solution of the equation

$$\frac{\partial^2 u}{\partial t^2} - a^2\left[\frac{\partial^2 u}{\partial r^2} + \frac{1}{r}\frac{\partial u}{\partial r}\right] = 0 \quad (0 \le r < l, 0 \le \varphi < 2\pi, t > 0)$$

with the conditions

$$u(r, \varphi, 0) = 0, \quad \frac{\partial u}{\partial t}(r, \varphi, 0) = A(l-r), \quad \frac{\partial u}{\partial r} + hu\bigg|_{r=1} = 0.$$

The oscillations of the membrane are radial since the initial functions do not depend on the polar angle φ; thus only terms with $n = 0$ are not zero.

The boundary condition of the problem is of *mixed type*, in which case eigenvalues $\mu_m^{(0)}$ are given by the roots of the eigenvalue equation $\mu J_0'(\mu) + h J_0(\mu) = 0$. The eigenfunctions are

$$V_{0m}^{(1)}(r,\varphi) = J_0\left(\mu_m^{(0)}r / l\right).$$

The oscillations of the membrane are radial since the initial functions do not depend on the polar angle φ; thus only terms with $n = 0$ are not zero.

The initial displacement is zero so that the coefficients $a_{0m} = 0$. The coefficients b_{0m} are given by equation (11.99), which result in

$$b_{0m} = \frac{2\pi}{||V_{0m}^{(1)}||^2} \int_0^l A(l-r) J_0(\mu_m^{(0)}r / l)\, r\, dr.$$

FIGURE 11.21   Graph of the membrane in Example 9 at (a) $t = 2$, (b) $t = 10$.

They are calculated in the program **Waves**. Their complete analytical calculation can be found in library example 3 (Vibrating Circular Membrane)) of this program. The oscillation of the membrane is given by the series in Bessel functions of zeroth order:

$$u(r, \varphi, t) = \sum_{m=0}^{\infty} b_{0m} \sin \frac{a\mu_m^{(0)}t}{l} J_0\left(\frac{\mu_m^{(0)}}{l}r\right).$$

Figure 11.21 shows two snapshots of the solution at the times $t = 2$ and $t = 10$. This solution was obtained with the program **Waves** for the case $a^2 = 0.25$, $l = 2$, $h = 1$, and $A = 5$.

11.3.3 The Fourier Method for Nonhomogeneous Equations

In this section we use the Fourier method to *nonhomogeneous* equations in polar coordinates for the case of a circular membrane equation (11.62),

$$\frac{\partial^2 u}{\partial t^2} + 2\kappa\frac{\partial u}{\partial t} - a^2\left(\frac{\partial^2 u}{\partial r^2} + \frac{1}{r}\frac{\partial u}{\partial r} + \frac{1}{r^2}\frac{\partial^2 u}{\partial \varphi^2}\right) + \gamma u = f(r, \varphi, t),$$

where $f(r, \varphi, t)$ is a given function.

First find solutions satisfying the *homogeneous* boundary condition (11.66)

$$\alpha\frac{\partial u}{\partial r} + \beta\, u\Big|_{r=l} = 0$$

and *nonhomogeneous* initial conditions (11.64)

$$u|_{t=0} = \phi(r, \varphi), \quad \frac{\partial u}{\partial t}\Big|_{t=0} = \psi(r, \varphi).$$

We begin by searching for a solution in the form of the sum

$$u(r, \varphi, t) = u_1(r, \varphi, t) + u_2(r, \varphi, t), \tag{11.101}$$

where $u_1(r, \varphi, t)$ is the solution to the *homogeneous* equation with *homogeneous* boundary and *nonhomogeneous* initial conditions given by

$$\frac{\partial^2 u_1}{\partial t^2} + 2\kappa \frac{\partial u_1}{\partial t} - a^2 \left(\frac{\partial^2 u_1}{\partial r^2} + \frac{1}{r} \frac{\partial u_1}{\partial r} + \frac{1}{r^2} \frac{\partial^2 u_1}{\partial \varphi^2} \right) + \gamma \, u_1 = 0, \tag{11.102}$$

$$\alpha \frac{\partial u_1}{\partial r} + \beta \, u_1 \bigg|_{r=l} = 0, \tag{11.103}$$

$$u_1|_{t=0} = \phi(r, \varphi), \quad \frac{\partial u_1}{\partial t} \bigg|_{t=0} = \psi(r, \varphi), \tag{11.104}$$

and $u_2(r, \varphi, t)$ is the solution to the *nonhomogeneous* equation with *zero boundary and initial conditions* given by

$$\frac{\partial^2 u_2}{\partial t^2} + 2\kappa \frac{\partial u_2}{\partial t} - a^2 \left(\frac{\partial^2 u_2}{\partial r^2} + \frac{1}{r} \frac{\partial u_2}{\partial r} + \frac{1}{r^2} \frac{\partial^2 u_2}{\partial \varphi^2} \right) + \gamma u_2 = f(r, \varphi, t), \tag{11.105}$$

$$\alpha \frac{\partial u_2}{\partial r} + \beta u_2 \bigg|_{r=l} = 0, \tag{11.106}$$

$$u_2|_{t=0} = 0, \quad \frac{\partial u_2}{\partial t} \bigg|_{t=0} = 0. \tag{11.107}$$

Physically the solution $u_1(r, \varphi, t)$ represents *free oscillations*, that is, oscillations that occur only due to an initial perturbation. The solution $u_2(r, \varphi, t)$ represents *forced oscillations*, that is, oscillations that result from the action of external periodic forces when initial perturbations are absent.

Methods for finding the solution $u_1(r, \varphi, t)$ for free oscillations were considered in the previous section; our task here need only be to find the solutions $u_2(r, \varphi, t)$ for forced oscillations. As in case of free oscillations we search for the solution $u_2(r, \varphi, t)$ in the form of the series

$$u_2(r, \varphi, t) = \sum_{n=0}^{\infty} \sum_{m=0}^{\infty} [T_{nm}^{(1)}(t) \cdot V_{nm}^{(1)}(r, \varphi) + T_{nm}^{(2)}(t) \cdot V_{nm}^{(2)}(r, \varphi)], \tag{11.108}$$

where $V_{nm}^{(1)}(r, \varphi)$ and $V_{nm}^{(2)}(r, \varphi)$ are eigenfunctions of the corresponding homogeneous boundary value problem, and $T_{nm}^{(1)}(t)$, $T_{nm}^{(2)}(t)$ are, for the moment, unknown functions of time, t.

Zero boundary conditions given in equation (11.106) for the function $u_2(r, \varphi, t)$ are satisfied for any choice of $T_{nm}^{(1)}(t)$ and $T_{nm}^{(2)}(t)$ under the restriction of uniform convergence of the series because they are known to be satisfied by the functions $V_{nm}^{(1)}(r,\varphi)$ and $V_{nm}^{(2)}(r,\varphi)$. However, the functions $T_{nm}^{(1)}(t)$ and $T_{nm}^{(2)}(t)$ must also be selected so that the series (11.108) satisfies the nonhomogeneous equation (11.105) and the homogeneous initial conditions (11.107).

Substituting the series in equation (11.108) into equation (11.105) we obtain

$$\sum_{n=0}^{\infty}\sum_{m=0}^{\infty}[T_{nm}''^{(1)}(t)+2\kappa\ T_{nm}'^{(1)}(t)+(a^2\lambda_{nm}+\gamma)T_{nm}^{(1)}(t)]V_{nm}^{(1)}(r,\varphi)$$

$$+\sum_{n=0}^{\infty}\sum_{m=0}^{\infty}[T_{nm}''^{(2)}(t)+2\kappa\ T_{nm}'^{(1)}(t)+(a^2\lambda_{nm}+\gamma)T_{nm}^{(2)}(t)]V_{nm}^{(2)}(r,\varphi)=f(r,\varphi,t). \tag{11.109}$$

Next we expand the function $f(r,\varphi,t)$ in a Fourier series with functions $V_{nm}^{(1)}(r,\varphi)$ and $V_{nm}^{(2)}(r,\varphi)$ as the basis functions:

$$f(r,\varphi,t)=\sum_{n=0}^{\infty}\sum_{m=0}^{\infty}[f_{nm}^{(1)}(t)V_{nm}^{(1)}(r,\varphi)+f_{nm}^{(2)}(t)V_{nm}^{(2)}(r,\varphi)], \tag{11.110}$$

where

$$f_{nm}^{(1)}(t)=\frac{1}{\|V_{nm}^{(1)}\|^2}\int_0^{2\pi}\int_0^l f(r,\varphi,t)\ V_{nm}^{(1)}\ r\,dr\,d\varphi,$$

$$f_{nm}^{(2)}(t)=\frac{1}{\|V_{nm}^{(2)}\|^2}\int_0^{2\pi}\int_0^l f(r,\varphi,t)\ V_{nm}^{(2)}\ r\,dr\,d\varphi. \tag{11.111}$$

Comparing the series (11.110) with equation (11.109) for the same function $f(r,\varphi,t)$ we obtain the following differential equations, which will determine the functions $T_{nm}^{(1)}(t)$ and $T_{nm}^{(2)}(t)$:

$$T_{nm}''^{(1)}(t)+2\kappa\ T_{nm}'^{(1)}(t)+(a^2\lambda_{nm}+\gamma)T_{nm}^{(1)}(t)=f_{nm}^{(1)}(t),$$

$$T_{nm}''^{(2)}(t)+2\kappa\ T_{nm}'^{(2)}(t)+(a^2\lambda_{nm}+\gamma)T_{nm}^{(2)}(t)=f_{nm}^{(2)}(t). \tag{11.112}$$

The solution $u_2(r,\varphi,t)$ defined by the series (11.108) satisfies initial conditions (11.107), which imposes on the functions $T_{nm}^{(1)}(t)$ and $T_{nm}^{(2)}(t)$ the conditions

$$\begin{cases}T_{nm}^{(1)}(0)=0,\\[2mm]T_{nm}'^{(1)}(0)=0\end{cases}\quad\text{and}\quad\begin{cases}T_{nm}^{(2)}(0)=0,\\[2mm]T_{nm}'^{(2)}(0)=0\end{cases}\quad n,m=0,1,2,\ldots \tag{11.113}$$

As for the one-dimensional case, solutions of the Cauchy problems defined in equations (11.112) and (11.113) for functions $T_{nm}^{(1)}(t)$ and $T_{nm}^{(2)}(t)$ can be written as

$$T_{nm}^{(1)}(t) = \frac{1}{\omega_{nm}} \int_0^t f_{nm}^{(1)}(\tau)\, Y_{nm}(t-\tau)\, d\tau,$$

$$T_{nm}^{(2)}(t) = \frac{1}{\omega_{nm}} \int_0^t f_{nm}^{(2)}(\tau)\, Y_{nm}(t-\tau)\, d\tau,$$

where

$$Y_{nm}(t) = \begin{cases} e^{-\kappa t}\sin\omega_{nm}t, & \kappa^2 < a^2\lambda_{nm}+\gamma, \\ e^{-\kappa t}\sinh\omega_{nm}t, & \kappa^2 > a^2\lambda_{nm}+\gamma, \\ te^{-\kappa t}, & \kappa^2 = a^2\lambda_{nm}+\gamma, \end{cases}$$

and $\omega_{nm} = \sqrt{|a^2\lambda_{nm}+\gamma-\kappa^2|}$. Substituting expressions (11.111) for $f_{nm}^{(1)}(\tau)$ and $f_{nm}^{(2)}(\tau)$ we obtain

$$T_{nm}^{(1)}(t) = \frac{1}{\omega_{nm}\|V_{nm}^{(1)}\|^2} \int_0^t d\tau \int_0^{2\pi}\int_0^l f(r,\varphi,\tau)V_{nm}^{(1)}(r,\varphi)\,Y_{nm}(t-\tau)\,r\,dr\,d\varphi,$$

$$\text{(11.114)}$$

$$T_{nm}^{(2)}(t) = \frac{1}{\omega_{nm}\|V_{nm}^{(2)}\|^2} \int_0^t d\tau \int_0^{2\pi}\int_0^l f(r,\varphi,\tau)V_{nm}^{(2)}(r,\varphi)\,Y_{nm}(t-\tau)\,r\,dr\,d\varphi.$$

Substituting these expressions for $T_{nm}^{(1)}(t)$ and $T_{nm}^{(2)}(t)$ in the series in equation (11.108) we obtain solutions to the boundary value problem defined in equations (11.105) through (11.107), assuming that equation (11.108) and the series obtained from it by twice differentiating term by term with respect to the variables r, φ and t converges uniformly. Thus the solution of the problem of forced oscillations with zero boundary conditions is

$$u(r,\varphi,t) = u_1(r,\varphi,t) + u_2(r,\varphi,t)$$

$$= \sum_{n=0}^{\infty}\sum_{m=0}^{\infty} \{[T_{nm}^{(1)}(t)+(a_{nm}\,y_{nm}^{(1)}(t)+b_{nm}\,y_{nm}^{(2)}(t))]\cdot V_{nm}^{(1)}(r,\varphi) \qquad \text{(11.115)}$$

$$+[T_{nm}^{(2)}(t)+(c_{nm}\,y_{nm}^{(1)}(t)+d_{nm}\,y_{nm}^{(2)}(t))]\cdot V_{nm}^{(2)}(r,\varphi)\},$$

where coefficients $T_{nm}^{(1)}(t)$ and $T_{nm}^{(2)}(t)$ are defined by equations (11.114), and a_{nm}, b_{nm}, c_{nm}, and d_{nm} were obtained previously in the discussion in Section 11.3.1.

11.3.4 Forced Radial Oscillations of a Circular Membrane

In the case of radial oscillations the solution of the nonhomogeneous equation becomes simpler. For radial oscillations the initial displacement, initial velocity, and function f do not depend on φ and are thus functions of r and t only:

$$u(r,\varphi,t)\big|_{t=0} = \phi(r), \quad \frac{\partial u}{\partial t}(r,\varphi,t)\bigg|_{t=0} = \psi(r), \quad \text{and} \quad f(r,\varphi,t) \equiv f(r,t).$$

For radial oscillations the solution does not contain the functions $V_{nm}^{(2)}(r,\varphi)$ and only the coefficients $f_{0m}^{(1)}$ are nonzero:

$$f_{0m}^{(1)}(t) = \frac{1}{\|V_{0m}^{(1)}\|^2} \int_0^l \int_0^{2\pi} f(r,t) J_0\left(\frac{\mu_m^{(0)}}{l} r\right) r\,dr\,d\varphi = \frac{2\pi}{\|V_{0m}^{(1)}\|^2} \int_0^l f(r,t) J_0\left(\frac{\mu_m^{(0)}}{l} r\right) r\,dr.$$

Substituting these functions into the first of the formulas (11.114) gives $T_{0m}^{(1)}(t)$ and the double series (11.115) reduces to a single series:

$$u(r,\varphi,t) = \sum_{m=0}^{\infty} \left\{ T_{0m}^{(1)}(t) + \left[a_{0m} y_{0m}^{(1)}(t) + b_{0m} y_{0m}^{(2)}(t)\right] \right\} \cdot J_0\left(\frac{\mu_m^{(0)}}{l} r\right).$$

Example 10

Find the transverse oscillations of a homogeneous circular membrane of radius l with a rigidly fixed edge where a variable pressure

$$p = p_0 \cos \omega t$$

acts on one side of the membrane. Assume that initial deviations and initial velocities are absent and that the reaction of the environment is negligibly small.

Solution. The boundary problem modeling the evolution of such oscillations leads to the equation

$$\frac{\partial^2 u}{\partial t^2} - a^2 \left[\frac{\partial^2 u}{\partial r^2} + \frac{1}{r}\frac{\partial u}{\partial r}\right] = \frac{p_0}{\rho} \cos \omega\, t \quad (0 \le r < l, 0 \le \varphi < 2\pi, t > 0)$$

under zero initial and boundary conditions given by

$$u(r,\varphi,0) = 0, \quad \frac{\partial u}{\partial t}(r,\varphi,0) = 0, \quad u(l,\varphi,t) = 0.$$

In this example the oscillations are radial because the initial conditions are homogeneous and the external pressure is a function of t only. From the previous discussion we have that the eigenvalues $\mu_m^{(0)}$ are roots of equation $J_0(\mu) = 0$; the eigenfunctions are

$$V_{0m}^{(1)}(r,\phi) = J_0\left(\frac{\mu_m^{(0)}}{l} r\right).$$

Consequently the solution $u(r, \varphi, t)$ is defined by the series

$$u(r, \varphi, t) = \sum_{m=0}^{\infty} T_{0m}^{(1)}(t) \cdot J_0\left(\frac{\mu_m^{(n)}}{l}r\right),$$

where

$$T_{0m}^{(1)}(t) = \frac{1}{\omega_{0m}} \int_0^t f_{0m}^{(1)}(p) \sin\omega_{0m}(t-p)\ dp.$$

They are calculated in the program **Waves**. Their complete analytical calculation can be found in library example 3 (Vibrating Circular Membrane)) of this program.

The program **Waves** calculates these coefficients; thus the reader can consider these two formulas as the solution. The way to calculate the coefficients $T_{0m}^{(1)}(t)$ in analytical form can be found in library example 5 (Vibrating Circular Membrane) of this program. As the result, the deflection of the membrane as a function of time is described by the series

$$u(r, \phi, t) = \frac{2p_0}{\rho} \sum_{m=0}^{\infty} \frac{\cos\omega_{0m}t - \cos\omega t}{(\omega^2 - \omega_{0m}^2)\mu_m^{(0)} J_1(\mu_m^{(0)})} J_0\left(\frac{\mu_m^{(0)}}{l}r\right).$$

Figure 11.22 shows two snapshots of the solution at the times $t = 7$ and $t = 10.5$. This solution was obtained with the program **Waves** for $a^2 = 1$, $l = 2$, $p_0 = 0.25$, and $\omega = 3$.

11.3.5 The Fourier Method for Equations with Nonhomogeneous Boundary Conditions

Consider now the general boundary value problem for equations describing the forced oscillations of a circular membrane with nonhomogeneous boundary and initial conditions given by equations (11.63) through (11.65). We have

$$\frac{\partial^2 u}{\partial t^2} + 2\kappa\frac{\partial u}{\partial t} - a^2\left(\frac{\partial^2 u}{\partial r^2} + \frac{1}{r}\frac{\partial u}{\partial r} + \frac{1}{r^2}\frac{\partial^2 u}{\partial \varphi^2}\right) + \gamma u = f(r, \varphi, t),$$

(a)   (b)

FIGURE 11.22   Graph of the membrane in Example 10 at (a) $t = 7$, (b) $t = 10.5$.

$$\alpha \frac{\partial u}{\partial r} + \beta \left. u \right|_{r=l} = g(\varphi, t),$$

$$u|_{t=0} = \phi(r, \varphi), \quad \left. \frac{\partial u}{\partial t} \right|_{t=0} = \psi(r, \varphi).$$

It is not possible to apply the Fourier method directly to this problem because the boundary conditions are nonhomogeneous. However, we may reduce the problem to one with zero boundary conditions.

We search for the solution of the problem in the form of the sum

$$u(r, \varphi, t) = v(r, \varphi, t) + w(r, \varphi, t),$$

where $v(r, \varphi, t)$ is a new, unknown function, and the function $w(r, \varphi, t)$ is chosen so that it satisfies the given nonhomogeneous boundary condition

$$\alpha \frac{\partial w}{\partial r} + \beta \left. w \right|_{r=l} = g(\varphi, t)$$

and has the necessary number of continuous derivatives in r, φ, and t.

For the function $v(r, \varphi, t)$ we obtain following boundary value problem:

$$\frac{\partial^2 v}{\partial t^2} + 2\kappa \frac{\partial v}{\partial t} - a^2 \left(\frac{\partial^2 v}{\partial r^2} + \frac{1}{r} \frac{\partial v}{\partial r} + \frac{1}{r^2} \frac{\partial^2 v}{\partial \varphi^2} \right) + \gamma v = f^*(r, \varphi, t),$$

$$\alpha \frac{\partial v}{\partial r}(l, \varphi, t) + \beta \, v(l, \varphi, t) = 0,$$

$$v(r, \varphi, t)|_{t=0} = \phi^*(r, \varphi), \quad \left. \frac{\partial v}{\partial t}(r, \varphi, t) \right|_{t=0} = \psi^*(r, \varphi)$$

where

$$f^*(r, \varphi, t) = f(r, \varphi, t) - \frac{\partial^2 w}{\partial t^2} - 2\kappa \frac{\partial w}{\partial t} + a^2 \left(\frac{\partial^2 w}{\partial r^2} + \frac{1}{r} \frac{\partial w}{\partial r} + \frac{1}{r^2} \frac{\partial^2 w}{\partial \varphi^2} \right) - \gamma w, \quad (11.116)$$

$$\phi^*(r, \varphi) = \phi(r, \varphi) - w(r, \varphi, 0), \quad (11.117)$$

$$\psi^*(r, \varphi) = \psi(r, \varphi) - \frac{\partial w}{\partial t}(r, \varphi, 0). \quad (11.118)$$

Solutions to this problem were considered in previous section.

Reading Exercise

Verify equations (11.116) through (11.118).

The function $w(r, \varphi, t)$ can be chosen in different forms, the simplest of which is

$$w(r,\varphi,t) = (c_0 + c_1 r^2)g(\varphi,t), \tag{11.119}$$

where c_0 and c_1 are constants. These constants must be chosen so that function $w(r, \varphi, t)$ satisfies the given boundary conditions.

1) For the case of Dirichlet boundary conditions $\alpha = 0$ and $\beta = 1$:

    a) The boundary condition is $u(l, \varphi, t) = g(\varphi, t)$.

    The auxiliary function (with $c_0 = 0$, $c_1 = 1/l^2$) is given by

$$w(r, \varphi, t) = \frac{r^2}{l^2} g(\varphi,t). \tag{11.120}$$

    b) The boundary condition is $u(l, \varphi, t) = g(t)$ or $u(l, \varphi, t) = g_0 = \text{const}$.

    The auxiliary function (with $c_0 = 1, c_1 = 0$) is given by

$$w(r, \varphi, t) = g(t) \text{ or } w(r, \varphi, t) = g_0. \tag{11.121}$$

2) For the case of Neumann boundary conditions $\alpha = 1$ and $\beta = 0$:

$$\frac{\partial u}{\partial r}(l,\varphi,t) = g(\varphi,t).$$

The auxiliary function is (with $c_0 = 0, c_1 = 1/2l$)

$$w(r,\varphi,t) = \frac{r^2}{2l} \cdot g(\varphi,t) + C, \tag{11.122}$$

where C is an arbitrary constant.

3) For the case of mixed boundary conditions $\alpha = 1$ and $\beta = h$:

$$\frac{\partial}{\partial r}u(l,\varphi,t) + h\,u(l, \varphi, t) = g(\varphi, t).$$

The auxiliary function is (with $c_0 = 0$, $c_1 = 1/(2l + hl^2)$)

$$w(r, \varphi, t) = \frac{r^2}{l(2 + hl)} g(\varphi, t). \tag{11.123}$$

Reading Exercise

We leave it to the reader to check that so chosen function $w(r, \varphi, t)$ in equations (11.120) through (11.123) satisfies the boundary condition

$$\alpha \frac{\partial w}{\partial r}(l, \varphi, t) + \beta \, w(l, \varphi, t) = g(\varphi, t). \tag{11.124}$$

We next consider examples of oscillating membrane problems where we need to find solutions with nonhomogeneous boundary conditions.

Example 11

Find the oscillations of a circular membrane ($0 \leq r \leq l$) in an environment without resistance and with zero initial conditions where the motion is caused by movement at its edge described by

$$u(l, \varphi, t) = A \sin \omega t, \, t \geq 0.$$

Solution. The boundary problem modeling the evolution of such oscillations is given by the equation

$$\frac{\partial^2 u}{\partial t^2} - a^2 \left[\frac{\partial^2 u}{\partial r^2} + \frac{1}{r} \frac{\partial u}{\partial r} \right] = 0,$$

$$0 \leq r < l, \, 0 \leq \varphi < 2\pi, \, t > 0$$

with zero initial conditions

$$u(r, \varphi, 0) = 0, \quad \frac{\partial u}{\partial t}(r, \varphi, 0) = 0$$

and boundary condition

$$u(l, \varphi, t) = A \sin \omega t.$$

Here $\alpha = 0$ and $\beta = 1$ so, using equation (11.121), we may immediately write

$$w(r, \varphi, t) = A \sin \omega t.$$

Recall that for this initial boundary value problem we are searching for the solution in the form of the sum

$$u(r, \varphi, t) = w(r, \varphi, t) + v(r, \varphi, t).$$

FIGURE 11.23   Graph of membrane in Example 11 at (a) $t = 2$ and (b) $t = 20$.

For the function $v(r, \varphi, t)$ we obtain from the boundary conditions and the above definitions given in equations (11.116) through (11.118),

$$f^*(r, \varphi, t) = A\omega^2 \sin \omega t,$$

$$\phi^*(r, \varphi) = 0, \quad \psi^*(r, \varphi) = -A\omega.$$

The function $f^*(r, \varphi, t)$ does not depend on the polar angle φ; thus the function $v(r, \varphi, t)$ defines radial oscillations of the membrane only. The solution $v(r, \varphi, t)$ is thus given by the series

$$v(r, \varphi, t) = \sum_{m=0}^{\infty} [T_{0m}^{(1)}(t) + b_{0m} \sin \omega_{0m} t] \cdot J_0\left(\frac{\mu_m^{(0)}}{l} r\right),$$

where $\omega_{0m} = a\sqrt{\lambda_{0m}} = a(\mu_m^{(0)}/l)$ and $\mu_m^{(0)}$ are the roots of equation $J_0(\mu) = 0$.

The program **Waves** calculates coefficients $T_{0m}^{(1)}(t)$ and b_{0m}. The way to calculate the coefficients $T_{0m}^{(1)}(t)$ in analytical form can be found in library example 6 (Vibrating Circular Membrane) of this program. As the result, the deflection of the membrane as a function of time is described by the series

$$u(r, \varphi, t) = w(r, \varphi, t) + v(r, \varphi, t)$$

$$= A\sin \omega t + \sum_{m=0}^{\infty} [T_{0m}^{(1)}(t) + b_{0m} \sin \omega_{0m} t] J_0\left(\frac{\mu_m^{(0)}}{l} r\right)$$

$$= A\sin \omega t + 2A\omega \sum_{m=0}^{\infty} \frac{[\omega_{0m} \sin \omega_{0m} t - \omega \sin \omega t]}{(\mu_m^{(0)})^2 J_1(\mu_m^{(0)})(\omega^2 - \omega_{0m}^2)} J_0\left(\frac{\mu_m^{(0)}}{l} r\right).$$

Figure 11.23 shows two snapshots of the solution at the times $t = 2$ and $t = 20$. This solution was obtained with the program **Waves** for $a^2 = 1$, $l = 2$, $A = 0.2$, and $\omega = 3$.

Problems

In problems 1 through 24 we consider transverse oscillations of a rectangular membrane $(0 \leq x \leq l_x, 0 \leq y \leq l_y)$, in problems 25 through 47 a homogeneous *circular* membrane of radius l. Solve these problems analytically, which means the following: formulate the equation and initial and boundary conditions, obtain the eigenvalues and eigenfunctions, and write the formulas for coefficients of the series expansion and the expression for the solution of the problem. If the integrals in the coefficients are not easy to evaluate, leave it for the program, which evaluates them numerically. Then study the problem in detail with the program **Waves**. Obtain the pictures of several eigenfunctions and screenshots of solution in different moments of time. Directions on how to use the program are found in Appendix 5. The reader should assign values for the parameters such as the values of surface mass density,ρ, tension, T, coefficient of resistance of the environment, κ, and so on. In some problems it might be interesting to consider realistic values for the parameters. Remember that the oscillations should remain small.

The program **Waves** allows the user to see the oscillation of individual harmonics. The problems discussed refer to membranes, but it should be remembered that many other, similar physical problems are described by the same two-dimensional hyperbolic equation—for example, oscillations of the surface of a liquid (analogous to a membrane with free boundaries), axial oscillations of a gas in a cylindrical tube, and so on.

Note. When you choose the parameters of the problem and coefficients of the functions (initial and boundary conditions, and external forces), do not forget that the amplitudes of oscillations should remain small.

In problems 1 through 5 external forces and resistance of the embedding medium are absent. Find the transverse oscillations of the membrane.

Using the program **Waves** investigate the behavior of the membrane for each of the given problems. Choose the value of wave speed squared a^2 such that the initial period of vibrations decreases (1) twice, (2) four times, (3) five times. Obtain the same results by changing the lengths of membrane along axes for a constant value of a^2.

1. The membrane is fixed along its edges. The initial conditions are

$$\varphi(x, y) = A \sin \frac{\pi x}{l_x} \sin \frac{\pi y}{l_y}, \quad \psi(x,y) = 0.$$

2. The edge at $x = 0$ of membrane is free and other edges are fixed. The initial conditions are

$$\varphi(x, y) = A \cos \frac{\pi x}{2l_x} \sin \frac{\pi y}{l_y}, \quad \psi(x,y) = 0.$$

3. The edge at $y = 0$ of the membrane is free and other edges are fixed. The initial conditions are

$$\varphi(x, y) = A \sin \frac{\pi x}{l_x} \cos \frac{\pi y}{2l_y}, \quad \psi(x,y) = 0.$$

4. The edges $x = 0$ and $y = 0$ of membrane are fixed, and edges $x = l_x$ and $y = l_y$ are free. The initial conditions are

$$\varphi(x,y) = 0, \quad \psi(x, y) = Axy \left(l_x - \frac{x}{2} \right) \left(l_y - \frac{y}{2} \right).$$

5. The edges $x = 0$, $x = l_x$ and $y = l_y$ of membrane are fixed, and the edge $y = 0$ is elastically constrained with coefficient of elasticity $h = 1$. The initial conditions are

$$\varphi(x,y) = Ax\left(1 - \frac{x}{l_x}\right)\sin\frac{\pi y}{l_y}, \quad \psi(x,y) = 0.$$

In problems 6 through 10 external forces are absent and the resistance of the embedding environment is proportional to velocity with proportionality constant κ. Find the transverse oscillations of the membrane.

Using the program **Waves** investigate the behavior of the membrane for each of the given problems. Choose a coefficient of resistance, κ, so that oscillations decay to zero (with a precision of 5%) during (a) three periods of oscillation; (b) five periods of oscillation; (c) ten periods of oscillation.

6. The edges $x = 0$, $y = 0$, and $y = l_y$ of membrane are fixed and the edge $x = l_x$ is attached elastically with coefficient of elasticity $h = 1$. The initial conditions are

$$\varphi(x,y) = 0, \quad \psi(x,y) = Axy(l_y - y).$$

7. The edges $x = 0$ and $y = 0$ of membrane are fixed, the edge $y = l_y$ is free, and the edge $x = l_x$ is attached elastically with coefficient of elasticity $h = 1$. The initial conditions are

$$\varphi(x,y) = 0, \quad \psi(x,y) = Ax\left(1 - \frac{x}{l_x}\right)\sin\frac{\pi y}{l_y}.$$

8. The edges $x = l_x$ and $y = l_y$ are fixed, the edge $x = 0$ is free, and the edge $y = 0$ is attached elastically with coefficient of elasticity $h = 1$. The initial conditions are

$$\varphi(x,y) = Ay\left(1 - \frac{y}{l_y}\right)\sin\frac{\pi x}{l_x}, \quad \psi(x,y) = 0.$$

9. The edges $y = 0$ and $y = l_y$ are free, the edge $x = 0$ is fixed, and the edge $x = l_x$ is attached elastically with coefficient of elasticity $h = 1$. The initial conditions are

$$\varphi(x,y) = Axy^2\left(1 - \frac{x}{l_x}\right), \quad \psi(x,y) = 0.$$

10. The edges $x = 0$ and $y = l_y$ are free, the edge $x = l_x$ is attached elastically with coefficient of elasticity $h = 1$, and the edge $y = 0$ is attached elastically with coefficient of elasticity $h = 1$. The initial conditions are

$$\varphi(x,y) = Ay\left[1 - (x/l_x)^2\right], \quad \psi(x,y) = 0.$$

In problems 11 through 15 at the initial instant of time, $t = 0$, the membrane is set in motion by a blow that applies an impulse, I, at the point (x_0, y_0) $(0 < x_0 < l_x, 0 < y_0 < l_y)$.

For the following cases, simulate free transverse oscillations of the membrane assuming the initial displacement is zero, external forcing is absent, and the environment causes a resistance proportional to velocity ($\kappa > 0$).

11. The membrane is fixed along its edges.

12. The edge $x = l_x$ of membrane is free and the other edges are fixed.

13. The edges $x = 0$ and $y = l_y$ of the membrane are free and the edges $x = l_x$ and $y = 0$ are fixed.

14. The edges $x = l_x$, $y = 0$, and $y = l_y$ of the membrane are fixed and the edge $x = 0$ is attached elastically with coefficient of elasticity $h = 1$.

15. The edges $y = 0$ and $y = l_y$ are free, the edge $x = l_x$ is fixed, and the edge $x = 0$ is attached elastically with coefficient of elasticity $h = 1$.

In problems 16 through 20 assume that a force density $f(x, y, t)$ acts, the initial displacement from equilibrium as well as the initial velocity are zero, and resistance of the embedding medium is absent ($\kappa = 0$).

Solve the equations of motion with the given boundary conditions for each case given below. Using the program **Waves** investigate the behavior of the membrane.

16. The membrane is fixed along its boundaries. The external force is

$$f(x, y, t) = A x \sin \frac{2\pi y}{l_y} (1 - \sin \omega t).$$

17. The edge of the membrane at $x = 0$ is free, and other edges are fixed. The external force is

$$f(x, y, t) = A(l_x - x) \sin \frac{\pi y}{l_y} \sin \omega t.$$

18. The edges $x = l_x$ and $y = l_y$ of the membrane are free and edges $x = 0$ and $y = 0$ are fixed. The external force is

$$f(x, y, t) = A x \sin \frac{\pi y}{2 l_y} \cos \omega t.$$

19. The edges $x = 0$, $y = 0$, and $y = l_y$ of the membrane are fixed and the edge $x = l_x$ is attached elastically with elasticity coefficient $h = 1$. The external force is

$$f(x, y, t) = A x y \cos \frac{\pi y}{2 l_y} (1 - \sin \omega t).$$

20. The edges $x = 0$ and $y = l_y$ are attached elastically with elasticity coefficient $h = 1$, the edge $x = l_x$ is fixed, and the edge $y = 0$ is free. The external force is

$$f(x,y,t) = A\sin\frac{\pi x}{l_x}\cos\frac{\pi y}{2l_y}\sin\omega t.$$

In problems 21 through 25 the boundary of the membrane is driven. Assume the initial velocities are zero and the resistance of embedding material is absent ($\kappa = 0$). The initial shape of the membrane is $u(x,y,0) = \phi(x,y)$, and nonhomogeneous terms in the boundary condition are given.

Solve the equations of motion with the given boundary conditions for each case given below. Using the program **Waves** investigate the behavior of the membrane.

21. The motion of the edge of the membrane is given by

$$g_1(y,t) = A\sin\frac{\pi y}{l_y}\cos\omega t$$

and the other edges are fixed. Initially the membrane has the shape

$$\varphi(x,y) = A\left(1 - \frac{x}{l_x}\right)\sin\frac{\pi y}{l_y}.$$

22. The motion of the edge $y = l_y$ of the membrane is given by

$$g_4(x,t) = A\sin\frac{\pi x}{l_x}\sin\omega t$$

and the other edges are fixed. Initially the membrane has the shape

$$\varphi(x,y) = Axy\cos\frac{\pi x}{2l_x}.$$

23. Edges $x = 0$ and $x = l_x$ are fixed, the edge $y = 0$ is free, and the edge $y = l_y$ moves as

$$g_4(x,t) = A\sin\frac{\pi x}{l_x}\cos\omega t.$$

Initially the membrane has the shape

$$\varphi(x,y) = Ax(l_x - x)\sin\frac{\pi y}{l_y}.$$

24. The edges $x = 0$ and $y = l_y$ are fixed and the edge $y = 0$ is subject to the action of a harmonic force causing displacements

$$g_3(x,t) = A\sin\frac{\pi x}{l_x}\cos\omega t.$$

Initially the membrane has the shape

$$\varphi(x, y) = Ax(y - l_y)\cos\frac{\pi x}{2l_x}.$$

In the following problems consider transverse oscillations of a homogeneous *circular* membrane of radius l, located in the x-y plane.

For all problems the displacement of the membrane, $u(r,\varphi,0) = \phi(r, \varphi)$, is given at some initial moment of time, $t = 0$, and the membrane is released with initial velocity $u_t(r, \varphi, 0) = \psi(r, \varphi)$.

In problems 25 through 30 external forces and resistance of the environment are absent. Find solutions and simulate the transverse oscillations of the membrane using the program **Waves** for each of the given problems.

Choose the value of wave speed squared a^2 so that the initial period of vibrations decreases (1) twice, (2) four times, (3) five times. Obtain the same results by changing the radius of the membrane l for a constant value of a^2.

25. The membrane is fixed along its contour. The initial conditions are

$$\phi(r,\varphi) = Ar(l^2 - r^2)\sin\varphi, \quad \psi(r,\varphi) = 0.$$

26. The membrane is fixed along its contour. The initial conditions are

$$\phi(r, \varphi) = 0, \quad \psi(r,\varphi) = Ar(l^2 - r^2)\cos 4\varphi.$$

27. The edge of the membrane is free. The initial conditions are

$$\phi(r,\varphi) = Ar\left(l - \frac{r}{2}\right)\sin 3\varphi, \quad \psi(r,\varphi) = 0.$$

28. The edge of the membrane is free. The initial conditions are

$$\phi(r,\varphi) = 0, \quad \psi(r,\varphi) = Ar\left(l - \frac{r}{2}\right)\cos 2\varphi.$$

29. The edge of the membrane is fixed elastically with coefficient $h = 1$. The initial conditions are

$$\phi(r,\varphi) = Ar(1-r)\sin\varphi, \quad \psi(r,\varphi) = 0.$$

30. The edge of the membrane is fixed elastically with coefficient $h = 1$. The initial conditions are

$$\phi(r, \varphi) = 0, \quad \psi(r, \varphi) = Ar(1-r)\cos\varphi.$$

In problems 31 through 35 external forces are absent but the coefficient of resistance of the environment (a force of resistance that is proportional to velocity) is κ. Find solutions and simulate the transverse oscillations of the membrane using the program **Waves** for each of the given problems.

Choose a coefficient of resistance, κ, so that oscillations decay to zero (with a precision of 5%) during (a) three periods of oscillation; (b) five periods of oscillation; (c) ten periods of oscillation.

31. The membrane is fixed along its contour. The initial conditions are

$$\phi(r, \varphi) = Ar^3(l^2 - r^2)\sin 3\varphi, \quad \psi(r, \varphi) = 0.$$

32. The membrane is fixed along its contour. The initial conditions are

$$\phi(r, \varphi) = 0, \quad \psi(r, \varphi) = Ar^2(l^2 - r^2)\sin 2\varphi.$$

33. The edge of the membrane is free. The initial conditions are

$$\phi(r, \varphi) = Ar\left(l - \frac{r}{2}\right)\sin 5\varphi, \quad \psi(r, \varphi) = 0.$$

34. The edge of the membrane is free. The initial conditions are

$$\phi(r, \varphi) = 0, \quad \psi(r, \varphi) = Ar\left(l - \frac{r}{2}\right)\cos 4\varphi.$$

35. The edge of the membrane is fixed elastically with coefficient $h = 1$. The initial conditions are

$$\phi(r, \varphi) = Ar(r-1)\sin 3\varphi, \quad \psi(r, \varphi) = 0.$$

In problems 36 through 38 the membrane is exited at time $t = 0$ by a sharp impact from a hammer, transferring to the membrane an impulse I at a point (r_0, φ_0) where $0 < r_0 < 1$ and $0 \le \varphi_0 < 2\pi$.

For the following situations, simulate free transverse oscillations of the membrane assuming the initial displacement is zero, external forcing is absent, and the environment causes a resistance proportional to velocity ($\kappa > 0$).

36. The membrane is fixed along its contour.

37. The edge of the membrane is free.

38. The edge of the membrane is attached elastically with elastic coefficient $h = 1$.

In problems 39 through 43 consider circular membranes of radius l. The initial displacement and initial velocity are zero. The resistance of environment is absent ($\kappa = 0$). Find the transversal vibrations of the membrane caused by the action of a varying external pressure $f(r, \varphi, t)$ on one side of the membrane surface.

39. The membrane is fixed along its contour. The external force is

$$f(r,\varphi,t) = A(\sin \omega t + \cos \omega t).$$

40. The membrane is fixed along its contour. The external force is

$$f(r, \varphi, t) = A(l-r)\sin\omega t.$$

41. The edge of the membrane is free. The external force is

$$f(r, \varphi, t) = Ar\cos\omega t.$$

42. The edge of the membrane is fixed elastically with coefficient $h = 1$. The external force is

$$f(r,\varphi,t) = A(l^2 - r^2)\sin\omega t.$$

43. The edge of the membrane is fixed elastically with coefficient $h = 1$. The external force is

$$f(r,\varphi,t) = A(l-r)\cos\omega t.$$

In problems 44 through 47 simulate the transverse oscillations of a membrane caused by its border being displaced according to the function $g(t)$ (the nonhomogeneous term in the boundary condition). Assume external forces, initial velocities, and resistance of the environment are absent ($\kappa = 0$). The initial displacement is given by $u(r, \varphi, 0) = \phi(r)$.

44. $g(t) = A\sin^2 \omega t$, $\phi(r) = Br(l-r)$

45. $g(t) = A(1 - \cos\omega t)$, $\phi(r) = B(l^2 - r^2)$

46. $g(t) = A\cos^2 \omega t$, $\phi(r) = Br\left(\dfrac{r}{2} - l\right)$

47. $g(t) = A(1 - \sin\omega t)$, $\phi(r) = Br(r - l)$

CHAPTER 12

One-Dimensional Parabolic Equations

In this chapter we shall consider a general class of equations known as parabolic equations and their solutions. We start in Section 12.1 with two physical examples: heat conduction and diffusion. More general properties of parabolic equations and their solutions are discussed in subsequent sections.

12.1 HEAT CONDUCTION AND DIFFUSION: BOUNDARY VALUE PROBLEMS

12.1.1 Heat Conduction

Heat may be defined as the flow of energy through a body due to a difference in temperature. This is a kinetic process at the molecular level and involves energy transfer due to molecular collisions. We introduce the equations that model heat transfer in solids where there is no macroscopic mass transfer.

Heat flow through a solid due to a temperature change is assumed to obey the linear heat flow equation established by Fourier:

$$\vec{q} = -\kappa \nabla T. \tag{12.1}$$

Here $\vec{q}$ is the heat flux (or current density), which is the heat (or energy) that flows through a cross-sectional surface area of the body per unit time. The quantity ∇T is the temperature gradient (the difference in temperature along a line parallel to the flux) and the coefficient κ is called the thermal conductivity.

Functions $\vec{q}$ and T are also connected by the *continuity equation*:

$$\rho c \frac{\partial T}{\partial t} + \operatorname{div} \vec{q} = 0, \tag{12.2}$$

where ρ is the mass density of medium and c is its heat capacity.

Equation (12.2) may be stated as follows:

> The amount of heat (or energy) obtained by a unit volume of a body during some unit time interval equals the negative of the divergence of the current density.

Using the relation

$$\operatorname{div} \vec{q} = \operatorname{div}\left[-\kappa\,\nabla T\right] = -\kappa\,\nabla^2 T,$$

equations (12.1) and (12.2) lead to the *heat conduction equation* given by

$$\frac{\partial T}{\partial t} = \chi \nabla^2 T. \tag{12.3}$$

The coefficient $\chi = \kappa/c\rho$ is called the thermal diffusivity.

In a steady state—for instance, with the help of some external heat source—the temperature in the solid becomes time independent and the heat equation reduces to *Laplace's equation* given by

$$\nabla^2 T = 0. \tag{12.4}$$

If Q is the rate at which heat is added (or removed) per unit time and unit volume, the heat conduction equation becomes

$$\frac{\partial T}{\partial t} = \chi \nabla^2 T + \frac{Q}{\rho c}. \tag{12.5}$$

12.1.2 Diffusion Equation

Diffusion is a mixing process that occurs when one substance is introduced into a second substance. The introduced quantity, by means of molecular or atomic transport, spreads from locations where its concentration is higher to locations where the concentration is lower. Examples include the diffusion of perfume molecules into a room when the bottle is opened and the diffusion of neutrons in a nuclear reactor. Given sufficient time, diffusion will lead to an equalizing of the concentration of the introduced substance. We may imagine situations, however, where equilibrium is not reached. This could occur, for example, by continually adding more of the introduced substance at one location and/or removing it at another location.

For low concentration gradients, diffusion obeys Fick's law in which the current density of each component of a mixture is proportional to the concentration gradient of this component:

$$\vec{I} = -D\nabla q. \tag{12.6}$$

Here ∇q is the concentration gradient, I is the current density, and D is the coefficient of diffusion. The case $I = $ constant is the case of steady-state diffusion where a substance is introduced and removed at the same rate.

Let us derive an equation that describes changes of concentration of the introduced substance. The continuity equation

$$\frac{\partial q}{\partial t} + \operatorname{div} \vec{I} = 0, \tag{12.7}$$

is a statement of conservation of mass:

Any increase in the amount of molecules in some volume must equal the amount of molecules entering through the surface enclosing this volume.

Substituting $\vec{I}$ from Equation (12.6) we obtain the diffusion equation

$$\frac{\partial q}{\partial t} = D\nabla^2 q. \tag{12.8}$$

If the introduced substance is being created (or destroyed)—for example, in chemical reactions—we may describe this action by some function f of time and coordinates on the right side of the continuity equation. The diffusion equation then takes the form

$$\frac{\partial q}{\partial t} = D\nabla^2 q + f. \tag{12.9}$$

If we now compare equations (12.8) and (12.9) to equations (12.3) and (12.5), we see they are identical mathematical equations known as *parabolic equations*.

We may also allow the particles of the diffusing substance to be unstable in the sense that they may disappear (such as an unstable gas or a gas being absorbed) or multiply (as with neutron diffusion). If the rates of these processes at each point in space are proportional to the concentration, the process is described by

$$\frac{\partial q}{\partial t} = D\nabla^2 q + \beta q, \tag{12.10}$$

where β is coefficient of disintegration ($\beta < 0$) or multiplication ($\beta > 0$). If we consider the possibility of heat exchange with the environment (in the case that the lateral surfaces of the rod are not insulated), the term βq will appear in heat equation too (for the details see book [1]).

We may also consider the possibility of a diffusing substance participating in the motion of the material in which it is diffusing (which can be a fluid or gas). Suppose the fluid and diffusing substance flow along the x-axis with velocity v, or in the simplest case, a sample of fluid containing the introduced substance moves along the x-axis.

Selecting the element $(x, x + \Delta x)$ and considering the amount of substance that flows through cross sections at x and $x + \Delta x$ due to diffusion as well as fluid motion, we arrive to an equation that includes the first derivative of u with respect to x in addition to the second derivative:

$$\frac{\partial q}{\partial t} = D\frac{\partial^2 q}{\partial x^2} - v\frac{\partial q}{\partial x}. \tag{12.11}$$

We start by considering *the boundary value problem* on the bounded interval $[0,l]$ for the *one-dimensional* heat conduction equation (we use "heat terminology" just for convenience) in its most general form

$$\frac{\partial u}{\partial t} = a^2\frac{\partial^2 u}{\partial x^2} + \xi\frac{\partial u}{\partial x} - \gamma u + f(x,t). \tag{12.12}$$

The initial condition is

$$u(x,0) = \varphi(x), \, 0 < x < l, \tag{12.13}$$

where $\varphi(x)$ is a given function.

12.1.3 Boundary Conditions

1. Dirichlet's condition The temperature at the end, $x = a$ (here $a = 0$ or l), of the bar changes by a specified law given by

$$u(a,t) = g(t), \tag{12.14}$$

where $g(t)$ is a known function of time t. In particular, g is constant if the end of bar is maintained at a steady temperature.

2. *Neumann's condition*—The heat current is given at the end, $x = a$, of the bar in which case

$$q = -\kappa\frac{\partial u}{\partial x}.$$

This may be written as

$$\left.\frac{\partial u}{\partial x}\right|_{x=a} = g(t), \tag{12.15}$$

where $g(t)$ is a known function.

If one of the ends of the bar is insulated, then the coefficient of heat exchange equal zero and the boundary condition at this end takes the form

$$\left.\frac{\partial u}{\partial x}\right|_{x=a} = 0. \tag{12.16}$$

3. *Mixed condition*—In this case the ends of the bar are exchanging heat with the environment, which has a temperature, θ. Actual heat exchange in real physical situations is very complicated, but we may simplify the problem by assuming that it obeys Newton's law, $q = H(u - \theta)$, where H is a positive constant.

It may also be the case that the external environments at the ends of the bar are different. In this case boundary conditions at the ends become

$$\kappa S \Delta t \left. \frac{\partial u}{\partial x} \right|_{x=0} = H_1 [u|_{x=0} - \theta_1],$$

$$-\kappa S \Delta t \left. \frac{\partial u}{\partial x} \right|_{x=l} = H_2 [u|_{x=l} - \theta_2],$$

(12.17)

where the temperatures of the environment at the left and right ends, θ_1 and θ_2, are considered to be known functions of time. In the simplest case they are constants. Thus, in the case of free heat exchange, the boundary conditions at the $x = a$ end of the rod has the form

$$\frac{\partial u}{\partial x} \pm hu \bigg|_{x=a} = g(t),$$

(12.18)

with given function $g(t)$ and $h = $ const.

In general, the boundary conditions are

$$P_1[u] \equiv \alpha_1 \frac{\partial u}{\partial x} + \beta_1 u \bigg|_{x=0} = g_1(t),$$

$$P_2[u] \equiv \alpha_2 \frac{\partial u}{\partial x} + \beta_2 u \bigg|_{x=l} = g_2(t), \quad t>0.$$

(12.19)

At the end, $x = 0$, when $\alpha_1 = 0$ we have Dirichlet's condition, when $\beta_1 = 0$ we have Neumann's condition, and when both constants α_1 and β_1 are not zero we have the mixed condition (clearly the constants α_1 and β_1 cannot be zero simultaneously). The same holds at the end $x = l$. As we have discussed in Chapter 4, physical limitations most often lead to the restrictions $\alpha_1 / \beta_1 < 0$ and $\alpha_2 / \beta_2 > 0$ for the signs of the coefficients in boundary condition (12.19).

Now let us show how equation (12.12), with the help of a proper substitution, can be reduced to the equation without the term with $\partial u / \partial x$.

Substituting

$$u(x,t) = e^{\mu x} v(x,t)$$

where $\mu = -\xi/(2a^2)$ into equation (12.12) yields the equation

$$\frac{\partial v}{\partial t} = a^2 \frac{\partial^2 v}{\partial x^2} - \tilde{\gamma}v + \tilde{f}(x,t),$$

where

$$\tilde{\gamma} = \gamma + \frac{\xi^2}{4a^2}, \quad \tilde{f}(x,t) = e^{-\mu x} f(x,t).$$

The initial condition for the function $v(x,t)$ has the form

$$v(x,0) = \tilde{\varphi}(x),$$

with $\tilde{\varphi}(x) = e^{-\mu x} \varphi(x)$. Boundary conditions for the function $v(x,t)$ are as follows:

$$P_1[v] \equiv \alpha_1 \frac{\partial v}{\partial x} + \tilde{\beta}_1 v \Big|_{x=0} = g_1(t), \quad P_2[v] \equiv \alpha_2 \frac{\partial v}{\partial x} + \tilde{\beta}_2 v \Big|_{x=l} = \tilde{g}_2(t),$$

where

$$\tilde{\beta}_1 = \beta_1 + \mu \cdot \alpha_1, \quad \tilde{\beta}_2 = \beta_2 + \mu \cdot \alpha_2, \quad \tilde{g}_2(t) = e^{-\mu l} \cdot g_2(t)$$

(check these results as homework). Keeping in mind this result, in future calculations we will consider parabolic equation without the term with $\partial u/\partial x$.

We have focused here on the heat conduction equation because the associated terminology is more concrete and intuitively fruitful than that for diffusion. But because the diffusion and heat conduction equations have identical forms, the solutions to diffusion problems can be obtained by a trivial replacement of D and q by χ and T. The boundary condition in equation (12.14) corresponds to the concentration maintained at the ends, condition (12.15) corresponds to an impenetrable end, and condition (12.18) corresponds to a semi-permeable end (when diffusion through this end is similar to that described by Newton's law for heat exchange). An analogue from chemistry is the case of a reaction on the boundary of a body when the speed of reaction, that is, the speed of creation or absorption of one of the chemical components, is proportional to the concentration of this component.

Let us briefly notice the *uniqueness of the solution* of the heat conduction equation under the conditions in equations (12.12), (12.13), and (12.19) and the continuous dependence of this solution on the right-hand terms of the boundary and initial conditions. This material and important *the principle of the maximum* can be found in book [1].

12.2 THE FOURIER METHOD FOR HOMOGENEOUS EQUATIONS

Let us first find the solution of the *homogeneous equation*

$$\frac{\partial u}{\partial t} = a^2 \frac{\partial^2 u}{\partial x^2} - \gamma u, \quad 0 < x < l, \ t > 0, \tag{12.20}$$

which satisfies the initial condition

$$u(x,t)\big|_{t=0} = \varphi(x), \tag{12.21}$$

and has homogeneous boundary conditions

$$P_1[u] = \alpha_1 \frac{\partial u}{\partial x} + \beta_1 u \bigg|_{x=0} = 0, \quad P_2[u] = \alpha_2 \frac{\partial u}{\partial x} + \beta_2 u \bigg|_{x=l} = 0. \tag{12.22}$$

The Fourier method of separation of variables supposes that a solution of equation (12.20) can be found as a product of two functions, one depending only on x, the second depending only on t:

$$u(x,t) = X(x)\,T(t). \tag{12.23}$$

Substituting equation (12.23) into equation (12.20) we obtain

$$X(x)[T'(t) + \gamma T(t)] - a^2 X''(x) T(t) = 0.$$

The variables can be separated and denoting a *separation constant as* $-\lambda$, we obtain

$$\frac{T'(t) + \gamma T(t)}{a^2 T(t)} \equiv \frac{X''(x)}{X(x)} = -\lambda. \tag{12.24}$$

Thus equation (12.24) gives two ordinary linear homogeneous differential equations, a first-order equation for function $T(t)$,

$$T'(t) + (a^2 \lambda + \gamma) T(t) = 0, \tag{12.25}$$

and a second-order equation for function $X(x)$,

$$X''(x) + \lambda\, X(x) = 0. \tag{12.26}$$

To find the allowed values of λ we apply the boundary conditions. Homogenous boundary condition (12.22), imposed on $u(x,t)$, yields homogeneous boundary conditions on the function $X(x)$ given by

$$\alpha_1 X' + \beta_1 X\big|_{x=0} = 0, \quad \alpha_2 X' + \beta_2 X\big|_{x=l} = 0. \tag{12.27}$$

Thus we obtain the Sturm-Liouville boundary problem for eigenvalues, λ, and the corresponding eigenfunctions, $X(x)$. As we know from Chapter 4 and previous discussion in this chapter, there exist an *infinite* sets of real non-negative discrete spectrum of eigenvalues $\{\lambda_n\}$ and corresponding set of eigenfunctions $\{X_n(x)\}$ (clearly $\lambda = 0$ is also possible if $\beta_1 = \beta_2 = 0$).

As we obtained in Chapter 4, the eigenvalues of the Sturm-Liouville problem stated in equations (12.26) and (12.27) are

$$\lambda_n = \left(\frac{\mu_n}{l}\right)^2, \tag{12.28}$$

where μ_n is the nth non-negative root of the equation

$$\tan\mu = \frac{(\alpha_1\beta_2 - \alpha_2\beta_1)l\mu}{\mu^2\alpha_1\alpha_2 + l^2\beta_1\beta_2}. \tag{12.29}$$

The corresponding eigenfunctions can be written as

$$X_n(x) = \frac{1}{\sqrt{\alpha_1^2\lambda_n + \beta_1^2}}\left[\alpha_1\sqrt{\lambda_n}\cos\sqrt{\lambda_n}\,x - \beta_1\sin\sqrt{\lambda_n}\,x\right]. \tag{12.30}$$

Now consider equation (12.25). It is a linear first-order differential equation and the general solution with $\lambda = \lambda_n$ is

$$T_n(t) = C_n e^{-(a^2\lambda_n + \gamma)t}, \tag{12.31}$$

where C_n is an arbitrary constant. Non-negative values of λ_n are required so that the solution cannot grow to infinity with time. Now we have that each function

$$u_n(x,t) = T_n(t)X_n(x) = C_n e^{-(a^2\lambda_n + \gamma)t}X_n(x)$$

is a solution of equation (12.20) satisfying boundary conditions (12.22). To satisfy the initial conditions (12.21), we compose the series

$$u(x,t) = \sum_{n=1}^{\infty} C_n e^{-(a^2\lambda_n + \gamma)t}X_n(x). \tag{12.32}$$

If this series converges uniformly as well as the series obtained by differentiating twice by x and once by t, the sum gives a solution to equation (12.20) and satisfies the boundary conditions (12.22). The initial condition in equation (12.21) gives

$$u|_{t=0} = \varphi(x) = \sum_{k=1}^{\infty} C_k X_k(x), \tag{12.33}$$

where we have expanded the function $\varphi(x)$ in a series of the eigenfunctions of the boundary value problem given by equations (12.26) and (12.27).

Assuming uniform convergence of the series (12.32) we can find the coefficients C_n. Multiplying both sides of equation (12.33) by $X_n(x)$, integrating from 0 to l, and imposing the orthogonality condition of the functions $X_n(x)$, we obtain

$$C_n = \frac{1}{\| X_n \|^2} \int_0^l \varphi(x) X_n(x)\, dx. \tag{12.34}$$

If the series (12.32) and the series obtained from it by differentiating by t and twice differentiating by x are uniformly convergent, then by substituting these values of coefficients, C_n, into the series (12.32) we obtain the solution of the problem stated in equations (12.20) through (12.22).

Equation (12.32) gives a solution for *free heat exchange* (heat exchange without sources of heat within the body). It can be considered as the decomposition of an unknown function, $u(x,t)$, into a Fourier series over an orthogonal set of functions $\{X_n(x)\}$. This series converges well for "reasonable" assumptions about initial and boundary conditions.

Example 1

Let zero temperature be maintained on both the ends, $x = 0$ and $x = l$, of a uniform isotropic bar of length l with a heat-insulated lateral surface. Initially the temperature distribution inside the bar is given by

$$u(x,0) = \varphi(x) = \begin{cases} \dfrac{x}{l} u_0 & \text{for } 0 < x \le \dfrac{l}{2}, \\[2mm] \dfrac{l-x}{l} u_0 & \text{for } \dfrac{l}{2} < x < l, \end{cases}$$

where $u_0 = $ const. There are no sources of heat inside the bar. Find the temperature distribution for the interior of the bar for time $t > 0$.

Solution. The problem is described by the equation

$$\frac{\partial u}{\partial t} = a^2 \frac{\partial^2 u}{\partial x^2}, \quad 0 < x < l, \quad t > 0$$

with initial and boundary conditions (these conditions are consistent)

$$u(x,0) = \varphi(x), \ u(0,t) = u(l,t) = 0.$$

The boundary conditions of the problem are Dirichlet homogeneous boundary conditions; therefore eigenvalues and eigenfunctions of problem are

$$\lambda_n = \left(\frac{n\pi}{l}\right)^2, \ X_n(x) = \sin\frac{n\pi x}{l}, \ ||X_n||^2 = \frac{l}{2}, \ n = 1,2,\ldots$$

Equation (12.34) gives

$$C_n = \frac{2}{l}\int_0^l \varphi(x)\sin\frac{n\pi x}{l}\,dx = \frac{4u_0}{n^2\pi^2}\sin\frac{n\pi}{2} = \begin{cases} 0, & n = 2k, \\ \dfrac{4u_0}{(2k-1)^2\pi^2}(-1)^k, & n = 2k-1. \end{cases}$$

Hence the distribution of temperature inside bar for some moment is described by the series

$$u(x,t) = \frac{4u_0}{\pi^2}\sum_{k=1}^{\infty}\frac{(-1)^k}{(2k-1)^2}e^{-\frac{a^2(2k-1)^2\pi^2}{l^2}t}\sin\frac{(2k-1)\pi x}{l}.$$

The series (as well as the series in several of the following examples) converges rather rapidly because it has coefficients that decrease as $1/k^2$.

Figure 12.1 shows the spatial-time-dependent solution $u(x, t)$ for Example 1. This solution was obtained with the program **Heat** for the case when $l=10$, $u_0 = 5$, and $a^2 = 0.25$. All parameters are dimensionless.

The bold black line represents the initial temperature; the bold gray line is the temperature at time $t = 100$. The gray lines in between show the temperature evolution within the period of time from 0 until 100.

As a Reading Exercise, choose values of parameters in real physical units and estimate how much time it takes to reach equilibrium for different materials and for samples of different size. Make sure that the answers are physically reasonable. *Hint*: Clearly the

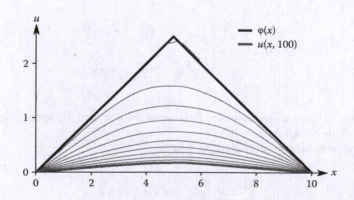

FIGURE 12.1   Solution $u(x, t)$ for Example 1.

approach to equilibrium (the time for temperature fluctuations to decay) is governed by the factor $\exp[-(a^2\lambda_n + \gamma)t]$.

Example 2

Consider the case when the ends, $x = 0$ and $x = l$, of a bar are thermally insulated from the environment. The lateral surface is also insulated. In this case the derivatives of temperature with respect to x on the ends of the bar equal zero. Initially the temperature is distributed as in the previous example:

$$\varphi(x) = \begin{cases} \dfrac{x}{l} u_0 & \text{for } 0 < x \le \dfrac{l}{2}, \\[2mm] \dfrac{l-x}{l} u_0 & \text{for } \dfrac{l}{2} < x < l, \end{cases}$$

where $u_0 = $ const. Sources of heat are absent. Find the temperature distribution inside the bar for $t>0$.

Solution. We are to solve the equation

$$\frac{\partial u}{\partial t} = a^2 \frac{\partial^2 u}{\partial x^2}, \quad 0 < x < l, \ t > 0$$

with initial and boundary conditions

$$u(x,0) = \varphi(x), \quad \frac{\partial u}{\partial x}(0,t) = \frac{\partial u}{\partial x}(l,t) = 0.$$

Notice that the initial and boundary conditions in this case are not consistent (they are contradictory); in such situations we can only obtain a *generalized solution*. The boundary conditions of the problem are Neumann homogeneous boundary conditions. We leave it to the reader as a Reading Exercise to obtain the formulas that define the problem (for all possible situations see Appendix 1). The eigenvalues and eigenfunctions obtained from these formulas are

$$\lambda_n = \left(\frac{n\pi}{l}\right)^2, \quad X_n(x) = \cos\frac{n\pi x}{l}, \quad \|X_n\|^2 = \begin{cases} l, & n = 0, \\ l/2, & n > 0, \end{cases} \quad n = 0,1,2,\ldots$$

Having applied equation (12.34), we obtain

$$C_0 = \frac{u_0}{l^2}\left[\int_0^{l/2} x\,dx + \int_{l/2}^{l}(l-x)\,dx\right] = \frac{u_0}{4},$$

$$C_n = \frac{2u_0}{l^2}\int_0^{l/2} x\cos\frac{n\pi x}{l}\,dx + \frac{2u_0}{l^2}\int_{l/2}^{l}(l-x)\cos\frac{n\pi x}{l}\,dx$$

$$= \begin{cases} 0 & \text{for } n = 2k-1, \\[3mm] \dfrac{u_0}{k^2\pi^2}\left[(-1)^k - 1\right] & \text{for } n = 2k, \end{cases} \quad k = 1,2,3,\ldots$$

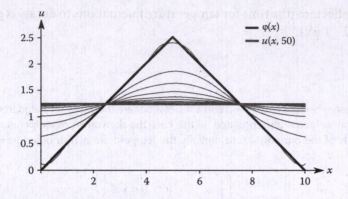

FIGURE 12.2   Solution $u(x, t)$ for Example 2.

Thus the only nonzero C_n are those for which $n = 2k$ with $k = 2m + 1$, $m = 0,1,2,...$, that is, $n = 4m + 2$; so we have

$$C_{4m+2} = -\frac{2u_0}{(2m+1)^2\pi^2}.$$

Finally, the temperature distribution inside the bar for some moment is expressed by the series

$$u(x,t) = \frac{u_0}{4} - \frac{2u_0}{\pi^2} \sum_{m=0}^{\infty} \frac{1}{(2m+1)^2} e^{-\frac{(4m+2)^2 a^2\pi^2}{l^2}t} \cos\frac{(4m+2)\pi x}{l}.$$

At $x = l/4$ and $x = 3l/4$ all cosine terms equal zero; hence at these points $u = u_0/4$ for any $t \geq 0$. It is also clear that

$$\int_0^l u(x,t)\,dx = \frac{1}{4}u_0 l.$$

Notice that this area (the definite integral) is proportional to the amount of energy (heat) in the bar. The insulated ends of the bar correspond to a graph of $u(x,t)$, which has horizontal tangents at $x=0$ and $x = l$. As $t \to \infty$ the first term of the series dominates. From this and from physical considerations we conclude that $u \to u_0/4$ as $t \to \infty$.

Figure 12.2 shows the spatial-time-dependent solution $u(x,t)$ for Example 2. This solution was obtained with the program **Heat** for the case when $l=10$, $u_0 = 5$, and $a^2 = 0.25$.

Example 3

Consider the situation where heat flux is governed by Newton's law of cooling and a constant temperature environment occurs at each end of a uniform isotropic bar of length l ($0 \leq x \leq l$) with insulated lateral surface. The initial temperature of the bar is equal to $u_0 = $ const. Internal sources of heat in the bar are absent. Find the temperature distribution inside the bar for $t > 0$.

Solution. We need to solve the equation

$$\frac{\partial u}{\partial t} = a^2 \frac{\partial^2 u}{\partial x^2}, \quad 0 < x < l, t > 0$$

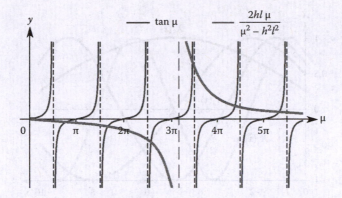

FIGURE 12.3 Graphical solution of the eigenvalue equation for Example 3.

for the initial condition

$$u(x,0) = u_0$$

and boundary conditions

$$\frac{\partial u}{\partial x}(0,t) - hu(0,t) = 0, \quad \frac{\partial u}{\partial x}(l,t) + hu(l,t) = 0,$$

where $h > 0$ is the coefficient of heat exchange with environment. Obviously, as in the previous example, we can only obtain a generalized solution.

The boundary conditions of the problem are mixed homogeneous boundary conditions, so eigenvalues are

$$\lambda_n = \left(\frac{\mu_n}{l}\right)^2, \quad n = 1,2,3,...,$$

where μ_n is the nth root of the equation $\tan\mu = 2hl\mu/(\mu^2 - h^2l^2)$. Figure 12.3 shows curves of the two functions, $y = \tan\mu$ and $y = 2hl\mu/(\mu^2 - h^2l^2)$, plotted on the same set of axes. The eigenvalues are the squares of the values of μ at the intersection points of these curves divided by length l.

Each eigenvalue corresponds to an eigenfunction

$$X_n(x) = \frac{1}{\sqrt{\lambda_n + h^2}}\left[\sqrt{\lambda_n}\cos\sqrt{\lambda_n}\,x + h\sin\sqrt{\lambda_n}\,x\right]$$

with the norm

$$\|X_n\|^2 = \frac{l}{2} + \frac{h}{\lambda_n + h^2}.$$

Figure 12.4 shows first four eigenfunctions of the given boundary value problem.

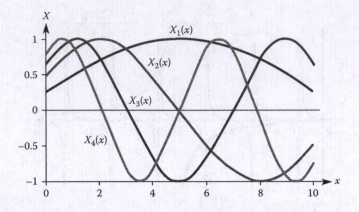

FIGURE 12.4   Eigenfunctions $X_1(x)$ through $X_4(x)$ for Example 3.

Applying equation (12.34) we obtain

$$C_n = \frac{1}{\|X_n\|^2} \int_0^l \frac{u_0}{\sqrt{\lambda_n + h^2}} \left[\sqrt{\lambda_n} \cos \sqrt{\lambda_n}\, x + h \sin \sqrt{\lambda_n}\, x \right] dx$$

$$= \frac{2u_0 \sqrt{\lambda_n + h^2}}{l(\lambda_n + h^2) + 2h} \left[\sin \sqrt{\lambda_n}\, l - \frac{h}{\sqrt{\lambda_n}} (\cos \sqrt{\lambda_n}\, l - 1) \right].$$

Hence the temperature distribution inside the bar for some moment of time is expressed by the series

$$u(x,t) = \sum_{n=1}^{\infty} C_n e^{-a^2 \lambda_n t} \frac{1}{\sqrt{\lambda_n + h^2}} \left[\sqrt{\lambda_n} \cos \sqrt{\lambda_n}\, x + h \sin \sqrt{\lambda_n}\, x \right].$$

Figure 12.5 shows the spatial time-dependent solution $u(x,t)$ for Example 3. This solution was obtained with the program **Heat** for the case when $l = 10$, $u_0 = 5$, and $a^2 = 0.25$.

In this example the boundary and initial conditions do not match each other; as a result at $t = 0$ the temperature, $u(x,0)$, given by the solution in the form of an eigenfunction expansion

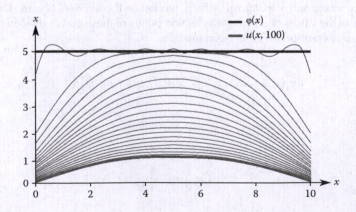

FIGURE 12.5   Solution $u(x, t)$ for Example 3.

does not converge uniformly. The convergence is poor at points close to the ends of the rod and the solution appears to have unphysical oscillations of temperature initially. Increasing the number of terms in the series smoothes these oscillations at all points except the ends. These oscillations do not occur physically and, in fact, they disappear from the solution for any finite time $t > 0$ in which case $u(x,t)$ converges rapidly to a physically reasonable result.

12.3 THE FOURIER METHOD FOR NONHOMOGENEOUS EQUATIONS

Consider the *nonhomogeneous* linear equation

$$\frac{\partial u}{\partial t} = a^2 \frac{\partial^2 u}{\partial x^2} - \gamma u + f(x,t), \qquad (12.35)$$

where $f(x,t)$ is a known function, with initial condition

$$u(x,t)|_{t=0} = \varphi(x) \qquad (12.36)$$

and *homogeneous boundary conditions*

$$P_1[u] \equiv \alpha_1 u_x + \beta_1 u|_{x=0} = 0, \quad P_2[u] \equiv \alpha_2 u_x + \beta_2 u|_{x=l} = 0. \qquad (12.37)$$

To start, let us express the function $u(x,t)$ as the sum of two functions:

$$u(x,t) = u_1(x,t) + u_2(x,t),$$

where $u_1(x,t)$ satisfies the *homogeneous equation* with the given (homogeneous) boundary conditions and the specified (nonhomogeneous) initial condition:

$$\frac{\partial u_1}{\partial t} = a^2 \frac{\partial^2 u_1}{\partial x^2} - \gamma u_1,$$

$$u_1(x,t)|_{t=0} = \varphi(x),$$

$$P_1[u_1] \equiv \alpha_1 u_{1x} + \beta_1 u_1|_{x=0} = 0, \quad P_2[u_1] \equiv \alpha_2 u_{1x} + \beta_2 u_1|_{x=l} = 0.$$

The function $u_2(x,t)$ satisfies the nonhomogeneous equation with *zero boundary and initial conditions*:

$$\frac{\partial u_2}{\partial t} = a^2 \frac{\partial^2 u_2}{\partial x^2} - \gamma u_2 + f(x,y), \qquad (12.38)$$

$$u_2(x,t)|_{t=0} = 0, \qquad (12.39)$$

$$P_1[u_2] \equiv \alpha_1 u_{2x} + \beta_1 u_2|_{x=0} = 0, \quad P_2[u_2] \equiv \alpha_2 u_{2x} + \beta_2 u_2|_{x=l} = 0. \qquad (12.40)$$

The methods for finding $u_1(x,t)$ have been discussed in the previous section; therefore here we concentrate on finding the solutions $u_2(x,t)$. As for the case of free heat exchange inside the bar, let us expand function $u_2(x,t)$ as a series

$$u_2(x,t) = \sum_{n=1}^{\infty} T_n(t)X_n(x), \qquad (12.41)$$

where $X_n(x)$ are eigenfunctions of the corresponding homogeneous boundary value problem and $T_n(t)$ are unknown functions of t.

Boundary conditions in equation (12.40) for $u_2(x,t)$ are valid for any choice of functions $T_n(t)$ (when the series converge uniformly) because they are valid for the functions $X_n(x)$. Substituting the series (12.41) into equation (12.38) we obtain

$$\sum_{n=1}^{\infty}[T_n'(t)+(a^2\lambda_n+\gamma)T_n(t)]X_n(x) = f(x,t). \qquad (12.42)$$

Using the completeness property we can expand the function $f(x,t)$, as function of x, into a Fourier series of the functions $X_n(x)$ on the interval $(0,l)$ such that

$$f(x,t) = \sum_{n=1}^{\infty} f_n(t)X_n(x). \qquad (12.43)$$

Using the orthogonality property of the functions $X_n(x)$ we find that

$$f_n(t) = \frac{1}{\|X_n\|^2} \int_0^l f(x,t)X_n(x)dx. \qquad (12.44)$$

Comparing the two expansions in equations (12.42) and (12.43) for the same function $f(x,t)$ we obtain a differential equations for the functions $T_n(t)$:

$$T_n'(t)+(a^2\lambda_n+\gamma)T_n(t) = f_n(t). \qquad (12.45)$$

In order that $u_2(x,t)$ given by equation (12.41) satisfy the initial condition (12.39) it is necessary that the functions $T_n(t)$ obey the condition

$$T_n(0) = 0. \qquad (12.46)$$

The solution to the ordinary differential equation of the fist order, equation (12.45), with initial condition (12.46) can be represented in the integral form

$$T_n(t) = \int_0^t f_n(\tau)e^{-(a^2\lambda_n+\gamma)(t-\tau)}d\tau, \qquad (12.47)$$

or

$$T_n(t) = \int\limits_0^t f_n(\tau) Y_n(t-\tau)\,d\tau, \quad \text{where} \quad Y_n(t-\tau) = e^{-(a^2\lambda_n+\gamma)(t-\tau)}.$$

Thus the solution of the nonhomogeneous heat conduction problem for a bar with boundary conditions equal to zero has the form

$$u(x,t) = u_1(x,t) + u_2(x,t) = \sum_{n=1}^{\infty}\left[T_n(t) + C_n e^{-(a^2\lambda_n+\gamma)t}\right] X_n(x), \tag{12.48}$$

where functions $T_n(t)$ are defined by equation (12.47) and coefficients C_n have been found earlier when we considered the homogeneous heat equation.

Example 4

A point-like heat source with power Q = const. is located at x_0 $(0 < x_0 < l)$ in a uniform isotropic bar with insulated lateral surfaces. The initial temperature of the bar is zero. Temperatures at the ends of the bar are maintained at zero. Find the temperature inside the bar for $t > 0$.

Solution. The boundary value problem modeling heat propagation for this example is

$$\frac{\partial u}{\partial t} = a^2 \frac{\partial^2 u}{\partial x^2} + \frac{Q}{c\rho}\delta(x-x_0),$$

$$u(x,0) = 0, \quad u(0,t) = u(l,t) = 0,$$

where $\delta(x-x_0)$ is the delta function.

The boundary conditions are Dirichlet homogeneous boundary conditions, so the eigenvalues and eigenfunctions of the problem are

$$\lambda_n = \left(\frac{n\pi}{l}\right)^2, \quad X_n(x) = \sin\frac{n\pi x}{l}, \quad \|X_n\|^2 = \frac{l}{2}, \quad (n=1,2,3,\ldots).$$

In the case of homogeneous initial conditions, $\varphi(x) = 0$, we have $C_n = 0$ and the solution $u(x,t)$ is defined by the series

$$u(x,t) = \sum_{n=1}^{\infty} T_n(t)\sin\frac{n\pi x}{l},$$

where f_n and T_n are defined by equations (12.44) and (12.47):

$$f_n(t) = \frac{2}{l}\int\limits_0^l \frac{Q}{c\rho}\delta(x-x_0)\sin\frac{n\pi x}{l}\,dx = \frac{2Q}{lc\rho}\sin\frac{n\pi x_0}{l}\,dx,$$

$$T_n(t) = \frac{2Ql}{c\rho a^2 n^2\pi^2}(1-e^{-n^2a^2\pi^2 t/l^2})\sin\frac{n\pi x_0}{l}.$$

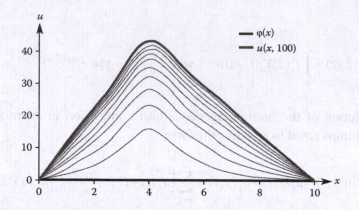

FIGURE 12.6   Solution $u(x,t)$ for Example 4.

Substituting the expression for $T_n(t)$ into the general formulas we obtain the solution of the problem:

$$u(x,t) = \frac{2Ql}{cpa^2\pi^2} \sum_{n=1}^{\infty} \frac{1}{n^2}(1 - e^{-n^2a^2\pi^2t/l^2})\sin\frac{n\pi x_0}{l}\sin\frac{n\pi x}{l}.$$

Figure 12.6 shows the spatial-time-dependent solution $u(x,t)$ for Example 4. This solution was obtained with the program **Heat** for the case when $l = 10$, $Q/cp = 5$, $x_0 = 4$, and $a^2 = 0.25$.

12.4 THE FOURIER METHOD FOR NONHOMOGENEOUS EQUATIONS WITH NONHOMOGENEOUS BOUNDARY CONDITIONS

Now consider the general boundary problem for heat conduction, equation (12.35), given by

$$\frac{\partial u}{\partial t} = a^2\frac{\partial^2 u}{\partial x^2} - \gamma u + f(x,t)$$

with nonhomogeneous initial (equation (12.36)) and boundary conditions

$$u(x,t)|_{t=0} = \varphi(x),$$

$$P_1[u] \equiv \alpha_1 u_x + \beta_1 u|_{x=0} = g_1(t), \quad P_2[u] \equiv \alpha_2 u_x + \beta_2 u|_{x=l} = g_2(t). \tag{12.49}$$

We cannot apply the Fourier method directly to obtain a solution of the problem because the boundary conditions are nonhomogeneous. However the problem can easily be reduced to a problem with boundary conditions equal to zero in the following way.

Let us search for the solution of the problem in the form

$$u(x,t)=v(x,t)+w(x,t), \tag{12.50}$$

where $v(x,t)$ is a new unknown function, and the function $w(x,t)$ is chosen so that it satisfies the given nonhomogeneous boundary conditions

$$P_1[w] \equiv \alpha_1 w_x + \beta_1 w|_{x=0} = g_1(t),$$

$$P_2[w] \equiv \alpha_2 w_x + \beta_2 w|_{x=l} = g_2(t).$$

For the function $v(x,t)$ we obtain the following boundary value problem:

$$\frac{\partial v}{\partial t} = a^2 \frac{\partial^2 v}{\partial x^2} - \gamma v + f^*(x,t),$$

$$v(x,t)|_{t=0} = \varphi^*(x),$$

$$P_1[v] \equiv \alpha_1 v_x + \beta_1 v|_{x=0} = 0, \quad P_2[v] \equiv \alpha_2 v_x + \beta_2 v|_{x=l} = 0,$$

where

$$f^*(x,t) = f(x,t) - \frac{\partial w}{\partial t} + a^2 \frac{\partial^2 w}{\partial x^2} - \gamma w,$$

$$\varphi^*(x) = \varphi(x) - w(x,0).$$

The solution of such a problem with homogeneous boundary conditions has been considered in the previous section.

The auxiliary function $w(x,t)$ is ambiguously defined. The simplest way to proceed is to use polynomials and construct it in the form

$$w(x,t) = P_1(x)g_1(t) + P_2(x)g_2(t),$$

where $P_1(x)$ and $P_2(x)$ are polynomials of the first or second order. Coefficients of these polynomials will be chosen so that the function $w(x,t)$ satisfies the given boundary conditions. We have the following possibilities.

Case I

If β_1 and β_2 in equation (12.49) are not zero simultaneously we may seek for the function $w(x,t)$ in the form

$$w(x,t) = (\gamma_1 + \delta_1 x)g_1(t) + (\gamma_2 + \delta_2 x)g_2(t).$$

Substituting this into boundary conditions (12.49), and taking into account that the derived system of equations must be valid for arbitrary t, we obtain coefficients γ_1, δ_1, γ_2, and δ_2 as

$$\gamma_1 = \frac{\alpha_2 + \beta_2 l_x}{\beta_1 \beta_2 l_x + \beta_1 \alpha_2 - \beta_2 \alpha_1}, \quad \delta_1 = \frac{-\beta_2}{\beta_1 \beta_2 l_x + \beta_1 \alpha_2 - \beta_2 \alpha_1},$$

$$\gamma_2 = \frac{-\alpha_1}{\beta_1 \beta_2 l_x + \beta_1 \alpha_2 - \beta_2 \alpha_1}, \quad \delta_2 = \frac{\beta_1}{\beta_1 \beta_2 l_x + \beta_1 \alpha_2 - \beta_2 \alpha_1}.$$

Reading Exercise

We leave it to the reader to obtain the results above.

Case II

If $\beta_1 = \beta_2 = 0$, that is, $\begin{cases} u_x(0,t) = g_1(t) \\ u_x(l,t) = g_2(t) \end{cases}$, the auxiliary function has the form

$$w(x,t) = \left[x - \frac{x^2}{2l} \right] \cdot g_1(t) + \frac{x^2}{2l} \cdot g_2(t).$$

Reading Exercise

Verify the statement above.

It is easily checked that, defined in such a way, the auxiliary functions $w(x,t)$ satisfy the boundary conditions in equation (12.37).

Reading Exercise

Prove the statement above.

Combining the different kinds of boundary conditions listed above, we obtain nine different auxiliary functions, which are listed in Appendix 2.

In the following examples we consider problems that involve the nonhomogeneous heat conduction equation with nonhomogeneous boundary conditions.

Example 5

Find the temperature change in a homogeneous isotropic bar of length l ($0 \leq x \leq l$) with a heat-insulated lateral surface during free heat exchange if the initial temperature is given by

$$u(x,0) = \varphi(x) = u_0 \frac{x^2}{l^2}.$$

The left end of the bar at $x = 0$ is insulated and at the right end temperature is held constant:

$$u(l,t) = u_0, \quad \text{where} \quad u_0 = \text{const} > 0.$$

Solution. The problem is described by the equation

$$\frac{\partial u}{\partial t} = a^2 \frac{\partial^2 u}{\partial x^2}$$

with the conditions

$$u(x,0) = \varphi(x) = u_0 x^2 / l^2,$$

$$\frac{\partial u}{\partial x}(0,t) = 0, \quad u(l,t) = u_0.$$

The solution of the problem will be of the form

$$u(x,t) = v(x,t) + w(x,t).$$

The auxiliary function can be easily obtained from the general formulas above (do this as a Reading Exercise; the answers are found in Appendix 2) and we find

$$w(x,t) = u_0.$$

This function corresponds to the steady-state regime as $t \to \infty$. The eigenvalues and eigenfunctions are easy to obtain (you can also find them in Appendix 1) and we leave it to the reader as a Reading Exercise to check that they are

$$\lambda_n = \left[\frac{(2n-1)\pi}{2l}\right]^2, \quad X_n(x) = \cos\frac{(2n-1)\pi x}{2l}, \quad \| X_n \|^2 = \frac{l}{2}, \quad (n = 1,2,3,\ldots).$$

Following the same logic as in previous problems, for the function $v(x,t)$ we obtain the conditions

$$f^*(x,t) = 0,$$

$$\varphi^*(x) = u_0 \frac{x^2}{l^2} - u_0 = u_0\left(\frac{x^2}{l^2} - 1\right).$$

We apply equation (12.34) to obtain

$$C_n = \frac{2}{l}\int_0^l \varphi^*(x)\cos\frac{(2n-1)\pi x}{2l}\,dx$$

$$= \frac{2}{l}\int_0^l u_0\left(\frac{x^2}{l^2} - 1\right)\cos\frac{(2n-1)\pi x}{2l}\,dx = -\frac{32u_0}{(2n-1)^3\pi^3}(-1)^n.$$

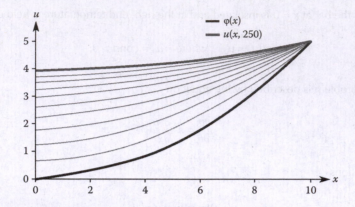

FIGURE 12.7   Solution $u(x,t)$ for Example 5.

With these coefficients we obtain the solution of the problem:

$$u(x,t) = u_0 - \frac{32u_0}{\pi^3} \sum_{n=1}^{\infty} \frac{(-1)^n}{(2n-1)} e^{-\frac{(2n-1)^2 a^2 \pi^2}{l^2} t} \cos \frac{(2n-1)\pi x}{2l}$$

Figure 12.7 shows the spatial-time-dependent solution $u(x,t)$ for Example 5. This solution was obtained with the program **Heat** for the case when $l = 10$, $u_0 = 5$, and $a^2 = 0.25$.

Example 6

The initial temperature of a homogeneous isotropic bar of length l $(0 \leq x \leq l)$ is

$$u(x,0) = u_0 = \text{const.}$$

There exists a steady heat flux from the environment into the ends of the bar, which is given by

$$\frac{\partial u}{\partial x}(0,t) = Q_1 = -\frac{q_1}{\kappa S}, \quad \frac{\partial u}{\partial x}(l,t) = Q_2 = \frac{q_2}{\kappa S}.$$

Convective heat transfer occurs with the environment through the lateral surface. Find the temperature $u(x,t)$ of the bar for $t > 0$.

Solution. The problem is described by the equation

$$\frac{\partial u}{\partial t} = a^2 \frac{\partial^2 u}{\partial x^2} - \gamma u, \quad a^2 = \frac{\kappa}{c\rho}, \quad \gamma = \frac{h}{c\rho}$$

with initial and boundary conditions

$$u(x,0) = \varphi(x) = u_0,$$

$$\frac{\partial u}{\partial x}(0,t) = Q_1, \quad \frac{\partial u}{\partial x}(l,t) = Q_2.$$

The boundary conditions of the problem are Neumann boundary conditions and the eigenvalues and eigenfunctions are, respectively,

$$\lambda_n = \left(\frac{n\pi}{l}\right)^2, \ X_n(x) = \cos\frac{n\pi x}{l}, \ ||X_n||^2 = \begin{cases} l, & n = 0, \\ l/2, & n > 0. \end{cases}$$

The auxiliary function is (see Appendix 2)

$$w(x,t) = \left(x - \frac{x^2}{2l}\right)Q_1 + \frac{x^2}{2l}Q_2 = xQ_1 + \frac{x^2}{2l}(Q_2 - Q_1).$$

We leave it to the reader as a Reading Exercise to obtain these auxiliary function and the eigenvalues and eigenfunctions.

The solution is

$$u(x,t) = w(x,t) + \sum_{n=0}^{\infty} \left[T_n(t) + C_n \cdot e^{-(a^2\lambda_n + \gamma)t}\right]\cos\frac{n\pi x}{l}.$$

To obtain the coefficients we find first

$$f^*(x,t) = \frac{a^2}{l}(Q_2 - Q_1) - \gamma\left[\left(x - \frac{x^2}{2l}\right)Q_1 + \frac{x^2}{2l}Q_2\right],$$

$$\varphi^*(x) = u_0 - \left(x - \frac{x^2}{2l}\right)Q_1 - \frac{x^2}{2l}Q_2.$$

Applying equations (12.34), (12.44), and (12.47), we obtain

$$C_0 = \frac{1}{l}\int_0^l \left[u_0 - \left(x - \frac{x^2}{2l}\right)Q_1 - \frac{x^2}{2l}Q_2\right]dx = u_0 - \frac{l}{6}(2Q_1 + Q_2),$$

$$C_n = \frac{2}{l}\int_0^l \left[u_0 - Q_1x + (Q_1 - Q_2)\frac{x^2}{2l}\right]\cos\frac{n\pi x}{l}dx = \frac{2l}{n^2\pi^2}\left[Q_1 - (-1)^n Q_2\right],$$

$$f_0(t) = \frac{1}{l}\int_l f^*(x,t)dx = \frac{a^2(Q_2 - Q_1)}{l} - \frac{\gamma l}{6}(2Q_1 + Q_2),$$

$$T_0(t) = \int_0^t f_0(\tau)\cdot e^{-\gamma(t-\tau)}\,d\tau = \frac{1}{\gamma}\cdot\left[\frac{a^2(Q_2 - Q_1)}{l} - \frac{\gamma l}{6}(2Q_1 + Q_2)\right]\cdot(1 - e^{-\gamma t}).$$

For $n > 0$,

$$f_n(t) = \frac{2}{l}\int_0^l f^*(x,t)\cos\frac{n\pi x}{l}\,dx = \frac{2l\gamma}{n^2\pi^2}\left[Q_1 - (-1)^n Q_2\right],$$

$$T_n(t) = \int_0^t f_n(\tau)\cdot e^{-(a^2\lambda_n + \gamma)(t-\tau)}\,d\tau = \frac{2l\gamma}{n^2\pi^2(a^2\lambda_n + \gamma)}\cdot\left[Q_1 - (-1)^n Q_2\right]\cdot\left(1 - e^{-(a^2\lambda_n + \gamma)t}\right).$$

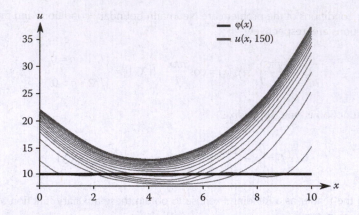

FIGURE 12.8   Solution $u(x,t)$ for Example 6.

Figure 12.8 shows the spatial-time-dependent solution $u(x,t)$ for Example 6. This solution was obtained with the program **Heat** for the case when $l = 10$, $\gamma = 0.02$, $u_0 = 10$, $Q_1 = -5$, $Q_2 = 10$, and $a^2 = 0.25$.

Example 7

Find the temperature distribution inside a thin homogeneous isotropic bar of length l ($0 \le x \le l$) with insulated lateral surface if the initial temperature is zero. The temperature is maintained at zero on the right end of the bar ($x = l$), and on the left it changes as governed by

$$u(0,t) = u_0 \cos \omega t,$$

where u_0, ω are known constants. There are no sources or absorbers of heat inside the bar.

Solution. The temperature is given by a solution of the equation

$$\frac{\partial u}{\partial t} = a^2 \frac{\partial^2 u}{\partial x^2}$$

with conditions

$$u(x,0) = 0,$$

$$u(0,t) = u_0 \cos \omega t, \quad u(l,t) = 0.$$

For these Dirichlet boundary conditions the eigenvalues and eigenfunctions of the problem are

$$\lambda_n = \left(\frac{n\pi}{l}\right)^2, \quad X_n(x) = \sin\frac{n\pi x}{l}, \quad n = 1, 2, 3, \ldots$$

The solution will be of the form

$$u(x,t) = v(x,t) + w(x,t).$$

The auxiliary function follows from the general case (also see Appendix 2):

$$w(x,t) = \left(1 - \frac{x}{l}\right) u_0 \cos \omega t.$$

Then

$$f^*(x,t) = u_0 \omega \left(1 - \frac{x}{l}\right) \sin \omega t \quad \text{and} \quad \varphi^*(x) = -u_0 \left(1 - \frac{x}{l}\right).$$

Applying equations (12.34), (12.44), and (12.47), we obtain

$$C_n = \frac{2}{l} \int_0^l \varphi^*(x) \sin \frac{n\pi x}{l} dx = -\frac{2u_0}{n\pi},$$

$$f_n(t) = \frac{2}{l} \int_0^l f^*(x) \sin \frac{n\pi x}{l} dx = \frac{2u_0\omega}{n\pi} \sin \omega t,$$

$$T_n(t) = \int_0^t f_n(\tau) e^{-a^2 \lambda_n (t-\tau)} d\tau = \frac{2u_0\omega}{n\pi} \int_0^t \sin \omega \tau \, e^{-a^2 \lambda_n (t-\tau)} d\tau$$

$$= \frac{2u_0\omega}{n\pi (a^4 \lambda_n^2 + \omega^2)} \left[a^2 \lambda_n \sin \omega t - \omega \cos \omega t + \omega e^{-a^2 \lambda_n t} \right].$$

Substituting the expressions for C_n and $T_n(t)$ into the general formulas, we obtain the solution:

$$u(x,t) = \left(1 - \frac{x}{l}\right) u_0 \cos \omega t + \sum_{n=1}^{\infty} \left[T_n(t) + C_n e^{-a^2 \lambda_n t} \right] \sin \frac{n\pi x}{l} = \left(1 - \frac{x}{l}\right) u_0 \cos \omega t +$$

$$+ \frac{2u_0}{\pi} \sum_{n=1}^{\infty} \frac{1}{n(a^4 \lambda_n^2 + \omega^2)} \left[a^2 \lambda_n \omega \sin \omega t - \omega^2 \cos \omega t + a^2 \lambda_n e^{-a^2 \lambda_n t} \right] \sin \frac{n\pi x}{l}.$$

Figure 12.9 shows the spatial-time-dependent solution $u(x,t)$ for Example 7. This solution was obtained with the program **Heat** for the case when $l = 10$, $\omega = 0.5$, $u_0 = 5$, and $a^2 = 4$.

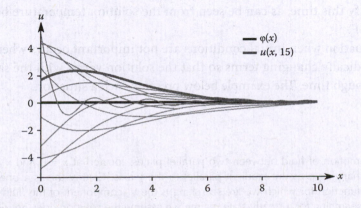

FIGURE 12.9   Solution $u(x, t)$ for Example 7.

Reading Exercise

Investigate the behavior of this solution, which has many important physical applications. This can be done with help of the program **Heat**. Try large values for l and ω. Observe that for these cases heat does not propagate far from the left end of the bar. Determine the behavior of the solution for $t\to\infty$. In the following section we will demonstrate a way to find an analytical solution for the steady-state regime for large t by separating the variables in the form $\mathrm{Re}\,X(x)e^{i\omega t}$.

Reading Exercise

Check that for $t\to\infty$ the solution obtained above is a sum of purely periodic harmonics and can be presented in the form

$$u(x,t) = \left(1 - \frac{x}{l}\right)u_0 \cos\omega t + \frac{2u_0}{\pi}\sum_{n=1}^{\infty}\frac{1}{n}\sin\delta_n \cdot \sin(\omega t - \delta_n)\cdot\sin\frac{n\pi x}{l},$$

where the phase shifts are given by $\delta_n = \tan^{-1}(\omega/a^2\lambda_n)$.

12.5 BOUNDARY PROBLEMS WITHOUT INITIAL CONDITIONS

Next we consider frequently encountered physical situations where knowledge of initial conditions is not important. For example, clearly the influence of the initial conditions decreases with time for cases where heat propagates through a body. If the moment of interest is long enough after the initial time, the temperature of a bar, for example, is for all purposes defined by the boundary conditions since the effects of the initial conditions have had time to decay. In this case we may suppose that after a long enough time the initial condition vanishes.

This brings us to boundary problems *without initial conditions*. This situation also frequently applies when *boundary conditions change periodically*—for example, in the previous Example 7. In these cases we may assume that after a large interval of time the temperature of a body varies periodically with the same frequency as the boundary condition. Generally this time can be estimated as $t \gg 1/\omega$. After this time the initial temperature can always be assumed to be equal to zero (even if it is not). In problems like Example 7 we can specify this time; as can be seen from the solution temperature becomes steady when $t \gg (l/a)^2$.

Another situation when initial conditions are not important occurs when the equation contains periodically changing terms so that the solution varies with the same frequency after a long enough time. The example below presents such a situation.

Example 8

Consider the motion of fluid between two parallel plates, located at $x = 0$ and $x = H$, under a periodically changing pressure gradient parallel to the y-axis. Clearly this is a one-dimensional problem; the function for which we are searching is the y-component of the fluid speed, $u(x,t)$. Since we are searching for a steady-state regime we assume the solution does not depend on the initial condition and formally we set $u(x,0) = 0$.

Solution. This problem is described by the equation that follows from the Navier-Stokes equation for fluid motion given by

$$\frac{\partial u}{\partial t} = a^2 \frac{\partial^2 u}{\partial x^2} + b \cos \omega t, \tag{12.51}$$

where $a^2 \equiv \nu$ is the coefficient of kinematic viscosity. The boundary conditions

$$u(0,t) = u(H,t) = 0$$

correspond to zero velocity at the plates and the initial condition is

$$u(x,0) = 0.$$

The eigenvalues and eigenfunctions of the problem we have seen a number of times before and are

$$\lambda_n = \left(\frac{n\pi}{H}\right)^2, \quad X_n(x) = \sin\frac{n\pi x}{H}, \quad \|X_n\|^2 = \frac{H}{2}, \quad n = 1,2,3,\ldots$$

The coefficient, $C_n = 0$, since $\varphi(x) = 0$. Applying equations (12.44) and (12.47) we obtain

$$f_n(t) = \frac{2}{H}\int_0^H b\,\cos\omega t \cdot \sin\frac{n\pi x}{H}\,dx = \frac{2b\cos\omega t}{n\pi}\left[1 - (-1)^n\right],$$

$$T_n = \frac{2b[1-(-1)^n]}{n\pi}\int_0^t \cos\omega\tau\; e^{-\frac{a^2 n^2 \pi^2}{H^2}(t-\tau)}\,d\tau = \frac{2bH^2}{n\pi}\frac{[1-(-1)^n]}{[a^4 n^4 \pi^4 + \omega^2 H^4]}$$

$$\times \left[a^2 n^2 \pi^2 \cos\omega t + \omega H^2 \sin\omega t - a^2 n^2 \pi^2 e^{-\frac{a^2 n^2 \pi^2}{H^2}t}\right]$$

Obviously from above we see that $T_{2k}(t) = 0$. Also

$$a^2 n^2 \pi^2 \cos\omega t + \omega H^2 \sin\omega t = \sqrt{a^4 n^4 \pi^4 + \omega^2 H^4}\,\sin(\omega t + \theta_n),$$

where $\theta_n = \arctan(a^2 n^2 \pi^2 / \omega H^2)$. Hence, the solution of the problem has the form

$$u(x,t) = \frac{4bH^2}{\pi}\sum_{k=1}^{\infty}\frac{1}{2k-1}\left\{\frac{\sin(\omega t + \theta_{2k-1})}{\sqrt{a^4(2k-1)^4\pi^4 + \omega^2 H^4}}\right.$$

$$\left. -\frac{a^2(2k-1)^2\pi^2}{[a^4(2k-1)^4\pi^4 + \omega^2 H^4]}e^{-\frac{a^2(2k-1)^2\pi^2}{H^2}t}\right\}\sin\frac{(2k-1)\pi x}{H}.$$

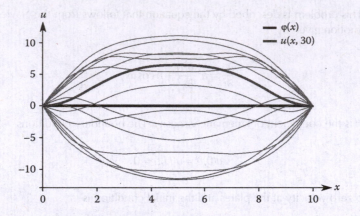

FIGURE 12.10   Solution $u(x, t)$ for Example 8.

Figure 12.10 shows the spatial-time-dependent solution $u(x,t)$ for Example 8. This solution was obtained with the program **Heat** for the case when $H = 10$, $\omega = 0.5$, $b = 5$, and $a^2 = \nu = 1$. Clearly as $t \rightarrow \infty$ we have

$$u(x,t) \rightarrow \frac{4bH^2}{\pi} \sum_{k=1}^{\infty} \frac{\sin(\omega t + \theta_{2k-1}) \sin \frac{(2k-1)\pi x}{H}}{(2k-1)\sqrt{a^4(2k-1)^4 \pi^4 + \omega^2 H^4}}.$$

We can also see that $u(0,t)=u(H,t)=0$ as it should be. Also $u(x,t)$ takes its maximum value at $x = H/2$. Notice that the initial conditions and the equation formally disagree but in this periodic problem the role of the initial condition becomes negligible for times $t \gg H^2/a^2$.

The curves shown in Figure 12.11 depict time traces of the fluid speed. A careful examination of the graphs shows that after time $t \approx 35$ the solution no longer depends on the initial condition.

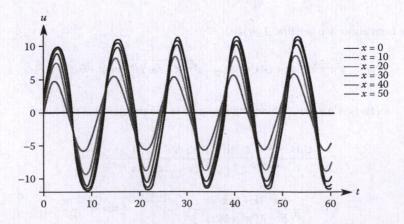

FIGURE 12.11   Time traces of fluid speed for Example 8.

In this and similar problems it is easy to obtain a steady-state (terminal) solution. From the physical point of view it follows that such a solution is periodic with frequency ω because in systems with dissipation (described by parabolic equations) the internal oscillations decay exponentially with time (which is exactly the opposite to the case of hyperbolic equations). This allows us to search for a solution in the form

$$u(x,t) = \text{Re}\{X(x)\exp(i\omega t)\}. \tag{12.52}$$

Notice that we cannot look for a time dependence for the solution in the form of an external force, $\cos\omega t$, only as we did for hyperbolic equations. The reason is that in cases where the equation has a first derivative in time the $\sin e$ and cosine functions mix and we have to take into account both (or resort to using the real part of an exponential using the Re operator). While performing intermediate operations in the following we may omit the symbol Re and wait to take the real part close to the final step. The function $X(x)$ is complex and the system response will be, as follows from the previous formula, shifted in phase relative to the external influence.

Substituting equation (12.52) into equation (12.51) we obtain the ordinary differential equation

$$X'' - \frac{i\omega}{a^2} X = -\frac{b}{a^2}. \tag{12.53}$$

First we solve the homogeneous equation,

$$X'' - \frac{i\omega}{a^2} X = 0. \tag{12.54}$$

The characteristic equation for this linear equation has two roots:

$$\frac{\sqrt{i\omega}}{a} = \pm \frac{\sqrt{\omega/2}}{a}(1+i). \tag{12.55}$$

Thus, a general solution to the homogeneous equation (12.54) is

$$X(x) = C_1 \exp[q(1+i)x] + C_2 \exp[-q(1+i)x], \quad q \equiv \frac{1}{a}\sqrt{\frac{\omega}{2}}. \tag{12.56}$$

A particular solution of the nonhomogeneous equation (12.53) is $(-ib/\omega)$, in which case we may write a solution of this equation which satisfies zero boundary conditions as

$$X(x) = -\frac{ib}{\omega}\left(1 - \frac{\cos\alpha\left(x - \frac{H}{2}\right)}{\cos\alpha\frac{H}{2}}\right), \quad \alpha = q(1-i). \tag{12.57}$$

From this we have

$$u(x,t) = \operatorname{Re} X \cos \omega t - \operatorname{Im} X \sin \omega t. \tag{12.58}$$

To obtain a final form we use the identity

$$\cos z = \cos(x + iy) = \cos x \cos iy + \sin x \sin iy = \cos x \cosh y + i \sin x \sinh y \tag{12.59}$$

to yield the result for $\operatorname{Re} X$ and $\operatorname{Im} X$ that, with equation (12.58), gives the final answer for the steady-state solution:

$$\operatorname{Re} X = -\frac{b}{\omega} \frac{\sin qx \sinh q(x-H) - \sin q(x-H)\sinh qx}{\cos qH + \cosh qH},$$

$$\operatorname{Im} X = -\frac{b}{\omega} \left(1 - \frac{\cos qx \cosh q(x-H) + \cos q(x-H)\cosh qx}{\cos qH + \cosh qH} \right). \tag{12.60}$$

Next we will consider two examples of solutions of *diffusion problems*.

Example 9

Let the pressure and temperature of air in a cylinder length l ($0 \le x \le l$) be equal to the atmospheric pressure. One end of the cylinder at $x = 0$ is opened at the instant $t = 0$, and the other, at $x = l$, remains closed. The concentration of some gas in the external environment is constant ($u_0 = $ const). Find the concentration of gas in the cylinder for $t > 0$ if at the instant $t = 0$ the gas begins to diffuse into the cylinder through the opened end.

Solution. This problem can be represented by the equation

$$\frac{\partial u}{\partial t} = a^2 \frac{\partial^2 u}{\partial x^2}, \quad a^2 = D,$$

under conditions

$$u(x,0) = 0, \quad u(0,t) = u_0, \quad DS\frac{\partial u}{\partial x}(l,t) = 0,$$

where D is the diffusion coefficient.

Clearly the eigenvalues and eigenfunctions of problem are (see Appendix 1)

$$\lambda_n = \left[\frac{(2n-1)\pi}{2l} \right]^2, \quad X_n(x) = \sin\frac{(2n-1)\pi x}{2l}, \quad \| X_n \|^2 = \frac{l}{2}, \quad n = 1, 2, 3, \ldots$$

and the solution will be of the form

$$u(x,t)=v(x,t)+w(x,t).$$

In general, for a specific problem an auxiliary function is easily obtained from general formulas for $w(x,t)$ found in Appendix 2. We may often guess what the function must look like based on the physical observation that we are searching for a terminal or steady-state solution. In the present case

$$w(x,t) = u_0.$$

We also have

$$f^*(x,t) = 0, \quad \varphi^*(x) = -u_0.$$

Applying formula (12.34), we obtain

$$C_n = -\frac{2u_0}{l} \int_0^l \sin\frac{(2n-1)\pi x}{2l}\,dx = -\frac{4u_0}{(2n-1)\pi}.$$

Substituting the expression for C_n into the general formula, we obtain the final solution:

$$u(x,t) = w(x,t) + \sum_{n=1}^{\infty} C_n e^{-\frac{a^2(2n-1)^2\pi^2}{4l^2}t} \sin\frac{(2n-1)\pi x}{2l}$$

$$= u_0 - \frac{4u_0}{\pi} \sum_{n=1}^{\infty} \frac{1}{2n-1} e^{-\frac{a^2(2n-1)^2\pi^2}{4l^2}t} \sin\frac{(2n-1)\pi x}{2l}.$$

Figure 12.12 shows the spatial-time-dependent solution $u(x,t)$ for Example 9. This solution was obtained with the program **Heat** for the case when $l = 10$, $u_0 = 10$, and $a^2 = D = 1$. The last example is a bit more sophisticated.

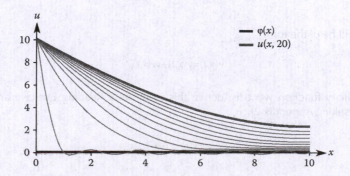

FIGURE 12.12 Solution $u(x,t)$ for Example 9.

Example 10

Let, as in the previous example, the pressure and temperature of air in a cylinder of length l $(0 \le x \le l)$ be equal to that of the atmosphere and the concentration of some gas in the external environment be constant, $u_0 = $ const. Both ends are closed by semi-permeable partitions. Gas diffuses through these ends into the cylinder from time $t = 0$. Find the concentration of diffusing gas in the cylinder for $t > 0$, assuming that the amount of diffused gas decreases through some chemical reaction and that the speed of dissociation at each point is proportional to the concentration of the gas at this point.

Solution. The problem is described by the equation

$$\frac{\partial u}{\partial t} = a^2 \frac{\partial^2 u}{\partial x^2} - \gamma u, \qquad a^2 = D,$$

where γ is the coefficient of decay. The initial condition

$$u(x,0) = 0,$$

and boundary conditions, corresponding to semi-permeable partitions (which allow some freedom for gas to diffuse through them) are

$$\frac{\partial u}{\partial x}(0,t) - h[u(0,t) - u_0] = 0, \qquad \frac{\partial u}{\partial x}(l,t) + h[u(l,t) - u_0] = 0.$$

Boundary conditions can be written in the form

$$\frac{\partial u}{\partial x} - hu \bigg|_{x-0} = -hu_0, \qquad \frac{\partial u}{\partial x} + hu \bigg|_{x-l} = hu_0.$$

Substituting $u(x,t) = e^{-\gamma t} v(x,t)$ into the above equations leads to the boundary value problem

$$\frac{\partial v}{\partial t} = a^2 \frac{\partial^2 v}{\partial x^2},$$

$$v(x,0) = 0, \qquad \frac{\partial v}{\partial x} - hv \bigg|_{x-0} = -hu_0 e^{\gamma t}, \qquad \frac{\partial v}{\partial x} + hv \bigg|_{x-l} = hu_0 e^{\gamma t}.$$

The solution will be of the form

$$v(x,t) = y(x,t) + w(x,t),$$

where an auxiliary function $w(x,t)$ is (derive this result as a Reading Exercise and verify your answer by consulting Appendix 2),

$$w(x,t) = u_0 e^{\gamma t}.$$

In this case we have

$$f^*(x,t) = -\gamma u_0 e^{\gamma t}, \qquad \varphi^*(x) = -u_0.$$

The eigenvalues and eigenfunctions of a problem with mixed boundary conditions are defined by formulas (see Appendix 1):

$$\lambda_n = \left(\frac{\mu_n}{l}\right)^2, \quad n = 1,2,3,...,$$

where μ_n is the nth root of the equation $\tan\mu = 2h/\mu/(\mu^2 - h^2 l^2)$,

$$X_n(x) = \frac{1}{\sqrt{\lambda_n + h^2}}\left[\sqrt{\lambda_n}\cos\sqrt{\lambda_n}x + h\sin\sqrt{\lambda_n}x\right], \quad ||X_n||^2 = \frac{l}{2} + \frac{h}{\lambda_n + h^2}.$$

Applying equations (12.34), (12.44), and (12.47) we obtain

$$C_n = \frac{1}{||X_n||^2}\int_0^l (-u_0)\frac{1}{\sqrt{\lambda_n + h^2}}\left[\sqrt{\lambda_n}\cos\sqrt{\lambda_n}x + h\sin\sqrt{\lambda_n}x\right]dx$$

$$= -\frac{2u_0\sqrt{\lambda_n + h^2}}{l(\lambda_n + h^2) + 2h}\left[\sin\sqrt{\lambda_n}l - \frac{h}{\sqrt{\lambda_n}}(\cos\sqrt{\lambda_n}l - 1)\right],$$

$$f_n(t) = \frac{1}{||X_n||^2}\int_0^l \left(-\gamma u_0 e^{\gamma t}\right)X_n(x)\,dx$$

$$= -e^{\gamma t}\frac{2\gamma u_0\sqrt{\lambda_n + h^2}}{l(\lambda_n + h^2) + 2h}\left[\sin\sqrt{\lambda_n}l - \frac{h}{\sqrt{\lambda_n}}(\cos\sqrt{\lambda_n}l - 1)\right],$$

$$T_n(t) = \int_0^t f_n(p)\,e^{-a^2\lambda_n(t-p)}dx = -\frac{2\gamma u_0\sqrt{\lambda_n + h^2}}{l(\lambda_n + h^2) + 2h}$$

$$\times\left[\sin\sqrt{\lambda_n}l - \frac{h}{\sqrt{\lambda_n}}(\cos\sqrt{\lambda_n}l - 1)\right]\int_0^t e^{-a^2\lambda_n(t-p)+\gamma t}\,dx$$

$$= -\frac{2\gamma u_0\sqrt{\lambda_n + h^2}}{\left[l(\lambda_n + h^2) + 2h\right]\left(a^2\lambda_n + \gamma\right)}$$

$$\times\left[\sin\sqrt{\lambda_n}l - \frac{h}{\sqrt{\lambda_n}}(\cos\sqrt{\lambda_n}l - 1)\right](e^{\gamma t} - e^{-a^2\lambda_n t}).$$

Substituting the expression for C_n and $T_n(t)$ in the general formula, we obtain the solution

$$u(x,t) = e^{-\gamma t}w(x,t) + e^{-\gamma t}\sum_{n=1}^{\infty}\left[T_n(t) + C_n e^{-a^2\lambda_n t}\right]\frac{\sqrt{\lambda_n}\cos\sqrt{\lambda_n}x + h\sin\sqrt{\lambda_n}x}{\sqrt{\lambda_n + h^2}}$$

$$= u_0 - 2u_0\sum_{n=1}^{\infty}\frac{[\sin\sqrt{\lambda_n}l - (h/\sqrt{\lambda_n})(\cos\sqrt{\lambda_n}l - 1)]}{[l(\lambda_n + h^2) + 2h](a^2\lambda_n + \gamma)}$$

$$\times\left[\gamma + a^2\lambda_n e^{-(a^2\lambda_n + \gamma)t}\right](\sqrt{\lambda_n}\cos\sqrt{\lambda_n}x + h\sin\sqrt{\lambda_n}x).$$

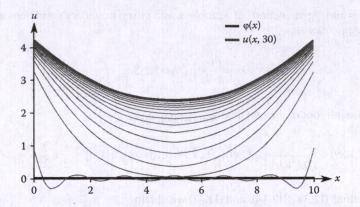

FIGURE 12.13   Solution $u(x, t)$ for Example 10.

Figure 12.13 shows the spatial-time-dependent solution $u(x,t)$ for Example 10. This solution was obtained with the program **Heat** for the case when $l = 10$, $u_0 = 5$, $h=1$, $\gamma = 0.05$, and $a^2 = D = 1$.

Reading Exercise

Find the asymptotes of this solution for small and large γ and h. Simulate this problem with the program **Heat** for different values of these parameters.

This problem is also useful in showing how it is possible to find the analytical solution at $t \to \infty$ in a closed form, rather than a series. Let us use the notation $U(x)$ for the part of the solution that is independent of time. The equation for this function is

$$U'' = k^2 U, \quad k^2 = \frac{\gamma}{a^2}$$

and the boundary conditions are

$$U'(0) - h[U(0) - u_0] = 0 \quad \text{and} \quad U'(l) + h[U(l) - u_0] = 0.$$

The solution to this boundary value problem for an ordinary differential equation that is symmetric about the central line is

$$U = A \cosh k\left(\frac{l}{2} - x\right).$$

The constant A is determined from either initial condition:

$$A\left[k \sinh \frac{kl}{2} + h \cosh k \frac{l}{2} \right] = h u_0.$$

Finally, the asymptotic solution as $t \to \infty$ is

$$U(x) = \frac{h u_0 \cosh k(l/2 - x)}{k \sinh kl/2 + h \cosh kl/2}.$$

Problems

Problems 1 through 30 involve the temperature distribution inside a homogeneous isotropic rod (or bar) of length l ($0 \le x \le l$).

Solve these problems analytically, which means the following: formulate the equation and initial and boundary conditions, obtain the eigenvalues and eigenfunctions, and write the formulas for coefficients of the series expansion and the expression for the solution of the problem. If the integrals in the coefficients are not easy to evaluate, leave it for the program, which evaluates them numerically. Then study the problem in detail with the program **Heat**. Obtain the pictures of several eigenfunctions and screenshots of solution in different moments of time. Directions on how to use the program are found in Appendix 5.

Using the program **Heat** the reader should assign the values of the parameters such as the values of the length of the rod l, mass density ρ, specific heat capacity c, coefficient of internal heat-conduction κ, coefficient of external (with the environment) heat-conduction γ, coefficients h_1 and h_2, and the parameters of functions in each problem. It is simpler to solve these problems assuming dimensionless values for the parameters, but using physical dimensions and realistic values of parameters can lead to insightful solutions. Also, by choosing alternative parameters the same problem can be made to represent diffusion and other related physical applications.

In problems 1 through 9 we consider rods that are thermally insulated over their lateral surfaces. In the initial time, $t = 0$, the temperature distribution is given by $u(x,0) = \varphi(x)$, $0 < x < l$. There are no heat sources or absorbers inside the rod.

Simulate the process of rod cooling with the **Heat** program for some values of parameters $a^2 = \kappa/(c\rho)$ and l. Vary the coefficient of thermal diffusivity a^2 in such a way that the cooling time decreased (with a precision of 5%) (1) twice and (2) four times. Obtain the same results changing the length of the rod l. (Use options "*Time traces of temperature at rod points*" or "*Evolution of rod temperature.*")

1. The ends of the rod are kept at zero temperature. The initial temperature of the rod is given as

$$\varphi_1(x) = 1, \quad \varphi_2(x) = x, \quad \varphi_3(x) = x(l - x).$$

2. The left end of the rod is kept at zero temperature and the right end is thermally insulated from the environment. The initial temperature of the rod is

$$\varphi_1(x) = x^2, \quad \varphi_2(x) = x, \quad \varphi_3(x) = x\left(l - \frac{x}{2}\right).$$

3. The left end of the rod is thermally insulated and the right end is kept at zero temperature. The initial temperature of the rod is

$$\varphi_1(x) = x, \quad \varphi_2(x) = 1, \quad \varphi_3(x) = l^2 - x^2.$$

4. Both ends of the rod are thermally insulated. The initial temperature of the rod is

$$\varphi_1(x) = x, \quad \varphi_2(x) = l^2 - x^2, \quad \varphi_3(x) = x^2\left(1 - \frac{2x}{3}\right).$$

5. The left end of the rod is kept at the zero temperature and the right end is subject to convective heat transfer with the environment. The initial temperature of the rod is

$$\varphi_1(x) = 1, \quad \varphi_2(x) = x, \quad \varphi_3(x) = x\left(1 - \frac{x}{2}\right).$$

6. The left end of the rod is subject to convective heat transfer with the environment, which has zero temperature, and the right end is kept at zero temperature. The initial temperature of the rod is

$$\varphi_1(x) = 1, \quad \varphi_2(x) = l - x, \quad \varphi_3(x) = \frac{x}{3}(l - 2x) + \frac{1}{3}.$$

7. The left end of the rod is thermally insulated and the right end is subject to convective heat transfer with the environment, which has a temperature of zero. The initial temperature of the rod is

$$\varphi_1(x) = 1, \quad \varphi_2(x) = x, \quad \varphi_3(x) = \left(1 - \frac{x^3}{3}\right).$$

8. The left end of the rod is subject to convective heat transfer with the environment (whose temperature is zero) and the right end is thermally insulated. The initial temperature of the rod is

$$\varphi_1(x) = 1, \quad \varphi_2(x) = x, \quad \varphi_3(x) = 1 - \frac{(x - 1)^3}{3}.$$

9. Both ends of the rod are subject to convective heat transfer with the environment, which has a temperature of zero. The initial temperature of the rod is

$$\varphi_1(x) = 1, \quad \varphi_2(x) = x, \quad \varphi_3(x) = \left(1 - \frac{x^3}{3}\right).$$

In problems 10 through 18 we consider rods whose lateral surfaces are subject to heat transfer according to Newton's law of cooling. The environment has zero temperature. The initial temperature of the rod is given as $u(x,0) = \varphi(x)$. Again there are no heat sources or absorbers inside the rod.

Study the meaning of the heat exchange coefficient γ (lateral heat exchange within a medium). Draw the graphs of evolution of rod temperature for the following values of γ: (a) $\gamma = 0$; (b) $\gamma = 0.05$; (c) $\gamma = 0.2$ (use the option "*Evolution of Rod Temperature*").

10. The ends of the rod are kept at a constant temperature; the left end has temperature $u(0,t) = u_1$ and the right end has temperature $u(l,t) = u_2$. The initial temperature of the rod is

$$\varphi_1(x) = x^2, \quad \varphi_2(x) = x, \quad \varphi_3(x) = x\left(l - \frac{x}{2}\right).$$

11. The left end of the rod is kept at a constant temperature $u(0,t) = u_1$, and a constant heat flow is supplied to the right end of the rod. The initial temperature of the rod is

$$\varphi_1(x) = x^2, \quad \varphi_2(x) = x, \quad \varphi_3(x) = x\left(l - \frac{x}{2}\right).$$

12. A constant heat flow is supplied to the left end of the rod from outside and the right end of the rod is kept at a constant temperature $u(l,t) = u_2$. The initial temperature of the rod is

$$\varphi_1(x) = x, \quad \varphi_2(x) = 1, \quad \varphi_3(x) = l^2 - x^2.$$

13. Constant heat flows are supplied to both ends of the rod. The initial temperature of the rod is

$$\varphi_1(x) = x, \quad \varphi_2(x) = l^2 - x^2, \quad \varphi_3(x) = x^2\left(1 - \frac{2x}{3}\right).$$

14. The left end of the rod is kept at a constant temperature $u(0,t) = u_1$ and the right end is subject to convective heat transfer with the environment, which has a constant temperature of u_0. The initial temperature of the rod is

$$\varphi_1(x) = 1, \quad \varphi_1(x) = 1, \quad \varphi_3(x) = x\left(l - \frac{x}{2}\right).$$

15. The left end of the rod is subjected to convective heat transfer with an environment that has a constant temperature of u_0; the right end is kept at the constant temperature $u(l,t) = u_2$. The initial temperature of the rod is

$$\varphi_1(x) = 1, \quad \varphi_2(x) = l - x, \quad \varphi_3(x) = \frac{x}{3}(l - 2x) + \frac{1}{3}.$$

16. A constant heat flow is supplied to the left end of the rod from outside and the right end of the rod is subject to convective heat transfer with an environment of constant temperature, u_0. The initial temperature of the rod is

$$\varphi_1(x)=1, \quad \varphi_2(x)=x, \quad \varphi_3(x)=\left(1-\frac{x^3}{3}\right).$$

17. The left end of the rod is subjected to convective heat transfer with an environment of constant temperature u_0, and a constant heat flow is supplied to the right end of the rod. The initial temperature of the rod is

$$\varphi_1(x)=1, \quad \varphi_2(x)=x, \quad \varphi_3(x)=1-\frac{(x-1)^3}{3}.$$

18. Both ends of the rod are subjected to convective heat transfer with an environment of constant temperature u_0. The initial temperature of the rod is

$$\varphi_1(x)=1, \quad \varphi_2(x)=x, \quad \varphi_3(x)=\left(1-\frac{x^3}{3}\right).$$

Problems 19 through 21 consider rods whose lateral surfaces are subjected to heat transfer according to Newton's law and the environment has a constant temperature θ. One internal source of heat acts at the point x_0 ($0 < x_0 < l$) inside the rod and the power of this source is Q.

19. The ends of the rod are kept at a constant temperatures—the left end has a temperature $u(0,t)=u_1$ and the right end has a temperature $u(l,t)=u_2$. The initial temperature of the rod is

$$\varphi_1(x)=x^2, \quad \varphi_2(x)=x, \quad \varphi_3(x)=x\left(l-\frac{x}{2}\right).$$

20. Constant heat flows, q_1 and q_2, are supplied to both ends of the rod from outside. The initial temperature of the rod is

$$\varphi_1(x)=x, \quad \varphi_2(x)=l^2-x^2, \quad \varphi_3(x)=x^2\left(1-\frac{2x}{3}\right).$$

21. At both ends of the rod a convective heat transfer occurs with the environment, which has a constant temperature of θ. The initial temperature of the rod is

$$\varphi_1(x)=1, \quad \varphi_2(x)=x, \quad \varphi_3(x)=\left(1-\frac{x^3}{3}\right).$$

In problems 22 through 24 we consider rods whose lateral surfaces are subjected to heat transfer according to Newton's law and the environment has a constant temperature θ. Internal heat sources and absorbers are active in the rod and their intensity (per unit mass of the rod) is given by $f(x,t)$. The initial temperature of the rod is zero, $u(x,0) = 0$.

22. The ends of the rod are kept at constant temperatures—the left end has temperature $u(0,t) = u_1$ and the right end has temperature $u(l,t) = u_2$. The intensities of heat sources and absorbers are

$$f_1(x,t) = A\sin\omega t, \qquad f_2(x,t) = Ae^{-\alpha t}\sin\omega t,$$

$$f_3(x,t) = A\cos\omega t, \qquad f_4(x,t) = Ae^{-\alpha t}\cos\omega t,$$

$$f_5(x,t) = A\sin\omega t\cos\omega t, \qquad f_6(x,t) = Ae^{-\alpha t}(\sin\omega t + \cos\omega t).$$

23. The constant heat flows, q_1 and q_2, are supplied to both ends of the rod from outside. The intensities of heat sources and absorbers are

$$f_1(x,t) = A\sin\omega t, \qquad f_2(x,t) = Ae^{-\alpha t}\sin\omega t,$$

$$f_3(x,t) = A\cos\omega t, \qquad f_4(x,t) = Ae^{-\alpha t}\cos\omega t,$$

$$f_5(x,t) = A\sin\omega t\cos\omega t, \qquad f_6(x,t) = Ae^{-\alpha t}(\sin\omega t + \cos\omega t).$$

24. Both ends of the rod are subjected to convective heat transfer with the environment at constant temperature θ. The intensities of heat sources and absorbers are

$$f_1(x,t) = A\sin\omega t, \qquad f_2(x,t) = Ae^{-\alpha t}\sin\omega t,$$

$$f_3(x,t) = A\cos\omega t, \qquad f_4(x,t) = Ae^{-\alpha t}\cos\omega t,$$

$$f_5(x,t) = A\sin\omega t\cos\omega t, \qquad f_6(x,t) = Ae^{-\alpha t}(\sin\omega t + \cos\omega t).$$

In problems 25 through 30 we consider rods with thermally insulated lateral surfaces. The initial temperature of the rod is zero $u(x,0) = 0$. Generation (or absorption) of heat by internal sources is absent. Find the temperature change inside the rod for the following cases:

25. The left end of the rod is kept at the constant temperature $u(0,t) = u_1$ and the temperature of the right end changes according to $g_2(t) = A\cos\omega t$.

26. The temperature of the left end changes as $g_1(t) = A\cos\omega t$; the right end of the rod is kept at the constant temperature $u(l,t) = u_2$.

27. The left end of the rod is kept at the constant temperature $u(0,t) = u_1$, and the heat flow $g_2(t) = A\sin\omega t$ is supplied to the right end of the rod from outside.

28. The heat flow $g_1(t) = A\cos\omega t$ is supplied to the left end of the rod from outside while the right end of the rod is kept at the constant temperature $u(l,t) = u_2$.

29. The left end of the rod is kept at the constant temperature $u(0,t) = u_1$ and the right end is subjected to a convective heat transfer with the environment, which has a temperature that varies as $u_{md}(t) = A\sin\omega t$.

30. The left end of the rod is subjected to a convective heat transfer with the environment, which has a temperatures that varies as $u_{md}(t) = A\cos\omega t$, and the right end is kept at the constant temperature $u(l,t) = u_2$.

CHAPTER 13

Two-Dimensional Parabolic Equations

In this chapter we discuss parabolic equations for two-dimensional bounded medium. We consider rectangular domains and domains where there is circular symmetry. The presentation is very similar to that for two-dimensional hyperbolic equations (Chapter 11). As always the reader is reminded that the mathematics presented is not limited only to discussions of heat conduction but has broader applications such as diffusion.

13.1 HEAT CONDUCTION WITHIN A FINITE RECTANGULAR DOMAIN

Next we consider problems involving parabolic equations for finite, two-dimensional domains such as rectangular and circular plates. Let a heat-conducting, uniform rectangular plate be placed in the horizontal x-y plane with boundaries given by edges along $x = 0$, $x = l_x$, $y = 0$, and $y = l_y$. The plate is assumed to be thin enough that the temperature is the same at all points with the same x-y coordinates. As before we discuss the heat problem in order to have a specific example at hand, but it should be remembered that any other physical problem described by a two-dimensional hyperbolic equation can be solved using the methods discussed below.

Let $u(x, y)$ be the temperature of the plate at the point (x,y) at time t. The heat conduction within such a thin uniform rectangular plate is described by the equation

$$\frac{\partial u}{\partial t} = a^2 \left[\frac{\partial^2 u}{\partial x^2} + \frac{\partial^2 u}{\partial y^2} \right] + \xi_x \frac{\partial u}{\partial x} + \xi_y \frac{\partial u}{\partial y} - \gamma u + f(x,y,t), \tag{13.1}$$

$$0 < x < l_x, \quad 0 < y < l_y, \quad t > 0.$$

Here a^2, ξ_x, ξ_y, and γ are real constants. In terms of heat exchange, $a^2 = k/c\rho$ is the thermal diffusivity of the material; $\gamma = h/c\rho$ where h is the heat exchange coefficient (for lateral heat exchange with an external medium); and $f(x,y,t) = Q(x,y,t)/c\rho$ where Q is the density of heat source ($Q < 0$ for locations where the heat is absorbed). The terms with coefficients ξ_x and

ξ_y describe heat gain (or loss if they are negative) due to bulk motion of the surrounding medium. Clearly these coefficients will equal zero for solids but are nonzero for liquids or gases in which bulk movement (advection) of the medium occurs—for example, $(-\xi_x)$ and $(-\xi_y)$ represent the x and y components of the speed of the medium for a liquid or gas medium adjacent to the surface.

The *initial condition* defines the temperature distribution within the membrane at time zero:

$$u(x,y,0) = \varphi(x,y). \tag{13.2}$$

The *boundary conditions* describe the thermal conditions at the boundary at any time, t. The boundary conditions can be written in a general form as

$$P_1[u] \equiv \alpha_1 \frac{\partial u}{\partial x} + \beta_1 u \Big|_{x=0} = g_1(y,t), \quad P_2[u] \equiv \alpha_2 \frac{\partial u}{\partial x} + \beta_2 u \Big|_{x=l_x} = g_2(y,t),$$

$$\tag{13.3}$$

$$P_3[u] \equiv \alpha_3 \frac{\partial u}{\partial x} + \beta_3 u \Big|_{y=0} = g_3(x,t), \quad P_4[u] \equiv \alpha_4 \frac{\partial u}{\partial x} + \beta_4 u \Big|_{y=l_y} = g_4(x,t),$$

where $g_1(y,t)$, $g_2(y,t)$, $g_3(x,t)$, and $g_4(x,t)$ are known functions of time and respective variable and $\alpha_1, \beta_1,\ldots, \alpha_4, \beta_4$ are real constants. As has been discussed previously, physical arguments lead to the sign restrictions $\alpha_1/\beta_1 < 0$ and $\alpha_3/\beta_3 < 0$.

As before, there are three main types of boundary conditions (here and in the following we denote $a = 0$ or l_x and $b = 0$ or l_y):

Case I. Boundary condition of the first type (*Dirichlet condition*) where we are given the temperature along the y- or x-edge:

$$u(a,y,t) = g(y,t) \text{ or } u(x,b,t) = g(x,t).$$

We may also have zero temperature at the edges in which case $g(y,t) \equiv 0$ or $g(x,t) \equiv 0$.

Case II. Boundary condition of the second type (*Neumann condition*) where we are given the heat flow along the y- or x-edge:

$$u_x(a,y,t) = g(y,t) \text{ or } u_y(x,b,t) = g(x,t).$$

We may also have a thermally insulated edge in which case $g(y,t) \equiv 0$ or $g(y,t) \equiv 0$.

Case III. Boundary condition of the third type (*mixed condition*) where there is heat exchange with a medium along the y- or x- edge given by

$$u_x(a,y,t) \pm hu(a,y,t) = g(y,t) \text{ or } u_y(x,b,t) \pm hu(x,b,t) = g(x,t).$$

We assume that h is a positive constant in which case the positive sign should be chosen in two previous formulas when $a = l_x$, $b = l_y$ and negative when $a = 0$.

The above boundary value problem can be reduced to the boundary value problem

$$\frac{\partial v}{\partial t} = a^2 \left[\frac{\partial^2 v}{\partial x^2} + \frac{\partial^2 v}{\partial y^2} \right] - \tilde{\gamma} v + \tilde{f}(x, y, t), \tag{13.4}$$

$$v(x, y, 0) = \tilde{\phi}(x, y), \tag{13.5}$$

and

$$P_1[v] \equiv \alpha_1 \frac{\partial v}{\partial x} + \tilde{\beta}_1 v \Big|_{x=0} = \tilde{g}_1(y, t), \quad P_2[v] \equiv \alpha_2 \frac{\partial v}{\partial x} + \tilde{\beta}_2 v \Big|_{x=l_x} = \tilde{g}_2(y, t),$$

$$P_3[v] \equiv \alpha_3 \frac{\partial v}{\partial x} + \tilde{\beta}_3 v \Big|_{y=0} = \tilde{g}_3(x, t), \quad P_4[v] \equiv \alpha_4 \frac{\partial v}{\partial x} + \tilde{\beta}_4 v \Big|_{y=l_y} = \tilde{g}_4(x, t) \tag{13.6}$$

with the help of the substitution

$$u(x, y, t) = e^{\mu x + v y} v(x, y, t), \tag{13.7}$$

where

$$\mu = -\frac{\xi_x}{2a^2} \quad \text{and} \quad v = -\frac{\xi_y}{2a^2}.$$

Here

$$\tilde{\gamma} = \gamma + \frac{\xi_x^2 + \xi_y^2}{4a^2},$$

$$\tilde{f}(x, y, t) = e^{-(\mu x + v y)} f(x, y, t),$$

$$\tilde{\phi}(x, y) = e^{-(\mu x + v y)} \phi(x, y),$$

$$\tilde{\beta}_1 = \beta_1 + \mu \alpha_1, \quad \tilde{g}_1(y, t) = e^{-v y} g_1(y, t),$$

$$\tilde{\beta}_2 = \beta_2 + \mu \alpha_2, \quad \tilde{g}_2(y, t) = e^{-(\mu l_x + v y)} g_2(y, t),$$

$$\tilde{\beta}_3 = \beta_3 + v \alpha_3, \quad \tilde{g}_3(x, t) = e^{-\mu x} g_3(x, t),$$

$$\tilde{\beta}_4 = \beta_4 + v \alpha_4, \quad \tilde{g}_4(x, t) = e^{-(\mu x + v l_y)} g_4(x, t).$$

Reading Exercise:

Make the substitution given in equation (13.7) and verify the results above.

From the above results we see that we need only consider the simpler problem given by equations (13.1) through (13.3) with the condition $\xi_x = \xi_y = 0$. This general boundary value problem for the rectangular membrane cannot be solved via the method of Fourier expansions directly since the boundary conditions are nonhomogeneous. However, this problem can be easily reduced to a boundary value problem with zero boundary conditions.

First we express the solution to the problem defined by equations (13.1) through (13.3) as the sum of two functions:

$$u(x,y,t) = v(x,y,t) + w(x,y,t), \tag{13.8}$$

where $v(x,y,t)$ is a new, unknown function and $w(x,y,t)$ is an auxiliary function satisfying the boundary conditions. We shall seek an auxiliary function, $w(x,y,t)$, in the form

$$w(x,y,t) = g_1(y,t)\overline{X} + g_2(y,t)\overline{\overline{X}} + g_3(x,t)\overline{Y} + g_4(x,t)\overline{\overline{Y}}$$

$$+ A(t)\overline{X}\,\overline{Y} + B(t)\overline{\overline{X}}\,\overline{Y} + C(t)\overline{X}\,\overline{\overline{Y}} + D(t)\overline{\overline{X}}\,\overline{\overline{Y}}, \tag{13.9}$$

where $\overline{X}(x)$, $\overline{\overline{X}}(x)$, $\overline{Y}(y)$, and $\overline{\overline{Y}}(y)$ are polynomials of first or second order. The coefficients of these polynomials will be adjusted in such a way that function $w(x,y,t)$ satisfies the boundary conditions given in equations (13.3).

The function $v(x,y,t)$ represents heat conduction when heat sources are present within the plate and *boundary conditions are zero*:

$$\frac{\partial v}{\partial t} = a^2\left[\frac{\partial^2 v}{\partial x^2} + \frac{\partial^2 v}{\partial y^2}\right] - \gamma v + f^*(x,t), \tag{13.10}$$

$$v(x,y,0) = \varphi^*(x,y),$$

$$P_1[v]\big\|_{x=0} = 0, \qquad P_2[v]\big\|_{x=l_x} = 0,$$

$$P_3[v]\big\|_{y=0} = 0, \qquad P_4[v]\big\|_{y=l_y} = 0,$$

where

$$f^*(x,y,t) = f(x,y,t) - \frac{\partial w}{\partial t} + a^2\left(\frac{\partial^2 w}{\partial x^2} + \frac{\partial^2 w}{\partial y^2}\right) - \gamma w,$$

$$\varphi^*(x,y) = \varphi(x,y) - w(x,y,0).$$

This problem can be solved using the Fourier expansion method. The solution, $v(x,y,t)$, can be expressed as the sum of two functions

$$v(x,y,t) = u_1(x,y,t) + u_2(x,y,t), \tag{13.11}$$

where $u_1(x,y,t)$ is the solution of the *homogeneous* equation with the *specified initial conditions* and $u_2(x,y,t)$ is the solution of a *nonhomogeneous* equation with *zero initial conditions*. For both functions, u_1 and u_2, *the boundary conditions are zero* (i.e., homogeneous).

The solution $u_1(x,y,t)$ represents the case of *free heat exchange*, that is, heat neither generated within nor lost from the plate, but only transferred by conduction. The solution $u_2(x,y,t)$ represents the *non-free heat exchange*, that is, the diffusion of heat due to generation (or absorption) of heat by internal sources when the initial distribution of temperature is zero. We find expressions for these two solutions in the following subsections.

13.1.1 The Fourier Method for the Homogeneous Heat Conduction Equation (Free Heat Exchange)

Let us first find the solution of the homogeneous equation

$$\frac{\partial u_1}{\partial t} = a^2 \left[\frac{\partial^2 u_1}{\partial x^2} + \frac{\partial^2 u_1}{\partial y^2} \right] - \gamma u_1, \quad 0 < x < l_x, \ 0 < y < l_y, \ t > 0 \tag{13.12}$$

with the initial condition

$$u_1(x,y,0) = \varphi(x,y), \tag{13.13}$$

and zero boundary conditions given by

$$P_1[u_1]\big|_{x=0} = 0, \qquad P_2[u_1]\big|_{x=l_x} = 0,$$
$$\tag{1.14}$$
$$P_3[u_1]\big|_{y=0} = 0, \qquad P_4[u_1]\big|_{y=l_y} = 0.$$

This, then, describes the case of *free heat exchange* within the plate.

Let us separate time and spatial variables:

$$u_1(x,y,t) = T(t) \, V(x,y). \tag{13.15}$$

We leave it to the reader as a Reading Exercise to obtain (*a*) the following equation for the function $T(t)$:

$$T'(t) + (\alpha^2 \lambda + \gamma)T(t) = 0 \tag{13.16}$$

and (*b*) the boundary value problem with homogeneous boundary conditions for the function $V(x,y)$, defined by

$$\frac{\partial^2 V}{\partial x^2} + \frac{\partial^2 V}{\partial y^2} + \lambda V(x,y) = 0, \tag{13.17}$$

$$\alpha_1 V_x(0,y) + \beta_1 V(0,y) = 0, \qquad \alpha_2 V_x(l_x,y) + \beta_2 V(l_x,y) = 0,$$
$$\tag{13.18}$$
$$\alpha_3 V_y(x,0) + \beta_3 V(x,0) = 0, \qquad \alpha_4 V_y(0,l_y) + \beta_4 V(0,l_y) = 0,$$

where λ is the constant of separation of variables.

Next we can again separate variables:

$$V(x,y) = X(x)Y(y).$$

As a Reading Exercise, obtain following one-dimensional boundary value problems:

$$X''(x) + \lambda_x X(x) = 0,$$
$$\alpha_1 X'(0) + \beta_1 X(0) = 0, \tag{13.19}$$
$$\alpha_2 X'(l_x) + \beta_2 X(l_x) = 0,$$

and

$$Y''(y) + \lambda_y Y(y) = 0,$$
$$\alpha_3 Y'(0) + \beta_3 Y(0) = 0, \tag{13.20}$$
$$\alpha_4 Y'(l_y) + \beta_4 Y(l_y) = 0,$$

where the separation of variables constants, λ_x and λ_y, are connected by the relation

$$\lambda_x + \lambda_y = \lambda.$$

For example, the boundary condition

$$\alpha_1 V_x(0,y) + \beta_1 V(0,y) = \alpha_1 X'(0)Y(y) + \beta_1 X(0)Y(y)$$
$$= [\alpha_1 X'(0) + \beta_1 X(0)]Y(y) = 0$$

leads to

$$\alpha_1 X'(0) + \beta_1 X(0) = 0,$$

assuming $Y(y) \neq 0$ (i.e., we seek nontrivial solutions only).

If λ_{xn}, λ_{ym}, $X_n(x)$, and $Y_m(y)$ are eigenvalues and eigenfunctions, respectively of the boundary value problems for $X(x)$ and $Y(y)$, then

$$\lambda_{nm} = \lambda_{xn} + \lambda_{ym} \tag{13.21}$$

and

$$V_{nm}(x,y) = X_n(x) Y_m(y) \tag{13.22}$$

are eigenvalues and eigenfunctions of the boundary value problem for $V(x,y)$. The functions $V_{nm}(x, y)$ are orthogonal and their norms are the products

$$\|V_{nm}\|^2 = \|X_n\|^2 \cdot \|Y_m\|^2 .$$

Eigenvalues and eigenfunctions of the boundary value problem depend on the types of boundary conditions. Combining different types of boundary conditions one can obtain nine different types of boundary value problems for the solution $X(x)$ and nine different types for the solution $Y(y)$ (see Appendix 1). Notice that equation (13.17) written in the form

$$\nabla^2 V(x,y) = \lambda V(x,y),$$

with the Laplace's operator

$$\nabla^2 \equiv \frac{\partial^2}{\partial x^2} + \frac{\partial^2}{\partial y^2}$$

allows us to conclude that functions V_{nm} are eigenfunctions, and λ_{nm} are the eigenvalues of this operator for the boundary conditions (13.18).

In general the eigenvalues can be written as follows:

$$\lambda_{xn} = \left(\frac{\mu_{xn}}{l_x} \right)^2, \ \lambda_{ym} = \left(\frac{\mu_{ym}}{l_y} \right)^2, \tag{13.23}$$

where μ_{xn} is the nth root of the equation

$$\tan \mu_x = \frac{(\alpha_1\beta_2 - \alpha_2\beta_1) \, l_x \mu_x}{\mu_x^2 \alpha_1\alpha_2 + l_x^2 \beta_1\beta_2}, \tag{13.24}$$

and μ_{ym} is the mth root of the equation

$$\tan \mu_y = \frac{(\alpha_3\beta_4 - \alpha_4\beta_3) \, l_y \mu_y}{\mu_y^2 \alpha_3\alpha_4 + l_y^2 \beta_3\beta_4}. \tag{13.25}$$

Similarly, as was obtained in the previous chapters, the eigenfunctions are

$$X_n(x) = \frac{1}{\sqrt{\alpha_1^2 \lambda_{xn} + \beta_1^2}} \left[\alpha_1 \sqrt{\lambda_{xn}} \cos \sqrt{\lambda_{xn}} \, x - \beta_1 \sin \sqrt{\lambda_{xn}} \, x \right],$$

$$\tag{13.26}$$

$$Y_m(y) = \frac{1}{\sqrt{\alpha_3^2 \lambda_{ym} + \beta_3^2}} \left[\alpha_3 \sqrt{\lambda_{ym}} \cos \sqrt{\lambda_{ym}} \, y - \beta_3 \sin \sqrt{\lambda_{ym}} \, y \right].$$

Now, having the eigenvalues, λ_{nm}, and eigenfunctions, $V_{nm}(x,y)$, of the boundary value problem, we may obtain the solution $u_1(x,y,t)$. The solution of the differential equation

$$T'_{nm}(t)+(a^2\lambda_{nm}+\gamma)T_{nm}(t)=0 \tag{13.27}$$

is

$$T_{nm}(t)=C_{nm}e^{-(a^2\lambda_{nm}+\gamma)t} \tag{13.28}$$

from which we see that the function $u_1(x,y,t)$ can be composed as the infinite series

$$u_1(x,y,t)=\sum_{n=1}^{\infty}\sum_{m=1}^{\infty}T_{nm}(t)\,V_{nm}(x,y). \tag{13.29}$$

Using that fact that this function must satisfy the initial condition (13.2) and using the orthogonality condition for functions $V_{nm}(x,y)$ we find coefficients C_{nm}:

$$C_{nm}=\frac{1}{\|V_{nm}\|^2}\int_0^{l_x}\int_0^{l_y}\varphi(x,y)V_{nm}(x,y)\,dx\,dy. \tag{13.30}$$

It is left to the reader as a Reading Exercise to verify equation (13.30) using the initial conditions (13.13) and the orthogonality condition.

Example 1

The initial temperature distribution within a thin uniform rectangular plate ($0 \le x \le l_x$, $0 \le y \le l_y$) with thermally insulated lateral faces is

$$u(x,y,0) = Axy(l_x - x)(l_y - y), \quad A = \text{const.}$$

Find the distribution of temperature within the plate at any later time if its boundary is kept a constant zero temperature. Generation (or absorption) of heat by internal sources is absent.

Solution. The problem may be modeled by the solution of the equation

$$\frac{\partial u}{\partial t} = a^2\left[\frac{\partial^2 u}{\partial x^2} + \frac{\partial^2 u}{\partial y^2}\right],$$

$$0 < x < l_x, \ 0 < y < l_y, \ t > 0$$

under the conditions

$$u(x,y,0) = \varphi(x,y) = Axy(l_x - x)(l_y - y),$$

$$u(0,y,t) = u(l_x,y,t) = u(x,0,t) = u(x,l_y,t) = 0.$$

The boundary conditions of the problem are Dirichlet homogeneous boundary conditions; therefore eigenvalues and eigenfunctions of problem are

$$\lambda_{nm} = \lambda_{xn} + \lambda_{ym} = \pi^2 \left(\frac{n^2}{l_x^2} + \frac{m^2}{l_y^2} \right), \quad n,m = 1,2,3,\ldots$$

$$\|V_{nm}\|^2 = \|X_n\|^2 \cdot \|Y_m\|^2 = \frac{l_x l_y}{4}.$$

Applying the equation (13.30), we obtain

$$C_{nm} = \frac{4}{l_x l_y} \int_0^{l_x} \int_0^{l_y} Axy(l_x - x)(l_y - y) \sin \frac{\pi n x}{l_x} \sin \frac{\pi m y}{l_y} \, dx \, dy$$

$$= \begin{cases} \dfrac{64 A l_x^2 l_y^2}{\pi^2 n^2 m^2}, & \text{if } n \text{ and } m \text{ are odd}, \\[2mm] 0, & \text{if } n \text{ or } m \text{ are even}. \end{cases}$$

Hence, the distribution of temperatures inside the rectangular plate at some instant of time is described by the series

$$u(x,y,t) = \frac{64 A l_x^2 l_y^2}{\pi^6} \sum_{n,m=1}^{\infty} \frac{e^{-\lambda_{nm} a^2 t}}{(2n+1)^2 (2m+1)^2} \sin \frac{(2n+1)\pi x}{l_x} \sin \frac{(2m+1)\pi y}{l_y}.$$

We suggest the reader to compare this example and its result with Example 1 from Chapter 11.

Figure 13.1 shows two snapshots of the solution at the times $t = 0$ and $t = 5$. This solution was obtained with the program **Heat** for the case $a^2 = 0.25$, $l_x = 4$, $l_y = 6$, and $A = 0.01$. Directions for using the program Heat are found in the appendix at the end of this chapter.

13.1.2 The Fourier Method for the Nonhomogeneous Heat Conduction Equation

As mentioned previously, the solution $u_2(x,y,t)$ represents the *non-free heat exchange* within the plate—that is, the diffusion of heat due to generation (or absorption) of heat by internal sources (or sinks) for the case of an initial distribution of temperatures equal to zero.

(a)   (b)

FIGURE 13.1   Surface graph of plate temperature at (a) $t = 0$, (b) $t = 5$ for Example 1.

The solution to the general problem of heat conduction in a plate consists of the sum of the *free heat exchange* solutions, $u_1(x,y,t)$, found in the previous section and the solutions, $u_2(x,y,t)$, which will be discussed in this section.

The function $u_2(x,y,t)$ is a solution of the *nonhomogeneous* equation

$$\frac{\partial u_2}{\partial t} = a^2 \left[\frac{\partial^2 u_2}{\partial x^2} + \frac{\partial^2 u_2}{\partial y^2} \right] - \gamma u_2 + f(x,y,t) \qquad (13.31)$$

with *zero initial and boundary conditions*. As above, we can separate time and spatial variables to obtain a general solution to equation (13.31) in the form

$$u_2(x,y,t) = \sum_{n=1}^{\infty} \sum_{m=1}^{\infty} T_{nm}(t)\, V_{nm}(x,y), \qquad (13.32)$$

where $V_{nm}(x,y)$ are eigenfunctions of the corresponding homogeneous boundary value problem given in equations (13.12) and (13.14), with solutions given by equation (13.29). Here $T_{nm}(t)$ are, as yet, unknown functions of t. Zero boundary conditions for $u_2(x,y,t)$ given by

$$P_1[u_2]\big|_{x=0} = 0, \qquad P_2[u_2]\big|_{x=l_x} = 0,$$

$$P_3[u_2]\big|_{y=0} = 0, \qquad P_4[u_2]\big|_{y=l_y} = 0.$$

are valid for any choice of functions $T_{nm}(t)$ (assuming the series converge uniformly) because they are valid for the functions $V_{nm}(x,y)$. We leave it to the reader to obtain these results as a Reading Exercise.

We now determine the functions $T_{nm}(t)$ in such a way that the series (13.32) satisfies the nonhomogeneous equation (13.31) and the homogeneous initial condition. Substituting the series (13.32) into equation (13.31) we obtain

$$\sum_{n=1}^{\infty} \sum_{m=1}^{\infty} \left[T'_{nm}(t) + (a^2 \lambda_{nm} + \gamma) T_{nm}(t) \right] V_{nm}(x,y) = f(x,y,t). \qquad (13.33)$$

We can expand the function $f(x,y,t)$ in a Fourier series of the functions $V_{nm}(x,y)$ in the rectangular region $[0,l_x;0,l_y]$ such that

$$f(x,y,t) = \sum_{n=1}^{\infty} \sum_{m=1}^{\infty} f_{nm}(t)\, V_{nm}(x,y), \qquad (13.34)$$

where

$$f_{nm}(t) = \frac{1}{\|V_{nm}\|^2} \int_0^{l_x} \int_0^{l_y} f(x,y,t)\, V_{nm}(x,y)\, dx\, dy. \qquad (13.35)$$

Comparing the two expansions in equations (13.33) and (13.34) for the same function $f(x,y,t)$ we obtain differential equations for the functions $T_{nm}(t)$:

$$T'_{nm}(t)+(a^2\lambda_{nm}+\gamma)T_{nm}(t)=f_{nm}(t). \tag{13.36}$$

In order that the solution represented by the series $u_2(x,y,t)$ given in equation (13.32) satisfy the zero temperature initial condition it is necessary that the functions $T_{nm}(t)$ obey the condition

$$T_{nm}(0)=0. \tag{13.37}$$

As we proved before, the solution of the ordinary differential equation (13.36) with initial condition (13.37) may be written in the integral form

$$T_{nm}(t)= \int_0^t f_{nm}(\tau)\, e^{-(a^2\lambda_{nm}+\gamma)(t-\tau)}\, d\tau. \tag{13.38}$$

Thus the solution of the nonhomogeneous heat conduction problem for a thin uniform rectangular plate with boundary conditions equal to zero has the form

$$u(x,y,t)=u_1(x,y,t)+u_2(x,y,t)= \sum_{n=1}^{\infty}\sum_{m=1}^{\infty}[T_{nm}(t)+C_{nm}e^{-(a^2\lambda_{nm}+\gamma)t}]\,V_{nm}(x,y),$$

where the functions $T_{nm}(t)$ are defined by equation (13.38) and the coefficients C_{nm} have been found earlier in equation (13.30).

Example 2

Find the temperature, $u(x,y,t)$, of a thin rectangular plate ($0 \le x \le l_x$, $0 \le y \le l_y$) if its boundary is kept at constant zero temperature, the initial temperature distribution within the plate is zero, and one internal source of heat acts at the point (x_0, y_0) in the plate. The value of this source is

$$Q(t) = A\sin\omega t.$$

Assume the plate is thermally insulated over its lateral faces.

Solution. The problem is expressed as

$$\frac{\partial u}{\partial t}=a^2\left[\frac{\partial^2 u}{\partial x^2}+\frac{\partial^2 u}{\partial y^2}\right]+\frac{A}{c\rho}\sin\omega t\cdot\delta(x-x_0)\delta(y-y_0),$$

under the conditions

$$u(x,y,0)=0,$$

$$u(0,y,t)=u(l_x,y,t)=u(x,0,t)=u(x,l_y,t)=0.$$

The boundary conditions of the problem are Dirichlet homogeneous boundary conditions; therefore eigenvalues and eigenfunctions of problem are

$$\lambda_{nm}=\lambda_{xn}+\lambda_{ym}=\pi^2\left(\frac{n^2}{l_x^2}+\frac{m^2}{l_y^2}\right),\quad n,m=1,2,3,\ldots$$

$$\|V_{nm}\|^2=\|X_n\|^2\cdot\|Y_m\|^2=\frac{l_x l_y}{4}.$$

The initial condition is equal to zero, in which case $C_{nm} = 0$. Applying equation (13.35), we obtain

$$f_{nm}(t) = \frac{4}{l_x l_y} \int_0^{l_x} \int_0^{l_y} \frac{A}{c\rho} \sin \omega t \cdot \delta(x - x_0)\delta(y - y_0)\sin\frac{\pi n x}{l_x}\sin\frac{\pi m y}{l_y}\, dx\, dy$$

$$= \frac{4A}{c\rho\, l_x l_y}\sin\omega t \sin\frac{\pi n x_0}{l_x}\sin\frac{\pi m y_0}{l_y}.$$

We also have from the above formulas that

$$T_{nm}(t) = \int_0^t f_{nm}(\tau)e^{-\lambda_{nm}a^2(t-\tau)}\, d\tau = \frac{4A}{c\rho l_x l_y}\sin\frac{\pi n x_0}{l_x}\sin\frac{\pi m y_0}{l_y}$$

$$\times \frac{1}{\omega^2 + \left(a^2\lambda_{nm}\right)^2}\left[a^2\lambda_{nm}\sin\omega t - \omega\cos\omega t + \omega e^{-\lambda_{nm}a^2 t}\right]$$

so that finally we obtain

$$u(x,y,t) = \frac{4A}{c\rho l_x l_y}\sum_{n,m=1}^{\infty}\frac{1}{\omega^2 + \left(a^2\lambda_{nm}\right)^2}\left[a^2\lambda_{nm}\sin\omega t - \omega\cos\omega t + \omega e^{-\lambda_{nm}a^2 t}\right]$$

$$\times \sin\frac{\pi n x_0}{l_x}\sin\frac{\pi m y_0}{l_y}\sin\frac{\pi n x}{l_x}\sin\frac{\pi m y}{l_y}.$$

Figure 13.2 shows snapshots of the solution at times $t = 0.6$ and $t = 6.2$. This solution was obtained with the program **Heat** for the case, $a^2 = 0.25$, $l_x = 4$, $l_y = 6$, $A/c\rho = 120$, $\omega = 5$, $x_0 = 2$, and $y_0 = 3$.

Reading Exercise

Find a stationary solution (periodic in time) of this problem and compare it with the solution $u(x,y,t)$ obtained in this example when $t \rightarrow \infty$. *Hint*: Search for a stationary solution in the form $\text{Re}[F(x,y)\exp(i\omega t)]$; then, for a complex function, $F(x,y)$, you will obtain the Helmholtz equation with zero boundary and initial conditions. A similar problem was discussed for the one-dimensional heat equation.

(a)   (b)

FIGURE 13.2 Surface graph of plate temperature at (a) $t = 0.6$, (b) $t = 6.2$ for Example 2.

Example 3

A heat-conducting, thin, uniform rectangular plate ($0 \leq x \leq l_x$, $0 \leq y \leq l_y$) is thermally insulated over its lateral faces. One side of the plate, at $x = 0$, is thermally insulated and the rest of the boundary is kept at constant zero temperature. The initial temperature distribution within the plate is zero.

Let heat be generated throughout the plate with the intensity of internal sources (per unit mass of the membrane) given by

$$Q(x,y,t) = A\left(l_x - x\right)\sin\frac{\pi y}{l_y}\sin t.$$

Find the distribution of temperature within the plate when $t > 0$.

Solution. The problem involves finding the solution of the equation

$$\frac{\partial u}{\partial t} = a^2\left[\frac{\partial^2 u}{\partial x^2} + \frac{\partial^2 u}{\partial y^2}\right] + \frac{A}{C\rho}\left(l_x - x\right)\sin\frac{\pi y}{l_y}\sin t,$$

under the conditions

$$u(x,y,0) = 0,$$

$$\frac{\partial u}{\partial x}(0,y,t) = 0, \quad u(l_x,y,t) = 0, \quad u(x,0,t) = u(x,l_y,t) = 0.$$

Eigenvalues and eigenfunctions of problem are given by

$$\lambda_{nm} = \lambda_{xn} + \lambda_{ym} = \pi^2\left[\frac{(2n-1)^2}{4l_x^2} + \frac{m^2}{l_y^2}\right], \quad n,m = 1,2,3,\dots$$

$$V_{nm}(x,y) = X_n(x)Y_m(y) = \cos\frac{(2n-1)\pi x}{2l_x}\sin\frac{m\pi y}{l_y},$$

$$\|V_{nm}\|^2 = \|X_n\|^2 \cdot \|Y_m\|^2 = \frac{l_x l_y}{4}.$$

The three-dimensional picture shown in Figure 13.3 depicts two eigenfunctions (chosen as examples), $V_{11}(x,y)$ and $V_{61}(x,y)$ for the given problem. The values of λ are $\lambda_{11} = 0.428$ and $\lambda_{61} = 18.934$.

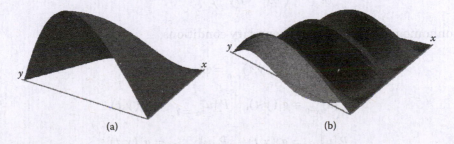

(a)　　　　　　　　　　　　　　　　(b)

FIGURE 13.3　Eigenfunctions (a) $V_{11}(x,y)$ and (b) $V_{61}(x,y)$ for Example 3.

FIGURE 13.4   Surface graph of plate temperature at (a) $t = 3$ and (b) $t = 6$ for Example 3.

The initial condition is equal to zero, in which case $C_{nm} = 0$. Applying equation (13.35), we obtain

$$f_{n1}(t) = \frac{4}{l_x l_y} \int_0^{l_x} \int_0^{l_y} f(x,y,t)\cos\frac{(2n-1)\pi x}{2l_x}\sin\frac{\pi y}{l_y}\,dx\,dy = \frac{8Al_x}{\pi^2(2n-1)^2}\sin(t),$$

and $f_{nm} = 0$, if $m \neq 1$. Thus we have

$$T_{n1}(t) = \frac{1}{||V_{n1}||^2}\int_0^t f_{n1}(\tau)e^{-\lambda_{n1}a^2(t-\tau)}\,d\tau$$

$$= \frac{8Al_x}{\pi^2(2n-1)^2[1+(a^2\lambda_{n1})^2]}\{a^2\lambda_{n1}\sin(t) - \cos(t) + e^{-\lambda_{n1}a^2 t}\}$$

$$T_{nm} = 0, \ \text{if } m \neq 1,$$

and, finally,

$$u(x,y,t) = \frac{8Al_x}{\pi^2}\sin\frac{\pi y}{l_y}\sum_{n=1}^{\infty}\frac{a^2\lambda_{n1}\sin t - \cos t + e^{-\lambda_{n1}a^2 t}}{(2n-1)^2[\omega^2 + (a^2\lambda_{n1})^2]}\cos\frac{(2n-1)\pi x}{2l_x}.$$

Figure 13.4 shows two snapshots of the solution at the times $t=3$ and $t=6$. This solution was obtained with the program **Heat** for the case $a^2 = 0.25$, $l_x = 4$, $l_y = 6$, and $A=0.1$.

13.1.3 The Fourier Method for the Nonhomogeneous Heat Conduction Equation with Nonhomogeneous Boundary Conditions

Now consider the general boundary problem for heat conduction given by

$$\frac{\partial u}{\partial t} = a^2\left(\frac{\partial^2 u}{\partial x^2} + \frac{\partial^2 u}{\partial y^2}\right) - \gamma u + f(x,y,t), \tag{13.39}$$

with nonhomogeneous initial and boundary conditions

$$u(x,y,t)\big|_{t=0} = \varphi(x,y),$$

$$P_1[u]\big|_{x=0} = g_1(y,t), \quad P_2[u]\big|_{x=l_x} = g_2(y,t),$$

$$P_3[u]\big|_{y=0} = g_3(x,t), \quad P_4[u]\big|_{y=l_y} = g_4(x,t).$$

As was mentioned in the previous section, to deal with homogeneous boundary conditions we introduce an auxiliary function, $w(x,y,t)$, and search for a solution of the problem in the form

$$u(x,y,t) = v(x,y,t) + w(x,y,t),$$

where $v(x,y,t)$ is a new unknown function, and the function $w(x,y,t)$ is chosen so that it satisfies the given nonhomogeneous boundary conditions.

Case I. Let the boundary value functions *satisfy consistency conditions* (i.e., the boundary conditions take the same values at the corners of the domain), in which case we have

$$P_1[g_3(x,t)]_{x=0} = P_3[g_1(y,t)]_{y=0}, \quad P_1[g_4(x,t)]_{x=0} = P_4[g_1(y,t)]_{y=l_y},$$

$$P_2[g_3(x,t)]_{x=l_x} = P_3[g_2(y,t)]_{y=0}, \quad P_2[g_4(x,t)]_{x=l_x} = P_4[g_2(y,t)]_{y=l_y}.$$

(13.40)

In this case we seek an auxiliary function $w(x,y,t)$ in the form of equation (13.9) with the coefficients of the first- or second-order polynomials $\overline{X}(x)$, $\overline{\overline{X}}(x)$, $\overline{Y}(y)$, and $\overline{\overline{Y}}(y)$ adjusted in such a way that the function $w(x,y,t)$ satisfies the boundary conditions.

Functions $\overline{X}(x)$ and $\overline{\overline{X}}(x)$ can be chosen in such a way that

$$P_1[\overline{X}(0)] = 1, \quad P_2[\overline{X}(l_x)] = 0,$$

$$P_1[\overline{\overline{X}}(0)] = 0, \quad P_2[\overline{\overline{X}}(l_x)] = 1.$$

(13.41)

If $\beta_1 \neq 0$ or $\beta_2 \neq 0$, then the functions $\overline{X}(x)$ and $\overline{\overline{X}}(x)$ are polynomials of the first order

$$\overline{X}(x) = \gamma_1 + \delta_1 x, \quad \overline{\overline{X}}(x) = \gamma_2 + \delta_2 x.$$

The coefficients $\gamma_1, \delta_1, \gamma_2, \delta_2$ of these polynomials are defined uniquely and depend on the types of boundary conditions:

$$\gamma_1 = \frac{\alpha_2 + \beta_2 l_x}{\beta_1 \beta_2 l_x + \beta_1 \alpha_2 - \beta_2 \alpha_1}, \quad \delta_1 = \frac{-\beta_2}{\beta_1 \beta_2 l_x + \beta_1 \alpha_2 - \beta_2 \alpha_1},$$

$$\gamma_2 = \frac{-\alpha_1}{\beta_1 \beta_2 l_x + \beta_1 \alpha_2 - \beta_2 \alpha_1}, \quad \delta_2 = \frac{\beta_1}{\beta_1 \beta_2 l_x + \beta_1 \alpha_2 - \beta_2 \alpha_1}.$$

(13.42)

Reading Exercise

Derive the above expressions.

If $\beta_1 = \beta_2 = 0$, then the functions $\overline{X}(x)$ and $\overline{\overline{X}}(x)$ are polynomials of the second order and we have

$$\overline{X}(x) = x - \frac{x^2}{2l_x}, \quad \overline{\overline{X}}(x) = \frac{x^2}{2l_x}. \tag{13.43}$$

The problem for $Y(y)$ (and its solution) is the same as the problem for $X(x)$. Functions $\overline{Y}(y)$ and $\overline{\overline{Y}}(y)$ can be chosen in such a way that

$$P_3[\overline{Y}(0)] = 1, \quad P_3[\overline{Y}(l_y)] = 0,$$

$$P_4[\overline{\overline{Y}}(0)] = 0, \quad P_4[\overline{\overline{Y}}(l_y)] = 1. \tag{13.44}$$

If $\beta_3 \neq 0$ or $\beta_4 \neq 0$, then the functions $\overline{Y}(y)$ and $\overline{\overline{Y}}(y)$ are polynomials of the first order and we may write

$$\overline{Y}(y) = \gamma_3 + \delta_3 y \quad \overline{\overline{Y}}(y) = \gamma_4 + \delta_4 y.$$

The coefficients $\gamma_3, \delta_3, \gamma_4, \delta_4$ of these polynomials are defined uniquely and depend on the types of boundary conditions:

$$\gamma_3 = \frac{\alpha_4 + \beta_4 l_y}{\beta_3 \beta_4 l_y + \beta_3 \alpha_4 - \beta_4 \alpha_3}, \quad \delta_3 = \frac{-\beta_4}{\beta_3 \beta_4 l_y + \beta_3 \alpha_4 - \beta_4 \alpha_3},$$

$$\gamma_4 = \frac{-\alpha_3}{\beta_3 \beta_4 l_y + \beta_3 \alpha_4 - \beta_4 \alpha_3}, \quad \delta_4 = \frac{\beta_3}{\beta_3 \beta_4 l_y + \beta_3 \alpha_4 - \beta_4 \alpha_3}. \tag{13.45}$$

If $\beta_3 = \beta_4 = 0$, then the functions $\overline{Y}(y)$ and $\overline{\overline{Y}}(y)$ are polynomials of the second order and we have

$$\overline{Y}(y) = y - \frac{y^2}{2l_y}, \quad \overline{\overline{Y}}(y) = \frac{y^2}{2l_y}. \tag{13.46}$$

Coefficients $A(t)$, $B(t)$, $C(t)$, and $D(t)$ of the auxiliary function $w(x,y,t)$ are defined from the boundary conditions:

At the edge at $x = 0$ we have

$$P_1[w]_{x=0} = g_1(y,t) + \{P_1[g_3(x,t)]_{x=0} + A(t)\}\overline{Y} + \{P_1[g_4(x,t)]_{x=0} + B(t)\}\overline{\overline{Y}}.$$

At the edge at $x = l_x$ we have

$$P_2[w]_{x=l_x} = g_2(y,t) + \{P_2[g_3(x)]_{x=l_x} + C(t)\}\overline{Y} + \{P_2[g_4(x,t)]_{x=l_x} + D(t)\}\overline{\overline{Y}}.$$

At the edge at $y = 0$ we have

$$P_3[w]_{y=0} = g_3(x,t) + \{P_3[g_1(y,t)]_{y=0} + A(t)\}\overline{X} + \{P_3[g_2(y,t)]_{y=0} + C(t)\}\overline{\overline{X}}.$$

At the edge at $y = l_y$ we have

$$P_4[w]_{y=l_y} = g_4(x,t) + \{P_4[g_1(y,t)]_{y=l_y} + B(t)\}\overline{X} + \{P_4[g_2(y,t)]_{y=l_y} + D(t)\}\overline{\overline{X}}.$$

The conformity conditions for the boundary value functions are valid, so the coefficients are

$$A(t) = -P_1[g_3(x,t)]_{x=0} = -P_3[g_1(y,t)]_{y=0},$$

$$B(t) = -P_1[g_4(x,t)]_{x=0} = -P_4[g_1(y,t)]_{y=l_y},$$

$$C(t) = -P_2[g_3(x,t)]_{x=l_x} = -P_3[g_2(y,t)]_{y=0},$$

$$D(t) = -P_2[g_4(x,t)]_{x=l_x} = -P_4[g_2(y,t)]_{y=l_y}.$$

$$(13.47)$$

Reading Exercise

Verify the expressions above for the coefficients.

It is easy to verify (which is left as a Reading Exercise for the reader) that the auxiliary function, $w(x,y,t)$, satisfies the given boundary conditions:

$$P_1[w]_{x=0} = g_1(y,t), \quad P_2[w]_{x=l_x} = g_2(y,t),$$

$$P_3[w]_{y=0} = g_3(x,t), \quad P_4[w]_{y=l_y} = g_4(x,t).$$

$$(13.48)$$

Case II. Suppose the boundary value functions *do not satisfy consistency conditions.* We shall search for an auxiliary function as the sum of two functions:

$$w(x,y,t) = w_1(x,y,t) + w_2(x,y,t). \tag{13.49}$$

Function $w_1(x,y,t)$ is an auxiliary function satisfying the consistent boundary conditions

$$P_1[w_1]_{x=0} = g_1(y,t), \quad P_2[w_1]_{x=l_x} = g_2(y,t),$$

$$P_3[w_1]_{y=0} = -A(t)\overline{X}(x) - C(t)\overline{\overline{X}}(x),$$

$$P_4[w_1]_{y=l_y} = -B(t)\overline{X}(x) - D(t)\overline{\overline{X}}(x),$$

(13.50)

where

$$A(t) = -P_3[g_1(y,t)]_{y=0}, \quad B(t) = -P_4[g_1(y,t)]_{y=l_y},$$

$$C(t) = -P_3[g_2(y,t)]_{y=0}, \quad D(t) = -P_4[g_2(y,t)]_{y=l_y}.$$

Such a function was constructed above. In this case it has the form

$$w_1(x,y,t) = g_1(y,t)\overline{X} + g_2(y,t)\overline{\overline{X}}.$$

Reading Exercise

Verify the above expression.

The function $w_2(x,y,t)$ is a particular solution of the equation

$$\frac{\partial^2 w_2}{\partial x^2} + \frac{\partial^2 w_2}{\partial y^2} = 0$$

(13.51)

(this is the *Laplace* equation) with the following boundary conditions:

$$P_1[w_2]_{x=0} = 0, \quad P_2[w_2]_{x=l_x} = 0,$$

$$P_3[w_2]_{y=0} = g_3(x,t) + A(t)\overline{X}(x) + C(t)\overline{\overline{X}}(x),$$

(13.52)

$$P_4[w_2]_{y=l_y} = g_4(x,t) + B(t)\overline{X}(x) + D(t)\overline{\overline{X}}(x).$$

The solution to this problem has the form

$$w_2(x,y,t) = \sum_{n=1}^{\infty} \{A_n(t)Y_{1n}(y) + B_n(t)Y_{2n}(y)\}X_n(x),$$

(13.53)

where λ_{xn} and $X_n(x)$ are eigenvalues and eigenfunctions of the Sturm-Liouville problem

$$X'' + \lambda X = 0, \ 0 < x < l_x$$

$$P_1[X]|_{x=0} = P_2[X]|_{x=l_x} = 0.$$

(13.54)

The coefficients $A_n(t)$ and $B_n(t)$ are defined by the formulas

$$A_n(t) = \frac{1}{\|X_n\|^2} \int_0^{l_x} \left[g_4(x,t) + B(t)\overline{X}(x) + D(t)\overline{\overline{X}}(x) \right] X_n(x)\,dx,$$

$$(13.55)$$

$$B_n(t) = \frac{1}{\|X_n\|^2} \int_0^{l_x} \left[g_3(x,t) + A(t)\overline{X}(x) + C(t)\overline{\overline{X}}(x) \right] X_n(x)\,dx.$$

Thus we see that eigenvalues and eigenfunctions of this boundary value problem depend on the types of boundary conditions (see Appendix 3 for a detailed account).

Having the eigenvalues, λ_{xn}, we obtain a similar equation for $Y(y)$ given by

$$Y'' - \lambda_{xn} Y = 0, \ 0 < y < l_y.$$

$$(13.56)$$

We shall choose fundamental system $\{Y_1, Y_2\}$ of solutions in such a way that

$$P_3[Y_1(0)] = 0, \quad P_3[Y_1(l_y)] = 1,$$

$$(13.57)$$

$$P_4[Y_2(0)] = 1, \quad P_4[Y_2(l_y)] = 0.$$

Two particular solutions of the previous equation (13.56) are $\exp(\pm\sqrt{\lambda_n}\,y)$ but for the future analysis it is more convenient to choose two linearly independent functions $Y_1(y)$ and $Y_2(y)$ in the form

$$Y_1(y) = a \sinh\sqrt{\lambda_n}\,y + b \cosh\sqrt{\lambda_n}\,y,$$

$$(13.58)$$

$$Y_2(y) = c \sinh\sqrt{\lambda_n}\,(l_y - y) + d \cosh\sqrt{\lambda_n}\,(l_y - y).$$

The values of coefficients a, b, c, and d depend on the types of boundary conditions $P_3[u]_{y=0}$ and $P_4[u]_{y=l_y}$ (see Appendix 3 for details). It can be verified that the auxiliary function

$$w(x,y,t) = w_1(x,y,t) + w_2(x,y,t)$$

$$(13.59)$$

$$= w_1(x,y,t) + \sum_{n=1}^{\infty} \{ A_n(t)Y_{1n}(y) + B_n(t)Y_{2n}(y) \} X_n(x)$$

satisfies the given boundary conditions.

It is important to notice that the separation of an unknown function, $u(x,y,t)$, into two functions, $u(x,y,t)=v(x,y,t)+w(x,y,t)$, has a transparent physical meaning. The function $w(x,y,t)$ is a solution to the Laplace equation

$$\frac{\partial^2 w}{\partial x^2} + \frac{\partial^2 w}{\partial y^2} = 0 \tag{13.60}$$

with nonhomogeneous boundary conditions, whereas the function $v(x,y,t)$ describes the relaxation of the temperature in the plate to the equilibrium. In particular for the simple case where the boundary conditions are stationary (i.e., the edges experience no change in temperature), the function w is the final, equilibrium temperature of the body as $t \to \infty$.

Example 4

A heat-conducting thin uniform rectangular plate ($0 \leq x \leq l_x$, $0 \leq y \leq l_y$) is thermally insulated over its lateral faces. The edge at $y=0$ of the plate is kept at the constant temperature $u = u_1$; the edge $y = l_y$ is kept at the constant temperature $u = u_2$; and the remaining boundary is thermally insulated. The initial temperature distribution within the plate is

$$u(x,y,0) = u_0 = \text{const}.$$

Find the temperature $u(x,y,t)$ of the plate at any later time, if generation (or absorption) of heat by internal sources is absent.

Solution. The problem may be resolved by solving the equation

$$\frac{\partial u}{\partial t} = a^2 \left[\frac{\partial^2 u}{\partial x^2} + \frac{\partial^2 u}{\partial y^2} \right],$$

under the conditions

$$u(x,y,0) = \varphi(x,y) = u_0,$$

$$\frac{\partial u}{\partial x}(0,y,t) = \frac{\partial u}{\partial x}(l_x,y,t) = 0, \ u(x,0,t) = u_1, \ u(x,l_y,t) = u_2.$$

The solution to this problem can be expressed as the sum of two functions

$$u(x,y,t) = w(x,y,t) + v(x,y,t),$$

where $v(x,y,t)$ is a new, unknown function and $w(x,y,t)$ is an auxiliary function satisfying the boundary conditions.

The boundary value functions *satisfy the conforming conditions*, that is,

$$g_1\big|_{y=0} = \frac{\partial g_3}{\partial x}\bigg|_{x=0} = 0, \ \ g_1\big|_{y=l_y} = \frac{\partial g_4}{\partial x}\bigg|_{x=0} = 0,$$

$$g_2\big|_{y=0} = \frac{\partial g_3}{\partial x}\bigg|_{x=l_x} = 0, \ g_2\big|_{y=l_y} = \frac{\partial g_4}{\partial x}\bigg|_{x=l_x} = 0.$$

An auxiliary function satisfying the given boundary condition has the form

$$w(x,y,t) = g_1(y,t)\overline{X} + g_2(y,t)\overline{\overline{X}} + g_3(x,t)\overline{Y} + g_4(x,t)\overline{\overline{Y}}$$

$$+ A(t)\overline{X}\overline{Y} + B(t)\overline{X}\overline{\overline{Y}} + C(t)\overline{\overline{X}}\overline{Y} + D(t)\overline{\overline{X}}\overline{Y},$$

where

$$\overline{X}(x) = x - \frac{x^2}{2l_x}, \ \overline{\overline{X}}(x) = \frac{x^2}{2l_x}, \ \overline{Y}(y) = 1 - \frac{y}{l_y}, \ \overline{\overline{Y}}(y) = \frac{y}{l_y},$$

$$A(t) = B(t) = C(t) = D(t) = 0$$

in which case we have

$$w(x,y,t) = u_1 + (u_2 - u_1)\frac{y}{l_y}.$$

Reading Exercise

Derive the above result for function w.

Given this expression for $w(x,y,t)$ we see that the separation of the function $u(x,y,t)$ into functions $w(x,y,t)$ and $v(x,y,t)$ is a separation into a stationary solution corresponding to the boundary conditions and the solution describing the relaxation of the temperature to the stationary state.

The relaxation process to a steady state described by the function $v(x,y,t)$ is the solution to the boundary value problem with zero boundary conditions where the stationary solution is described by

$$f^*(x,y,t) = -\frac{\partial w}{\partial t} + a^2\left(\frac{\partial^2 w}{\partial x^2} + \frac{\partial^2 w}{\partial y^2}\right) = 0,$$

$$\varphi^*(x,y) = u_0 - u_1 - (u_2 - u_1)\frac{y}{l_y}.$$

Eigenvalues and eigenfunctions of the problem can be easily obtained:

$$\lambda_{nm} = \lambda_{xn} + \lambda_{ym} = \pi^2\left[\frac{n^2}{l_x^2} + \frac{m^2}{l_y^2}\right], \ \ n = 0,1,2,... \ m = 1,2,3,...$$

$$V_{nm}(x,y) = X_n(x)Y_m(y) = \cos\frac{n\pi x}{l_x}\sin\frac{m\pi y}{l_y},$$

$$\|V_{nm}\|^2 = \|X_n\|^2 \ \|Y_m\|^2 = \begin{cases} l_x l_y/2, & n = 0 \\ l_x l_y/4, & n > 0. \end{cases}$$

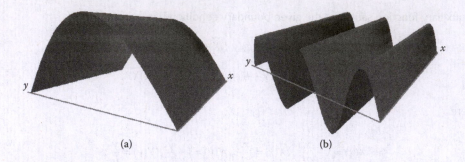

FIGURE 13.5   Eigenfunctions (a) $V_{01}(x,y)$ and (b) $V_{05}(x,y)$ for the free surface in Example 4.

The three-dimensional picture shown in Figure 13.5 depicts the two eigenfunctions, $V_{01}(x,y)$ and $V_{05}(x,y)$, chosen as examples for the given problem. The values of λ are $\lambda_{01} = 0.274$ and $\lambda_{05} = 6.854$.

Applying equation (13.57), we obtain

$$C_{nm} = \frac{1}{\|v_{nm}\|^2} \int_0^{l_x} \int_0^{l_y} \left[u_0 - u_1 - (u_2 - u_1)\frac{y}{l_y} \right] \cos\frac{n\pi x}{l_x} \sin\frac{m\pi y}{l_y}\, dx\, dy.$$

From this formula we have

$$C_{0m} = \frac{2}{l_y} \int_0^{l_x} \left[u_0 - u_1 - (u_2 - u_1)\cdot\frac{y}{l_y} \right] \sin\frac{m\pi y}{l_y}\, dx\, dy$$

$$= \frac{2}{m\pi} \left\{ (u_0 - u_1)\left[1 - (-1)^m\right] + (u_2 - u_1)(-1)^m \right\},$$

for $n > 0$, $C_{nm} = 0$. In Figure 13.5 we see that the eigenfunctions with index $n = 0$ do not depend on x. And as we just obtained, the temperature distribution does not depend on x at all. This result could be anticipated from the very beginning, since the initial and boundary conditions do not depend on x. In other words, the solution is actually one-dimensional for this problem. Hence, the distribution of temperature inside the rectangular plate for some instant of time is described by the series

$$u(x,y,t) = u_1 + (u_2 - u_1)\frac{y}{l_y} + \sum_{m=1}^{\infty} C_{0m} e^{-\lambda_{0m} a^2 t} \sin\frac{m\pi y}{l_y}.$$

Figure 13.6 shows two snapshots of the solution at the times $t = 0$ and $t = 10$. This solution was obtained with the program **Heat** for the case when $a^2 = 0.25$, $l_x = 4$, $l_y = 6$, $u_0 = 10$, $u_1 = 20$, and $u_2 = 50$.

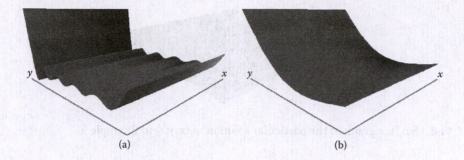

FIGURE 13.6   Surface graph of temperature at (a) $t = 0$ and (b) $t = 10$ for Example 4.

It is interesting to compare the first and second pictures in Figure 13.6. The first one is very rough; the second is smooth. The explanation of the roughness is that at $t = 0$ the initial and boundary conditions do not match (the solution is nonphysical). But at any nonzero time the temperature distribution very quickly becomes smooth, as it should be for real, physical situations. Here we see that this method of solution works quite well in approximating a real, physical situation.

Example 5

A heat-conducting thin uniform rectangular plate ($0 \le x \le l_x$, $0 \le y \le l_y$) is thermally insulated over its lateral faces. The edge at $y = 0$ is thermally insulated; the edge at $y = l_y$ is kept at constant zero temperature; the edge at $x = 0$ is kept at constant temperature $u = u_1$; and the edge at $x = l_x$ is kept at the temperature

$$u(l_y, x, t) = \cos\frac{3\pi y}{2l_y} e^{-t}.$$

The initial temperature distribution within the plate is $u(x, y, 0) = u_0 = \text{const}$. Find the temperature $u(x, y, t)$ of the plate at any later time, if generation (or absorption) of heat by internal sources is absent.

Solution. The problem depends on the solution of the equation

$$\frac{\partial u}{\partial t} = a^2 \left[\frac{\partial^2 u}{\partial x^2} + \frac{\partial^2 u}{\partial y^2} \right]$$

FIGURE 13.7   Surface graph of the auxiliary function $w_1(x, y, t)$ at (a) $t = 0$ and (b) $t = 3$ for Example 5.

FIGURE 13.8 Surface graph of the particular solution $w_2(x, y, t)$ to Example 5.

under the conditions

$$u(x,y,0) = \varphi(x,y) = u_0,$$

$$u(0,y,t) = u_1, \quad u(l_x,y,t) = \cos\frac{3\pi y}{2l_y}\cdot e^{-t}, \quad \frac{\partial u}{\partial y}(x,0,t) = 0, \quad u(x,l_y,t) = 0.$$

The solution to this problem can be expressed as the sum of two functions

$$u(x,y,t) = w(x,y,t) + v(x,y,t),$$

where $v(x,y,t)$ is a new, unknown function and $w(x,y,t)$ is an auxiliary function satisfying the boundary conditions.

The boundary value functions *do not satisfy the conforming conditions* at point $(0,l_y)$, that is,

$$g_1\big|_{y=l_y} = u_1 \neq g_4\big|_{x=0} = 0.$$

So, in this case we shall search for an auxiliary function as the sum of two functions

$$w(x,y,t) = w_1(x,y,t) + w_2(x,y,t),$$

where $w_1(x,y,t)$ is an auxiliary function satisfying the conforming boundary value functions

$$P_1[w_1]_{x=0} = g_1(y,t) = u_1, \quad P_2[w_1]_{x=l_x} = g_2(y,t) = \cos\frac{3\pi y}{2l_y}\,e^{-t},$$

$$P_3[w_1]_{y=0} = \frac{\partial w_1}{\partial y}(x,0) = g_3(x,t) = 0, \quad P_4[w_1]_{y=l_y} = w_1(x,l_y) = u_1\left(1-\frac{x}{l_x}\right),$$

and $w_2(x,y,t)$ is a particular solution of the Laplace equation

$$\nabla^2 w_2(x,y,t) = \left[\frac{\partial^2 w_2}{\partial x^2} + \frac{\partial^2 w_2}{\partial y^2}\right] = 0$$

with the following boundary conditions

$$P_1[w_2]_{x=0} = w_2(0,y,t) = 0, \quad P_2[w_2]_{x=l_x} = w_2(l_x,y,t) = 0,$$

$$P_3[w_2]_{y=0} = \frac{\partial u}{\partial y}(x,0,t) = 0, \quad P_4[w_2]_{y=l_y} = w_2(x,l_y,t) = u_1\left(\frac{x}{l_x} - 1\right).$$

The auxiliary function, $w_1(x,y,t)$, is

$$w_1(x,y,t) = u_1\left(1 - \frac{x}{l_x}\right) + \cos\frac{3\pi y}{2l_y}e^{-t}\frac{x}{l_x} = u_1 + \frac{x}{l_x}\left(\cos\frac{3\pi y}{2l_y}e^{-t} - u_1\right).$$

The particular solution, $w_2(x,y,t)$, of the problem has the form

$$w_2(x,y,t) = \sum_{n=1}^{\infty}\{A_n(t)Y_{1n}(y) + B_n(t)Y_{2n}(y)\}\ X_n(x),$$

where λ_{xn} and $X_n(x)$ are eigenvalues and eigenfunctions of the respective Sturm-Liouville problem:

$$\lambda_{xn} = \left[\frac{n\pi}{l_x}\right]^2,\ \ X_n(x) = \sin\frac{n\pi x}{l_x},\ \ \|X_n\|^2 = \frac{l_x}{2},\ \ n = 1,2,3,\ldots$$

The functions $Y_{1n}(y)$ and $Y_{2n}(y)$ for the given boundary conditions are (see Appendix 3)

$$Y_{1n}(y) = \frac{\cosh\sqrt{\lambda_n}\,y}{\cosh\sqrt{\lambda_n}\,l_y},\ \ Y_{2n}(y) = -\frac{\sinh\sqrt{\lambda_n}\,(l_y - y)}{\sqrt{\lambda_n}\,\cosh\sqrt{\lambda_n}\,l_y}.$$

Coefficients $A_n(t)$ and $B_n(t)$ are given by

$$A_n = \frac{2}{l_x}\int_0^{l_x} u_1\left(\frac{x}{l_x} - 1\right)\sin\frac{n\pi x}{l_x}\,dx = -\frac{2u_1}{n\pi},\ \ B_n = 0.$$

Thus,

$$w_2(x,y,t) = -\frac{2u_1}{\pi}\sum_{n=1}^{\infty}\frac{1}{n}\frac{\cosh\sqrt{\lambda_n}\,y}{\cosh\sqrt{\lambda_n}\,l_y}\sin\frac{n\pi x}{l_x}.$$

The function $v(x,y,t)$ is the solution to the boundary value problem with zero boundary conditions where

$$f^*(x,y,t) = \frac{x}{l_x}\cos\frac{3\pi y}{2l_y}e^{-t}\left[1 - a^2\left(\frac{3\pi}{2l_y}\right)^2\right],$$

$$\varphi^*(x,y) = u_0 - u_1 - \frac{x}{l_x}\left(\cos\frac{3\pi y}{2l_y}e^{-t} - u_1\right) + \frac{2u_1}{\pi}\sum_{n=1}^{\infty}\frac{1}{n}\frac{\cosh\sqrt{\lambda_n}\,y}{\cosh\sqrt{\lambda_n}\,l_y}\sin\frac{n\pi x}{l_x}.$$

Therefore, the eigenvalues and eigenfunctions of problem are

$$\lambda_{nm} = \lambda_{xn} + \lambda_{ym} = \pi^2\left[\frac{n^2}{l_x^2} + \frac{(2m-1)^2}{4l_y^2}\right],\ \ n,m = 1,2,3,\ldots$$

$$V_{nm}(x,y) = X_n(x)Y_m(y) = \sin\frac{n\pi x}{l_x}\cos\frac{(2m-1)\pi y}{2l_y},$$

$$\|V_{nm}\|^2 = \|X_n\|^2\cdot\|Y_m\|^2 = \frac{l_x l_y}{4}.$$

(As before, these eigenvalues and eigenfunctions can be found in Appendix 1 or easily derived by the reader.)

FIGURE 13.9   Surface graph of the plate temperature at (a) $t = 0$ and (b) $t = 2$ for Example 5.

Applying the equations (13.30) and (13.35), we obtain

$$C_{nm} = \frac{4}{l_x l_y} \int_0^{l_x} \int_0^{l_y} \varphi^*(x,y) \sin\frac{n\pi x}{l_x} \cos\frac{(2m-1)\pi y}{2l_y} \, dx \, dy,$$

$$f_{nm}(t) = \frac{4}{l_x l_y} \int_0^{l_x} \int_0^{l_y} f^*(x,y,t) \sin\frac{n\pi x}{l_x} \cos\frac{(2m-1)\pi y}{2l_y} \, dx \, dy.$$

Thus we have

$$T_{nm}(t) = \int_0^t f_{nm}(\tau) e^{-a^2 \lambda_{nm}(t-\tau)} \, d\tau$$

and

$$u(x,y,t) = \left[w_1(x,y,t) + w_2(x,y,t) \right] + v(x,y,t)$$

$$= u_1 + \frac{x}{l_x}\left(\cos\frac{3\pi y}{2l_y} e^{-t} - u_1 \right) - \frac{2u_1}{\pi} \sum_{n=1}^{\infty} \frac{1}{n} \frac{\cosh\sqrt{\lambda_n}\, y}{\cosh\sqrt{\lambda_n}\, l_y} \sin\frac{n\pi x}{l_x}$$

$$+ \sum_{n=1}^{\infty} \sum_{m=1}^{\infty} \left[T_{nm}(t) + C_{nm} e^{-\lambda_{nm} a^2 t} \right] \sin\frac{n\pi x}{l_x} \cos\frac{(2m-1)\pi y}{2l_y}.$$

Figure 13.9 shows two snapshots of the solution at the times $t=0$ and $t = 2$. This solution was obtained with the program **Heat** for the case when $a^2 = 1$, $l_x = 4$, $l_y = 6$, $u_0 = -1$, and $u_1 = 1$.

13.2 HEAT CONDUCTION WITHIN A CIRCULAR DOMAIN

Let a uniform circular plate be placed in the horizontal x-y plane and bounded at the circular periphery by a radius of length l. The plate is assumed to be thin enough so that the temperature is the same at all points with the same x-y coordinates.

For a circular domain it is convenient to write the heat conduction equation in polar coordinates where the Laplace operator is

$$\nabla^2 u = \frac{\partial^2 u}{\partial r^2} + \frac{1}{r}\frac{\partial u}{\partial r} + \frac{1}{r^2}\frac{\partial^2 u}{\partial \varphi^2}$$

and the heat conduction is described by the equation

$$\frac{\partial u}{\partial t} = a^2 \left(\frac{\partial^2 u}{\partial r^2} + \frac{1}{r}\frac{\partial u}{\partial r} + \frac{1}{r^2}\frac{\partial^2 u}{\partial \varphi^2} \right) - \gamma u + f(r,\varphi,t), \tag{13.61}$$

$$0 \le r < l, 0 \le \varphi < 2\pi, t > 0.$$

Here $u(r,\varphi,t)$ is the temperature of the membrane at point (r,φ) at time t.

The *initial condition* defines the temperature distribution within the membrane at time zero and may be expressed in generic form as

$$u(r,\varphi,0) = \phi(r,\varphi). \tag{13.62}$$

The *boundary condition* describes the thermal condition around the boundary at any time t. In general, the boundary condition can be written as follows:

$$P[u]_{r=l} \equiv \alpha \frac{\partial u}{\partial r} + \beta u \bigg|_{r=l} = g(\varphi,t). \tag{13.63}$$

It is obvious that function $g(\varphi,t)$ must be a singled-valued periodic function in φ of period 2π, that is,

$$g(\varphi + 2\pi,t) = g(\varphi,t).$$

Again we consider three main types of boundary conditions:

1. Boundary condition of the first type (Dirichlet condition), $u(l,\varphi,t) = g(\varphi,t)$, where the temperature at the boundary is given or is zero in which case $g(\varphi,t) \equiv 0$.

2. Boundary condition of the second type (Neumann condition), $u_r(l,\varphi,t) = g(\varphi,t)$, in which case the heat flow at the boundary is given or the boundary is thermally insulated and $g(\varphi,t) \equiv 0$.

3. Boundary condition of the third type (mixed condition), $u_r(l,\varphi,t) + h\, u(l,\varphi,t) = g(\varphi,t)$, where the conditions of heat exchange with a medium are specified (here $h = $ const).

In the case of nonhomogeneous boundary condition we introduce an auxiliary function, $w(r,\varphi,t)$, satisfying the given boundary condition and express the solution to the problem as the sum of two functions:

$$u(r,\varphi,t) = v(r,\varphi,t) + w(r,\varphi,t),$$

where $v(r,\varphi,t)$ is a new, unknown function *with* zero boundary condition. The construction of function $w(r,\varphi,t)$ will be discussed in Section 13.2.3 and function $w(r,\varphi,t)$ will be sought in the form

$$w(r,\varphi,t) = (c_0 + c_2 r^2)\, g(\varphi,t)$$

with constants c_0 and c_2 to be adjusted to satisfy the boundary conditions.

We may express the solution $v(r,\varphi,t)$ as the sum of two functions

$$v(r,\varphi,t) = u_1(r,\varphi,t) + u_2(r,\varphi,t),$$

where $u_1(r,\varphi,t)$ is the solution of the homogeneous equation (free heat exchange) with the given initial conditions and zero boundary conditions and $u_2(r,\varphi,t)$ is the solution of the nonhomogeneous equation (heat exchange involving internal sources) with zero initial and boundary conditions.

13.2.1 The Fourier Method for the Homogeneous Heat Conduction Equation

Let us find the solution of the homogeneous equation

$$\frac{\partial u_1}{\partial t} = a^2 \left(\frac{\partial^2 u_1}{\partial r^2} + \frac{1}{r}\frac{\partial u_1}{\partial r} + \frac{1}{r^2}\frac{\partial^2 u_1}{\partial \varphi^2} \right) - \gamma u_1, \tag{13.64}$$

with the initial condition

$$u_1(r,\varphi,0) = \phi(r,\varphi) \tag{13.65}$$

and zero boundary conditions

$$P[u_1]_{r=l} \equiv \alpha \frac{\partial u_1}{\partial r} + \beta u_1 \bigg|_{r=l} = 0. \tag{13.66}$$

As was done previously in Chapter 11, we can separate the variables so that

$$u_1(r,\varphi,t) = T(t)V(r,\varphi). \tag{13.67}$$

Substituting this in equation (13.64) and separating variables we obtain

$$\frac{T''(t) + \gamma T(t)}{a^2 T(t)} \equiv \frac{1}{V(r,\varphi)} \left[\frac{\partial^2 V}{\partial r^2} + \frac{1}{r}\frac{\partial V}{\partial r} + \frac{1}{r^2}\frac{\partial^2 V}{\partial \varphi^2} \right] = -\lambda,$$

where λ is a separation of variables constant (we know that a choice of minus sign before λ is convenient). Thus the function $T(t)$ is the solution of the ordinary linear homogeneous differential equation of first order

$$T'(t) + (a^2\lambda + \gamma)T(t) = 0, \tag{13.68}$$

and $V(r,\varphi)$ is the solution to the following boundary value problem:

$$\frac{\partial^2 V}{\partial r^2} + \frac{1}{r}\frac{\partial V}{\partial r} + \frac{1}{r^2}\frac{\partial^2 V}{\partial \varphi^2} + \lambda V = 0, \tag{13.69}$$

$$\alpha \frac{\partial V}{\partial r}(l,\varphi) + \beta V(l,\varphi) = 0. \tag{13.70}$$

Two restrictions on $V(r,\varphi)$ are that it be bounded, $|V(r,\varphi)|<\infty$, and that it be periodic in φ: $V(r,\varphi) = V(r,\varphi + 2\pi)$.

From this we see that the boundary value problem leads to solving a homogeneous equation with homogeneous boundary conditions.

Again, we can separate the variables by writing

$$V(r,\varphi) = R(r)\Phi(\varphi). \qquad (13.71)$$

Substituting equation (13.71) into equation (13.69), we obtain

$$\frac{r^2 R''(r)+rR'(r)}{R(r)}+\lambda r^2 = -\frac{\Phi''(\varphi)}{\Phi(\varphi)} = \nu,$$

where ν is another separation of variables constant.

This yields the following two boundary value problems, each of which is one-dimensional:

$$\Phi''(\varphi)+\nu\,\Phi(\varphi)=0,$$

$$\Phi(\varphi) = \Phi(\varphi+2\pi), \qquad (13.72)$$

$$\Phi'(\varphi) = \Phi'(\varphi+2\pi)$$

and

$$\frac{d^2 R}{dr^2}+\frac{1}{r}\frac{dR}{dr}+\left(\lambda-\frac{\nu}{r^2}\right)R=0,$$

$$\left.\alpha\frac{\partial R}{\partial r}+\beta\,R\right|_{r=l} = 0, \quad |R(0)|<\infty. \qquad (13.73)$$

Nontrivial periodical solutions for $\Phi(\varphi)$ exist only when $\nu=n^2$ (see discussion in Chapter 11) where n is an integer number, $n=0,1,2,...$ Function $\Phi_n(\varphi)$ can be presented in the form

$$\Phi_n(\varphi)= D_{1n}\cos n\varphi + D_{2n}\sin n\varphi. \qquad (13.74)$$

As we already saw in Chapter 11, eigenvalues and eigenfunctions of the boundary value problem for function $R_n(r)$ obeying the equations (13.73) have the form

$$\lambda_{nm} =\left(\frac{\mu_m^{(n)}}{l}\right)^2, \quad R_{nm}(r)= J_n\!\left(\frac{\mu_m^{(n)}}{l}r\right)\;((n,m=0,1,2,...)), \qquad (13.75)$$

where $\mu_m^{(n)}$ is the mth *positive* root of the equation

$$\alpha\mu J_n'(\mu)+\beta l J_n(\mu)=0 \qquad (13.76)$$

and $J_n(\mu)$ is the Bessel function of the first kind.

Collecting the above results we may write the eigenfunctions of the given boundary value problem as

$$V_{nm}^{(1)}(r,\varphi) = J_n\left(\frac{\mu_m^{(n)}}{l}r\right)\cdot\cos n\varphi \quad \text{and} \quad V_{nm}^{(2)}(r,\varphi) = J_n\left(\frac{\mu_m^{(n)}}{l}r\right)\cdot\sin n\varphi. \tag{13.77}$$

Functions $V_{nm}^{(1,2)}(r,\varphi)$ are eigenfunctions *of the Laplace's operator* in polar coordinates in the domain $0 \le r \le l$, $0 \le \varphi < 2\pi$, and λ_{nm} are the corresponding eigenvalues.

Using the above eigenvalues, λ_{nm}, we can write the solution of the differential equation

$$T'_{nm}(t)+(a^2\lambda_{nm}+\gamma)T_{nm}(t)=0 \tag{13.78}$$

as

$$T_{nm}(t)=C_{nm}e^{-(a^2\lambda_{nm}+\gamma)t}. \tag{13.79}$$

From this we see that the general solution for function u_1 is

$$u_1(r,\varphi,t)=\sum_{n=0}^{\infty}\sum_{m=0}^{\infty}[a_{nm}V_{nm}^{(1)}(r,\varphi)+b_{nm}V_{nm}^{(2)}(r,\varphi)]e^{-(a^2\lambda_{nm}+\gamma)t}. \tag{13.80}$$

The coefficients a_{nm} and b_{nm} are defined using the function that expresses the initial condition and the orthogonality property of functions $V_{nm}^{(1)}(r,\varphi)$ and $V_{nm}^{(2)}(r,\varphi)$:

$$a_{nm}=\frac{1}{\|V_{nm}^{(1)}\|^2}\int_0^l\int_0^{2\pi}\phi(r,\varphi)\,V_{nm}^{(1)}(r,\varphi)\,r\,dr\,d\varphi, \tag{13.81}$$

$$b_{nm}=\frac{1}{\|V_{nm}^{(2)}\|^2}\int_0^l\int_0^{2\pi}\phi(r,\varphi)\,V_{nm}^{(2)}(r,\varphi)\,r\,dr\,d\varphi. \tag{13.82}$$

The norms of eigenfunctions $V_{nm}^{(1)}(r,\varphi)$ and $V_{nm}^{(2)}(r,\varphi)$ can be found in Section 11.3.1.

Example 6

The initial temperature distribution within a very long (infinite) cylinder of radius l is

$$\phi(r,\varphi)=u_0\left(1-\frac{r^2}{l^2}\right), \quad u_0=\text{const},$$

Find the distribution of temperature within the cylinder if its surface is kept at constant zero temperature. Generation (or absorption) of heat by internal sources is absent.

Solution. The boundary value problem modeling the process of the cooling of an infinite cylinder is

$$\frac{\partial u}{\partial t} = a^2 \left[\frac{\partial^2 u}{\partial r^2} + \frac{1}{r} \frac{\partial u}{\partial r} \right],$$

$$u(r, \varphi, 0) = u_0 \left(1 - \frac{r^2}{l^2} \right), \quad u(l, \varphi, t) = 0.$$

The initial temperature does not depend on the polar angle φ; thus, only terms with $n = 0$ are not zero and solution includes only functions $V_{0m}^{(1)}(r, \varphi)$ given by the first of equations (13.77). The solution $u(r, \varphi, t)$ is therefore given by the series

$$u(r, \varphi, t) = \sum_{m=0}^{\infty} a_{0m} e^{-a^2 \lambda_{0m} t} J_0 \left(\frac{\mu_m^{(0)}}{l} r \right).$$

The boundary condition of the problem is of *Dirichlet type*, so eigenvalues are given by the equation

$$J_n(\mu) = 0;$$

therefore $\mu_m^{(n)}$ are the roots of this equation.

The coefficients a_{0m} are given by equation (13.81):

$$a_{0m} = \frac{2\pi}{\| V_{0m}^{(1)} \|^2} \int_0^l u_0 \left(1 - \frac{r^2}{l^2} \right) J_0 \left(\mu_m^{(0)} r/l \right) r \, dr.$$

The program **Heat** calculates these coefficients; thus the last two formulas can be considered as the final result. For completeness we present the analytical result for a_{0m} —see library example 1 (Heat Conduction within a Circular Membrane):

$$a_{0m} = \frac{8 u_0}{\left(\mu_m^{(0)} \right)^3 J_1 \left(\mu_m^{(0)} \right)}.$$

Collecting the above results, we find that the distribution of temperature within the cylinder is given by the series in Bessel functions of zeroth order:

$$u(r, \varphi, t) = 8 u_0 \sum_{m=0}^{\infty} \frac{e^{-a^2 \lambda_{0m} t}}{\left(\mu_m^{(0)} \right)^3 J_1 \left(\mu_m^{(0)} \right)} J_0 \left(\frac{\mu_m^{(0)}}{l} r \right).$$

On examination of the solution we see that, due to the exponential nature of the coefficients, the final temperature of the cylinder after a long time will be zero. This is due to dissipation of energy to the surrounding space and could have been anticipated from the physical configuration of the problem.

Figure 13.10 shows two snapshots of the solution at the times $t=0$ and $t=3$. This solution was obtained with the program **Heat** for the case when $a^2 = 0.25$, $l=2$, and $u_0 = 100$.

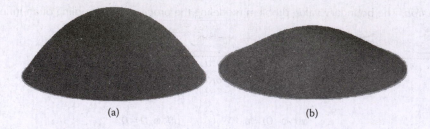

FIGURE 13.10 Plate temperature at (a) $t = 0$ and (b) $t = 3$ for Example 6.

Example 7

The initial temperature distribution within a very long (infinite) cylinder of radius l is

$$u(r, \varphi, 0) = u_0 = \text{const}.$$

Find the distribution of temperature within the cylinder if it is subjected to convective heat transfer according to Newton's law at its surface and the temperature of the medium is zero.

Solution. The boundary value problem modeling the process of the cooling of an infinite cylinder is

$$\frac{\partial u}{\partial t} = a^2 \left[\frac{\partial^2 u}{\partial r^2} + \frac{1}{r} \frac{\partial u}{\partial r} \right],$$

$$u(r, \varphi, 0) = u_0,$$

$$\frac{\partial u}{\partial r}(l, \varphi, t) + hu(l, \varphi, t) = 0.$$

The boundary condition of the problem is of *mixed type* so eigenvalues of problem $\lambda_{nm} = (\mu_m^{(n)}/l)^2$ are given by the roots of the equation

$$\mu J_n'(\mu) + hl J_n(\mu) = 0.$$

The eigenfunctions are

$$V_{nm}^{(1)} = J_n\left(\frac{\mu_m^{(n)}}{l} r \right) \cos n\varphi, \quad V_{nm}^{(2)} = J_n\left(\frac{\mu_m^{(n)}}{l} r \right) \sin n\varphi.$$

The coefficients a_{nm} and b_{nm} are given by equations (13.81) and (13.82):

$$a_{nm} = \frac{u_0}{|| V_{nm}^{(1)} ||^2} \int_0^{2\pi} \cos n\varphi \, d\varphi \int_0^l r J_n\left(\frac{\mu_m^{(n)}}{l} r \right) dr,$$

$$b_{nm} = \frac{u_0}{|| V_{nm}^{(2)} ||^2} \int_0^{2\pi} \sin n\varphi \, d\varphi \int_0^l r J_n\left(\frac{\mu_m^{(n)}}{l} r \right) dr.$$

(a)   (b)

FIGURE 13.11   Plate temperature at (a) $t = 0$ and (b) $t = 5$ for Example 7.

The initial temperature does not depend on the polar angle φ; thus, only terms with $n = 0$ are not zero. Obviously $b_{nm} = 0$ for all n. Coefficient a_{0m} can be obtained with the program **Heat**. For completeness show how to calculate it analytically. First, find

$$\| V_{0m}^{(1)} \|^2 = \pi l^2 \frac{l^2 h^2 + \left(\mu_m^{(0)}\right)^2}{l^2 h^2} \left[J_0'\left(\mu_m^{(0)}\right)\right]^2,$$

or, taking into account the relation $\mu J_0'(\mu) + h l J_0(\mu) = 0$, we may write

$$\| V_{0m}^{(1)} \|^2 = \pi \left[l^2 h^2 + \left(\mu_m^{(0)}\right)^2\right] \frac{l^2}{\left(\mu_m^{(0)}\right)^2} J_0^2\left(\mu_m^{(0)}\right).$$

From this we have

$$a_{0m} = \frac{2 u_0 \mu_m^{(0)}}{\left[\left(\mu_m^{(0)}\right)^2 + h^2 l^2\right] J_0^2\left(\mu_m^{(0)}\right)} J_1\left(\mu_m^{(0)}\right).$$

Using the above relations we may write the distribution of temperature within the cylinder as the series in Bessel functions of zero order:

$$u(r,\varphi,t) = 2 u_0 \sum_{m=0}^{\infty} \frac{\mu_m^{(0)} J_1\left(\mu_m^{(0)}\right) e^{-a^2 \lambda_{0m} t}}{\left[\left(\mu_m^{(0)}\right)^2 + h^2 l^2\right] J_0^2\left(\mu_m^{(0)}\right)} J_0\left(\frac{\mu_m^{(0)}}{l} r\right).$$

Figure 13.11 shows two snapshots of the solution at the times $t = 0$ and $t = 5$. This solution was obtained with the program **Heat** for the case when $a^2 = 0.25$, $l = 2$, $u_0 = 100$, and $h = 2$. As in the previous example, dissipation of energy to the environment brings the final temperature of the plate to zero after very long time periods.

13.2.2 The Fourier Method for the Nonhomogeneous Heat Conduction Equation

The solution $u_2(r,\varphi,t)$ represents the *non-free heat exchange* within the plate—that is, the diffusion of heat due to generation (or absorption) of heat by internal sources when the initial distribution of temperature is zero. The function $u_2(r,\varphi,t)$ is, thus, the solution of the *nonhomogeneous* equation

$$\frac{\partial u_2}{\partial t} = a^2 \left(\frac{\partial^2 u_2}{\partial r^2} + \frac{1}{r} \frac{\partial u_2}{\partial r} + \frac{1}{r^2} \frac{\partial^2 u_2}{\partial \varphi^2} \right) - \gamma u_2 + f(r,\varphi,t) \tag{13.83}$$

with *initial and boundary conditions equal to zero*.

After the separation of variables the general solution to this equation clearly is

$$u_2(r,\varphi,t) = \sum_{n=0}^{\infty}\sum_{m=0}^{\infty}\left[T_{nm}^{(1)}(t)V_{nm}^{(1)}(r,\varphi) + T_{nm}^{(2)}(t)V_{nm}^{(2)}(r,\varphi)\right], \tag{13.84}$$

where $V_{nm}^{(1)}(r,\varphi)$ and $V_{nm}^{(2)}(r,\varphi)$ are eigenfunctions of the corresponding homogeneous boundary value problem and $T_{nm}^{(1)}(t)$ and $T_{nm}^{(2)}(t)$ are unknown functions of t.

Reading Exercise

Provide the steps leading to equation (13.84).

Zero boundary conditions for $u_2(r,\varphi,t)$ may be expressed as

$$P[u_2]_{r=1} \equiv \alpha\frac{\partial u_2}{\partial r} + \beta_{u_2}\bigg|_{r=l} = 0$$

and are valid for any choice of functions $T_{nm}^{(1)}(t)$ and $T_{nm}^{(2)}(t)$ because they are valid for the functions $V_{nm}^{(1)}(r,\varphi)$ and $V_{nm}^{(2)}(r,\varphi)$.

Substitution of the series in equation (13.84) into the equation (13.83) allows one to obtain the equation

$$\sum_{n=0}^{\infty}\sum_{m=0}^{\infty}\left[\frac{dT_{nm}^{(1)}}{dt} + (a^2\lambda_{nm} + \gamma)T_{nm}^{(1)}(t)\right]V_{nm}^{(1)}(r,\varphi)$$

$$+ \sum_{n=0}^{\infty}\sum_{m=0}^{\infty}\left[\frac{dT_{nm}^{(2)}}{dt} + (a^2\lambda_{nm} + \gamma)T_{nm}^{(2)}(t)\right]V_{nm}^{(2)}(r,\varphi) = f(r,\varphi,t).$$

Next expand function $f(r,\varphi,t)$ within a circle of radius l as

$$f(r,\varphi,t) = \sum_{n=0}^{\infty}\sum_{m=0}^{\infty}[f_{nm}^{(1)}(t)V_{nm}^{(1)}(r,\varphi) + f_{nm}^{(2)}(t)V_{nm}^{(2)}(r,\varphi)]$$

where, using the orthogonality of this set of functions, the coefficients are

$$f_{nm}^{(1)}(t) = \frac{1}{\|V_{nm}^{(1)}\|^2}\int_0^l\int_0^{2\pi}f(r,\varphi,t)\,V_{nm}^{(1)}(r,\varphi)\,r\,dr\,d\varphi, \tag{13.85}$$

$$f_{nm}^{(2)}(t) = \frac{1}{\|V_{nm}^{(1)}\|^2}\int_0^l\int_0^{2\pi}f(r,\varphi,t)\,V_{nm}^{(2)}(r,\varphi)\,r\,dr\,d\varphi. \tag{13.86}$$

Comparing the expansions, we obtain differential equations for determining the functions $T_{nm}^{(1)}(t)$ and $T_{nm}^{(2)}(t)$:

$$\frac{dT_{nm}^{(1)}}{dt} + (a^2\lambda_{nm} + \gamma)T_{nm}^{(1)}(t) = f_{nm}^{(1)}(t),$$

$$\frac{dT_{nm}^{(2)}}{dt} + (a^2\lambda_{nm} + \gamma)T_{nm}^{(2)}(t) = f_{nm}^{(2)}(t).$$

(13.87)

In addition, these functions are necessarily subject to the initial conditions:

$$T_{nm}^{(1)}(0) = 0 \quad \text{and} \quad T_{nm}^{(2)}(0) = 0.$$

(13.88)

Solving the ordinary differential equations (13.87) with initial condition (13.88), we can present the functions $T_{nm}^{(1)}(t)$ and $T_{nm}^{(2)}(t)$ (as we did in Chapter 12) in the form of integral relations given by

$$T_{nm}^{(1)}(t) = \int_0^t f_{nm}^{(1)}(\tau)\, e^{-(a^2\lambda_{nm} + \gamma)(t-\tau)}\, d\tau,$$

(13.89)

$$T_{nm}^{(2)}(t) = \int_0^t f_{nm}^{(2)}(\tau)\, e^{-(a^2\lambda_{nm} + \gamma)(t-\tau)}\, d\tau.$$

(13.90)

Thus we have that the solution of the nonhomogeneous equation with zero initial and boundary conditions can be written as

$$u(x, y, t) = u_1(x, y, t) + u_2(x, y, t)$$

$$= \sum_{n=0}^{\infty}\sum_{m=0}^{\infty}\left\{\left[T_{nm}^{(1)}(t) + a_{nm}e^{-(a^2\lambda_{nm}+\gamma)t}\right]V_{nm}^{(1)}(x, y)\right.$$

(13.91)

$$\left. + \left[T_{nm}^{(2)}(t) + b_{nm}e^{-(a^2\lambda_{nm}+\gamma)t}\right]V_{nm}^{(2)}(x, y)\right\},$$

where functions $T_{nm}^{(1)}(t)$ and $T_{nm}^{(2)}(t)$ are defined by equations (13.89) and (13.90) and coefficients a_{nm} and b_{nm} were found in previous section (formulas (13.81), (13.82)).

Example 8

Find the temperature within a thin circular plate of radius l if its boundary is kept at constant zero temperature, the initial temperature distribution within the plate is zero, and one internal source of heat acts at the point (r_0, φ_0) of the plate where ($0 \le r_0 < l$, $0 \le \varphi_0 < 2\pi$). Assume that the value of this source is

$$Q(t) = A\sin\omega t.$$

The plate is thermally insulated over its lateral faces.

Solution. The problem is expressed as

$$\frac{\partial u}{\partial t} = a^2 \left[\frac{\partial^2 u}{\partial r^2} + \frac{1}{r}\frac{\partial u}{\partial r} \right] + \frac{A}{c\rho}\delta(r - r_0)\delta(\varphi - \varphi_0)\sin\omega t$$

under the conditions

$$u(r,\varphi,0) = 0, \ u(l,\varphi,t) = 0.$$

The boundary condition of the problem is of *Dirichlet type*, so eigenvalues are given by the equation $J_n(\mu) = 0$, the eigenfunctions are given by formulas (13.77), and the eigenfunctions' squared norms can be calculated using equation (11.86) and (11.87) from Chapter 11:

$$\left\|V_{nm}^{(1)}\right\|^2 = \left\|V_{nm}^{(2)}\right\|^2 = \varepsilon_n \pi \frac{l^2}{2}\left[J_n'\left(\mu_m^{(n)}\right)\right]^2, \quad \varepsilon_n = \begin{cases} 2, & \text{if } n = 0, \\ 1, & \text{if } n > 0. \end{cases}$$

The initial temperature of the plate is zero, so we have $a_{nm} = 0$ and $b_{nm} = 0$ for all n, m.
Let us next find $f_{nm}^{(1)}(t)$ and $f_{nm}^{(2)}(t)$:

$$f_{nm}^{(1)}(t) = \frac{A}{c\rho}\frac{2}{\sigma_n \pi l^2 \left[J_n'\left(\mu_m^{(n)}\right)\right]^2}\sin\omega t \cos n\varphi_0 J_n\left(\frac{\mu_m^{(n)}}{l}r_0\right),$$

$$f_{nm}^{(2)}(t) = \frac{A}{c\rho}\frac{2}{\sigma_n \pi l^2 \left[J_n'\left(\mu_m^{(n)}\right)\right]^2}\sin\omega t \sin n\varphi_0 J_n\left(\frac{\mu_m^{(n)}}{l}r_0\right).$$

From this we have

$$T_{nm}^{(1)}(t) = \frac{A}{c\rho}\frac{2}{\sigma_n \pi l^2 \left[J_n'\left(\mu_m^{(n)}\right)\right]^2}\cos n\varphi_0 J_n\left(\frac{\mu_m^{(n)}}{l}r_0\right)I(t),$$

$$T_{nm}^{(2)}(t) = \frac{A}{c\rho}\frac{2}{\sigma_n \pi l^2 \left[J_n'\left(\mu_m^{(n)}\right)\right]^2}\sin n\varphi_0 J_n\left(\frac{\mu_m^{(n)}}{l}r_0\right)I(t),$$

where we have introduced

$$I(t) = \frac{1}{\left[\omega^2 + \left(a^2\lambda_{nm}\right)^2\right]}\left\{a^2\lambda_{nm}\sin\omega t - \omega\cos\omega t + \omega e^{-\lambda_{nm}a^2 t}\right\}.$$

Therefore the evolution of temperature within the plate is described by series

$$u(r,\varphi,t) = \sum_{n=0}^{\infty}\sum_{m=0}^{\infty}\left[T_{nm}^{(1)}\cos(n\varphi) + T_{nm}^{(2)}\sin(n\varphi)\right]J_n\left(\frac{\mu_m^{(n)}}{l}r\right)$$

$$= \frac{2A}{c\rho\pi l^2}\sum_{n=0}^{\infty}\sum_{m=0}^{\infty}\frac{I(t)}{\sigma_n\left[J_n'\left(\mu_m^{(n)}\right)\right]^2}J_n\left(\frac{\mu_m^{(n)}}{l}r_0\right)J_n\left(\frac{\mu_m^{(n)}}{l}r\right)\cos n(\varphi - \varphi_0).$$

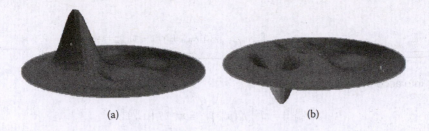

(a)   (b)

FIGURE 13.12   Temperature of plate at (a) $t = 0.5$ and (b) $t = 1$ for Example 8.

Figure 13.12 shows two snapshots of the solution at the times $t = 0.5$ and $t = 1$. This solution was obtained with the program **Heat** for the case when $a^2 = 0.25$, $l=2$, $r_0 = 1$, $\varphi_0 = 1$, $A/c\rho=100$, and $\omega = 5$.

Reading Exercise

Discuss the above result for the equilibrium (final) state when the source of heat is placed in the center of the plate at $r_0 = 0$. Show that in this case the problem can be reduced to a one-dimensional ordinary differential equation.

Example 9

A heat-conducting, thin, uniform, circular plate of radius l is thermally insulated over its lateral faces. The boundary of the plate is kept at constant zero temperature; the initial temperature distribution within the plate is zero. Let heat be generated throughout the plate with the intensity of internal sources (per unit mass of the membrane) given by

$$Q(t) = A\cos\omega t.$$

Find the distribution of temperature within the plate when $t > 0$.
   Solution. The problem may be expressed by the equation

$$\frac{\partial u}{\partial t} = a^2\left[\frac{\partial^2 u}{\partial r^2} + \frac{1}{r}\frac{\partial u}{\partial r}\right] + A\cos\omega t,$$

under the conditions

$$u(r,\varphi,0) = 0, \; u(l,\varphi,t) = 0.$$

The boundary condition of the problem is of *Dirichlet type*, so eigenvalues $\mu_m^{(n)}$ are given by equation $J_n(\mu) = 0$. The initial temperature of the plate is zero and the intensity of internal sources depends only on time t, so the solution $u(r,t)$ includes only functions $V_{0m}^{(1)} = J_0\left(\mu_m^{(0)}r/l\right)$ and is given by the series

$$u(r,t) = \sum_{m=0}^{\infty} T_{0m}^{(1)}(t)J_0\left(\frac{\mu_m^{(n)}}{l}r\right),$$

where

$$T_{0m}^{(1)}(t) = \int_0^t f_{0m}^{(1)}(\tau)\, e^{-(a^2\lambda_{0m} + \gamma)(t - \tau)}\, d\tau.$$

Taking into account that

$$\left\| V_{0m}^{(1)} \right\|^2 = \pi l^2 \left[J_0'\left(\mu_m^{(0)}\right) \right]^2 = \pi l^2 \left[J_1\left(\mu_m^{(0)}\right) \right]^2,$$

$$\int_0^l J_0\left(\frac{\mu_m^{(0)}}{l} r\right) r\, dr = \frac{l^2}{\left[\mu_m^{(0)}\right]^2} \int_0^{\mu_m^{(0)}} x J_0(x)\, dx = \frac{l^2}{\mu_m^{(0)}} J_1\left(\mu_m^{(0)}\right),$$

we find $f_{0m}^{(1)}(t)$ in the form

$$f_{0m}^{(1)}(t) = \frac{A\cos\omega t}{\left\| V_{0m}^{(1)} \right\|^2}\, 2\pi \int_0^l J_0\left(\frac{\mu_m^{(0)}}{l} r\right) r\, dr = \frac{2A}{\mu_m^{(0)} J_1\left(\mu_m^{(0)}\right)}\cos\omega t.$$

Then

$$T_{0m}^{(1)}(t) = \frac{2A}{\mu_m^{(0)} J_1\left(\mu_m^{(0)}\right)} \int_0^t \cos\omega\tau \cdot e^{-a^2\lambda_{0m}(t - \tau)}\, d\tau$$

$$= \frac{2A}{\mu_m^{(0)} J_1\left(\mu_m^{(0)}\right)\left[\omega^2 + \left(a^2\lambda_{0m}\right)^2\right]} \left[a^2\lambda_{0m}\cos\omega t + \omega\sin\omega t - a^2\lambda_{0m}e^{-a^2\lambda_{0m}t} \right].$$

Therefore the evolution of temperature within the plate is described by series

$$u(r,t) = \sum_{m=0}^{\infty} T_{0m}^{(1)}(t) J_0\left(\frac{\mu_m^{(n)}}{l} r\right).$$

Figure 13.13 shows two snapshots of the solution at the times $t = 1$ and $t = 3$. This solution was obtained with the program **Heat** for the case when $a^2 = 0.25$, $l = 2$, $A = 100$, and $\omega = 5$.

Reading Exercise

Discuss the above result for the equilibrium (final) state. Show that in this case the problem can be reduced to a one-dimensional ordinary differential equation.

(a)   (b)

FIGURE 13.13   Temperature of plate at (a) $t = 1$ and (b) $t = 3$ for Example 9.

13.2.3 The Fourier Method for the Nonhomogeneous Heat Conduction Equation with Nonhomogeneous Boundary Conditions

Finally we consider the general boundary problem for the heat conduction, equation (13.61), given by

$$\frac{\partial u}{\partial t} = a^2 \left(\frac{\partial^2 u}{\partial r^2} + \frac{1}{r} \frac{\partial u}{\partial r} + \frac{1}{r^2} \frac{\partial^2 u}{\partial \varphi^2} \right) - \gamma u + f(r, \varphi, t),$$

with nonhomogeneous initial condition (13.62) and boundary condition (13.63) given by

$$u(r, \varphi, 0) = \phi(r, \varphi),$$

$$P[u]_{r=l} \equiv \alpha \frac{\partial u}{\partial r} + \beta u \bigg|_{r=l} = g(\varphi, t).$$

To reduce the problem with nonhomogeneous boundary condition to the problem with homogeneous boundary condition, we introduce an auxiliary function $w(r, \varphi, t)$ that satisfies the given nonhomogeneous boundary condition. As always, we will search for the solution of the problem as

$$u(r, \varphi, t) = v(r, \varphi, t) + w(r, \varphi, t),$$

where $v(r, \varphi, t)$ is a function satisfying the homogeneous boundary condition, and will seek the auxiliary function $w(r, \varphi, t)$ in the form

$$w(r, \varphi, t) = (c_0 + c_1 r + c_2 r^2) g(\varphi, t).$$

The constants will be adjusted to satisfy the boundary condition. Because

$$\frac{1}{r} \frac{\partial w}{\partial r}(r, \varphi, t) = \left(\frac{c_1}{r} + 2c_2 \right) g(\varphi, t)$$

and $r=0$ is a regular point, the coefficient $c_1 \equiv 0$ and the auxiliary function reduces to

$$w(r, \varphi, t) = (c_0 + c_2 r^2) g(\varphi, t),$$

where c_0 and c_2 are real constants.

Below we present auxiliary functions for different types of boundary conditions.

Case I. For $\alpha = 0$ and and

a) boundary condition $u(l, \varphi, t) = g(\varphi, t)$ we have the auxiliary function $w(r, \varphi, t) = (r^2 / l^2) g(\varphi, t)$;

b) boundary condition $u(l, \varphi, t) = g(t)$ or $u(l, \varphi, t) = g_0 = \text{const}$. we have the auxiliary function $w(r, \varphi, t) = g(t)$ or $w(r, \varphi, t) = g_0$.

Case II. For $\alpha = 1$ and $\beta = 0$ the boundary condition $\partial u/\partial r(l,\varphi,t) = g(\varphi,t)$ we have the auxiliary function $w(r,\varphi,t) = (r^2/2l)g(\varphi,t) + C$, where C is an arbitrary constant.

Case III. For $\alpha = 1$ and $\beta = h$ the boundary condition is $\partial u/\partial r(l,\varphi,t) + h\, u(l,\varphi,t) = g(\varphi,t)$ and we have the auxiliary function $w(r,\varphi,t) = [r^2/l(2+lh)]g(\varphi,t)$.

It is easy to verify (we leave it to the reader as a Reading Exercise) that such auxiliary functions $w(r,\varphi,t)$ satisfy the given boundary condition

$$P[w] \equiv \alpha \frac{\partial w}{\partial r}(r,\varphi,t) + \beta w(r,\varphi,t)\Big|_{r=l} = g(\varphi,t).$$

Example 10

The initial temperature distribution within a very long (infinite) cylinder of radius l is

$$u(r,\varphi,0) = u_0 = \text{const}.$$

Find the distribution of temperature within the cylinder if a constant heat flow,

$$\frac{\partial u}{\partial r}(l,\varphi,t) = Q = \frac{q}{\kappa},$$

is supplied to the surface of the cylinder from the outside starting at time $t = 0$. Generation (or absorption) of heat by internal sources is absent.

Solution. The problem depends on the solution of the equation

$$\frac{\partial u}{\partial t} = a^2 \left[\frac{\partial^2 u}{\partial r^2} + \frac{1}{r}\frac{\partial u}{\partial r} \right]$$

under the conditions

$$u(r,\varphi,0) = u_0, \quad \frac{\partial u}{\partial r}(l,\varphi,t) = Q = \frac{q}{\kappa}.$$

An auxiliary function satisfying the given boundary condition is

$$w(r,\varphi,t) = \frac{Q}{2l}r^2.$$

The solution to the problem should be sought in the form

$$u(r,\varphi,t) = w(r,\varphi,t) + v(r,\varphi,t)$$

where the function $v(r,\varphi,t)$ is the solution to the boundary value problem for

$$f^*(r,\varphi,t) = \frac{2a^2}{l}Q, \quad \phi^*(r,\varphi) = u_0 - \frac{Q}{2l}r^2.$$

These functions do not depend on the polar angle φ, which is why the solution for function $v(r,t)$ contains only Bessel function of zeroth order:

$$v(r,t) = \sum_{m=0}^{\infty} \left[T_{0m}^{(1)}(t) + a_{0m}e^{-a^2\lambda_{0m}t} \right] J_0\left(\frac{\mu_m^{(0)}}{l}r \right).$$

The boundary condition of the problem is of *Neumann type*, so eigenvalues $\mu_m^{(n)}$ are given by the equation

$$J_n'(\mu) = 0.$$

The eigenfunctions and their norms are

$$V_{0m}^{(1)} = J_0\left(\frac{\mu_m^{(0)}}{l}r \right), \quad \left\| V_{0m}^{(1)} \right\|^2 = \pi\, l^2 J_0^2\left(\mu_m^{(0)} \right), \quad \left\| V_{00}^{(1)} \right\|^2 = \pi l^2,$$

in which case we have

$$a_{0m} = \frac{1}{\left\| V_{0m}^{(1)} \right\|^2} \int_0^{2\pi}\int_0^l \left(u_0 - \frac{Q}{2l}r^2 \right) J_0\left(\frac{\mu_m^{(0)}}{l}r \right) r\,dr\,d\varphi = -\frac{2Ql}{\left(\mu_m^{(0)} \right)^2 J_0\left(\mu_m^{(0)} \right)},$$

$$a_{00} = \frac{1}{\left\| V_{00}^{(1)} \right\|^2} \int_0^{2\pi}\int_0^l \left(u_0 - \frac{Q}{2l}r^2 \right) r\,dr\,d\varphi = u_0 - \frac{Ql}{4}.$$

Applying the equation (13.85) we obtain

$$f_{0m}^{(1)}(t) = \frac{2a^2Q}{l\left\| V_{0m}^{(1)} \right\|^2} \int_0^{2\pi} d\varphi \int_0^l J_0\left(\frac{\mu_m^{(0)}}{l}r \right) r\,dr = \frac{4a^2Q}{l\mu_m^{(0)}J_0^2\left(\mu_m^{(0)} \right)} J_1\left(\mu_m^{(0)} \right) = 0,$$

$$f_{00}^{(1)}(t) = \frac{2a^2Q}{l\left\| V_{00}^{(1)} \right\|^2} \int_0^{2\pi} d\varphi \int_0^l r\,dr = \frac{2a^2Q}{l},$$

and

$$T_{0m}^{(1)}(t) = \int_0^t f_{0m}^{(1)}(\tau)e^{-a^2\lambda_{0m}(t-\tau)}\,d\tau = 0, \quad T_{00}^{(1)}(t) = \int_0^t f_{00}^{(1)}(\tau)\,d\tau = \frac{2a^2Q}{l}t.$$

Hence, the distribution of temperature within the cylinder at some instant of time is described by the series (see formula (13.91))

$$u(r,t) = \frac{Q}{2l}r^2 + u_0 - \frac{Ql}{4} + \frac{2a^2Q}{l}t - 2Ql \sum_{m=1}^{\infty} \frac{e^{-a^2\lambda_{0m}t}}{\left(\mu_m^{(0)} \right)^2 J_0\left(\mu_m^{(0)} \right)} J_0\left(\frac{\mu_m^{(0)}}{l}r \right).$$

Figure 13.14 shows two snapshots of the solution at the times $t=0$ and $t=3$. This solution was obtained with the program **Heat** for the case when $a^2 = 0.25$, $l=2$, $u_0 = 10$, and $Q = 2$.

(a)   (b)

FIGURE 13.14   Plate temperature at (a) $t = 0$ and (b) $t = 3$ for Example 10.

Reading Exercise

Discuss the role and the origin of each term in the solution to Example 10.

Problems

In problems 1 through 21 we consider a heat conduction within a *rectangular* membrane ($0 \leq x \leq l_x$, $0 \leq y \leq l_y$), in problems 22 through 42 within a *circular* membrane of radius l. Solve these problems analytically, which means the following: formulate the equation and initial and boundary conditions, obtain the eigenvalues and eigenfunctions, and write the formulas for coefficients of the series expansion and the expression for the solution of the problem. If the integrals in the coefficients are not easy to evaluate, leave it for the program, which evaluates them numerically. Then study the problem in detail with the program **Heat**. Obtain the pictures of several eigenfunctions and screenshots of solution in different moments of time. Directions on how to use the program are found in Appendix 5.

If the boundary conditions for a rectangular domain do not satisfy consistency conditions, to find the auxiliary function w formulas from section 13.1.3 (Case II) and formulas from Heat Help System (section "*Heat Conduction within a Thin Uniform Rectangular Membrane*," "*Auxiliary Functions*") should be used.

In problems 1 through 5 we consider rectangular membranes that are thermally insulated over their lateral surfaces. In the initial time, $t=0$, the temperature distribution is given by $u(x,y,0)=\varphi(x,y)$. There are no heat sources or absorbers inside the membrane. Find the distribution of temperature within the membrane at any later time.

Simulate the process of membrane cooling with the **Heat** program for some value of parameter a^2. Vary the coefficient of thermal diffusivity a^2 in such a way that the cooling time decreased (with a precision of 5%) (1) twice and (2) four times. Also investigate changing the size of the membrane determined by l_x and l_y under fixed thermal diffusivity a^2.

1. The boundary of the membrane is kept at constant zero temperature. The initial temperature of the membrane is given as

$$\varphi(x,y) = A \sin\frac{\pi x}{l_x} \sin\frac{\pi y}{l_y}.$$

2. The edge at $x=0$ of membrane is thermally insulated and other edges are kept at zero temperature. The initial temperature of the membrane is given as

$$\varphi(x, y) = A\cos\frac{\pi x}{2l_x}\sin\frac{\pi y}{l_y}.$$

3. The edge at $y = 0$ of the membrane is thermally insulated and other edges are kept at zero temperature. The initial temperature of the membrane is given as

$$\phi(x, y) = A\sin\frac{\pi x}{l_x}\cos\frac{\pi y}{2l_y}.$$

4. The edges $x = 0$ and $y = 0$ of the membrane are kept at zero temperature, and the edges $x = l_x$ and $y = l_y$ are thermally insulated. The initial temperature of the membrane is given as

$$\varphi(x, y) = Axy(l_x - x)(l_y - y).$$

5. The edges $x = 0$, $x = l_x$, and $y = l_y$ of membrane are kept at zero temperature, and the edge $y = 0$ is subjected to convective heat transfer with the environment, which has a temperature of zero. The initial temperature of the membrane is given as

$$\varphi(x, y) = Axy(l_x - x)(l_y - y).$$

In problems 6 through 10 we consider rectangular plate ($0 \le x \le l_x$, $0 \le y \le l_y$), which is thermally insulated over its lateral surfaces. The initial temperature distribution within the plate is zero, and one internal source of heat acts at the point (x_0, y_0) of the plate. The value of this source is $Q(t)$. Find the temperature within the plate.

6. The edges $x = 0$, $y = 0$, and $y = l_y$ of the plate are kept at zero temperature and the edge $x = l_x$ is subjected to convective heat transfer with the environment, which has a temperature of zero. The value of the internal source is

$$Q(t) = A\cos\omega t.$$

7. The edges $x = 0$ and $y=0$ of the plate are kept at zero temperature, the edge $y = l_y$ is thermally insulated, and the edge $x = l_x$ is subjected to convective heat transfer with the environment, which has a temperature of zero. The value of the internal source is

$$Q(t) = A\sin\omega t.$$

8. The edges $x = l_x$ and $y = l_y$ of the plate are kept at zero temperature, the edge $x=0$ is thermally insulated, and the edge $y = 0$ is subjected to convective heat transfer with the environment, which has a temperature of zero. The value of the internal source is

$$Q(t) = A(\sin\omega t + \cos\omega t).$$

9. The edges $y = 0$ and $y = l_y$ are thermally insulated, the edge $x=0$ is kept at zero temperature, and the edge $x = l_x$ is subjected to convective heat transfer with the environment, which has a temperature of zero. The value of the internal source is

$$Q(t) = Ae^{-t} \sin \omega t.$$

10. The edges $x = 0$ and $y = l_y$ are thermally insulated, the edge $x = l_x$ is kept at zero temperature, and the edge $y = 0$ is subjected to convective heat transfer with the environment, which has a temperature of zero. The value of the internal source is

$$Q(t) = Ae^{-t} \cos \omega t.$$

11. Find the heat distribution in a thin rectangular plate, $0 \le x \le l_x$, $0 \le y \le l_y$, if it is subjected to heat transfer according to Newton's law at its edges. The temperature of the medium is $u_{md} = $ const, the initial temperature of the plate is zero, and there is a constant source of heat, Q, uniformly distributed over the plate.

Hint: The problem is formulated as follows:

$$\frac{\partial u}{\partial t} = a^2 \left[\frac{\partial^2 u}{\partial x^2} + \frac{\partial^2 u}{\partial y^2} \right] + Q, \ 0 < x < l_x, \ 0 < y < l_y, \ t > 0,$$

$$u(x, y, 0) = 0,$$

$$\frac{\partial u}{\partial x} - h(u - u_{md}) \bigg|_{x=0} = 0, \quad \frac{\partial u}{\partial x} + h(u - u_{md}) \bigg|_{x=l_x} = 0,$$

$$\frac{\partial u}{\partial y} - h(u - u_{md}) \bigg|_{y=0} = 0, \quad \frac{\partial u}{\partial y} + h(u - u_{md}) \bigg|_{y=l_y} = 0.$$

12. An infinitely long rectangular cylinder, $0 \le x \le l_x$, $0 \le y \le l_y$, with the central axis along the z-axis is placed in a coil. At $t=0$ a current in the coil turns on and the coil starts to generate an oscillation magnetic field outside the cylinder directed along the z-axis:

$$u(x, y, t) = H_0 \sin \omega t, \quad H_0 = \text{const}, \quad 0 < t < \infty.$$

Find the magnetic field inside the cylinder.

Hint: The problem is formulated as follows:

$$\frac{\partial u}{\partial t} = a^2 \left[\frac{\partial^2 u}{\partial x^2} + \frac{\partial^2 u}{\partial y^2} \right], 0 < x < l_x, 0 < y < l_y, t > 0,$$

$$u(x,y,0) = 0,$$

$$u|_{x=0} = u|_{x=l_x} = u|_{y=0} = u|_{y=l_y} = H_0 \sin \omega t.$$

For problems 13 through 15 consider a rectangular plate $(0 \le x \le l_x, 0 \le y \le l_y)$ that is thermally insulated over its lateral surfaces. The edges of the plate are kept at the temperatures described by the function of $u(x,y,t)|_\Gamma$, given below. The initial temperature distribution within the plate is $u(x,y,0)=u_0=$ const. Find heat distribution in the plate if there are no heat sources or absorbers inside the plate.

13. $u|_{x=0}=u|_{y=0}=u_1,\quad u|_{x=l_x}=u|_{y=l_y}=u_2$

14. $u|_{x=0}=u|_{x=l_x}=u_1,\quad u|_{y=0}=u|_{y=l_y}=u_2$

15. $u|_{x=0}=u|_{x=l_x}=u|_{y=0}=u_1,\quad u|_{y=l_y}=u_2$

For problems 16 through 18 a thin homogeneous plate with sides of length π lies in the x-y plane. The edges of the plate are kept at the temperatures described by the function of $u(x,y,t)|_\Gamma$, given below. Find the temperature in the plate if initially the temperature has a constant value A and there are no heat sources or absorbers inside the plate.

16. $u|_{x=0}=u|_{x=\pi}=0,\quad u|_{y=0}=u|_{y=\pi}=x^2-x$

17. $u|_{x=0}=u|_{x=\pi}=0,\quad u|_{y=0}=x^2,\quad u|_{y=\pi}=0.5x$

18. $u|_{x=0}=0,\quad u|_{x=\pi}=\cos y,\quad u|_{y=0}=u|_{y=\pi}=0$

In problems 19 through 21 an infinitely long rectangular cylinder has its central axis along the z-axis and its cross section is a rectangle with sides of length π. The sides of the cylinder are kept at the temperature described by functions $u(x,y,t)|_\Gamma$, given below. Find the temperature within the cylinder if initially the temperature is $u(x,y,0)=Axy$ and there are no heat sources or absorbers inside the cylinder.

19. $u|_{x=0}=3y^2,\quad u|_{x=\pi}=0,\quad u|_{y=0}=u|_{y=\pi}=0$

20. $u|_{x=0}=u|_{x=\pi}=y^2,\quad u|_{y=\pi}=0,\quad u|_{y=0}=x$

21. $u|_{x=0}=u|_{x=\pi}=\cos 2y,\quad u|_{y=0}=u|_{y=\pi}=0$

In problems 22 through 24 we consider circular membranes of radius l that are thermally insulated over their lateral surfaces. In the initial time, $t=0$, the temperature distribution is given by $u(r,\varphi,0)=\phi(r,\varphi)$. There are no heat sources or absorbers inside the membrane. Find the distribution of temperature within the membrane at any later time.

Simulate the process of membrane cooling with the **Heat** program for some value of parameter a^2. Vary the coefficient of thermal diffusivity a^2 in such a way that the cooling time decreased (with a precision of 5%) (1) twice and (2) four times. Also investigate changing the radius of the membrane under fixed thermal diffusivity a^2.

22. The boundary of the membrane is kept at constant zero temperature. The initial temperature of the membrane is given as

$$\phi(r,\varphi)=Ar(l^2-r^2)\sin\varphi.$$

23. The boundary of the membrane is thermally insulated. The initial temperature of the membrane is given as

$$\phi(r,\varphi) = u_0 r \cos 2\varphi.$$

24. The boundary of the membrane is subjected to convective heat transfer with the environment, which has a temperature of zero. The initial temperature of the membrane is given as

$$\phi(r,\varphi) = u_0(1 - r^2/l^2).$$

In problems 25 through 27 we consider a very long (infinite) cylinder of radius l. The initial temperature distribution within the cylinder is given by $u(r,\varphi,0)=\phi(r,\varphi)$. There are no heat sources or absorbers inside the cylinder. Find the distribution of temperature within the cylinder at any later time.

25. The surface of the cylinder is kept at constant temperature $u = u_0$. The initial temperature distribution within the cylinder is given by

$$\phi(r,\varphi) = Ar(l^2 - r^2)\sin\varphi.$$

26. The constant heat flow $\partial u/\partial r(l,\varphi,t)=Q$ is supplied to the surface of the cylinder from outside. The initial temperature distribution within the cylinder is given by

$$\phi(r,\varphi) = Ar\left(l - \frac{r}{2}\right)\sin 3\varphi.$$

27. The surface of the cylinder is subjected to convective heat transfer with the environment, which has a temperature $u = u_{md}$. The initial temperature distribution within the cylinder is given by

$$\phi(r,\varphi) = u_0 \sin 4\varphi.$$

In problems 28 through 30 we consider a circular plate of radius l that is thermally insulated over its lateral surfaces. The initial temperature distribution within the plate is zero, and one internal source of heat acts at the point (r_0,φ_0) of the plate. The value of this source is $Q(t)$. Find the temperature within the plate.

28. The edge of the plate is kept at zero temperature. The value of internal source is

$$Q(t) = A\cos\omega t.$$

29. The edge of the plate is thermally insulated. The value of internal source is

$$Q(t) = A\sin\omega t.$$

30. The edge of the plate is subjected to convective heat transfer with the environment, which has a temperature of zero. The value of internal source is

$$Q(t) = A(\sin \omega t + \cos \omega t).$$

In problems 31 through 36 we consider a circular membrane of radius l that is thermally insulated over its lateral surfaces. The initial temperature distribution within the membrane is zero. A heat is generated throughout the membrane; the intensity of internal sources (per unit mass of the membrane) is $Q(t)$. Find the temperature distribution within the membrane.

31. The edge of the membrane is kept at zero temperature. The intensity of the internal sources (per unit mass of the membrane) is

$$Q(t) = A\cos \omega t.$$

32. The edge of the membrane is kept at zero temperature. The intensity of the internal sources (per unit mass of the membrane) is

$$Q(t) = A(l-r)\sin \omega t.$$

33. The edge of the membrane is thermally insulated. The intensity of the internal sources (per unit mass of the membrane) is

$$Q(t) = A\sin \omega t.$$

34. The edge of the membrane is thermally insulated. The intensity of the internal sources (per unit mass of the membrane) is

$$Q(t) = A(l^2 - r^2)\sin\omega t.$$

35. The edge of the membrane is subjected to convective heat transfer with the environment, which has a temperature of zero. The intensity of the internal sources (per unit mass of the membrane) is

$$Q(t) = A(\sin \omega t + \cos \omega t).$$

36. The edge of the membrane is subjected to convective heat transfer with the environment, which has a temperature of zero. The intensity of the internal sources (per unit mass of the membrane) is

$$Q(t) = A(l-r)\cos \omega t.$$

In problems 37 through 39 we consider a very long (infinite) cylinder of radius l. At time $t = 0$ the constant magnetic field, parallel to the cylinder axis, is instantly established outside the cylinder. The external magnetic field strength is $H(t)$. Find the magnetic field strength within the cylinder, if the initial condition is zero.

37. $H(t) = H_0 \sin^2 \omega t$

38. $H(t) = H_0 \cos \omega t$

39. $H(t) = H_0(1 - \sin \omega t)$

In problems 40 through 42 we consider a very long (infinite) cylinder of radius l. The initial temperature of the cylinder is $u_0 = \text{const}$. Find the temperature distribution within the cylinder.

40. The surface of the cylinder is kept at the temperature described by the function $u(l, \varphi, t) = A \sin \omega t$.

41. The heat flow at the surface is governed by $\partial u / \partial r (l, \varphi, t) = A \cos \omega t$.

42. The temperature exchange with the environment with zero temperature is governed according to Newton's law $\partial u / \partial r + u|_{r=l} = A(1 - \cos \omega t)$.

CHAPTER 14

Elliptic Equations

14.1 ELLIPTIC DIFFERENTIAL EQUATIONS AND RELATED PHYSICAL PROBLEMS

When studying different stationary (time-independent) processes very often we meet *Laplace's equation*

$$\nabla^2 u = 0. \tag{14.1}$$

A function is said to be *harmonic* in a certain region if it satisfies Laplace's equation there and its first and second derivatives are continuous in that region. For example, the functions $2xy$, $x^2 - y^2$ and $e^{-x} \cos y$ satisfy the two-dimensional Laplace's equation

$$\frac{\partial^2 u}{\partial x^2} + \frac{\partial^2 u}{\partial y^2} = 0 \tag{14.2}$$

in the entire x, y-plane.

The nonhomogeneous equation

$$\nabla^2 u = -f \tag{14.3}$$

with a given function f of the coordinates is called *Poisson's equation*.

Laplace's and Poisson's partial differential equations are of *elliptic type*.

Let us present the Laplace's operator in different coordinate systems. Cylindrical coordinates r, φ, z are related with Cartesian coordinates as

$$x = r\cos\varphi, \quad y = r\sin\varphi, \quad z = z$$

the Laplacian is

$$\nabla^2 = \frac{1}{r}\frac{\partial}{\partial r}\left(r\frac{\partial}{\partial r}\right) + \frac{1}{r^2}\frac{\partial^2}{\partial \varphi^2} + \frac{\partial^2}{\partial z^2}. \tag{14.4}$$

For the particular case of cylindrical coordinates, a polar coordinate system r, φ with no dependence on z we have

$$x = r\cos\varphi, \quad y = r\sin\varphi$$

and the Laplacian is

$$\nabla^2 = \frac{1}{r}\frac{\partial}{\partial r}\left(r\frac{\partial}{\partial r}\right) + \frac{1}{r^2}\frac{\partial^2}{\partial \varphi^2}.$$

For spherical coordinates, r, ϑ, φ related with Cartesian coordinates as

$$x = r\sin\theta\cos\varphi, \quad y = r\sin\theta\sin\varphi, \quad z = r\cos\theta,$$

the Laplacian is

$$\nabla^2 = \frac{1}{r^2}\frac{\partial}{\partial r}\left(r^2\frac{\partial}{\partial r}\right) + \frac{1}{r^2\sin\theta}\frac{\partial}{\partial \theta}\left(\sin\theta\frac{\partial}{\partial \theta}\right) + \frac{1}{r^2\sin^2\theta}\frac{\partial^2}{\partial \varphi^2}. \tag{14.5}$$

Consider several physical problems.

1. If the heat distribution in a body is maintained in equilibrium, $\partial T/\partial t = 0$, the homogeneous heat equation (12.3), Chapter 12, reduces to Laplace's equation,

$$\nabla^2 T = 0. \tag{14.6}$$

If a medium contains a heat source or heat absorber, Q, and the temperature is time independent, the heat conduction equation (12.5), Chapter 12, becomes

$$\nabla^2 T = -\frac{Q}{\rho c\chi}, \tag{14.7}$$

which is a particular example of Poisson's equation (14.3).

2. The diffusion equation (12.9), Chapter 12, for stationary diffusion, $\partial q/\partial t = 0$, becomes

$$\nabla^2 q = -f/D, \tag{14.8}$$

where q is the concentration of a material diffusing through a medium, D is the diffusion coefficient, and f is a source or sink of the diffusing material; it also is Poisson's equation (or Laplace's equation when $f = 0$).

3. The electrostatic potential due to a point charge q is

$$\phi = \frac{q}{r},$$

(14.9)

where r is the distance from the charge to the point where the electric field is measured.

For continuous charge distribution with charge density ρ, the potential ϕ is related with ρ:

$$\nabla^2 \phi = -4\pi\rho.$$

(14.10)

Equation (14.10) is Poisson's equation for an electrostatic potential. In regions that do not contain electric charges—for instance, at points outside of a charged body—$\rho = 0$ and the potential that the body creates obeys Laplace's equation

$$\nabla^2 \phi = 0.$$

(14.11)

Reading Exercise:

Prove that if r is the distance from a fixed point, the scalar potential $\phi = q/r$ ($r \neq 0$) satisfies Laplace's equation (for simplicity let charge q be placed at the origin).

Hints:

i) Try a "physical" proof, that is, ϕ is the potential from a point charge; thus it should be a solution of the equation describing an electric potential for this case.

ii) Find $\nabla^2(r^{-1})$ in Cartesian coordinates.

$$r = \sqrt{x^2 + y^2 + z^2}, \quad \frac{\partial}{\partial x}\frac{1}{r} = \frac{x}{r^3}, \quad \frac{\partial^2}{\partial x^2}\frac{1}{r} = -3\frac{x^2}{r^4} + \frac{1}{r^3}, \text{ etc.}$$

From here $\nabla^2(r^{-1}) = 0$.

iii) Repeat for cylindrical coordinates. Hint: you cannot just substitute 1/r in equation (14.1) because in this equation r is the distance to the axis, not to the origin.

iv) Repeat for spherical coordinates.

The last two results can be easily obtained using the expressions for the operator ∇^2 in these coordinate systems and noticing that function ϕ depends only on r.

14.2 BOUNDARY CONDITIONS

First we note that problems described by elliptic equations do not contain time; therefore they do not need initial conditions. Physically it is also clear that the Laplace's and Poisson's equations by themselves are not sufficient to determine, for example, the temperature in all points of a body, or the electric potential outside the conductor. We have to know the heat regime at the body's surface or the charge distribution on the surface of the conductor to solve these problems and these constitute boundary conditions. From physical reasoning it is clear that, for instance, if the temperature distribution on the surface of a body is known, the solution of such a boundary value problem that consists of Laplace's or Poisson's equation together with a boundary condition should exist and be unique.

Boundary conditions can be set in several ways and in fact in our discussion above of various physical problems we have already met several kinds of boundary conditions. Below we categorize three primary kinds of boundary conditions, which correspond to the three different heat regimes on the surface (here we use temperature terminology but the arguments apply equally to other physical systems).

Consider some volume, V, bounded by a surface S. The boundary value problem for a stationary distribution of temperature, $u(x, y, z)$, inside the body is stated in the following way:

Find the function $u(x, y, z)$ inside the volume V satisfying the equation

$$\nabla^2 u = -f(x, y, z).$$

and satisfying one of the following kinds of boundary conditions:

1. $u = f_1$ on S   (boundary value problem of the first kind)

2. $\dfrac{\partial u}{\partial n} = f_2$ on S   (boundary value problem of the second kind)

3. $\dfrac{\partial u}{\partial n} + h(u - f_3) = 0$ on S ($h > 0$)   (boundary value problem of the third kind)

where f_1, f_2, f_3, and h are known functions and $\partial u / \partial n$ is the derivative in the direction of the outward normal to the surface, S.

We may use the above formulation in two distinct ways. If we want to find a solution, for example, the temperature, *inside* the volume V for the bounded region, we have what is referred to as an *interior boundary value problem*. If instead we need to find, for example, the temperature outside of a heater, or the electrostatic potential for an unbounded region outside the charged volume V, we have an *exterior boundary value problem*.

The physical sense of each of these boundary conditions is clear. The first boundary value problem when the *surface temperature is prescribed* is called *Dirichlet's problem*. The second boundary value problem when the *flux across the surface is prescribed* is called *Neumann's problem*. The third boundary value problem is called *the mixed problem*. This boundary condition corresponds to the well-known Newton's law of cooling, which governs the heat flux from the surface into the ambient medium.

Obviously a stationary temperature distribution is possible only if the net heat flow across the boundary is equal to zero. From here it follows that for the *interior Neumann's problem* the function f_2 should obey the additional requirement:

$$\iint\limits_{S} f_2 \, dS = 0. \tag{14.12}$$

In a similar way a boundary value problem can be formulated for the two-dimensional case when an area is bounded by a closed contour L. In this case the requirement that the net heat flow through the boundary for the interior problem is equal to zero becomes

$$\oint_L f_2 \, dl = 0. \tag{14.13}$$

Let us demonstrate the solution of a boundary value problem for a one-dimensional case. When the function $u(x)$ depends only on one variable, Laplace's equation becomes an ordinary differential equation and the solution is trivial.

Example 1

Solve the one-dimensional Laplace's equation in Cartesian coordinates, $d^2u/dx^2 = 0$ and apply Dirichlet boundary conditions to find a simple solution.

Solution. Integrating the equation gives $u = ax + b$. Dirichlet's problem with boundary conditions $u(x = 0) = u_1$ and $u(x = l) = u_2$ gives the solution as $u(x) = (u_2 - u_1)\, x/l + u_1$.

Reading Exercise

In Cartesian coordinates obtain a solution of a one-dimensional Neumann's problem for Laplace's equation.

Reading Exercise

Solve Dirichlet's problem in the case of axial symmetry with boundary conditions $u(r = a) = u_1$, $u(r = b) = u_2$. The result will give a solution to the problem of a stationary distribution of heat between two cylinders with common axis when cylinders' surfaces are kept at constant temperatures. The same solution also gives the electric potential between two equipotential cylindrical surfaces. (The solution is a harmonic function between the surfaces with the axis, $r = 0$, excluded.)

Hint: Use Laplace's operator in cylindrical coordinates when there is no dependence on φ and z.

Reading Exercise

Solve Dirichlet's problem in the case of spherical symmetry with boundary conditions $u(r = a) = u_1$, $u(r = b) = u_2$. The result will give a solution to the problem of a static distribution of heat between two spheres with a common center when the surfaces are kept at constant temperatures. It also gives the electric potential between two equipotential spherical surfaces. As in the previous case, the solution is a harmonic function between the surfaces, and the center at $r = 0$ is excluded.

Hint: Use Laplace's operator in spherical coordinates for the case of no dependence on φ and θ.

Let us formulate the *principle of the maximum*:

Theorem

If the function u is harmonic in some domain bounded by the surface S, it reaches its maximal (minimal) value at the boundary.

We will not prove this theorem; just notice that physical reasoning of this principal are clear in many cases—for instance, a stationary distribution of electric charges in a conductor is always on its surface.

14.3 THE BVP FOR LAPLACE'S EQUATION IN A RECTANGULAR DOMAIN

Boundary value problems for Laplace's equation in a rectangular domain can be solved with the method of separation of variables. We begin with the *Dirichlet problem* defined by

$$\nabla^2 u = 0 \quad (0 < x < l_x, \, 0 < y < l_y), \tag{14.14}$$

$$u(x, y)\big|_{x=0} = g_1(y), \quad u(x, y)\big|_{x=l_x} = g_2(y),$$

$$u(x, y)\big|_{y=0} = g_3(x), \quad u(x, y)\big|_{y=l_y} = g_4(x). \tag{14.15}$$

Let us split the problem in equations (14.14) through (14.15) into two parts, each of which has *homogeneous (zero) boundary conditions in one variable*. To proceed we introduce

$$u(x, y) = u_1(x, y) + u_2(x, y), \tag{14.16}$$

where $u_1(x, y)$ and $u_2(x, y)$ are the solutions to the following problems on a rectangular boundary:

$$\nabla^2 u_1 = 0, \tag{14.17}$$

$$u_1(x, y)\big|_{x=0} = u_1(x, y)\big|_{x=l_x} = 0, \tag{14.18}$$

$$u_1(x, y)\big|_{y=0} = g_3(x), \quad u_1(x, y)\big|_{y=l_y} = g_4(x), \tag{14.19}$$

and

$$\nabla^2 u_2 = 0, \tag{14.20}$$

$$u_2(x, y)\big|_{y=0} = u_2(x, y)\big|_{y=l_y} = 0, \tag{14.21}$$

$$u_2(x, y)\big|_{x=0} = g_1(y), \quad u_2(x, y)\big|_{x=l_x} = g_2(y). \tag{14.22}$$

First we consider the problem for the function $u_1(x, y)$ and search for the solution in the form

$$u_1(x, y) = X(x)Y(y). \tag{14.23}$$

Substituting equation (14.23) into the Laplace equation and separating the variables yields

$$\frac{X''(x)}{X(x)} \equiv -\frac{Y''(y)}{Y(y)} = -\lambda \tag{14.24}$$

where we take $\lambda > 0$ for the further solution.

From here we obtain equations for $X(x)$ and $Y(y)$. With the homogeneous boundary conditions in equation (14.18) we obtain the one-dimensional Sturm-Liouville problem for $X(x)$ given by

$$X'' + \lambda X = 0, \quad 0 < x < l_x,$$
$$X(0) = X(l_x) = 0.$$

If $X(x)$ is not equal to zero identically, the solution to this problem is

$$X_n = \sin\sqrt{\lambda_{xn}}\, x, \quad \lambda_{xn} = \left(\frac{\pi n}{l_x}\right)^2, \quad n = 1,2,3,\ldots \tag{14.25}$$

With these eigenvalues, λ_{xn}, we obtain an equation for $Y(y)$ from equation (14.24):

$$Y'' - \lambda_{xn} Y = 0, \quad 0 < y < l_y. \tag{14.26}$$

A general solution to this equation can be written in an obvious way as

$$Y_n = C_n^{(1)} \exp\left(\sqrt{\lambda_{xn}}\, y\right) + C_n^{(2)} \exp\left(-\sqrt{\lambda_{xn}}\, y\right). \tag{14.27}$$

Such a form of solution does not fit well the purposes of the further analysis. It is more suitable to take a fundamental system of solution $\left\{Y_n^{(1)}, Y_n^{(2)}\right\}$ of equation (14.26) in the way that function $Y_n^{(1)}$ and $Y_n^{(2)}$ satisfy the homogeneous boundary condition, one at $y = 0$ and another at $y = l_y$:

$$Y_n^{(1)}(0) = 0, \quad Y_n^{(2)}(l_y) = 0.$$

It is convenient to chose the following conditions at two other boundaries:

$$Y_n^{(1)}(l_y)=1, \quad Y_n^{(2)}(0)=1.$$

As the result the proper fundamental solutions of equations (14.26) are

$$Y_n^{(1)} = \frac{\sinh\sqrt{\lambda_{xn}}\,y}{\sinh\sqrt{\lambda_{xn}}\,l_y} \quad \text{and} \quad Y_n^{(2)} = \frac{\sinh\sqrt{\lambda_{xn}}\left(l_y - y\right)}{\sinh\sqrt{\lambda_{xn}}\,l_y}. \tag{14.28}$$

It is easily verified that they both satisfy equation (14.26) and are linearly independent; thus they can serve as a fundamental set of particular solutions for this equation.

Thus, a general solution of equation (14.26) can be written as

$$Y_n = A_n Y_n^{(1)} + B_n Y_n^{(2)}. \tag{14.29}$$

Using the above relations we may write a general solution of the Laplace equation satisfying the homogeneous boundary conditions at the boundaries $x = 0$ and $x = l_x$ in equation (14.18), as a series in the functions $Y_n^{(1)}(y)$ and $Y_n^{(2)}(y)$:

$$u_1 = \sum_{n=1}^{\infty}\left[A_n Y_n^{(1)}(y) + B_n Y_n^{(2)}(y)\right]\sin\sqrt{\lambda_{xn}}\,x. \tag{14.30}$$

The coefficients of this series are determined from the boundary conditions (14.19):

$$u_1(x,y)\big|_{y=0} = \sum_{n=1}^{\infty} B_n \sin\sqrt{\lambda_{xn}}\,x = g_3(x),$$

$$u_1(x,y)\big|_{y=l_y} = \sum_{n=1}^{\infty} A_n \sin\sqrt{\lambda_{xn}}\,x = g_4(x). \tag{14.31}$$

We see from here that B_n and A_n are Fourier coefficients of functions $g_3(x)$ and $g_4(x)$ in the system of eigenfunctions $\left\{\sin\sqrt{\lambda_{xn}}\,x\right\}_1^{\infty}$:

$$B_n = \frac{2}{l_x}\int_0^{l_x} g_3(\xi)\sin\sqrt{\lambda_{xn}}\,\xi d\xi,$$

$$A_n = \frac{2}{l_x}\int_0^{l_x} g_4(\xi)\sin\sqrt{\lambda_{xn}}\,\xi d\xi. \tag{14.32}$$

This completes the solution of the problem given in equations (14.17) through (14.19).

Obviously, the solution of the similar problem given in equations (14.20) through (14.22) can be obtained from equations (14.30) and (14.32) by replacing y for x, l_y for l_x and $g_3(x)$, $g_4(x)$ for $g_1(y)$ and $g_2(y)$. Carrying out this procedure yields

$$u_2 = \sum_{n=1}^{\infty} \left[C_n X_n^{(1)}(x) + D_n X_n^{(2)}(x) \right] \sin \sqrt{\lambda_{yn}}\, y, \tag{14.33}$$

where

$$X_n^{(1)}(x) = \frac{\sinh \sqrt{\lambda_{yn}}\, x}{\sinh \sqrt{\lambda_{yn}}\, l_x}, \quad X_n^{(2)}(x) = \frac{\sinh \sqrt{\lambda_{yn}}\, (l_x - x)}{\sinh \sqrt{\lambda_{yn}}\, l_x}, \quad \lambda_{yn} = \left(\frac{\pi n}{l_y} \right)^2, \tag{14.34}$$

and

$$C_n = \frac{2}{l_y} \int_0^{l_y} g_2(\xi) \sin \sqrt{\lambda_{yn}}\, \xi\, d\xi, \quad D_n = \frac{2}{l_y} \int_0^{l_y} g_1(\xi) \sin \sqrt{\lambda_{yn}}\, \xi\, d\xi. \tag{14.35}$$

Finally, the solution to the problem (14.14) and (14.15) has the form

$$u(x, y) = u_1(x, y) + u_2(x, y),$$

where functions $u_1(x, y)$ and $u_2(x, y)$ are defined by formulas (14.30) and (14.33), respectively.

In the same way can be solved a BVP for Laplace's equation in a rectangular domain with *other types of boundary conditions*. The only difference is that the other fundamental solutions should be used. Fundamental systems of solutions for different types of boundary conditions are collected in Appendix 3.

Example 2

Find a steady-state temperature distribution inside a rectangular material that has boundaries maintained under the following conditions:

$$T(x,y)\big|_{x=0} = T_0 + (T_3 - T_0)\frac{y}{l_y}, \quad T(x,y)\big|_{x=l_x} = T_1 + (T_2 - T_1)\frac{y}{l_y},$$

and

$$T(x,y)\big|_{y=0} = T_0 + (T_1 - T_0)\frac{x}{l_x}, \quad T(x,y)\big|_{y=l_y} = T_3 + (T_2 - T_3)\frac{x}{l_x},$$

that is, at the corners of the rectangular the temperatures are T_0, T_1, T_2, T_3, and on the boundaries the temperature are linear functions.

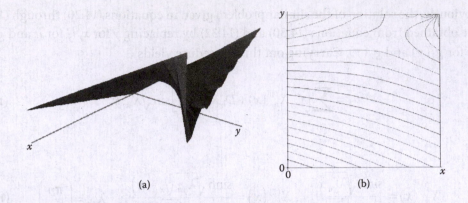

FIGURE 14.1   Surface plot (a) and contour map (b) for Example 2. The values of parameters $T_0 = 5$, $T_1 = 10$, $T_2 = 15$, $T_3 = 20$, $l_x = 4$, $l_y = 6$.

Solution. Introduce the function $u = T - T_0$ so that we measure the temperature relative to T_0. Then $g_1(y) = (T_3 - T_1)\dfrac{y}{l_y}$, and so on, and evaluating the integrals in equations (14.32) and (14.35), we obtain

$$A_n = \frac{2}{\pi n}\Big[T_3 - T_0 - (-1)^n T_2\Big], \qquad B_n = -2\frac{(-1)^n}{\pi n}(T_1 - T_0),$$

$$C_n = \frac{2}{\pi n}\Big[T_1 - T_0 - (-1)^n T_2\Big], \qquad D_n = -2\frac{(-1)^n}{\pi n}(T_3 - T_0).$$

These coefficients decay only as $1/n$; thus the series in equations (14.30) and (14.33) converge rather slowly. Figure 14.1, obtained with the program **Laplace**, shows the surface graph of temperature field $T(x, y)$ and lines of equal temperature (in program they are labeled with the temperature value) for Example 2. Directions for using the **Laplace** program are found in Appendix 5.

14.4 THE POISSON EQUATION WITH HOMOGENEOUS BOUNDARY CONDITIONS

Consider a boundary value problem for the Poisson equation

$$\frac{\partial^2 u}{\partial x^2} + \frac{\partial^2 u}{\partial y^2} = -f(x, y) \quad (0 < x < l_x, 0 < y < l_y) \tag{14.36}$$

with zero boundary conditions

$$P_1[u] \equiv \alpha_1 \frac{\partial u}{\partial x} + \beta_1 u \bigg|_{x=0} = 0, \quad P_2[u] \equiv \alpha_2 \frac{\partial u}{\partial x} + \beta_2 u \bigg|_{x=l_x} = 0,$$

$$P_3[u] \equiv \alpha_3 \frac{\partial u}{\partial x} + \beta_3 u \bigg|_{y=0} = 0, \quad P_4[u] \equiv \alpha_4 \frac{\partial u}{\partial x} + \beta_4 u \bigg|_{y=l_y} = 0. \tag{14.37}$$

The solution to the problem (14.36), (14.37) can be expanded in a series by eigenfunctions of the Sturm-Liouville problem for the Laplace operator over a rectangular domain

$$u(x,y) = \sum_{n=1}^{\infty} \sum_{m=1}^{\infty} C_{nm} V_{nm}(x,y), \tag{14.38}$$

where $V_{nm}(x,y)$ are eigenfunctions of the respective Laplace boundary value problem and coefficients are

$$C_{nm} = \frac{1}{\|V_{nm}\|^2} \int_0^{l_x} \int_0^{l_y} u(x,y) V_{nm}(x,y) \, dx \, dy. \tag{14.39}$$

Let us multiply the Poisson equation (14.36) by $V_{nm}(x,y)$ and integrate over the rectangle $[0, l_x; 0, l_y]$:

$$\int_0^{l_x} \int_0^{l_y} \left[\frac{\partial^2 u}{\partial x^2} + \frac{\partial^2 u}{\partial y^2} \right] \cdot V_{nm}(x,y) \, dx \, dy = -\int_0^{l_x} \int_0^{l_y} f(x,y) V_{nm}(x,y) \, dx \, dy. \tag{14.40}$$

Now substitute equation (14.38) into equation (14.40). Because $V_{nm}(x,y)$ are the eigenfunctions of the Laplacian we have

$$\nabla^2 V_{nm}(x,y) = -\lambda_{nm} V_{nm}(x,y), \tag{14.41}$$

where the eigenvalues λ_{nm} correspond to the boundary condition (14.37). The left sides of equations (14.40) become

$$-\sum_{i=1}^{\infty} \sum_{j=1}^{\infty} \lambda_{ij} \int_0^{l_x} \int_0^{l_y} C_{ij} \cdot V_{ij}(x,y) V_{nm}(x,y) \, dx \, dy.$$

Due to the orthogonality relation for the functions $V_{nm}(x,y)$, the only term in the sums that differs from zero is

$$-\lambda_{nm} C_{nm} \| V_{nm} \|^2.$$

Comparing with the right side in equation (14.40) we obtain

$$\lambda_{nm} C_{nm} = f_{nm}, \quad n, m = 1, 2, 3, \ldots, \tag{14.42}$$

where

$$f_{nm} = \frac{1}{\|V_{nm}\|^2} \int\limits_0^{l_x}\int\limits_0^{l_y} f(x,y)V_{nm}(x,y)dxdy. \tag{14.43}$$

From (14.42) and (14.43) the coefficients C_{nm} can be obtained.

In the case of boundary conditions of the first or third type (*Dirichlet condition* or *mixed condition*) eigenvalues $\lambda_{nm} \neq 0$ for all $n, m = 1,2,3, \ldots$; thus $C_{nm} = f_{nm}/\lambda_{nm}$ and the solution (14.38) is defined uniquely:

$$u(x,y) = \sum_{n=1}^{\infty}\sum_{m=1}^{\infty} \frac{f_{nm}}{\lambda_{nm}} V_{nm}(x,y). \tag{14.44}$$

In the case of boundary conditions of the second type (*Neumann conditions*) eigenvalue $\lambda_{00} = 0$ ($V_{00} = 1$) and all other eigenvalues are nonzero. Then there are two options. If

$$f_{00} = \int\limits_0^{l_x}\int\limits_0^{l_y} f(x,y)dx\,dy = 0,$$

then coefficient C_{00} is uncertain; the other coefficients are defined uniquely. The solution to the given problem exists but is determined only up to an arbitrary additive constant. The solution is

$$u(x,y) = \sum_{n=1}^{\infty}\sum_{m=1}^{\infty} \frac{f_{nm}}{\lambda_{nm}} V_{nm}(x,y) + \text{const.} \tag{14.45}$$

If

$$f_{00} = \int\limits_0^{l_x}\int\limits_0^{l_y} f(x,y)dx\,dy \neq 0,$$

then the solution to the given problem does not exist.

14.5 THE LAPLACE AND POISSON EQUATIONS WITH NONHOMOGENEOUS BOUNDARY CONDITIONS

Now consider the general boundary value problem for Poisson equation given by

$$\nabla^2 u = \frac{\partial^2 u}{\partial x^2} + \frac{\partial^2 u}{\partial y^2} = -f(x,y), \tag{14.46}$$

with nonhomogeneous boundary conditions

$$P_1[u] \equiv \alpha_1 \frac{\partial u}{\partial x} + \beta_1 u \bigg|_{x=0} = g_1(y), \quad P_2[u] \equiv \alpha_2 \frac{\partial u}{\partial x} + \beta_2 u \bigg|_{x=l_x} = g_2(y),$$

$$P_3[u] \equiv \alpha_3 \frac{\partial u}{\partial y} + \beta_3 u \bigg|_{y=0} = g_3(x), \quad P_4[u] \equiv \alpha_4 \frac{\partial u}{\partial y} + \beta_4 u \bigg|_{y=l_y} = g_4(x).$$

(14.47)

To deal with nonhomogeneous boundary conditions we introduce an auxiliary function, $w(x, y)$, and seek a solution of the problem in the form

$$u(x, y) = v(x, y) + w(x, y),$$

where $v(x, y)$ is a new unknown function, and the function $w(x, y)$ is chosen so that it satisfies the given nonhomogeneous boundary conditions (14.47). Then function $v(x, y)$ satisfies the homogeneous boundary conditions.

We shall seek an auxiliary function, $w(x, y)$, in the form

$$w(x, y) = g_1(y)\bar{X} + g_2(y)\bar{\bar{X}} + g_3(x)\bar{Y} + g_4(x)\bar{\bar{Y}}$$
$$+ A\bar{X}\bar{Y} + B\bar{X}\bar{\bar{Y}} + C\bar{\bar{X}}\bar{Y} + D\bar{\bar{X}}\bar{\bar{Y}},$$

(14.48)

where $\bar{X}(x)$, $\bar{\bar{X}}(x)$, $\bar{Y}(y)$, and $\bar{\bar{Y}}(y)$ are polynomials of first or second order. The coefficients of these polynomials will be adjusted in such a way that function $w(x, y)$ satisfies the boundary conditions given in equations (14.47). Notice, that function $w(x, y)$ depends only on the boundary conditions (14.47), that is, is the same for Poisson and Laplace equations.

14.5.1 Consistent Boundary Conditions

Let the boundary functions *satisfy consistency conditions* (i.e., the boundary functions take the same values at the corners of the domain), in which case we have

$$P_1[g_3(x)]_{x=0} = P_3[g_1(y)]_{y=0}, \quad P_1[g_4(x)]_{x=0} = P_4[g_1(y)]_{y=l_y},$$

$$P_2[g_3(x)]_{x=l_x} = P_3[g_2(y)]_{y=0}, \quad P_2[g_4(x)]_{x=l_x} = P_4[g_2(y)]_{y=l_y}.$$

(14.49)

Functions $\bar{X}(x)$ and $\bar{\bar{X}}(x)$ can be chosen in such a way that

$$P_1\left[\bar{X}(0)\right] = 1, \quad P_2\left[\bar{X}(l_x)\right] = 0,$$

$$P_1\left[\bar{\bar{X}}(0)\right] = 0, \quad P_2\left[\bar{\bar{X}}(l_x)\right] = 1.$$

(14.50)

If $\beta_1 \neq 0$ or $\beta_2 \neq 0$, then the functions $\bar{X}(x)$ and $\bar{\bar{X}}(x)$ are polynomials of the first order

$$\bar{X}(x) = \gamma_1 + \delta_1 x, \qquad \bar{\bar{X}}(x) = \gamma_2 + \delta_2 x. \tag{14.51}$$

The coefficients γ_1, δ_1, γ_2, δ_2 of these polynomials are defined uniquely and depend on the types of boundary conditions:

$$
\gamma_1 = \frac{\alpha_2 + \beta_2 l_x}{\beta_1 \beta_2 l_x + \beta_1 \alpha_2 - \beta_2 \alpha_1}, \quad \delta_1 = \frac{-\beta_2}{\beta_1 \beta_2 l_x + \beta_1 \alpha_2 - \beta_2 \alpha_1},
$$

$$
\gamma_2 = \frac{-\alpha_1}{\beta_1 \beta_2 l_x + \beta_1 \alpha_2 - \beta_2 \alpha_1}, \quad \delta_2 = \frac{\beta_1}{\beta_1 \beta_2 l_x + \beta_1 \alpha_2 - \beta_2 \alpha_1}. \tag{14.52}
$$

If $\beta_1 = \beta_2 = 0$, then the functions $\bar{X}(x)$ and $\bar{\bar{X}}(x)$ are polynomials of the second order and we have

$$\bar{X}(x) = x - \frac{x^2}{2 l_x}, \qquad \bar{\bar{X}}(x) = \frac{x^2}{2 l_x}. \tag{14.53}$$

The problem for $Y(y)$ (and its solution) is the same as the problem for $X(x)$. Functions $\bar{Y}(y)$ and $\bar{\bar{Y}}(y)$ can be chosen in such a way that

$$P_3[\bar{Y}(0)] = 1, \quad P_3[\bar{Y}(l_y)] = 0,$$

$$P_4[\bar{\bar{Y}}(0)] = 0, \quad P_4[\bar{\bar{Y}}(l_y)] = 1. \tag{14.54}$$

If $\beta_3 \neq 0$ or $\beta_4 \neq 0$, then the functions $\bar{Y}(y)$ and $\bar{\bar{Y}}(y)$ are polynomials of the first order and we may write

$$\bar{Y}(y) = \gamma_3 + \delta_3 y, \qquad \bar{\bar{Y}}(y) = \gamma_4 + \delta_4 y. \tag{14.55}$$

The coefficients, γ_3, δ_3, γ_4, δ_4 of these polynomials are defined uniquely and depend on the types of boundary conditions:

$$
\gamma_3 = \frac{\alpha_4 + \beta_4 l_y}{\beta_3 \beta_4 l_y + \beta_3 \alpha_4 - \beta_4 \alpha_3}, \quad \delta_3 = \frac{-\beta_4}{\beta_3 \beta_4 l_y + \beta_3 \alpha_4 - \beta_4 \alpha_3},
$$

$$
\gamma_4 = \frac{-\alpha_3}{\beta_3 \beta_4 l_y + \beta_3 \alpha_4 - \beta_4 \alpha_3}, \quad \delta_4 = \frac{\beta_3}{\beta_3 \beta_4 l_y + \beta_3 \alpha_4 - \beta_4 \alpha_3}. \tag{14.56}
$$

If $\beta_3 = \beta_4 = 0$, then the functions $\overline{Y}(y)$ and $\overline{\overline{Y}}(y)$ are polynomials of the second order and we have

$$\overline{Y}(y) = y - \frac{y^2}{2l_y}, \quad \overline{\overline{Y}}(y) = \frac{y^2}{2l_y}. \tag{14.57}$$

Coefficients A, B, C, and D of the auxiliary function $w(x, y)$ are defined from the boundary conditions.

At the edge : $x = 0 : P_1[w]_{x=0} = g_1(y) + (P_1[g_3(0)] + A)\overline{Y} + (P_1[g_4(0)] + B)\overline{\overline{Y}}.$

At the edge : $x = l_x : P_2[w]_{x=l_x} = g_2(y) + (P_2[g_3(l_x)] + C)\overline{Y} + (P_2[g_4(l_x)] + D)\overline{\overline{Y}}.$

At the edge : $y = 0 : P_3[w]_{y=0} = g_3(x) + (P_3[g_1(0)] + A)\overline{X} + (P_3[g_2(0)] + C)\overline{\overline{X}}.$

At the edge : $y = l_y : P_4[w]_{y=l_y} = g_4(x) + (P_4[g_1(l_y)] + B)\overline{X} + (P_4[g_2(l_y)] + D)\overline{\overline{X}}.$

The conformity conditions for the boundary value functions are valid, so the coefficients are

$$A = -P_3[g_1(y)]_{y=0}, \quad B = -P_4[g_1(y)]_{y=l_y},$$

$$C = -P_3[g_2(y)]_{y=0}, \quad D = -P_4[g_2(y)]_{y=l_y}. \tag{14.58}$$

It is easy to verify (which is left as a Reading Exercise) that the auxiliary function, $w(x, y)$, satisfies the given boundary conditions:

$$P_1[w]_{x=0} = g_1(y), \quad P_2[w]_{x=l_x} = g_2(y),$$

$$P_3[w]_{y=0} = g_3(x), \quad P_4[w]_{y=l_y} = g_4(x).$$

14.5.2 Inconsistent Boundary Conditions

Suppose the boundary functions *do not satisfy consistency conditions*. We shall search for an auxiliary function as the sum of two functions:

$$w(x, y) = w_1(x, y) + w_2(x, y). \tag{14.59}$$

Function $w_1(x, y)$ is an auxiliary function satisfying the consistent boundary conditions

$$P_1[w_1]_{x=0} = g_1(y), \quad P_2[w_1]_{x=l_x} = g_2(y),$$

$$P_3[w_1]_{y=0} = -A\overline{X}(x) - C\overline{\overline{X}}(x),$$

$$P_4[w_1]_{y=l_y} = -B\overline{X}(x) - D\overline{\overline{X}}(x), \tag{14.60}$$

where

$$A = -P_3[g_1(y)]_{y=0}, \quad B = -P_4[g_1(y)]_{y=l_y},$$

$$C = -P_3[g_2(y)]_{y=0}, \quad D = -P_4[g_2(y)]_{y=l_y}. \tag{14.61}$$

Such a function was constructed above. In this case it has the form

$$w_1(x,y) = g_1(y)\bar{X} + g_2(y)\bar{\bar{X}}. \tag{14.62}$$

The function $w_2(x,y)$ is a particular solution of the *Laplace problem* with the following boundary conditions:

$$P_1[w_2]_{x=0} = 0, \quad P_2[w_2]_{x=l_x} = 0,$$

$$P_3[w_2]_{y=0} = g_3(x) + A\bar{X}(x) + C\bar{\bar{X}}(x), \tag{14.63}$$

$$P_4[w_2]_{y=l_y} = g_4(x) + B\bar{X}(x) + D\bar{\bar{X}}(x).$$

This problem was considered in detail in Section 14.3. The solution of this problem has a form

$$w_2(x,y) = \sum_{n=1}^{\infty} \left[A_n Y_{1n}(y) + B_n Y_{2n}(y) \right] X_n(x), \tag{14.64}$$

where λ_{xn} and $X_n(x)$ are eigenvalues and eigenfunctions of the Sturm-Liouville problem

$$X'' + \lambda X = 0 \quad (0 < x < l_x),$$

$$P_1[X]\big|_{x=0} = P_2[X]\big|_{x=l_x} = 0. \tag{14.65}$$

The coefficients A_n and B_n are defined by the formulas

$$A_n = \frac{1}{\|X_n\|^2} \int_0^{l_x} \left[g_4(x) + B\bar{X}(x) + D\bar{\bar{X}}(x) \right] X_n(x)\,dx,$$

$$\tag{14.66}$$

$$B_n = \frac{1}{\|X_n\|^2} \int_0^{l_x} \left[g_3(x) + A\bar{X}(x) + C\bar{\bar{X}}(x) \right] X_n(x)\,dx.$$

Thus we see that eigenvalues and eigenfunctions of this boundary value problem depend on the types of boundary conditions (see Appendix 1 for a detailed account).

Having the eigenvalues, λ_{xn}, we obtain a similar equation for $Y(y)$ given by

$$Y'' - \lambda_{xn} Y = 0 \quad (0 < y < l_y). \tag{14.67}$$

We shall choose fundamental system $\{Y_1, Y_2\}$ of solutions in such a way that

$$P_3[Y_1(0)] = 0, \quad P_3[Y_1(l_y)] = 1,$$
$$P_4[Y_2(0)] = 1, \quad P_4[Y_2(l_y)] = 0. \tag{14.68}$$

Two particular solutions of the previous equation (14.65) are $\exp(\pm\sqrt{\lambda_n}\,y)$, but for the future analysis it is more convenient to choose two linearly independent functions $Y_1(y)$ and $Y_2(y)$ in the form

$$Y_1(y) = a \sinh \sqrt{\lambda_n}\, y + b \cosh \sqrt{\lambda_n}\, y,$$
$$Y_2(y) = c \sinh \sqrt{\lambda_n}\, (l_y - y) + d \cosh \sqrt{\lambda_n}\, (l_y - y). \tag{14.69}$$

The values of coefficients a, b, c, and d depend on the types of boundary conditions $P_3[w_2]_{y=0}$ and $P_4[w_2]_{y=l_y}$ (see Appendix 3 for details). It can be verified that the auxiliary function

$$w(x,y) = w_1(x,y) + w_2(x,y)$$
$$= w_1(x,y) + \sum_{n=1}^{\infty} [A_n Y_{1n}(y) + B_n Y_{2n}(y)] X_n(x) \tag{14.70}$$

satisfies the given boundary conditions when $n \to \infty$.

Example 3

Find the stationary distribution of temperature within a very long (infinite) parallelepiped of rectangular cross section ($0 \le x \le \pi$, $0 \le y \le \pi$) if the faces $y = 0$ and $y = \pi$ follow the temperature distributions

$$u(x,0) = \cos x \quad \text{and} \quad u(x,\pi) = \cos 3x$$

respectively and the constant heat flows are supplied to the faces $x = 0$ and $x = \pi$ from outside:

$$\frac{\partial u}{\partial x}(0,y) = \sin y \quad \text{and} \quad \frac{\partial u}{\partial x}(\pi,y) = \sin 5y.$$

Generation (or absorption) of heat by internal sources is absent.

Solution. The problem is expressed as Laplace equation

$$\frac{\partial^2 u}{\partial x^2} + \frac{\partial^2 u}{\partial y^2} = 0,$$

under the conditions

$$P_1[u]_{x=0} \equiv \frac{\partial u}{\partial x}(0, y) = \sin y \quad (\alpha_1 = 1, \ \beta_1 = 0 - \text{Neumann condition}),$$

$$P_2[u]_{x=\pi} \equiv \frac{\partial u}{\partial x}(\pi, y) = \sin 5y \quad (\alpha_2 = 1, \ \beta_2 = 0 - \text{Neumann condition}),$$

$$P_3[u]_{y=0} \equiv u(x, 0) = \cos x \quad (\alpha_3 = 0, \ \beta_3 = 1 - \text{Dirichlet condition}),$$

$$P_4[u]_{y=\pi} \equiv u(x, \pi) = \cos 3x \quad (\alpha_4 = 0, \ \beta_4 = 1 - \text{Dirichlet condition}).$$

The solution to this problem can be expressed as the sum of two functions

$$u(x, y) = w(x, y) + v(x, y),$$

where $v(x,y)$ is a new unknown function and $w(x,y)$ is an auxiliary function satisfying the boundary conditions.

The boundary functions *satisfy the consisting conditions*, that is (see formulas (14.49)),

$$P_1[\cos x]_{x=0} = \frac{\partial(\cos x)}{\partial x}\bigg|_{x=0} = 0 \quad = \quad P_3[\sin y]_{y=0} = \sin y\big|_{y=0} = 0,$$

$$P_1[\cos 3x]_{x=0} = \frac{\partial(\cos 3x)}{\partial x}\bigg|_{x=0} = 0 \quad = \quad P_4[\sin y]_{y=\pi} = \sin y\big|_{y=\pi} = 0,$$

$$P_2[\cos x]_{x=\pi} = \frac{\partial(\cos x)}{\partial x}\bigg|_{x=\pi} = 0 \quad = \quad P_3[\sin 5y]_{y=0} = \sin 5y\big|_{y=0} = 0,$$

$$P_2[\cos 3x]_{x=\pi} = \frac{\partial(\cos 3x)}{\partial x}\bigg|_{x=\pi} = 0 \quad = \quad P_4[\sin 5y]_{y=\pi} = \sin 5y\big|_{y=\pi} = 0.$$

Let's construct the auxiliary function (14.48) satisfying the given boundary condition. In our problem $\beta_1 = \beta_2 = 0$, so (see formulas (14.53))

$$\overline{X}(x) = x - \frac{x^2}{2l_x}, \quad \overline{\overline{X}}(x) = \frac{x^2}{2l_x}.$$

From formulas (14.55) and (14.56) we have $\gamma_3 = 1$, $\delta_3 = -1/l_y$, $\gamma_4 = 0$; $\delta_4 = 1/l_y$, so

$$\overline{Y}(y) = 1 - \frac{y}{\pi}, \quad \overline{\overline{Y}}(y) = \frac{y}{\pi}.$$

Coefficients (14.58) of auxiliary function (14.48) are zero:

$$A = -P_3[\sin y]_{y=0} = -\sin y\big|_{y=0} = 0,$$

$$B = -P_4[\sin y]_{y=\pi} = -\sin y\big|_{y=\pi} = 0,$$

$$C = -P_3[\sin 5y]_{y=0} = -\sin 5y\big|_{y=0} = 0,$$

$$D = -P_4[\sin 5y]_{y=\pi} = -\sin 5y\big|_{y=\pi} = 0.$$

So, the auxiliary function (14.48) has the form

$$w(x,y) = \sin y \cdot \bar{X} + \sin 5 \cdot \bar{\bar{X}} + \cos x \cdot \bar{Y} + \cos 3x \cdot \bar{\bar{Y}}$$

$$= \sin y \left(x - \frac{x^2}{2\pi} \right) + \sin 5y \frac{x^2}{2\pi} + \cos x \left(1 - \frac{y}{\pi} \right) + \cos 3x \frac{y}{\pi}.$$

Function $v(x, y)$ is the solution to the Poisson problem with zero boundary conditions

$$\frac{\partial^2 u}{\partial x^2} + \frac{\partial^2 u}{\partial y^2} = f^*(x,y),$$

where

$$f^*(x,y) = \frac{\partial^2 w}{\partial x^2} + \frac{\partial^2 w}{\partial y^2} = \frac{1}{2\pi}\Big[\sin y \cdot (x^2 - 2\pi x - 2) + \sin 5y \cdot (2 - 25x^2)$$

$$+ 2\cos x \cdot (y - \pi) - 18y \cdot \cos 3x\Big].$$

Eigenvalues and eigenfunctions are (see Appendix 1)

$$\lambda_{nm} = \lambda_{xn} + \lambda_{ym} = n^2 + m^2, \, n = 0,1,2,\ldots, \, m = 1,2,\ldots$$

$$V_{nm}(x,y) = X_n(x) \cdot Y_m(y) = \cos nx \cdot \sin my,$$

$$\|V_{nm}\|^2 = \|X_n\|^2 \cdot \|Y_m\|^2 = \begin{cases} \pi^2/2, & \text{if } n = 0, \\ \pi^2/4, & \text{if } n > 0. \end{cases}$$

The solution $v(x, y)$ is defined by the series

$$v(x,y) = \sum_{n=0}^{\infty} \sum_{m=1}^{\infty} C_{nm} \cos nx \cdot \sin my,$$

where $C_{nm} = \dfrac{f_{nm}}{\lambda_{nm}}$, $f_{nm} = \dfrac{1}{\|V_{nm}\|^2} \int_0^\pi \int_0^\pi f^*(x,y) \cdot \cos nx \cdot \sin my \, dxdy.$

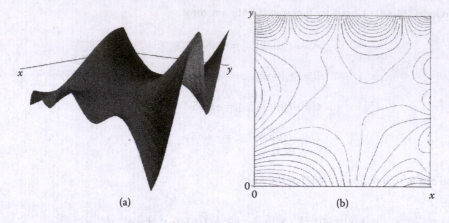

FIGURE 14.2 Surface plot (a) and lines of equal temperature (b) for Example 3. This solution was obtained with the program **Laplace**.

Thus,

$$C_{01} = \frac{f_{01}}{\lambda_{01}} = -\frac{1}{\pi}\left(\frac{\pi^2}{3} + 1\right), \quad C_{n1} = \frac{f_{n1}}{\lambda_{n1}} = \frac{2}{\pi n^2 (n^2 + 1)},$$

$$C_{05} = \frac{f_{05}}{\lambda_{05}} = \frac{1}{\pi}\left(\frac{1}{25} - \frac{\pi^2}{6}\right), \quad C_{n5} = \frac{f_{n5}}{\lambda_{n5}} = (-1)^{n+1}\frac{50}{\pi n^2 (n^2 + 25)},$$

$$C_{1m} = \frac{f_{1m}}{\lambda_{1m}} = -\frac{2}{\pi m(m^2 + 1)}, \quad C_{3m} = \frac{f_{3m}}{\lambda_{3m}} = (-1)^m \frac{18}{\pi m(m^2 + 9)},$$

$$C_{nm} = 0 \quad \text{in other cases.}$$

The final solution (see also Figure 14.2) is

$$u(x,y) = w(x,y) + v(x,y) = \sin y \cdot \left(x - \frac{x^2}{2\pi}\right) + \sin 5y \cdot \frac{x^2}{2\pi} + \cos x \cdot \left(1 - \frac{y}{\pi}\right) + \cos 3x \cdot \frac{y}{\pi}$$

$$+ \sum_{n=0}^{\infty}[C_{n1}\sin y + C_{n5}\sin 5y]\cdot \cos nx + \sum_{n=1}^{\infty}[C_{1m}\cos x + C_{3m}\cos 3x]\cdot \sin my.$$

Example 4

A heat-conducting thin uniform rectangular membrane ($0 \le x \le l_x, 0 \le y \le l_y$) is thermally insulated over its lateral faces. The bounds $x = 0$ and $y = 0$ are thermally insulated; the other bounds, $x = l_x$ and $y = l_y$, are held at fixed temperature $u = u_0$. One constant internal source of heat acts at the point (x_0, y_0) of the membrane. The value of this source is $Q = \text{const}$. Find the steady-state temperature distribution in the membrane.

Solution. The problem may be explored in the solution of the Poisson equation

$$\frac{\partial^2 u}{\partial x^2} + \frac{\partial^2 u}{\partial y^2} = -Q \cdot \delta(x - x_0)\delta(y - y_0)$$

under the conditions

$$P_1[u]_{x=0} \equiv \frac{\partial u}{\partial x}(0, y) = 0 \quad \left(\alpha_1 = 1, \ \beta_1 = 0 - \text{Neumann condition}\right),$$

$$P_2[u]_{x=l_x} \equiv u(l_x, y) = u_0 \quad \left(\alpha_2 = 0, \ \beta_2 = 1 - \text{Dirichlet condition}\right),$$

$$P_3[u]_{y=0} \equiv \frac{\partial u}{\partial y}(x, 0) = 0 \quad \left(\alpha_3 = 1, \ \beta_3 = 0 - \text{Neumann condition}\right),$$

$$P_4[u]_{y=l_y} \equiv u(x, l_y) = u_0 \quad \left(\alpha_4 = 0, \ \beta_4 = 1 - \text{Dirichlet condition}\right).$$

The solution to this problem can be expressed as the sum of two functions

$$u(x, y) = v(x, y) + w(x, y),$$

where $v(x, y)$ is a new unknown function and $w(x, y)$ is an auxiliary function satisfying the boundary conditions.

The boundary functions *satisfy the consisting conditions*, that is (see formulas (14.49)),

$$P_1[g_3]_{x=0} = \frac{\partial(g_3 \equiv 0)}{\partial x}\bigg|_{x=0} = 0 \quad = \quad P_3[g_1]_{y=0} = \frac{\partial(g_1 \equiv 0)}{\partial y}\bigg|_{y=0} = 0,$$

$$P_1[g_4]_{x=0} = \frac{\partial(g_4 \equiv u_0)}{\partial x}\bigg|_{x=0} = 0 \quad = \quad P_4[g_1]_{y=l_y} = g_1\big|_{y=l_y} = 0,$$

$$P_2[g_3]_{x=l_x} = g_3\big|_{x=l_x} = 0 \quad = \quad P_3[g_2]_{y=0} = \frac{\partial(g_2 \equiv u_0)}{\partial y}\bigg|_{y=0} = 0,$$

$$P_2[g_4]_{x=l_x} = g_4\big|_{x=l_x} = u_0 \quad = \quad P_4[g_2]_{y=l_y} = g_2\big|_{y=l_y} = u_0.$$

Lets construct the auxiliary function (14.48) satisfying the given boundary condition. In our problem $\gamma_1 = -l_x$, $\delta_1 = 1$, $\gamma_2 = 1$, $\delta_2 = 0$ and $\gamma_3 = -l_y$, $\delta_3 = 1$, $\gamma_4 = 1$, $\delta_4 = 0$ (see formulas (14.51), (14.52) and (14.55), (14.56)), so

$$\bar{X}(x) = x - l_x, \quad \bar{\bar{X}}(x) = 1, \quad \bar{Y}(y) = y - l_y, \quad \bar{\bar{Y}}(y) = 1.$$

Coefficients (14.58) of auxiliary function (14.48) are zeros:

$$A = -P_3[g_1]_{y=0} = 0, \quad B = -P_4[g_1]_{y=l_y} = 0,$$

$$C = -P_3[g_2]_{y=0} = 0, \quad D = -P_4[g_2]_{y=l_y} = u_0.$$

So, an auxiliary function (14.48) has the form

$$w(x,y) = u_0 \overline{\overline{X}} + u_0 \overline{\overline{Y}} + u_0 \overline{\overline{XY}} = 3u_0.$$

This function $w(x, y)$ is a harmonic function, that is, $\nabla^2 w(x,y) = 0$.

Function $v(x, y)$ is the solution to the Poisson problem with zero boundary conditions

$$\frac{\partial^2 v}{\partial x^2} + \frac{\partial^2 v}{\partial y^2} = f^*(x,y),$$

where $f^*(x,y) = f(x,y) + \dfrac{\partial^2 w}{\partial x^2} + \dfrac{\partial^2 w}{\partial y^2} = -Q \cdot \delta(x - x_0)\, \delta(y - y_0)$.

Eigenvalues and eigenfunctions are (see Appendix 1)

$$\lambda_{nm} = \lambda_{xn} + \lambda_{ym} = \pi^2\left(\frac{(2n-1)^2}{4l_x^2} + \frac{(2m-1)^2}{4l_y^2} \right), \quad n,m = 1,2,\ldots$$

$$V_{nm}(x,y) = X_n(x) \cdot Y_m(y) = \cos\frac{\pi(2n-1)x}{2l_x} \cos\frac{\pi(2m-1)y}{2l_y}$$

$$\|V_{nm}\|^2 = \|X_n\|^2 \cdot \|Y_m\|^2 = \frac{l_x l_y}{4}.$$

The solution $v(x, y)$ is defined by the series

$$v(x,y) = \sum_{n=1}^{\infty}\sum_{m=1}^{\infty} C_{nm} \cos\frac{(2n-1)\pi x}{2l_x} \cos\frac{(2m-1)\pi y}{2l_y},$$

where $C_{nm} = \dfrac{f_{nm}}{\lambda_{nm}}$ with

$$f_{nm} = \frac{1}{\|V_{nm}\|^2} \int_0^{l_x}\int_0^{l_y} f(x,y) \cdot \cos\frac{(2n-1)\pi x}{2l_x} \cos\frac{(2m-1)\pi y}{2l_y}\, dx dy$$

$$= \frac{4Q}{l_x l_y} \cos\frac{(2n-1)\pi x_0}{2l_x} \cos\frac{(2m-1)\pi y_0}{2l_y}$$

$$C_{nm} = \frac{f_{nm}}{\lambda_{nm}} = \frac{16Q l_x l_y}{\pi^2\left[(2n-1)^2 l_y^2 + (2m-1)^2 l_x^2\right]} \cos\frac{(2n-1)\pi x_0}{2l_x} \cos\frac{(2m-1)\pi y_0}{2l_y}.$$

The final solution (see also Figure 14.3) is

$$u(x,y) = w(x,y) + v(x,y) = 3u_0 + \sum_{n=1}^{\infty}\sum_{m=1}^{\infty} C_{nm} \cos\frac{(2n-1)\pi x}{2l_x} \cos\frac{(2m-1)\pi y}{2l_y}.$$

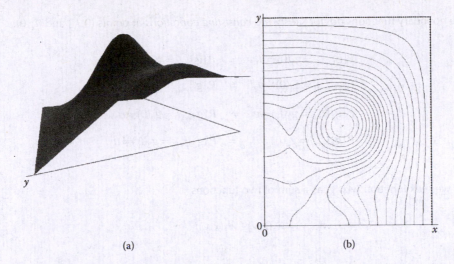

(a)

(b)

FIGURE 14.3 Surface plot (a) and lines of equal temperature (b) for Example 4. This solution was obtained with the program **Laplace** for the case $x_0 = 2$, $y_0 = 3$, $u_0 = 10$, $Q = 100$.

Example 5

Find the electrostatic potential in a charge-free rectangular domain $(0 \leq x \leq l_x, 0 \leq y \leq l_y)$, if one part of the bound ($x = 0$ and $y = 0$) is held at fixed potential $u = u_1$ and the other part ($x = l_x$ and $y = l_y$) at fixed potential $u = u_2$.

Solution. The problem is the Laplace equation

$$\frac{\partial^2 u}{\partial x^2} + \frac{\partial^2 u}{\partial y^2} = 0,$$

under the conditions

$$P_1[u]_{x=0} \equiv u(0, y) = u_1 \quad \left(\alpha_1 = 0, \ \beta_1 = 1 - \text{Dirichlet condition}\right),$$

$$P_2[u]_{x=l_x} \equiv u(l_x, y) = u_2 \quad \left(\alpha_2 = 0, \ \beta_2 = 1 - \text{Dirichlet condition}\right),$$

$$P_3[u]_{y=0} \equiv u(x, 0) = u_1 \quad \left(\alpha_3 = 0, \ \beta_3 = 1 - \text{Dirichlet condition}\right),$$

$$P_4[u]_{y=l_y} \equiv u(x, l_y) = u_2 \quad \left(\alpha_4 = 0, \ \beta_4 = 1 - \text{Dirichlet condition}\right).$$

The solution to this problem can be expressed as the sum of two functions

$$u(x, y) = v(x, y) + w(x, y),$$

where $v(x, y)$ is a new unknown function and $w(x, y)$ is an auxiliary function satisfying the boundary conditions.

The boundary functions *do not satisfy the consisting conditions* at points $(0, l_y)$ and $(l_x, 0)$:

$$P_1[g_3]_{x=0} = g_3(0) = u_1 \quad = \quad P_3[g_1]_{y=0} = g_1(0) = u_1,$$

$$P_1[g_4]_{x=0} = g_4(0) = u_2 \quad \neq \quad P_4[g_1]_{y=l_y} = g_1(l_y) = u_1,$$

$$P_2[g_3]_{x=l_x} = g_3(l_x) = u_1 \quad \neq \quad P_3[g_2]_{y=0} = g_2(0) = u_2,$$

$$P_2[g_4]_{x=l_x} = g_4(l_x) = u_2 \quad = \quad P_4[g_2]_{y=l_y} = g_2(l_y) = u_2.$$

Thus we seek function $w(x, y)$ as a sum of two functions

$$w(x, y) = w_1(x, y) + w_2(x, y),$$

where $w_1(x, y)$ is the auxiliary function satisfying the consisting boundary conditions (see (14.60))

$$P_1[w_1]_{x=0} = g_1(y), \quad P_2[w_1]_{x=l_x} = g_2(y),$$

$$P_3[w_1]_{y=0} = -A\bar{X}(x) - C\bar{\bar{X}}(x),$$

$$P_4[w_1]_{y=l_y} = -B\bar{X}(x) - D\bar{\bar{X}}(x),$$

and has the form (14.62)

$$w_1(x, y) = g_1(y)\bar{X} + g_2(y)\bar{\bar{X}}.$$

Like in the previous situations, we find functions $\bar{X}(x)$ and $\bar{\bar{X}}(x)$

$$\bar{X}(x) = 1 - \frac{x}{l_x}, \quad \bar{\bar{X}}(x) = \frac{x}{l_x},$$

and coefficients *A, B, C, D* (see (14.61))

$$A = -P_3[g_1(y)]_{y=0} = -u_1, \quad B = -P_4[g_1(y)]_{y=l_y} = -u_1,$$

$$C = -P_3[g_2(y)]_{y=0} = -u_2, \quad D = -P_4[g_2(y)]_{y=l_y} = -u_2.$$

Thus, the auxiliary function $w_1(x, y)$, which satisfies the consisting boundary conditions

$$P_1[w_1]_{x=0} = u_1, \quad P_2[w_1]_{x=l_x} = u_2,$$

$$P_3[w_1]_{y=0} = u_1 + (u_2 - u_1)\frac{x}{l_x}, \quad P_4[w_1]_{y=l_y} = u_1 + (u_2 - u_1)\frac{x}{l_x},$$

is

$$w_1(x,y) = u_1 + \left(u_2 - u_1\right)\frac{x}{l_x}.$$

Function $w_1(x,y)$ is harmonic, that is, $\nabla^2 w_1(x,y) = 0$.

The function $w_2(x,y)$ is a particular solution of the *Laplace problem* with the following boundary conditions (see (14.63)):

$$P_1[w_2]_{x=0} = 0, \quad P_2[w_2]_{x=l_x} = 0,$$

$$P_3[w_2]_{y=0} = g_3(x) + A\overline{X}(x) + C\overline{\overline{X}}(x) = \frac{x}{l_x}\left(u_1 - u_2\right),$$

$$P_4[w_2]_{y=l_y} = g_4(x) + B\overline{X}(x) + D\overline{\overline{X}}(x) = \left(u_2 - u_1\right)\left(1 - \frac{x}{l_x}\right).$$

The solution of this problem has the form (14.64):

$$w_2(x,y) = \sum_{n=1}^{\infty} \{A_n Y_{1n}(y) + B_n Y_{2n}(y)\} \cdot X_n(x),$$

where λ_{xn}, $X_n(x)$, eigenvalues and eigenfunctions of respective Sturm-Liouville problem,

$$\lambda_{xn} = \left[\frac{n\pi}{l_x}\right]^2, \quad X_n(x) = \sin\frac{n\pi x}{l_x}, \quad \|X_n\|^2 = \frac{l_x}{2}, \quad n = 1,2,\dots$$

and functions $Y_{1n}(y)$ and $Y_{2n}(y)$ for the given boundary functions are

$$Y_{1n}(y) = \frac{\sinh\sqrt{\lambda_n}\, y}{\sinh\sqrt{\lambda_n}\, l_y}, \quad Y_{2n}(y) = \frac{\sinh\sqrt{\lambda_n}\, (l_y - y)}{\sinh\sqrt{\lambda_n}\, l_y}$$

(see Appendixes 1 and 3).

Coefficients A_n and B_n are determined with formulas (14.66):

$$A_n = \frac{1}{\|X_n\|^2}\int_0^{l_x}\left[g_4(x) + B\overline{X} + D\overline{\overline{X}}\right]X_n(x)\,dx$$

$$= \frac{2}{l_x}\int_0^{l_x}\left(u_2 - u_1\right)\left(1 - \frac{x}{l_x}\right)\sin\frac{n\pi x}{l_x}\,dx = \frac{2}{n\pi}\left(u_2 - u_1\right),$$

$$B_n = \frac{1}{\|X_n\|^2}\int_0^{l_x}\left[g_3(x) + A\overline{X} + C\overline{\overline{X}}\right]X_n(x)\,dx$$

$$= \frac{2}{l_x}\int_0^{l_x}\frac{x}{l_x}\left(u_1 - u_2\right)\sin\frac{n\pi x}{l_x}\,dx = (-1)^n\frac{2}{n\pi}\left(u_2 - u_1\right).$$

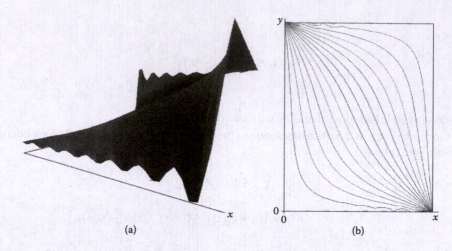

(a)   (b)

FIGURE 14.4 Surface plot (a) and lines of equal potential (b) for Example 5. This solution was obtained with the program **Laplace** for the case $u_1 = 10$, $u_2 = 20$.

So,

$$w_2(x,y) = 2(u_2 - u_1)\sum_{n=1}^{\infty}\frac{1}{n\pi \cdot \sinh\sqrt{\lambda_n}l_y}\left\{\sinh\sqrt{\lambda_n}y + (-1)^n \sinh\sqrt{\lambda_n}(l_y - y)\right\}\cdot \sin\frac{n\pi x}{l_x}.$$

Function $w_2(x,y)$ is a particular solution of the *Laplace problem* $\nabla^2 w_2(x,y) = 0$.
  Function $v(x, y)$ is the solution to the *Poisson problem with zero boundary conditions*, where

$$f^*(x,y) = \frac{\partial^2 w_1}{\partial x^2} + \frac{\partial^2 w_1}{\partial y^2} + \frac{\partial^2 w_2}{\partial x^2} + \frac{\partial^2 w_2}{\partial y^2} \equiv 0,$$

so, in this example,

$$v(x, y) \equiv 0.$$

The solution to the problem (see also Figure 14.4) is

$$u(x,y) = w_1(x,y) + w_2(x,y) = u_1 + (u_2 - u_1)\frac{x}{l_x}$$

$$+ 2(u_2 - u_1)\sum_{n=1}^{\infty}\frac{1}{n\pi \cdot \sinh\sqrt{\lambda_n}l_y}\left\{\sinh\sqrt{\lambda_n}y + (-1)^n \sinh\sqrt{\lambda_n}(l_y - y)\right\}\cdot \sin\frac{n\pi x}{l_x}.$$

14.6 LAPLACE'S EQUATION IN POLAR COORDINATES

In this and the following sections we consider two-dimensional problems that have a symmetry allowing the use of polar coordinates. Solutions to these problems contain simple trigonometric functions of the polar angle as well as power and logarithmic functions of the radius.

In polar coordinates, (r, φ), the Laplacian has the form

$$\nabla^2 = \frac{\partial^2}{\partial r^2} + \frac{1}{r}\frac{\partial}{\partial r} + \frac{1}{r^2}\frac{\partial^2}{\partial \varphi^2} \equiv \nabla_r^2 + \frac{1}{r^2}\nabla_\varphi^2. \qquad (14.71)$$

We begin by solving the Laplace's equation $\nabla^2 u = 0$ using the Fourier method of separation of variables. First we represent $u(r,\varphi)$ in the form

$$u(r,\varphi) = R(r)\Phi(\varphi). \qquad (14.72)$$

Substituting equation (14.72) into $\nabla^2 u = 0$ and separating the variables we have

$$\frac{r^2 \nabla_r R}{R} = -\frac{\nabla_\varphi^2 \Phi}{\Phi} \equiv \lambda.$$

Because the first term does not depend on the angular variable φ and the second does not depend on r, each term must equal a constant which we denoted as λ.

From here we obtain two separate equations for $R(r)$ and $\Phi(\varphi)$:

$$\Phi'' + \lambda\Phi = 0,$$

$$r^2 R'' + rR' - \lambda R = 0.$$

The equation for function $\Phi(\varphi)$ we have met several times before. The periodicity condition, $\Phi(\varphi + 2\pi) = \Phi(\varphi)$, leads to a discrete spectrum of eigenvalues:

$$\lambda_n = n^2, \quad n = 0,1,2,\ldots$$

thus the eigenfunctions are

$$\Phi = \Phi_n(\varphi) = \begin{cases} \cos n\varphi, \\ \sin n\varphi. \end{cases}$$

Negative values of n correspond to the same eigenfunctions and therefore need not be included in the list of eigenvalues.

The equation for $R(r)$

$$r^2 R'' + rR' - n^2 R = 0, \qquad (14.73)$$

is known as the Euler equation. The general solution to this equation is

$$R = R_n(r) = C_1 r^n + C_2 r^{-n}, \quad n \neq 0, \tag{14.74}$$

$$R_0(r) = C_1 + C_2 \ln r, \qquad n = 0.$$

Combining the above results for $\Phi_n(\varphi)$ and $R_n(r)$ we obtain the following particular and general solutions of Laplace's equation:

a) Under the condition that the solution be finite at $r = 0$ and infinite at $r \to \infty$, we have

$$u_n(r,\varphi) = r^n \begin{Bmatrix} \cos n\varphi \\ \sin n\varphi \end{Bmatrix}, \quad n = 0,1,\ldots$$

We can write a general solution for Laplace's problem for an *interior boundary value problem*, $0 \leq r \leq a$, as the expansion with these particular solutions $u_n(r,\varphi)$ as

$$u(r,\varphi) = \sum_{n=0}^{\infty} r^n \left(A_n \cos n\varphi + B_n \sin n\varphi \right).$$

The term with $n = 0$ is more conveniently written as $A_0 / 2$; thus we have

$$u(r,\varphi) = \frac{A_0}{2} + \sum_{n=1}^{\infty} r^n \left(A_n \cos n\varphi + B_n \sin n\varphi \right). \tag{14.75}$$

b) For the case that the solution is infinite at $r = 0$ and finite at $r \to \infty$ we have

$$u_n(r,\varphi) = \frac{1}{r^n} \begin{Bmatrix} \cos n\varphi \\ \sin n\varphi \end{Bmatrix}, \quad n = 0,1,\ldots$$

These functions may be used as solutions to the Laplace problem for regions outside of a circle. The general solution of Laplace's equation for such an *exterior boundary value problem*, $r \geq a$, limited (i.e., bounded) at infinity, can be written as

$$u(r,\varphi) = \frac{A_0}{2} + \sum_{n=1}^{\infty} \frac{1}{r^n} \left(A_n \cos n\varphi + B_n \sin n\varphi \right). \tag{14.76}$$

c) We also have a third set of solutions,

$$u_n(r,\varphi) = 1, \quad \ln r, \quad r^n \begin{Bmatrix} \cos n\varphi \\ \sin n\varphi \end{Bmatrix}, \quad \frac{1}{r^n} \begin{Bmatrix} \cos n\varphi \\ \sin n\varphi \end{Bmatrix}, \quad n = 1,2,\ldots$$

for the cases where the solution is unbounded as $r \to 0$, as well as $r \to \infty$. This set of functions is used to solve Laplace's equation for regions that form a circular ring, $a \le r \le b$.

14.7 LAPLACE'S EQUATION AND INTERIOR BVP FOR CIRCULAR DOMAIN

In this section we consider the first of the three cases presented in the previous section—solve the boundary value problem for a disk:

$$\nabla^2 u = 0 \quad \text{in} \quad 0 \le r < l, \tag{14.77}$$

with boundary condition

$$u(r, \varphi)\Big|_{r=l} = f(\varphi). \tag{14.78}$$

Applying the boundary condition (14.78) to formula (14.75) we obtain

$$\frac{A_0}{2} + \sum_{n=1}^{\infty} l^n \left(A_n \cos n\varphi + B_n \sin n\varphi \right) = f(\varphi) \tag{14.79}$$

From this we see that $l^n A_n$ and $l^n B_n$ are the Fourier coefficients of expansion of the function $f(\varphi)$ in the system (or basis) of trigonometric functions $\{\cos n\varphi, \sin n\varphi\}$. We may evaluate the coefficients using the formulas

$$A_n l^n = \frac{1}{\pi} \int_0^{2\pi} f(\varphi) \cos n\varphi \, d\varphi, \quad B_n l^n = \frac{1}{\pi} \int_0^{2\pi} f(\varphi) \sin n\varphi \, d\varphi, \quad n = 0,1,2,\dots \tag{14.80}$$

Thus, the solution of the *interior Dirichlet problem* for the Laplace equation is

$$u(r, \varphi) = \frac{A_0}{2} + \sum_{n=1}^{\infty} \left(\frac{r}{l} \right)^n \left[A_n \cos n\varphi + B_n \sin n\varphi \right]. \tag{14.81}$$

Example 6

Find the temperature distribution inside a circle if the boundary is kept at the temperature $T_0 = C_1 + C_2 \cos \varphi + C_3 \sin 2\varphi$.

Solution. It is obvious that for this particular case the series given by equation (14.80) reduces to three nonzero terms:

$$A_0 = 2C_1, \quad lA_1 = C_2, \quad l^2 B_2 = C_3.$$

In this case the solution given by equation (14.81) is

$$T = C_1 + C_2 \frac{r}{l} \cos\varphi + C_3 \left(\frac{r}{l}\right)^2 \sin 2\varphi.$$

Similarly we can obtain solutions of the *second and third boundary value problems* for Laplace's equation for a disc. We leave to the reader as Reading Exercises to check that the resulting formulas are correct for the following two cases of Neumann and mixed boundary conditions.

The *Neumann problem* for the Laplace equation with boundary condition

$$\left.\frac{\partial u}{\partial r}\right|_{r=l} = f(\varphi) \tag{14.82}$$

has the solution

$$u(r,\varphi) = \sum_{n=1}^{\infty} \frac{r^n}{nl^{n-1}} \left[A_n \cos n\varphi + B_n \sin n\varphi\right] + C, \tag{14.83}$$

where C is an arbitrarily function. We remind the reader that a solution of the interior Neumann problem can exist only under the condition

$$\int_{C_l} f \, dl = \int_0^{2\pi} f(\varphi) \, d\varphi = 0.$$

This condition is necessary and sufficient; the function $u(r,\varphi)$ in such a case is not unique because any arbitrary constant C can be added to this solution.

The *mixed problem* for the Laplace equation with boundary condition

$$\left.\frac{\partial u}{\partial r} + hu\right|_{r=l} = f(\varphi), \qquad h = \text{const} \tag{14.84}$$

has the solution

$$u(r,\varphi) = \frac{A_0}{2h} + \sum_{n=1}^{\infty} \frac{r^n}{(n+lh)l^{n-1}} \left[A_n \cos n\varphi + B_n \sin n\varphi\right]. \tag{14.85}$$

Clearly, in the case of homogeneous boundary conditions ($f(\varphi) = 0$) all three problems have only trivial solutions (i.e., equal to zero or for the Neumann problem, any constant).

Coefficients in the expansions in equations (14.81), (14.83), and (14.85) are determined using equation (14.80). Let us briefly discuss the convergence of the series in these expansions. If the function $f(\varphi)$ defining the boundary condition can be integrated absolutely, its Fourier coefficients are bounded and, as can be seen from the structure of these series, they converge in any interior point of the circle ($r<l$). The smoother the function $f(\varphi)$ is, the faster these series converge. The series can be differentiated term by term any number of times and the sums satisfy Laplace's equation, that is, they are harmonic functions. The same can be said for the problems discussed in the following sections.

An approximate value of $u(r,\varphi)$ can be obtained by keeping the first N terms. For an interior problem the absolute error of this approximation is

$$\varepsilon = \left| \sum_{n=N+1}^{\infty} \left(\frac{r}{l}\right)^n \left[A_n \cos n\varphi + B_n \sin n\varphi \right] \right| \le \sum_{n=N+1}^{\infty} \left(\frac{r}{l}\right)^n \left[|A_n| + |B_n| \right].$$

Because the terms in the series monotonically decrease, at $r<a$:

$$\sum_{n=N+1}^{\infty} \left(\frac{r}{l}\right)^n \left[|A_n| + |B_n| \right] \le \left[|A_{N+1}| + |B_{N+1}| \right] \sum_{n=N+1}^{\infty} \left(\frac{r}{l}\right)^n$$

$$= \left[|A_{N+1}| + |B_{N+1}| \right] \left(\frac{r}{l}\right)^{N+1} \frac{1}{1-r/l}.$$

The last expression was obtained as a sum of the geometric progression. From this we see that the values of ε are smaller for smaller values of r. Therefore, these series converge *faster* in any internal point of the domain than on the boundary.

Example 7

Let $u(r,\varphi)\big|_{r=a} = \sin(\varphi/2)$ at $0 \le \varphi < 2\pi$, $l = 10$. Keeping six terms in equation (14.81), find the temperature at several points $P(r,\varphi)$ of the disc: $P_1(0,\varphi)$, $P_2(2,\pi/18)$, $P_3(3,\pi/18)$.

Solution. The coefficients in (14.79) are $A_0 = 4/\pi$, $A_n = 4/\left[\pi\left(1-4n^2\right)\right]$ (for $n = 1,2,\ldots$), and $B_n = 0$; thus we have the expansion

$$u(r,\varphi) = \frac{2}{\pi} + \sum_{n=1}^{\infty} \left(\frac{r}{l}\right)^n \frac{4}{\pi(1-4n^2)} \cos n\varphi.$$

At point $P_1(0,0)$ the temperature is $u(0,0) = 2/\pi$ and the error is zero because $r = 0$. At point $P_2(2,\pi/18)$ keeping six terms in the partial sum and rounding off the result with accuracy, $\varepsilon_1 = 10^{-4}$, we obtain $u(2,\pi/18) = 0.5496$ and

$$\varepsilon \le \left| \left(\frac{2}{10}\right)^7 \frac{4}{\pi(1-4\cdot 7^2)} \frac{1}{1-\dfrac{2}{10}} \right| \approx 0.5 \cdot 10^{-5}.$$

The total error is $\varepsilon + \varepsilon_1 \approx 10^{-4}$.

Similarly, at point $P_3(3, \pi/18)$, $u(3, \pi/18) = 0.5031$, $\varepsilon \approx 10^{-4}$, and $\varepsilon + \varepsilon_1 \approx 2 \cdot 10^{-4}$.

Reading Exercise

Solve this example again but with discontinuous boundary conditions $u(r, \varphi)|_{r=a} = \sin(\varphi/2)$ for $0 < \varphi < \pi$ and $u(r, \varphi)|_{r=a} = \cos(\varphi/2)$ for $\pi < \varphi < 2\pi$.

Example 8

Find the temperature at the point $P_2(2, \pi/18)$ in the previous example with the accuracy 10^{-5}.

Solution. First we have to decide how many terms it is necessary to keep in the partial sum. Keeping only two, we obtain

$$\varepsilon \leq \left| \left(\frac{2}{10} \right)^3 \frac{4}{\pi(1 - 4 \cdot 3^2)} \frac{1}{1 - \dfrac{2}{10}} \right| \approx 10^{-3}.$$

This is insufficient accuracy. Including a few more terms we can find that $N = 6$ gives $\varepsilon \approx 0.5 \cdot 10^{-5}$, which is sufficient. Thus, we have to keep six terms and to round off the result with an accuracy of 10^{-5}. The result is $u(2, \pi/18) = 0.54956$.

It is important to highlight that in the case of a piecewise continuous boundary function $f(\varphi)$, the solutions of the boundary value problems exist and approach the values of $f(\varphi)$ continuously at points of continuity of this function. At a discontinuity point of the function $f(\varphi)$ the Fourier series, as was discussed in Chapter 9, converges to one-half the sum of left and right limits.

Example 9

Let an infinite homogenous cylinder with a circular surface of radius a be kept at a constant temperature

$$u(r, \varphi)|_{r=l} = \begin{cases} T_0, & 0 \leq \varphi < \pi, \\ -T_0, & \pi \leq \varphi < 2\pi \end{cases}$$

for any z. After a long period of time the temperature inside the cylinder will become constant, that is, the system reaches equilibrium. Find the temperature distribution inside the cylinder when this occurs.

Solution. This is the Dirichlet's interior problem for a circle. The solution to this problem is given by equation (14.81); formulas (14.80) give

$$A_n = 0, \quad B_{2k} = 0, \quad B_{2k+1} = \frac{4T_0}{\pi(2k+1)}, \quad l = 0,1,2....$$

Reading Exercise

Consider examples of discontinuous boundary conditions with big and small jumps at discontinuity points, for instance, $u(r,\varphi)|_{r=a} = A$ for $0 < \varphi < \pi$ and $u(r,\varphi)|_{r=a} = B$ for $\pi < \varphi < 2\pi$. Using different constant values of A and B, check that the number of terms necessary to keep the same accuracy at some given point P, depends on the value of the difference $|A - B|$.

Note that for the piecewise continuous boundary function, $f(\varphi)$, the series converges more slowly than for the continuous case. In the current example $|B_n| \sim n^{-1}$ at large n, whereas $|A_n| \sim n^{-2}$ in the previous example.

Along with analytical solutions, you should use the program **Laplace** to solve this and other problems and to make useful estimations.

14.8 LAPLACE'S EQUATION AND EXTERIOR BVP FOR CIRCULAR DOMAIN

In this section we consider the second of the three cases presented in Section 14.5. This problem is formulated as

$$\nabla^2 u = 0 \quad \text{for } r > l.$$

With the boundary condition $u|_{r=l} = f(\varphi)$ we directly obtain the solution of the *Dirichlet problem* as (14.51):

$$u(r,\varphi) = \frac{A_0}{2} + \sum_{n=1}^{\infty}\left(\frac{l}{r}\right)^n \left[A_n \cos n\varphi + B_n \sin n\varphi\right]. \tag{14.86}$$

The *Neumann problem* with boundary condition

$$\left.\frac{\partial u}{\partial r}\right|_{r=l} = f(\varphi) \tag{14.87}$$

has the solution

$$u(r,\varphi) = -\sum_{}^{\infty} \frac{1}{n}\frac{l^{n+1}}{r^n}\left[A_n \cos n\varphi + B_n \sin n\varphi\right] + C, \tag{14.88}$$

where C is an arbitrarily constant. We remind the reader again that the exterior Neumann problem for a plane has a solution only under the condition

$$\int_{C_l} f\, dl = \int_0^{2\pi} f(\varphi)\, d\varphi = 0 \tag{14.89}$$

and its solution has an arbitrary additive constant.

The *mixed problem*, with a boundary condition

$$\left.\frac{\partial u}{\partial r} - hu\right|_{r=l} = f(\varphi), \tag{14.90}$$

has the solution

$$u(r,\varphi) = -\frac{A_0}{2h} - \sum_{n=1}^{\infty} \frac{l^{n+1}}{(n+lh)r^n}\left[A_n \cos n\varphi + B_n \sin n\varphi\right] \tag{14.91}$$

Notice that different signs for h in equation (14.84) as compared to equation (14.91) are because of different directions of the vector normal to the boundary. For the interior problem this vector is directed outward; for the exterior problem it is inward.

Coefficients A_n and B_n in the series (14.86), (14.88), and (14.91) are the Fourier coefficients of function $f(\varphi)$ and are calculated with equations (14.80). We leave to the reader as useful Reading Exercises to prove the results in equations (14.86), (14.88), and (14.91).

14.9 POISSON'S EQUATION: GENERAL NOTES AND A SIMPLE CASE

Let us briefly discuss how to find a solution of Poisson's equation

$$\nabla^2 u = -f \tag{14.92}$$

with Dirichlet boundary condition:

$$u(r,\varphi)\big|_{r=l} = g(\varphi). \tag{14.93}$$

Either the interior or exterior problem can be considered. Note that, contrary to the solution of Laplace's equation, the solution of Poisson's equation is not zero even for homogeneous boundary conditions $g(\varphi) \equiv 0$.

The way to solve the problem (14.92), (14.93) is to solve the nonhomogeneous equation (Poisson's equation) without taking into consideration boundary conditions and then to add a solution of Laplace's equation in a way that the sum satisfies the boundary condition. In other words, the function u is presented as the sum of two functions, $u = u_p + u_0$, where u_p is a particular solution of Poisson's equation

$$\nabla^2 u_p = -f, \tag{14.94}$$

and u_0 is a solution of Laplace's equation

$$\nabla^2 u_0 = 0. \tag{14.95}$$

The function u should satisfy the necessary boundary condition from which follows the boundary condition for u_0:

$$u_0(r,\varphi)\big|_{r=l} = g(\varphi) - u_p. \tag{14.96}$$

The boundary value problem defined by equations (14.95) and (14.96) was considered in Sections 14.6 and 14.7; thus now we turn to the solution of equation (14.94). The question is how to find a particular solution of Poisson's equation that is finite in the center (for the interior problem) or at infinity (for the exterior one) irrespectively of the boundary condition at $r = l$. It should be emphasized that changing the type of boundary condition for Poisson's equation, (14.92), we need only to change the boundary condition for Laplace's equation (14.95).

Let us consider a common particular case: very often the inhomogeneous term $f(r,\varphi)$ has the form

$$f(r,\varphi) = r^m \cos n\varphi \tag{14.97}$$

(or perhaps $r^m \sin n\varphi$). Here m is an arbitrary real number, and n is an integer since the function $f(r,\varphi)$ should be periodic in φ; thus $f(r,\varphi + 2\pi) = f(r,\varphi)$. In this case a particular solution of Poisson's equation can be obtained using the method of undetermined coefficients. Note that $m > -2$ corresponds to an interior problem, while $m < -2$ to an exterior one. Indeed, the function $f(r,\varphi)$ can be infinite at $r = 0$ for the interior problem; we should only ensure that the integral

$$\int_S |f(r,\varphi)|\, dS$$

remains finite (for example, in electrostatics it means that the full charge inside the domain is finite). It is obvious that at $m > -2$ we have a finite value of

$$\int_0^l f(r,\varphi) r\, dr.$$

On the other hand, at $m < -2$ both $f(r \to \infty, \varphi)$ and the integral $\int_l^\infty f(r,\varphi) r\, dr$ remain finite.

The value $m = -2$ can be used only for a boundary problem involving solutions inside an annulus (ring).

In polar coordinates equation (14.92) is

$$\frac{\partial^2 u}{\partial r^2} + \frac{1}{r}\frac{\partial u}{\partial r} + \frac{1}{r^2}\frac{\partial^2 u}{\partial \varphi^2} = -f(r,\varphi)$$

and because of

$$\nabla^2 r^{m+2} \cos n\varphi = \left[\left(m+2 \right)^2 - n^2 \right] r^m \cos n\varphi,$$

the particular solution of equation

$$\nabla^2 u = -r^m \cos n\varphi$$

is

$$u_p = -\frac{r^{m+2} \cos n\varphi}{\left(m+2 \right)^2 - n^2}. \tag{14.98}$$

A difficulty occurs if $m + 2 = \pm n$, in which case we cannot apply equation (14.98). In this case we may seek the solution in the form

$$u_p = R(r) \cos n\varphi,$$

and obtain for $R(r)$ the equation

$$R'' + \frac{1}{r} R' - \frac{n^2}{r^2} R = -r^{\pm n - 2},$$

where the derivatives with respect to r are denoted by primes. The particular solution of this equation is

$$\mp \frac{r^{\pm n} \ln r}{2n} \quad \text{at } n \neq 0 \tag{14.99}$$

and

$$-\frac{\ln^2 r}{2} \quad \text{at } n = 0. \tag{14.100}$$

Recall that $m > -2$ corresponds to an interior problem, whereas n stays in the argument of $\cos n\varphi$, that is, there is no need to consider negative values of n. The result is the solution in equation (14.99), in which the upper sign can be used for the interior problem, $r \leq l$, and the lower sign for the exterior problem, $r \geq l$. The solution in equation (14.100) with $n = 0$ is the particular solution for a boundary value problem inside an annulus.

Example 10

Solve the boundary value problem for a disk given by

$$\nabla^2 u = -Axy, \quad r \le l,$$

$$u\big|_{r=l} = 0. \tag{14.101}$$

Solution. The function, f, on the right side of Poisson's equation is $f = Axy = Ar^2 \sin 2\varphi / 2$; thus we have the situation described by equation (14.97) with $m = n = 2$. Using the result of equation (14.98) we obtain a particular solution of equation (14.101) as

$$u_p(r,\varphi) = -\frac{A}{2 \cdot 12} r^4 \sin 2\varphi. \tag{14.102}$$

Taking the solution given by equation (14.102) into account, the boundary value problem of equations (14.95) and (14.96) takes the following form:

$$\nabla^2 u_0 = 0, \tag{14.103}$$

$$r = l: \quad u_0 = -u_p = \frac{A}{24} l^4 \sin 2\varphi. \tag{14.104}$$

Since the boundary condition (14.104) contains only one Fourier harmonic, we can conclude that $u_0(r,\varphi) = Cr^2 \sin 2\varphi$ (see Example 6 in Section 14.6) with $C = -Al^2/24$. Thus the function $u = u_p + u_0$ that satisfies the given boundary condition is

$$u(r,\varphi) = \frac{A}{24} r^2 \left(l^2 - r^2 \right) \sin 2\varphi. \tag{14.105}$$

14.10 POISSON'S INTEGRAL

It is possible to present a solution of Dirichlet's problem for Laplace's equation as an integral formula. Let us do this first for the interior problem for a circle. Substituting the formulas for Fourier coefficients in equation (14.80) into Equation (14.81) and switching the order of summation and integration, we obtain

$$u(r,\varphi) = \frac{1}{\pi} \int_0^{2\pi} f(\phi) \left\{ \frac{1}{2} + \sum_{n=1}^{\infty} \left(\frac{r}{l} \right)^n \left[\cos n\phi \cos n\varphi + \sin n\phi \sin n\varphi \right] \right\} d\phi$$

$$= \frac{1}{\pi} \int_0^{2\pi} f(\phi) \left\{ \frac{1}{2} + \sum_{n=1}^{\infty} \left(\frac{r}{l} \right)^n \cos n(\phi - \varphi) \right\} d\phi. \tag{14.106}$$

Since $t \equiv r/l < 1$, the expression in the parentheses can be transformed as follows:

$$Z \equiv \frac{1}{2} + \sum_{n=1}^{\infty} t^n \cos n(\phi - \varphi) = \frac{1}{2} + \frac{1}{2} \sum_{n=1}^{\infty} t^n \left[e^{in(\varphi - \phi)} + e^{-in(\varphi - \phi)} \right]$$

$$= \frac{1}{2} \left\{ 1 + \sum_{n=1}^{\infty} \left[\left(t e^{i(\varphi - \phi)} \right)^n + \left(t e^{-i(\varphi - \phi)} \right)^n \right] \right\}.$$

Using

$$\sum_{n=0}^{\infty} x^n = \frac{1}{1-x}, \quad \sum_{n=1}^{\infty} x^n = \sum_{n=0}^{\infty} x^n - 1 = \frac{x}{1-x},$$

we have

$$Z = \frac{1}{2} \left[1 + \frac{t e^{i(\varphi - \phi)}}{1 - t e^{i(\varphi - \phi)}} + \frac{t e^{-i(\varphi - \phi)}}{1 - t e^{-i(\varphi - \phi)}} \right] = \frac{1}{2} \frac{1 - t^2}{1 - 2t \cos(\varphi - \phi) + t^2}.$$

Therefore equation (14.106) becomes

$$u(r, \varphi) = \frac{1}{2\pi} \int_0^{2\pi} f(\phi) \frac{l^2 - r^2}{r^2 - 2lr \cos(\varphi - \phi) + l^2} \, d\phi. \qquad (14.107)$$

This formula gives the solution to the first boundary value problem inside a circle and is called the Poisson integral. The expression

$$u(r, \varphi, l, \phi) = \frac{l^2 - r^2}{r^2 - 2lr \cos(\varphi - \phi) + l^2} \qquad (14.108)$$

is called the Poisson kernel.

Expression (14.107) is not applicable at $r = l$, but its limit as $r \to l$ for any fixed value of φ is equal to $f(\varphi)$ because the series we used to obtain equation (14.107) is a continuous function in the closed region $r \le l$. Thus, the function defined by the formula

$$u(r, \varphi) = \begin{cases} \dfrac{1}{2\pi} \displaystyle\int_0^{2\pi} f(\phi) \dfrac{l^2 - r^2}{r^2 - 2lr \cos(\varphi - \phi) + l^2} \, d\phi, & \text{if } r < l, \\[4mm] f(\varphi), & \text{if } r = l \end{cases}$$

is a harmonic function satisfying the Laplace equation $\nabla^2 u = 0$ for $r < l$ and continuous in the closed region $r \le l$.

Similarly we obtain the solution to the exterior boundary value problem for a circle as

$$u(r,\varphi) = \begin{cases} \dfrac{1}{2\pi} \displaystyle\int_0^{2\pi} f(\phi) \dfrac{r^2 - l^2}{r^2 - 2lr\cos(\varphi-\phi) + l^2} \, d\phi, & \text{if } r > l, \\[4mm] f(\varphi), & \text{if } r = l. \end{cases} \tag{14.109}$$

Poisson's integrals cannot be evaluated analytically for an arbitrary function $f(\varphi)$; however, they are often very useful in certain applications. In particular they can be more useful for numerical calculations than the infinite series solution.

Example 11

Consider a stationary membrane's deflection from the equilibrium position. For the stationary case the membrane surface is described by the function $u = u(x, y)$, which satisfies the equation

$$\frac{\partial^2 u}{\partial x^2} + \frac{\partial^2 u}{\partial y^2} = 0.$$

If the membrane contour projection onto the x-y plane is a circle of radius l, we can consider this problem as an interior Dirichlet's problem for a circle.

Let the equation of the contour be given by function $u = f(\varphi)$, where f is a z-coordinate of the contour at angle φ. As an example consider a film fixed on a firm frame that has a circular projection onto the x-y plane with radius l and center at point O. The equation of the film contour in polar coordinates is $u = C\cos 2\varphi$ ($0 \le \varphi \le 2\pi$), $r = l$. Find the shape, $u(r,\varphi)$, of the film.

Solution. The solution is given by Poisson's integral:

$$u(r,\varphi) = \frac{1}{2\pi} \int_0^{2\pi} C\cos 2\phi \cdot \frac{l^2 - r^2}{r^2 - 2lr\cos(\phi - \varphi) + l^2} \, d\phi.$$

To evaluate this integral we use the substitution $\phi - \varphi = \zeta$. The limits of integration will not change because the integrand is periodic with period 2π (an integral in the limits from $-\varphi$ to $2\pi - \varphi$ is equal to the same integral in the limits from 0 to 2π). Thus we have

$$u(r,\varphi) = \frac{C(l^2 - r^2)}{2\pi} \int_0^{2\pi} \frac{\cos(2\zeta + 2\varphi)}{r^2 - 2lr\cos\zeta + l^2} \, d\zeta$$

$$= \frac{C(l^2 - r^2)}{2\pi} \left[\cos 2\varphi \int_0^{2\pi} \frac{\cos 2\zeta \, d\zeta}{r^2 - 2lr\cos\zeta + l^2} - \sin 2\varphi \int_0^{2\pi} \frac{\sin 2\zeta \, d\zeta}{r^2 - 2lr\cos\zeta + l^2} \right].$$

FIGURE 14.5   Shape of the film in Example 11.

The second of these integrals is equal to zero as the integral of the odd function on the interval $(0, 2\pi)$. So for this case

$$u(r, \varphi) = \frac{C(l^2 - r^2)}{2\pi} \cos 2\varphi \int_{-\pi}^{\pi} \frac{\cos 2\zeta}{r^2 - 2lr \cos \zeta + l^2} d\zeta.$$

With the substitution $\tan(\zeta/2) = v$ finally we obtain

$$u(r, \varphi) = \frac{C(l^2 - r^2)}{2\pi} \cos 2\varphi \frac{2\pi r^2}{l^2(l^2 - r^2)} = \frac{Cr^2}{l^2} \cos 2\varphi.$$

The function $C(r/l)^2 \cos 2\varphi$ is harmonic and takes the values $C\cos 2\varphi$ on the contour of the circle. It has a shape of a saddle, shown in Figure 14.5.

Reading Exercise

Solve this problem using the method of Section 14.8.

14.11 APPLICATION OF BESSEL FUNCTIONS FOR THE SOLUTION OF LAPLACE'S AND POISSON'S EQUATIONS IN A CIRCLE

As we already know from solving hyperbolic and parabolic equations, the eigenfunctions of the Laplacian in polar coordinates for Dirichlet, Neumann, and mixed interior problems can be expressed in terms of the Bessel functions

$$V_{nm}^{(1)} = J_n\left(\frac{\mu_m^{(n)}}{l} r\right) \cos n\varphi, \qquad V_{nm}^{(2)} = J_n\left(\frac{\mu_m^{(n)}}{l} r\right) \sin n\varphi. \tag{14.110}$$

Different types of boundary conditions lead to different eigenvalues $\mu_m^{(n)}$.

Therefore we may use functions (14.110) to solve Laplace's and Poisson's equations. On this way we can solve Poisson's equations for general form of functions $f(r, \varphi)$, not necessarily $r^{\pm n}\sin n\varphi$, like in Section 14.9. Note that the method discussed in the current section is realized in the program **Laplace**.

Let us consider the following interior boundary value problem for and Poisson's equation:

$$\nabla^2 u = \frac{\partial^2 u}{\partial r^2} + \frac{1}{r}\frac{\partial u}{\partial r} + \frac{1}{r^2}\frac{\partial^2 u}{\partial \varphi^2} = -f(r,\varphi), \quad 0 \le r < l,\, 0 \le \varphi < 2\pi, \tag{14.111}$$

$$\alpha\frac{\partial u(r,\varphi)}{\partial r} + \beta u(r,\varphi)\bigg|_{r=l} = g(\varphi), \tag{14.112}$$

$$u(r,\varphi) = u(r,\varphi + 2\pi), \quad |\alpha| + |\beta| \ne 0.$$

To separate variables when a boundary condition is nonhomogeneous, we should split function $u(r,\varphi)$ into two functions:

$$u(r,\varphi) = v(r,\varphi) + w(r,\varphi), \tag{14.113}$$

where the introduced auxiliary function $w(r,\varphi)$ must satisfy the nonhomogeneous boundary condition (14.112) and leaves the boundary condition for the function $v(r,\varphi)$ homogeneous.

The function $w(r,\varphi)$ satisfying the boundary condition (14.112) can be chosen in different ways; the only restriction is that it should be continuous and finite. Let us seek it in the form

$$w(r,\varphi) = (c_0 + c_1 r + c_2 r^2)\cdot g(\varphi).$$

Because

$$\frac{1}{r}\frac{\partial w(r,\varphi)}{\partial r} = \left(\frac{c_1}{r} + 2c_2\right)g(\varphi)$$

then $c_1 = 0$ and as the result

$$w(r,\varphi) = (c_0 + c_2 r^2)g(\varphi). \tag{14.114}$$

Case 1. For $\alpha = 0$, $\beta = 1$ we have:

a) Boundary condition $u(r,\varphi)\big|_{r=l} = g(\varphi)$ with auxiliary function

$$w(r,\varphi) = \frac{r^2}{l^2}g(\varphi); \tag{14.115}$$

b) Boundary condition $u(r,\varphi)\big|_{r=l} = g_0 = \text{const}$ with auxiliary function

$$w(r,\varphi) = g_0. \tag{14.116}$$

Case 2. For $\alpha = 1$, $\beta = 0$ we have the boundary condition $\dfrac{\partial u}{\partial r}(r,\varphi)\bigg|_{r=l} = g(\varphi)$ with auxiliary function

$$w(r,\varphi) = \frac{r^2}{2l}g(\varphi) + C, \tag{14.117}$$

where C is an arbitrary constant.

Case 3. For $\alpha = 1$, $\beta = h>0$ we have the boundary condition $\dfrac{\partial u}{\partial r}(r,\varphi) + hu(r,\varphi)\bigg|_{r=l} = g(\varphi)$ with auxiliary function

$$w(r,\varphi) = \frac{r^2}{l(2+hl)}g(\varphi). \tag{14.118}$$

It is easy to verify by the direct substitution that the above expressions for $w(r,\varphi)$ satisfy boundary condition (14.112).

For function $v(r, \varphi)$ we have the BVP with zero boundary condition

$$\nabla^2 v = \frac{\partial^2 v}{\partial r^2} + \frac{1}{r}\frac{\partial v}{\partial r} + \frac{1}{r^2}\frac{\partial^2 v}{\partial \varphi^2} = -f^*(r,\varphi), \tag{14.119}$$

$$\alpha\frac{\partial v(l,\varphi)}{\partial r} + \beta v(l,\varphi)\bigg| = 0, \tag{14.120}$$

with function $f^*(r,\varphi)$ defined as

$$f^*(r,\varphi) = f(r,\varphi) + \frac{\partial^2 w}{\partial r^2} + \frac{1}{r}\frac{\partial w}{\partial r} + \frac{1}{r^2}\frac{\partial^2 w}{\partial \varphi^2}.$$

Let us present the solution to the problem of equations (14.119) and (14.120) as a series in functions (14.110):

$$v(r,\varphi) = \sum_{n=0}^{\infty}\sum_{m=0}^{\infty}\left[A_{nm}V_{nm}^{(1)}(r,\varphi) + B_{nm}V_{nm}^{(2)}(r,\varphi)\right] \tag{14.121}$$

with $V_{nm}^{(1)}(r,\varphi), V_{nm}^{(2)}(r,\varphi)$ (14.110) satisfying the corresponding boundary condition. Because of the orthogonality, the coefficients in equation (14.121) can be obtained via the (unknown) function $v(r,\varphi)$ by multiplying (14.121) by $dS = rdrd\varphi$ and integrating over the disc's area:

$$A_{nm} = \frac{1}{\|V_{nm}^{(1)}\|^2} \int_0^{2\pi} \int_0^l v(r,\varphi)V_{nm}^{(1)}(r,\varphi)rdrd\varphi,$$

$$\tag{14.122}$$

$$B_{nm} = \frac{1}{\|V_{nm}^{(2)}\|^2} \int_0^{2\pi} \int_0^l v(r,\varphi)V_{nm}^{(2)}(r,\varphi)rdrd\varphi.$$

To find a final form for these functions we first multiply equation (14.119) by $V_{nm}^{(1)}(r,\varphi)$ and $V_{nm}^{(2)}(r,\varphi)$ and integrate over the circular domain:

$$\int_0^{2\pi} \int_0^l \nabla^2 v\, V_{nm}^{(1)}(r,\varphi)rdrd\varphi = -\int_0^{2\pi} \int_0^l f^*(r,\varphi)V_{nm}^{(1)}(r,\varphi)rdrd\varphi, \tag{14.123}$$

$$\int_0^{2\pi} \int_0^l \nabla^2 v\, V_{nm}^{(2)}(r,\varphi)rdrd\varphi = -\int_0^{2\pi} \int_0^l f^*(r,\varphi)V_{nm}^{(2)}(r,\varphi)rdrd\varphi. \tag{14.124}$$

Now substitute equation (14.121) into equations (14.123) and (14.124). Because $V_{nm}^{(1)}(r,\varphi)$ and $V_{nm}^{(2)}(r,\varphi)$ are the eigenfunctions of the Laplacian we have

$$\nabla^2 V_{nm}^{(1,2)}(r,\varphi) = -\lambda_{nm} V_{nm}^{(1,2)}(r,\varphi), \tag{14.125}$$

where the eigenvalues $\lambda_{nm} = (\mu_m^{(n)}/l)^2$ correspond to the boundary condition (14.120). The left sides of equations (14.123) ($p = 1$) and (14.124) ($p = 2$) become

$$-\sum_{i=0}^{\infty} \sum_{j=0}^{\infty} \lambda_{ij} \int_0^{2\pi} \int_0^l \left[A_{ij} V_{ij}^{(1)}(r,\varphi) + B_{ij} V_{ij}^{(2)}(r,\varphi) \right] V_{nm}^{(p)}(r,\varphi)rdrd\varphi.$$

Due to the orthogonality relation for the functions $V_{nm}^{(p)}(r,\varphi)$, the only term in the sums that differs from zero is

$$-\lambda_{nm} A_{nm} \|V_{nm}^{(p)}\|^2 \delta_{p1} \quad \text{or} \quad -\lambda_{nm} B_{nm} \|V_{nm}^{(p)}\|^2 \delta_{p2}.$$

Comparing with the right sides in equations (14.123) and (14.124) we obtain

$$\lambda_{nm} A_{nm} = f_{nm}^{(1)}, \quad \lambda_{nm} B_{nm} = f_{nm}^{(2)}, \quad n,m = 0,1,2,\ldots,$$

where

$$f_{nm}^{(p)} = \frac{1}{\| V_{nm}^{(p)} \|^2} \int_0^{2\pi} \int_0^a f^*(r,\varphi) V_{nm}^{(p)}(r,\varphi) r dr d\varphi. \tag{14.126}$$

From this the coefficients A_{nm}, B_{nm} can be obtained.

In the case of boundary conditions of the first or third type (Dirichlet's condition or mixed condition) the eigenvalues $\lambda_{nm} \neq 0$ for all $n, m = 0,1,2,\ldots$; in this case the solution is defined uniquely and has the form

$$v(r,\varphi) = \sum_{n=0}^{\infty} \sum_{m=0}^{\infty} \frac{1}{\lambda_{nm}} \left[f_{nm}^{(1)} V_{nm}^{(1)}(r,\varphi) + f_{nm}^{(2)} V_{nm}^{(2)}(r,\varphi) \right]. \tag{14.127}$$

In the case of boundary conditions of the second type (Neumann's condition) the eigenvalue $\lambda_{00} = 0$ ($V_{00}^{(1)} = 1$, $V_{00}^{(2)} = 0$) and all other eigenvalues are nonzero. There are two options. If

$$f_{00}^{(1)} = \int_0^{2\pi} \int_0^l f^*(r,\varphi) \, r dr d\varphi = 0,$$

then the coefficient A_{00} is undefined. A solution to the given problem exists but is determined only up to an arbitrary additive constant. The other coefficients are defined uniquely. The solution in this case is

$$v(r,\varphi) = \sum_{n=0}^{\infty} \sum_{m=0}^{\infty} \frac{1}{\lambda_{nm}} \left[f_{nm}^{(1)} V_{nm}^{(1)}(r,\varphi) + f_{nm}^{(2)} V_{nm}^{(2)}(r,\varphi) \right] + \text{const.} \tag{14.128}$$

If

$$\int_0^{2\pi} \int_0^l f^*(r,\varphi) \, r dr d\varphi \neq 0,$$

then the solution to the given problem does not exist.

Thus, the *general solution* of Poisson's problem in a circular domain with nonzero boundary condition has the form

$$u(r,\varphi) = w(r,\varphi) + v(r,\varphi)$$

$$= w(r,\varphi) + \sum_{n=0}^{\infty} \sum_{m=0}^{\infty} \left[A_{nm} V_{nm}^{(1)}(r,\varphi) + B_{nm} V_{nm}^{(2)}(r,\varphi) \right],$$

or

$$u(r,\varphi) = w(r,\varphi) + \sum_{n=0}^{\infty} \sum_{m=0}^{\infty} \left[A_{nm} \cos n\varphi + B_{nm} \sin n\varphi \right] J_n\left(\frac{\mu_m^{(n)}}{l} r \right), \tag{14.129}$$

where

$$A_{nm} = \frac{f_{nm}^{(1)}}{\lambda_{nm}} = \frac{1}{\lambda_{nm} \, \| V_{nm}^{(1)} \|^2} \int_0^{2\pi} \int_0^l f^*(r,\varphi) V_{nm}^{(1)}(r,\varphi) r \, dr d\varphi, \tag{14.130}$$

$$B_{nm} = \frac{f_{nm}^{(2)}}{\lambda_{nm}} = \frac{1}{\lambda_{nm} \, \| V_{nm}^{(2)} \|^2} \int_0^{2\pi} \int_0^l f^*(r,\varphi) V_{nm}^{(2)}(r,\varphi) r \, dr d\varphi \tag{14.131}$$

with $V_{nm}^{(p)}(r,\varphi)$ defined by equation (14.110) and the auxiliary function $w(r,\varphi)$ (14.114). Recall that the squared norms are different for different types of the BVP.

Example 12

Find a stationary temperature distribution in a thin circular plate of radius *l* if a boundary of the plate is kept at zero temperature and the plate contains a distributed source of heat with

$$Q(r,\varphi) = r^2 \cos 2\varphi.$$

Solution. The BVP is formulated as the Poisson equation

$$\frac{\partial^2 u}{\partial r^2} + \frac{1}{r} \frac{\partial u}{\partial r} + \frac{1}{r^2} \frac{\partial^2 u}{\partial \varphi^2} = -r^2 \cos 2\varphi \quad (0 \le r < l, \, 0 \le \phi < 2\pi)$$

with zero boundary condition

$$u(l,\varphi,t) = 0.$$

This is the Dirichlet BVP and the eigenvalues $\mu_m^{(n)}$ are positive roots of equation $J_n(\mu r / l) = 0$. From there, $\lambda_{nm} = \left(\mu_m^{(n)} / l \right)^2$.

Next find $f_{nm}^{(1)}$, $f_{nm}^{(2)}$ using formulas (14.126). Because the boundary condition is homogeneous, $f^*(r,\varphi) = f(r,\varphi)$; integrals in (14.126) contain

$$\int_0^{2\pi} \cos 2\varphi \cos n\varphi \, d\varphi = \begin{cases} 2\pi, & \text{if } n = 2, \\ 0, & \text{if } n \ne k, \end{cases} \quad \int_0^{2\pi} \cos 2\varphi \sin n\varphi \, d\varphi = 0,$$

thus $f_{nm}^{(1)} = 0$ for $n \ne 2$ and $f_{nm}^{(2)} = 0$ for other values of *n*.

To find $f_{2m}^{(1)}$ let us use the recurrence formula for Bessel functions

$$\int x^n J_{n-1}(x)\,dx = x^n J_n(x),$$

which gives

$$\int_0^l r^3 J_2\left(\frac{\mu_m^{(2)}}{l}r\right)dr = \frac{l^4}{\mu_m^{(2)}}J_3\left(\mu_m^{(2)}\right).$$

Thus

$$f_{2m}^{(1)} = \frac{1}{\left\|V_{2m}^{(1)}\right\|^2}\int_0^{2\pi}\int_0^l r^2\cos 2\varphi\cdot\cos 2\varphi\cdot J_2\left(\frac{\mu_m^{(2)}}{l}r\right)r\,dr\,d\varphi$$

$$= \frac{2}{\pi l^2\left[J_2'\left(\mu_m^{(2)}\right)\right]^2}\cdot 2\pi\cdot\frac{l^4}{\mu_m^{(2)}}J_3\left(\mu_m^{(2)}\right) = \frac{4l^2}{\mu_m^{(2)}\cdot\left[J_2'\left(\mu_m^{(2)}\right)\right]^2}\cdot J_3\left(\mu_m^{(2)}\right).$$

Therefore, the solution of the problem $u(r,\varphi)$ is the series

$$u(r,\varphi) = \sum_{m=0}^{\infty}\frac{l^2}{\left(\mu_m^{(2)}\right)^2}f_{2m}^{(1)}\cdot\cos 2\varphi\cdot J_2\left(\frac{\mu_m^{(2)}}{l}r\right).$$

See also Figures 14.6 and 14.7.

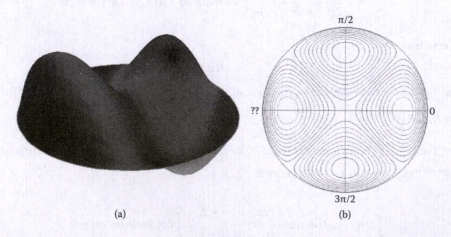

FIGURE 14.6　Surface plot (a) and lines of equal temperature (b) of the solution $u(r,\varphi)$ for Example 12.

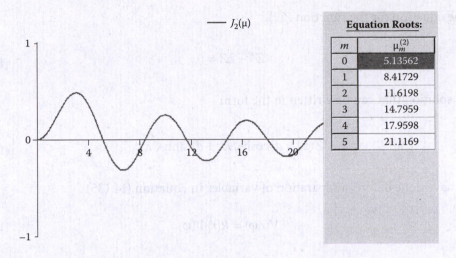

FIGURE 14.7   Graph of function $J_2(\mu)$ and table for the roots of equation $J_2(\mu)=0$ for Example 12. The graph and table are obtained with program **Laplace**.

14.12   THREE-DIMENSIONAL LAPLACE EQUATION FOR A CYLINDER

Up to now we did not discuss three-dimensional problems. In this section we will show that they can be solved in the way similar to two-dimensional ones.

Let us separate the variables in the *three-dimensional* Laplace equation

$$\nabla^2 u = 0, \tag{14.132}$$

inside a circular bounded cylinder, $r \le a$, $0 \le \varphi < 2\pi$, $0 \le z \le l$.

Let us represent the unknown function, u, in the following form:

$$u(r,\varphi,z) = V(r,\varphi)Z(z). \tag{14.133}$$

Substituting (14.133) into equation (14.132), after the separation of variables, we get

$$\frac{1}{Vr}\frac{\partial}{\partial r}\left(r\frac{\partial V}{\partial r}\right) + \frac{1}{Vr^2}\frac{\partial^2 V}{\partial \varphi^2} = -\frac{1}{Z}\frac{\partial^2 Z}{\partial z^2} = -\lambda, \tag{14.134}$$

where $\lambda > 0$ is a separation constant that will be determined from the conditions of existence of a nontrivial solution of the problem. As a result, we obtain the equation for the function $V(r,\varphi)$

$$\frac{1}{r}\frac{\partial}{\partial r}\left(r\frac{\partial V}{\partial r}\right) + \frac{1}{r^2}\frac{\partial^2 V}{\partial \varphi^2} + \lambda V = 0 \tag{14.135}$$

and the equation for the function $Z(z)$

$$Z'' - \lambda Z = 0. \tag{14.136}$$

with a solution that can be written in the form

$$Z(z) = d_1 \cosh \sqrt{\lambda} z + d_2 \sinh \sqrt{\lambda} z. \tag{14.137}$$

As we have done before, a separation of variables in equation (14.135)

$$V(r,\varphi) = R(r)\Phi(\varphi) \tag{14.138}$$

results in

$$\frac{1}{r} \frac{\partial}{\partial r}\left(r \frac{\partial R}{\partial r} \right) + \left(\lambda - \frac{\nu}{r^2} \right) R = 0 \tag{14.139}$$

and

$$\Phi'' + \nu\Phi = 0, \tag{14.140}$$

where ν is a separation constant. From periodicity condition, $\Phi(\varphi) = \Phi(\varphi + 2\pi)$, we have $\nu = n^2$, where $n = 0,1,2,...$, and two sets of eigenfunctions $\Phi_n(\varphi)$:

$$\Phi_n(\varphi) = \sin n\varphi \quad \text{and} \quad \Phi_n(\varphi) = \cos n\varphi. \tag{14.141}$$

Equation (14.139) for the function $R(r)$ is the Bessel equation, which bounded at $r = 0$ solutions are the Bessel functions

$$R(r) = J_n\left(\sqrt{\lambda} r \right). \tag{14.142}$$

This result explains our choice of the sign, $\lambda > 0$—only in this case the separation of variables leads to the Bessel equation for function $R(r)$. If $\lambda < 0$ the solutions of equation (14.139) for $R(r)$ give the modified Bessel functions, $I_n\left(\sqrt{-\lambda} r \right)$ (see book [1]). Also, when $\lambda < 0$, the solutions of equation (14.136) for function $Z(z)$ are periodic functions, $\sin\left(\sqrt{-\lambda} z \right)$ and $\cos\left(\sqrt{-\lambda} z \right)$, with the eigenvalues $\lambda = \lambda_m$ determined by the boundary conditions at $z = 0$ and $z = l$. The physics of the problems governs what sign of λ has to be chosen: in one case we expect oscillatory behavior of function $u(x, y, z)$ in z; in the other, the exponential behavior. In book [1] the reader can find solution of equation (14.134) for both signs of λ.

Consider the Dirichlet boundary value problem with zero boundary condition at the lateral surface

$$u\big|_{r=a} = 0 \tag{14.143}$$

and boundary conditions at the bottom and top surfaces

$$u\big|_{z=0} = g(r,\varphi), \quad u\big|_{z=l} = F(r,\varphi), \tag{14.144}$$

where $g(r,\varphi)$ and $F(r,\varphi)$ are given functions.

Assume that an expected solution is not periodic in z; thus $\lambda > 0$. The boundary condition at the lateral surface (14.143) results in $R(a) = 0$, which gives

$$J_n\left(\mu_m^{(n)}\right) = 0. \tag{14.145}$$

where $\mu_m^{(n)} = \sqrt{\lambda}a$, $m = 0,1,2\ldots$ numerates the roots of this equation.

Therefore, equation (14.135) gives the eigenvalues $\lambda_{nm} = \left(\mu_m^{(n)}/a\right)^2$ for Dirichlet BVP; the corresponding eigenfunctions are

$$V_{nm}^{(1)} = J_n\left(\frac{\mu_m^{(n)}}{a}r\right)\cos n\varphi, \quad V_{nm}^{(2)} = J_n\left(\frac{\mu_m^{(n)}}{a}r\right)\sin n\varphi. \tag{14.146}$$

The norms in the case of Dirichlet BVP are

$$\left\|V_{nm}^{(1)}\right\|^2 = \frac{a^2}{2}\left|J_n'\left(\mu_m^{(n)}\right)\right|^2 \pi\varepsilon_n \quad \text{and} \quad \left\|V_{nm}^{(2)}\right\|^2 = \frac{a^2}{2}\left|J_n'\left(\mu_m^{(n)}\right)\right|^2 \pi, \tag{14.147}$$

where $\varepsilon_n = 2$ for $n = 0$ and $\varepsilon_n = 1$ for $n \neq 0$.

Using the above results, the solution of the first BVP for equations (14.132) with zero boundary condition at the lateral surface can be represented as the series

$$u(r,\varphi,z) = \sum_{n=0}^{\infty}\sum_{m=0}^{\infty}\left\{\left[a_{1nm}V_{nm}^{(1)}(r,\varphi) + b_{1nm}V_{nm}^{(2)}(r,\varphi)\right]\cosh\left(\sqrt{\lambda_{nm}}z\right)\right.$$

$$\left. + \left[a_{2nm}V_{nm}^{(1)}(r,\varphi) + b_{2nm}V_{nm}^{(2)}(r,\varphi)\right]\sinh\left(\sqrt{\lambda_{nm}}z\right)\right\}. \tag{14.148}$$

From the boundary condition at $z = 0$ we have

$$\sum_{n=0}^{\infty}\sum_{m=0}^{\infty}\left[a_{1nm}V_{nm}^{(1)}(r,\varphi)+b_{1nm}V_{nm}^{(2)}(r,\varphi)\right]=g(r,\varphi)$$

where the coefficients a_{1nm} and b_{1nm} may be determined by expanding the function $g(r,\varphi)$ in a Fourier series in basis functions $V_{nm}^{(1)}(r,\varphi)$ and $V_{nm}^{(2)}(r,\varphi)$:

$$a_{1nm}=\frac{1}{\left\|V_{nm}^{(1)}\right\|^2}\int_0^a\int_0^{2\pi}g(r,\varphi)V_{nm}^{(1)}(r,\varphi)r\,dr\,d\varphi,$$

(14.149)

$$b_{1nm}=\frac{1}{\left\|V_{nm}^{(2)}\right\|^2}\int_0^a\int_0^{2\pi}g(r,\varphi)V_{nm}^{(2)}(r,\varphi)r\,dr\,d\varphi.$$

Analogously, we find the coefficients a_{2nm} and b_{2nm} using the boundary condition at $z = l$:

$$a_{2nm}=\frac{1}{\left\|V_{nm}^{(1)}\right\|^2}\int_0^a\int_0^{2\pi}F(r,\varphi)V_{nm}^{(1)}(r,\varphi)r\,dr\,d\varphi,$$

(14.150)

$$b_{2nm}=\frac{1}{\left\|V_{nm}^{(2)}\right\|^2}\int_0^a\int_0^{2\pi}F(r,\varphi)V_{nm}^{(2)}(r,\varphi)r\,dr\,d\varphi.$$

In a similar way the *three-dimensional* Laplace and Poisson equations can be solved in a cylindrical domain for other types of boundary conditions on the lateral surface. The only difference is that the equations (14.145) and (14.147) should be replaced by the proper ones for the corresponding types of the boundary conditions (it is shown in detail in Chapter 11).

Example 13

Find an expression for the potential of the electrostatic field inside a cylinder $r \leq a$, $0 \leq z \leq l$, the upper end and outside surfaces of which are grounded and the lower end of which is held at potential $A\sin 2\varphi$.

Solution. The problem is formulated as

$$\nabla^2 u = 0 \quad u\big|_{r=a}=0 \quad u\big|_{z=0}=A\sin 2\varphi, \quad u\big|_{z=l}=0.$$

Coefficients $a_{2mn} = b_{2mn} = 0$. Clearly all $a_{1nm} = 0$ because of orthogonality of function sin2φ and functions cos$n\varphi$ on $[0,2\pi]$. Among coefficients b_{1nm} only coefficients $b_{12m} \neq 0$:

$$b_{12m} = \frac{A}{\left\|V_{2m}^{(2)}\right\|^2} \int_0^a \int_0^{2\pi} J_2\left(\frac{\mu_m^{(2)}}{a}r\right) \sin 2\varphi \sin 2\varphi r \, dr \, d\varphi = \frac{A\pi}{\left\|V_{2m}^{(2)}\right\|^2} \int_0^a J_2\left(\frac{\mu_m^{(2)}}{a}r\right) r \, dr,$$

or

$$b_{12m} = \frac{2A}{a^2 \left|J_2'\left(\mu_m^{(2)}\right)\right|^2} \int_0^a J_2\left(\frac{\mu_m^{(2)}}{a}r\right) r \, dr \text{ —this integral can be calculated numerically.}$$

Thus,

$$u(r,\varphi,z) = \sin 2\varphi \cdot \sum_{m=0}^{\infty} b_{12m} J_2\left(\frac{\mu_m^{(2)}}{a}r\right) \cosh\left(\mu_m^{(2)} z / a\right).$$

Problems

In problems 1 through 20 we consider a Laplace and Poisson problems over a *rectangular* domains $(0 \leq x \leq l_x, 0 \leq y \leq l_y)$, in problems 21 through 38 over a *circular* domains of radius l. Solve these problems analytically, which means the following: formulate the equation and boundary conditions, obtain the eigenvalues and eigenfunctions, and write the formulas for coefficients of the series expansion and the expression for the solution of the problem. If the integrals in the coefficients are not easy to evaluate, leave it for the program, which evaluates them numerically. Then study the problem in detail with the program **Laplace**. Obtain the pictures of several eigenfunctions and screenshots of solution and of the auxiliary functions. Directions on how to use the program are found in Appendix 5.

In problems 1 through 5 we consider rectangular plates $(0 \leq x \leq l_x, 0 \leq y \leq l_y)$ that are thermally insulated over their lateral surfaces. There are no heat sources or absorbers inside the plates. Find the steady-state temperature distribution in the plates.

1. The sides $x = 0$, $y = 0$, and $y = l_y$ have a fixed temperature of zero and the side $x = l_x$ follows the temperature distribution $u(l_x, y) = \sin^2(\pi y / l_y)$.

2. The sides $x = 0$, $x = l_x$, and $y = 0$ have a fixed temperature of zero and the side $y = l_y$ follows the temperature distribution $u(x, l_y) = \sin^2(\pi x / l_x)$.

3. The sides $x = 0$ and $y = 0$ have a fixed temperature of zero, the side $x = l_x$ is thermally insulated, and the side $y = l_y$ follows the temperature distribution

$$u(x, l_y) = \sin(5\pi x / l_x).$$

4. The sides $x = 0$ and $x = l_x$ have a fixed temperature of zero, and the sides $y = 0$ and $y = l_y$ follow the temperature distributions

$$u(x,0) = \sin\frac{\pi x}{l_x} \quad \text{and} \quad u(x,l_y) = \sin\frac{3\pi x}{l_x}.$$

5. The sides $y = 0$ and $y = l_y$ have a fixed temperature of zero, and the constant heat flows

$$u_x(0,y) = u_x(l_x,y) = \sin\left(3\pi y / l_y\right)$$

are supplied to the sides $x = 0$ and $x = l_x$ of the plate from outside.

In problems 6 through 10 we consider a rectangular plate ($0 \le x \le l_x$, $0 \le y \le l_y$) that is thermally insulated over its lateral surfaces. One constant internal source of heat acts at the point (x_0, y_0) of the plate. The value of this source is $Q = \text{const}$. Find the steady-state temperature distribution in the plate.

6. The edges $x = 0$, $y = 0$, and $y = l_y$ of the plate are kept at zero temperature and the edge $x = l_x$ is subjected to convective heat transfer with the environment, which has a temperature of zero.

7. The edges $x = 0$ and $y = 0$ of the plate are kept at zero temperature, the edge $y = l_y$ is thermally insulated, and the edge $x = l_x$ is subjected to convective heat transfer with the environment, which has a temperature of zero.

8. The edges $x = l_x$ and $y = l_y$ of the plate are kept at zero temperature, the edge $x = 0$ is thermally insulated, and the edge $y = 0$ is subjected to convective heat transfer with the environment, which has a temperature of zero.

9. The edges $y = 0$ and $y = l_y$ are thermally insulated, the edge $x = 0$ is kept at zero temperature, and the edge $x = l_x$ is subjected to convective heat transfer with the environment, which has a temperature of zero.

10. The edges $x = 0$, $x = l_x$, and $y = l_y$ are thermally insulated and the edge $y = 0$ is subjected to convective heat transfer with the environment, which has a temperature of zero.

In problems 11 through 15 we consider a heat-conducting rectangular plate ($0 \le x \le l_x$, $0 \le y \le l_y$) thermally insulated over its lateral surfaces. Let heat be generated throughout the plate; the intensity of internal sources (per unit mass of the plate) is $Q(x, y)$. Find the steady-state temperature distribution in the plate.

11. Part of the plate bound ($x = 0$ and $x = l_x$) is thermally insulated, and the other part is subjected to convective heat transfer with a medium. The temperature of the medium is $u_{md} = \text{const}$. The intensity of internal sources (per unit mass of the plate) is

$$Q(x,y) = A\cos\frac{\pi x}{l_x}\cos\frac{\pi y}{l_y}.$$

12. Part of the plate bound ($y = 0$ and $y = l_y$) is thermally insulated, and the other part is subjected to convective heat transfer with a medium. The temperature of the medium is $u_{md} = \text{const}$. The intensity of internal sources (per unit mass of the plate) is

$$Q(x, y) = A \cos \frac{\pi x}{l_x} \cos \frac{\pi y}{l_y}.$$

13. Sides $x = 0$ and $x = l_x$ of the plate are thermally insulated, and sides $y = 0$ and $y = l_y$ are held at fixed temperatures $u(x, 0) = 0$ and $u(x, l_y) = \cos(5\pi x / l_x)$. The intensity of internal sources (per unit mass of the plate) is

$$Q(x, y) = A x \sin \frac{\pi y}{l_y}.$$

14. Sides $y = 0$ and $y = l_y$ of the plate are thermally insulated, and sides $x = 0$ and $x = l_x$ are held at fixed temperatures $u(0, y) = 0$ and $u(l_x, y) = \cos(3\pi y / l_y)$.

$$Q(x, y) = A x \cos \frac{\pi y}{l_y}.$$

15. Sides $y = 0$ and $y = l_y$ of the plate are thermally insulated, side $x = 0$ is held at fixed temperature $u = u_1$, and side $x = l_x$ is subjected to convective heat transfer with a medium. The temperature of the medium is zero. The intensity of internal sources (per unit mass of the plate) is

$$Q(x, y) = A x y.$$

In problems 16 through 20 an infinitely long rectangular cylinder has its central axis along the z-axis and its cross section is a rectangular with sides of length π. The sides of the cylinder are kept at an electric potential described by functions $u(x, y)|_\Gamma$ given below. Find the electric potential within the cylinder.

16. $u\big|_{x=0} = u\big|_{x=\pi} = y^2$, $u\big|_{y=0} = x$, $u\big|_{y=\pi} = 0$

17. $u\big|_{x=0} = y$, $u\big|_{x=\pi} = y^2$, $u\big|_{y=0} = u\big|_{y=\pi} = 0$

18. $u\big|_{x=0} = 0$, $u\big|_{x=\pi} = y^2$, $u\big|_{y=0} = 0$, $u\big|_{y=\pi} = \cos x$

19. $u\big|_{x=0} = u\big|_{x=\pi} = \cos 2y$, $u\big|_{y=0} = u\big|_{y=\pi} = 0$

20. $u\big|_{x=0} = \cos 3y$, $u\big|_{x=\pi} = 0$, $u\big|_{y=0} = x^2$, $u\big|_{y=\pi} = 0$

In problems 21 through 23 we consider a circular plate of radius l that is thermally insulated over its lateral surfaces. The circular periphery of the plate is kept at the temperature described by functions of the polar angle $u(l,\varphi) = g(\varphi)$, given below. Find the steady-state temperature distribution in the plate.

21. $g(\varphi) = \cos 3\varphi$

22. $g(\varphi) = \cos\dfrac{\varphi}{2} + \dfrac{\varphi}{\pi}$

23. $g(\varphi) = \cos\dfrac{\varphi}{2} + \sin\dfrac{\varphi}{2}$

In problems 24 and 26 a thin homogeneous circular plate of radius l is electrically insulated over its lateral surfaces. The boundary of the plate is kept at an electric potential described by functions of polar angle $u(l,\varphi) = g(\varphi)$, given below. Find an electric potential in the plate.

24. $g(\varphi) = \sin 4\varphi$

25. $g(\varphi) = \sin\dfrac{\varphi}{2} + \dfrac{\pi}{2}$

26. $g(\varphi) = 2\cos\varphi - 3\sin\varphi$

In problems 27 through 29 we consider a very long (infinite) cylinder of radius l. The constant heat flow $Q(\varphi)$ is supplied to the surface of the cylinder from outside. Find the steady-state temperature distribution in the cylinder.

27. $Q(\varphi) = 3\sin\varphi + 2\sin^3\varphi$

28. $Q(\varphi) = 4\cos^3\varphi + 2\sin\varphi$

29. $Q(\varphi) = 5\sin\varphi - \cos\varphi$

In problems 30 through 32 we consider a very long (infinite) cylinder of radius l. At the surface of the cylinder there is a heat exchange with the medium. The medium temperature is $u_{md}(\varphi)$. Find the steady-state temperature distribution in the cylinder.

30. $u_{md}(\varphi) = \sin\varphi + \cos 4\varphi$

31. $u_{md}(\varphi) = 1 + 4\cos^2\varphi$

32. $u_{md}(\varphi) = 2\sin^2\varphi + 1$

In problems 33 through 35 we consider a circular plate of radius l that is thermally insulated over its lateral surfaces. One constant internal source of heat acts at the point (r_0, φ_0) of the plate. The value of this source is $Q = $ const. Find the steady-state temperature distribution in the plate.

33. The edge of the plate is kept at zero temperature.

34. The edge of the plate is subjected to convective heat transfer with the environment, which has a temperature of zero.

35. The edge of the plate is subjected to convective heat transfer with the environment, which has a temperature of $u_{md} = $ const.

In problems 36 through 38 we consider a circular plate of radius l that is thermally insulated over its lateral surfaces. The contour of the plate is maintained at zero temperature. A uniformly distributed source of heat with power $Q(r, \varphi)$ is acting in the plate. Find the steady-state temperature distribution in the plate.

36. $Q(r, \varphi) = r \cos 2\varphi$

37. $Q(r, \varphi) = r^2 \sin \varphi$

38. $Q(r, \varphi) = r^2 (\cos 3\varphi + \sin 3\varphi)$

Appendix 1: Eigenvalues and Eigenfunctions of One-Dimensional Sturm-Liouville Boundary Value Problem for Different Types of Boundary Conditions

The one-dimensional Sturm-Liouville boundary value problem for eigenvalues and eigenfunctions is formulated as follows:

Find values of parameter λ for which there exist nontrivial (not identically equal to zero) solutions of the boundary value problem:

$$X'' + \lambda X = 0, \quad 0 < x < l,$$

$$P_1[X] \equiv \alpha_1 X' + \beta_1 X \big|_{x=0} = 0, \quad |\alpha_1| + |\beta_1| \neq 0,$$

$$P_2[X] \equiv \alpha_2 X' + \beta_2 X \big|_{x=l} = 0, \quad |\alpha_2| + |\beta_2| \neq 0.$$

The eigenfunctions of this Sturm-Liouville problem are

$$X_n(x) = \frac{1}{\sqrt{\alpha_1^2 \lambda_n + \beta_1^2}} \left[\alpha_1 \sqrt{\lambda_n} \cos \sqrt{\lambda_n} \, x - \beta_1 \sin \sqrt{\lambda_n} \, x \right].$$

These eigenfunctions are orthogonal. Their square norms are

$$\|X_n\|^2 = \int_0^l X_n^2(x) dx = \frac{1}{2} \left[l + \frac{(\beta_2 \alpha_1 - \beta_1 \alpha_2)(\lambda_n \alpha_1 \alpha_2 - \beta_1 \beta_2)}{(\lambda_n \alpha_1^2 + \beta_1^2)(\lambda_n \alpha_2^2 + \beta_2^2)} \right].$$

The eigenvalues are

$$\lambda_n = \left(\frac{\mu_n}{l}\right)^2$$

where μ_n is the nth root of the equation

$$\tan\mu = \frac{(\alpha_1\beta_2 - \alpha_2\beta_1)l\mu}{\mu^2\alpha_1\alpha_2 + l^2\beta_1\beta_2}.$$

Below we consider all possible cases of boundary conditions.

1. Boundary conditions ($\alpha_1 = 0$, $\beta_1 = -1$, $\alpha_2 = 0$, $\beta_2 = 1$):

$$\begin{cases} X(0) = 0, & \text{— Dirichlet condition,} \\ X(l) = 0, & \text{— Dirichlet condition.} \end{cases}$$

Eigenvalues: $\lambda_n = \left[\dfrac{\pi n}{l}\right]^2$, $\quad n = 1, 2, 3, \dots$.

Eigenfunctions: $X_n(x) = \sin\dfrac{\pi n}{l}x$, $\quad \|X_n\|^2 = \dfrac{l}{2}$.

2. Boundary conditions ($\alpha_1 = 0$, $\beta_1 = -1$, $\alpha_2 = 1$, $\beta_2 = 0$):

$$\begin{cases} X(0) = 0, & \text{— Dirichlet condition,} \\ X'(l) = 0, & \text{— Neumann condition.} \end{cases}$$

Eigenvalues: $\lambda_n = \left[\dfrac{\pi(2n+1)}{2l}\right]^2$, $\quad n = 0, 1, 2, \dots$.

Eigenfunctions: $X_n(x) = \sin\dfrac{\pi(2n+1)}{2l}x$, $\quad \|X_n\|^2 = \dfrac{l}{2}$.

3. Boundary conditions ($\alpha_1 = 0$, $\beta_1 = -1$, $\alpha_2 = 1$, $\beta_2 = h_2$):

$$\begin{cases} X(0) = 0 & \text{—Dirichlet condition,} \\ X'(l) + h_2 X(l) = 0 & \text{— mixed condition.} \end{cases}$$

Eigenvalues: $\lambda_n = \left[\dfrac{\mu_n}{l}\right]^2$, $\quad n = 0,1,2, \ldots$, where μ_n is nth root of the equation $\tan\mu = -\dfrac{\mu}{h_2 l}$.

Eigenfunctions: $\quad X_n(x) = \sin\sqrt{\lambda_n}\,x$, $\quad \|X_n\|^2 = \dfrac{1}{2}\left(l + \dfrac{h_2}{\lambda_n + h_2^2}\right)$.

4. Boundary conditions ($\alpha_1 = 1$, $\beta_1 = 0$, $\alpha_2 = 0$, $\beta_2 = 1$):

$$\begin{cases} X'(0) = 0, & -\text{Neumann condition,} \\ X(l) = 0, & -\text{Dirichlet condition.} \end{cases}$$

Eigenvalues: $\lambda_n = \left[\dfrac{\pi(2n+1)}{2l}\right]^2$, $\quad n = 0,1,2, \ldots$.

Eigenfunctions: $\quad X_n(x) = \cos\dfrac{\pi(2n+1)}{2l}x$, $\quad \|X_n\|^2 = \dfrac{l}{2}$.

5. Boundary conditions ($\alpha_1 = 1$, $\beta_1 = 0$, $\alpha_2 = 1$, $\beta_2 = 0$):

$$\begin{cases} X'(0) = 0, & -\text{Neumann condition,} \\ X'(l) = 0, & -\text{Neumann condition.} \end{cases}$$

Eigenvalues: $\lambda_n = \left[\dfrac{\pi n}{l}\right]^2$, $\quad n = 0,1,2, \ldots$.

Eigenfunctions: $\quad X_n(x) = \cos\dfrac{\pi n}{l}x$, $\quad \|X_n\|^2 = \begin{cases} l, & n = 0, \\ l/2, & n > 0. \end{cases}$

6. Boundary conditions ($\alpha_1 = 1$, $\beta_1 = 0$, $\alpha_2 = 1$, $\beta_2 = h_2$):

$$\begin{cases} X'(0) = 0, & -\text{Neumann condition,} \\ X'(l) + h_2 X(l) = 0, & -\text{mixed condition.} \end{cases}$$

Eigenvalues: $\lambda_n = \left[\dfrac{\mu_n}{l}\right]^2$, $\quad n = 0,1,2, \ldots$, where μ_n is nth root of the equation $\tan\mu = \dfrac{h_2 l}{\mu}$.

Eigenfunctions: $\quad X_n(x) = \cos\sqrt{\lambda_n}\,x$, $\quad \|X_n\|^2 = \dfrac{1}{2}\left(l + \dfrac{h_2}{\lambda_n + h_2^2}\right)$.

7. Boundary conditions ($\alpha_1 = 1$, $\beta_1 = -h_1$, $\alpha_2 = 0$, $\beta_2 = 1$):

$$\begin{cases} X'(l) - h_1 X(l) = 0, & -\text{mixed condition,} \\ X(0) = 0, & -\text{Dirichlet condition.} \end{cases}$$

Eigenvalues: $\lambda_n = \left[\dfrac{\mu_n}{l} \right]^2$, $n = 0, 1, 2, \ldots$, where μ_n is nth root of the equation $\tan \mu = -\dfrac{\mu}{h_1 l}$.

Eigenfunctions: $X_n(x) = \sin \sqrt{\lambda_n}\,(l - x)$, $\|X_n\|^2 = \dfrac{1}{2}\left(l + \dfrac{h_1}{\lambda_n + h_1^2} \right)$.

8. Boundary conditions ($\alpha_1 = 1$, $\beta_1 = -h_1$, $\alpha_2 = 1$, $\beta_2 = 0$):

$$\begin{cases} X'(0) - h_1 X(0) = 0, & -\text{mixed condition,} \\ X'(l) = 0, & -\text{Neumann condition.} \end{cases}$$

Eigenvalues: $\lambda_n = \left[\dfrac{\mu_n}{l} \right]^2$, $n = 0, 1, 2, \ldots$, where μ_n is nth root of the equation $\tan \mu = \dfrac{h_1 l}{\mu}$.

Eigenfunctions: $X_n(x) = \cos \sqrt{\lambda_n}\,(l - x)$, $\|X_n\|^2 = \dfrac{1}{2}\left(l + \dfrac{h_1}{\lambda_n + h_1^2} \right)$.

9. Boundary conditions ($\alpha_1 = 1$, $\beta_1 = -h_1$, $\alpha_2 = 1$, $\beta_2 = h_2$):

$$\begin{cases} X'(0) - h_1 X(0) = 0, & -\text{mixed condition,} \\ X'(l) + h_2 X(l) = 0, & -\text{mixed condition.} \end{cases}$$

Eigenvalues: $\lambda_n = \left[\dfrac{\mu_n}{l} \right]^2$, $n = 0, 1, 2, \ldots$, where μ_n is nth root of the equation $\tan \mu = \dfrac{(h_1 + h_2)l\mu}{\mu^2 - h_1 h_2 l^2}$.

Eigenfunctions: $X_n(x) = \dfrac{1}{\sqrt{\lambda_n + h_1^2}}\left[\sqrt{\lambda_n} \cos \sqrt{\lambda_n}\, x + h_1 \sin \sqrt{\lambda_n}\, x \right]$,

$\|X_n\|^2 = \dfrac{1}{2}\left(l + \dfrac{(h_1 + h_2)(\lambda_n + h_1 h_2)}{(\lambda_n + h_1^2)(\lambda_n + h_2^2)} \right)$.

Appendix 2: Auxiliary Functions, $w(x,t)$, for Different Types of Boundary Conditions

In the case of nonhomogeneous boundary conditions

$$P_1[u] \equiv \alpha_1 \frac{\partial u}{\partial x} + \beta_1 u \bigg|_{x=0} = g_1(t), \quad |\alpha_1| + |\beta_1| \neq 0,$$

$$P_2[u] \equiv \alpha_2 \frac{\partial u}{\partial x} + \beta_2 u \bigg|_{x=l} = g_2(t), \quad |\alpha_2| + |\beta_2| \neq 0.$$

the solution to the boundary value problem can be expressed as the sum of two functions

$$u(x,t) = v(x,t) + w(x,t),$$

where $w(x,t)$ is an auxiliary function satisfying the boundary conditions and $v(x,t)$ is a solution of the boundary value problem with zero boundary conditions.

We seek an auxiliary function $w(x,t)$ in a form

$$w(x,t) = g_1(t)\overline{X}(x) + g_2(t)\overline{\overline{X}}(x),$$

where $\overline{X}(x)$ and $\overline{\overline{X}}(x)$ are polynomials of first or second order. The coefficients of these polynomials are adjusted to satisfy the boundary conditions.

Functions $\overline{X}(x)$ and $\overline{\overline{X}}(x)$ should be chosen in such a way that

$$P_1\left[\overline{X}(0)\right] = 1, \quad P_2\left[\overline{X}(l)\right] = 0,$$
$$P_1\left[\overline{\overline{X}}(0)\right] = 0, \quad P_2\left[\overline{\overline{X}}(l)\right] = 1.$$

If $\beta_1 \neq 0$ or $\beta_2 \neq 0$, then functions $\overline{X}(x)$ and $\overline{\overline{X}}(x)$ are polynomials of the first order

$$\overline{X}(x) = \gamma_1 + \delta_1 x, \quad \overline{\overline{X}}(x) = \gamma_2 + \delta_2 x,$$

Coefficients γ_1, δ_1, γ_2, δ_2 of these polynomials are defined uniquely and depend on the types of boundary conditions

$$\gamma_1 = \frac{\alpha_2 + \beta_2 l}{\beta_1\beta_2 l + \beta_1\alpha_2 - \beta_2\alpha_1}, \qquad \delta_1 = \frac{-\beta_2}{\beta_1\beta_2 l + \beta_1\alpha_2 - \beta_2\alpha_1},$$

$$\gamma_2 = \frac{-\alpha_1}{\beta_1\beta_2 l + \beta_1\alpha_2 - \beta_2\alpha_1}, \qquad \delta_2 = \frac{\beta_1}{\beta_1\beta_2 l + \beta_1\alpha_2 - \beta_2\alpha_1}.$$

If $\beta_1 = \beta_2 = 0$, then functions $\overline{X}(x)$ and $\overline{\overline{X}}(x)$ are polynomials of the second order

$$\overline{X}(x) = x - \frac{x^2}{2l}, \quad \overline{\overline{X}}(x) = \frac{x^2}{2l}.$$

1. Boundary conditions ($\alpha_1 = 0$, $\beta_1 = -1$, $\alpha_2 = 0$, $\beta_2 = 1$):

$$\begin{cases} u(0,t) = g_1(t), & -\text{Dirichlet condition,} \\ u(l,t) = g_2(t), & -\text{Dirichlet condition.} \end{cases}$$

Auxiliary function: $\quad w(x,t) = \left[1 - \dfrac{x}{l}\right] \cdot g_1(t) + \dfrac{x}{l} \cdot g_2(t).$

2. Boundary conditions ($\alpha_1 = 0$, $\beta_1 = -1$, $\alpha_2 = 1$, $\beta_2 = 0$):

$$\begin{cases} u(0,t) = g_1(t), & -\text{Dirichlet condition,} \\ u_x(l,t) = g_2(t), & -\text{Neumann condition.} \end{cases}$$

Auxiliary function: $w(x,t) = g_1(t) + x g_2(t).$

3. Boundary conditions ($\alpha_1 = 0$, $\beta_1 = -1$, $\alpha_2 = 1$, $\beta_2 = h_2$):

$$\begin{cases} u(0,t) = g_1(t), & -\text{Dirichlet condition,} \\ u_x(l,t) + h_2 u(l,t) = g_2(t), & -\text{mixed condition.} \end{cases}$$

Auxiliary function: $w(x,t) = \left[1 - \dfrac{h_2}{1 + h_2 l} x\right] \cdot g_1(t) + \dfrac{x}{1 + h_2 l} \cdot g_2(t).$

4. Boundary conditions ($\alpha_1 = 1$, $\beta_1 = 0$, $\alpha_2 = 0$, $\beta_2 = 1$):

$$\begin{cases} u_x(0,t) = g_1(t), & -\text{Neumann condition}, \\ u(l,t) = g_2(t), & -\text{Dirichlet condition}. \end{cases}$$

Auxiliary function: $w(x,t) = (x - l)g_1(t) + g_2(t)$.

5. Boundary conditions ($\alpha_1 = 1$, $\beta_1 = 0$, $\alpha_2 = 1$, $\beta_2 = 0$):

$$\begin{cases} u_x(0,t) = g_1(t), & -\text{Neumann condition}, \\ u_x(l,t) = g_2(t), & -\text{Neumann condition}. \end{cases}$$

Auxiliary function: $w(x,t) = \left[x - \dfrac{x^2}{2l} \right] \cdot g_1(t) + \dfrac{x^2}{2l} \cdot g_2(t)$.

6. Boundary conditions ($\alpha_1 = 1$, $\beta_1 = 0$, $\alpha_2 = 1$, $\beta_2 = h_2$):

$$\begin{cases} u_x(0,t) = g_1(t), & -\text{Neumann condition}, \\ u_x(l,t) + h_2 u(l,t) = g_2(t), & -\text{mixed condition}. \end{cases}$$

Auxiliary function: $w(x,t) = \left[x - \dfrac{1 + h_2 l}{h_2} \right] \cdot g_1(t) + \dfrac{1}{h_2} \cdot g_2(t)$.

7. Boundary conditions ($\alpha_1 = 1$, $\beta_1 = -h_1$, $\alpha_2 = 1$, $\beta_2 = h_2$):

$$\begin{cases} u_x(0,t) - h_1 u(0,t) = g_1(t), & -\text{mixed condition}, \\ u(l,t) = g_2(t), & -\text{Dirichlet condition}. \end{cases}$$

Auxiliary function: $w(x,t) = \dfrac{x - l}{1 + h_1 l} \cdot g_1(t) + \dfrac{1 + h_1 x}{1 + h_1 l} \cdot g_2(t)$.

8. Boundary conditions ($\alpha_1 = 1$, $\beta_1 = -h_1$, $\alpha_2 = 1$, $\beta_2 = h_2$):

$$\begin{cases} u_x(0,t) - h_1 u(0,t) = g_1(t), & -\text{mixed condition}, \\ u_x(l,t) = g_2(t), & -\text{Neumann condition}. \end{cases}$$

Auxiliary function: $w(x,t) = -\dfrac{1}{h_1} \cdot g_1(t) + \left[x + \dfrac{1}{h_1} \right] \cdot g_2(t).$

9. Boundary conditions ($\alpha_1 = 1$, $\beta_1 = -h_1$, $\alpha_2 = 1$, $\beta_2 = h_2$):

$$\begin{cases} u_x(0,t) - h_1 u(0,t) = g_1(t), & - \text{mixed condition}, \\ u_x(l,t) + h_2 u(l,t) = g_2(t), & - \text{mixed condition}. \end{cases}$$

Auxiliary function: $w(x,t) = \dfrac{h_2(x-l)-1}{h_1 + h_2 + h_1 h_2 l} \cdot g_1(t) + \dfrac{1 + h_1 x}{h_1 + h_2 + h_1 h_2 l} \cdot g_2(t).$

Appendix 3: Eigenfunctions of Sturm-Liouville Boundary Value Problem for the Laplace Equation in a Rectangular Domain for Different Types of Boundary Conditions

Let the function $u(x,y)$ be a particular solution of the Laplace problem with the following boundary conditions

$$P_1[u]_{x=0} = 0, \qquad P_2[u]_{x=l_x} = 0,$$

$$P_3[u]_{y=0} = g_3(x), \quad P_4[u]_{y=l_y} = g_4(x).$$

The solution of this problem has a form

$$u(x,y) = \sum_{n=1}^{\infty} \{A_n Y_{1n}(y) + B_n Y_{2n}(y)\} \cdot X_n(x),$$

where $\lambda_{xn}, X_n(x)$ are eigenvalues and eigenfunctions of Sturm-Liouville problem for an interval

$$X'' + \lambda X = 0, \quad 0 < x < l_x,$$

$$P_1[X]\|_{x=0} = P_2[X]\|_{x=l_x} = 0.$$

(The solution of this boundary value problem depends on the types of boundary conditions $P_1[u]$ and $P_2[u]$.)

Taking into account the eigenvalues λ_{xn} we obtain an equation for $Y(y)$:

$$Y'' - \lambda_{xn}Y = 0, \quad 0 < y < l_y.$$

The functions $Y_{1n}(y)$ and $Y_{2n}(y)$ are sought in the form

$$Y_{1n}(y) = a \cdot \sinh\sqrt{\lambda_{xn}}\,y + b \cdot \cosh\sqrt{\lambda_{xn}}\,y,$$

$$Y_{2n}(y) = c \cdot \sinh\sqrt{\lambda_{xn}}\,(l_y - y) + d \cdot \cosh\sqrt{\lambda_{xn}}\,(l_y - y).$$

The values of coefficients a, b, c, d are adjusted to satisfy the boundary conditions $P_3[u]_{y=0}$ and $P_4[u]_{y=l_y}$ and should be chosen in such a way that

$$P_3[Y_1(0)] = 0, \quad P_3[Y_1(l_y)] = 1,$$

$$P_4[Y_2(0)] = 1, \quad P_4[Y_2(l_y)] = 0.$$

1. Boundary conditions $\begin{cases} P_3[u] \equiv u|_{y=0} = g_3(x) & \text{(Dirichlet condition)}, \\ P_4[u] \equiv u|_{y=l_y} = g_4(x) & \text{(Dirichlet condition)}. \end{cases}$

Fundamental system:

$$Y_{1n}(y) = \frac{\sinh\sqrt{\lambda_{xn}}\,y}{\sinh\sqrt{\lambda_{xn}}\,l_y}, \quad Y_{2n}(y) = \frac{\sinh\sqrt{\lambda_{xn}}\,(l_y - y)}{\sinh\sqrt{\lambda_{xn}}\,l_y}.$$

If $\lambda_{x0} = 0$, $X_0(x) \equiv 1$ then

$$Y_{1n}(y) = \frac{y}{l_y}, \quad Y_{2n}(y) = 1 - \frac{y}{l_y}.$$

2. Boundary conditions $\begin{cases} P_3[u] \equiv u|_{y=0} = g_3(x) & \text{(Dirichlet condition)}, \\ P_4[u] \equiv \dfrac{\partial u}{\partial y}\Big|_{y=l_y} = g_4(x) & \text{(Neumann condition)}. \end{cases}$

Fundamental system:

$$Y_{1n}(y) = \frac{\sinh\sqrt{\lambda_{xn}}\,y}{\sqrt{\lambda_{xn}}\cosh\sqrt{\lambda_{xn}}\,l_y}, \quad Y_{2n}(y) = \frac{\cosh\sqrt{\lambda_{xn}}\,(l_y - y)}{\cosh\sqrt{\lambda_{xn}}\,l_y}.$$

If $\lambda_{x0} = 0$, $X_0(x) \equiv 1$ then

$$Y_{1n}(y) = y, \quad Y_{2n}(y) = 1.$$

3. Boundary conditions $\begin{cases} P_3[u] \equiv u|_{y=0} = g_3(x) & \text{(Dirichlet condition)}, \\ P_4[u] \equiv \dfrac{\partial u}{\partial y} + h_4 u \Big|_{y=l_y} = g_4(x) & \text{(mixed condition)}. \end{cases}$

Fundamental system:

$$Y_{1n}(y) = \frac{\sinh \sqrt{\lambda_{xn}}\, y}{h_4 \sinh \sqrt{\lambda_{xn}}\, l_y + \sqrt{\lambda_{xn}} \cosh \sqrt{\lambda_{xn}}\, l_y},$$

$$Y_{2n}(y) = \frac{h_4 \sinh \sqrt{\lambda_{xn}}\,(l_y - y) + \sqrt{\lambda_{xn}} \cosh \sqrt{\lambda_{xn}}\,(l_y - y)}{h_4 \sinh \sqrt{\lambda_{xn}}\, l_y + \sqrt{\lambda_{xn}} \cosh \sqrt{\lambda_{xn}}\, l_y}.$$

If $\lambda_{x0} = 0$, $X_0(x) \equiv 1$ then

$$Y_{1n}(y) = \frac{y}{1 + h_4 l_y}, \qquad Y_{2n}(y) = 1 - \frac{h_4}{1 + h_4 l_y}\, y.$$

4. Boundary conditions $\begin{cases} P_3[u] \equiv \dfrac{\partial u}{\partial y} \Big|_{y=0} = g_3(x) & \text{(Neumann condition)}, \\ P_4[u] \equiv u|_{y=l_y} = g_4(x) & \text{(Dirichlet condition)}. \end{cases}$

Fundamental system:

$$Y_{1n}(y) = \frac{\cosh \sqrt{\lambda_{xn}}\, y}{\cosh \sqrt{\lambda_{xn}}\, l_y}, \qquad Y_{2n}(y) = -\frac{\sinh \sqrt{\lambda_{xn}}\,(l_y - y)}{\sqrt{\lambda_{xn}} \cosh \sqrt{\lambda_{xn}}\, l_y}.$$

If $\lambda_{x0} = 0$, $X_0(x) \equiv 1$ then

$$Y_{1n}(y) = 1, \qquad Y_{2n}(y) = y - l_y.$$

5. Boundary conditions $\begin{cases} P_3[u] \equiv \dfrac{\partial u}{\partial y} \Big|_{y=0} = g_3(x) & \text{(Neumann condition)}, \\ P_4[u] \equiv \dfrac{\partial u}{\partial y} \Big|_{y=l_y} = g_4(x) & \text{(Neumann condition)}. \end{cases}$

Fundamental system:

$$Y_{1n}(y) = \frac{\cosh \sqrt{\lambda_{xn}}\, y}{\sqrt{\lambda_{xn}} \sinh \sqrt{\lambda_{xn}}\, l_y}, \qquad Y_{2n}(y) = -\frac{\cosh \sqrt{\lambda_{xn}}\,(l_y - y)}{\sqrt{\lambda_{xn}} \sinh \sqrt{\lambda_{xn}}\, l_y}.$$

If $\lambda_{x0} = 0$, $X_0(x) \equiv 1$ then

$$Y_{1n}(y) = \frac{1}{2l_y} y^2, \quad Y_{2n}(y) = y - \frac{1}{2l_y} y^2.$$

6. Boundary conditions $\begin{cases} P_3[u] \equiv \dfrac{\partial u}{\partial y}\bigg|_{y=0} = g_3(x) & \text{(Neumann condition)}, \\[4mm] P_4[u] \equiv \dfrac{\partial u}{\partial y} + h_4 u\bigg|_{y=l_y} = g_4(x) & \text{(mixed condition)}. \end{cases}$

Fundamental system:

$$Y_{1n}(y) = \frac{\cosh\sqrt{\lambda_{xn}}\, y}{\sqrt{\lambda_{xn}}\, \sinh\sqrt{\lambda_{xn}}\, l_y + h_4 \cosh\sqrt{\lambda_{xn}}\, l_y},$$

$$Y_{2n}(y) = -\frac{h_4 \sinh\sqrt{\lambda_{xn}}\,(l_y - y) + \sqrt{\lambda_{xn}}\, \cosh\sqrt{\lambda_{xn}}\,(l_y - y)}{\sqrt{\lambda_{xn}}\left[\sqrt{\lambda_{xn}}\, \sinh\sqrt{\lambda_{xn}}\, l_y + h_4 \cosh\sqrt{\lambda_{xn}}\, l_y\right]}.$$

If $\lambda_{x0} = 0$, $X_0(x) \equiv 1$ then

$$Y_{1n}(y) = \frac{y}{h_4}, \quad Y_{2n}(y) = y - \frac{1 + h_4 l_y}{h_4}.$$

7. Boundary conditions $\begin{cases} P_3[u] \equiv \dfrac{\partial u}{\partial y} - h_3 u\bigg|_{y=0} = g_3(x) & \text{(mixed condition)}, \\[4mm] P_4[u] \equiv u\big|_{y=l_y} = g_4(x) & \text{(Dirichlet condition)}. \end{cases}$

Fundamental system:

$$Y_{1n}(y) = \frac{h_3 \sinh\sqrt{\lambda_{xn}}\, y + \sqrt{\lambda_{xn}}\, \cosh\sqrt{\lambda_{xn}}\, y}{h_3 \sinh\sqrt{\lambda_{xn}}\, l_y + \sqrt{\lambda_{xn}}\, \cosh\sqrt{\lambda_{xn}}\, l_y},$$

$$Y_{2n}(y) = -\frac{\sinh\sqrt{\lambda_{xn}}\,(l_y - y)}{h_3 \sinh\sqrt{\lambda_{xn}}\, l_y + \sqrt{\lambda_{xn}}\, \cosh\sqrt{\lambda_{xn}}\, l_y}.$$

If $\lambda_{x0} = 0$, $X_0(x) \equiv 1$ then

$$Y_{1n}(y) = \frac{1 + h_3 y}{1 + h_3 l_y}, \quad Y_{2n}(y) = \frac{y - l_y}{1 + h_3 l_y}.$$

8. Boundary conditions
$$\begin{cases} P_3[u] \equiv \dfrac{\partial u}{\partial y} - h_3 u \Big|_{y=0} = g_3(x) \quad \text{(mixed condition)}, \\[3mm] P_4[u] \equiv \dfrac{\partial u}{\partial y} \Big|_{y=l_y} = g_4(x) \qquad \text{(Neumann condition)}. \end{cases}$$

Fundamental system:

$$Y_{1n}(y) = \frac{h_3 \sinh \sqrt{\lambda_{xn}}\, y + \sqrt{\lambda_{xn}} \cosh \sqrt{\lambda_{xn}}\, y}{\sqrt{\lambda_{xn}} \left[\sqrt{\lambda_{xn}} \sinh \sqrt{\lambda_{xn}}\, l_y + h_3 \cosh \sqrt{\lambda_{xn}}\, l_y \right]},$$

$$Y_{2n}(y) = -\frac{\cosh \sqrt{\lambda_{xn}}\,(l_y - y)}{\sqrt{\lambda_{xn}} \sinh \sqrt{\lambda_{xn}}\, l_y + h_3 \cosh \sqrt{\lambda_{xn}}\, l_y}.$$

If $\lambda_{x0} = 0$, $X_0(x) \equiv 1$ then

$$Y_{1n}(y) = \frac{1 + h_3 y}{1 + h_3 l_y}, \quad Y_{2n}(y) = \frac{y - l_y}{1 + h_3 l_y}.$$

9. Boundary conditions
$$\begin{cases} P_3[u] \equiv \dfrac{\partial u}{\partial y} - h_3 u \Big|_{y=0} = g_3(x) \quad \text{(mixed condition)}, \\[3mm] P_4[u] \equiv \dfrac{\partial u}{\partial y} + h_4 u \Big|_{y=l_y} = g_4(x) \quad \text{(mixed condition)} \end{cases}$$

Fundamental system:

$$Y_{1n}(y) = \frac{h_3 \sinh \sqrt{\lambda_{xn}}\, y + \sqrt{\lambda_{xn}} \cosh \sqrt{\lambda_n}\, y}{(\lambda_{xn} + h_3 h_4) \sinh \sqrt{\lambda_{xn}}\, l_y + \sqrt{\lambda_{xn}}\,(h_3 + h_4) \cosh \sqrt{\lambda_{xn}}\, l_y},$$

$$Y_{2n}(y) = -\frac{h_4 \sinh \sqrt{\lambda_{xn}}\,(l_y - y) + \sqrt{\lambda_{xn}} \cosh \sqrt{\lambda_{xn}}\,(l_y - y)}{(\lambda_{xn} + h_3 h_4) \sinh \sqrt{\lambda_{xn}}\, l_y + \sqrt{\lambda_{xn}}\,(h_3 + h_4) \cosh \sqrt{\lambda_{xn}}\, l_y}.$$

If $\lambda_{x0} = 0$, $X_0(x) \equiv 1$ then

$$Y_1(y) = \frac{1 + h_3 y}{h_3 + h_4 + h_3 h_4 l_y}, \quad Y_2(y) = \frac{h_4(y - l_y) - 1}{h_3 + h_4 + h_3 h_4 l_y}.$$

Appendix 4: A Primer on the Matrix Eigenvalue Problems and the Solution of the Selected Examples in Section 5.2

Let A be a given $n \times n$ real matrix and consider the matrix equation

$$AX = \lambda X, \tag{A4.1}$$

where X is an unknown column n-vector and λ is an unknown scalar. To solve equation (A4.1) is to determine all solution vectors X and constants λ that satisfy equation (A4.1). If we think of A as matrix operator, then we are looking for vectors X such that the result of an operator action on X is the multiplication of X by a scalar λ.

The trivial solution of equation (A4.1) is $X = 0$ (for any λ). This solution is of no interest. A value of λ for which equation (A4.1) has a solution $X \neq 0$ is called an *eigenvalue* of the matrix A. The corresponding nonzero solutions X are called the *eigenvectors* of A corresponding to that eigenvalue λ. The set of all the eigenvalues of A is called the spectrum of A. The spectrum consists of at least one eigenvalue and at most n numerically different eigenvalues. Furthermore, some or all eigenvalues may be complex and the complex eigenvalues occur in complex conjugate pairs.

The problem of determining the eigenvalues and eigenvectors of a matrix is called an algebraic (or matrix) eigenvalue problem. As if to match the name, the *analytical* solution of any such problem starts with the determination of eigenvalues, which is then followed by the solution for eigenvectors. (Note that numerical methods may be designed to determine eigenvectors first.) To begin with, equation (A4.1) is written in the equivalent form $(A - \lambda I)X = 0$, where I is the $n \times n$ identity matrix (that is, a matrix with all 1's on the main diagonal, and zeros elsewhere) and 0 is the zero (column) n-vector. This defines the homogeneous algebraic system of size n for the unknown components of vector X. It is known from linear algebra that the condition for the existence of a unique nontrivial solution of such a system is that its determinant is equal to zero; thus

$$\det(A - \lambda I) = 0. \tag{A4.2}$$

Equation (A4.2) is called the *characteristic equation* of A. Since A and I are given, the left-hand side of this equation, as we will see, defines an algebraic polynomial of nth degree for λ. This polynomial is called the *characteristic polynomial* of A. $A-\lambda I$ is called the *characteristic matrix* of A. Solving the characteristic equation gives all eigenvalues of A (the spectrum). Next, equation (A4.1) is written in the form

$$(A-\lambda I)X=0, \tag{A4.3}$$

the eigenvalues are substituted one by one, and for each eigenvalue the corresponding eigenvector, or eigenvectors, are found by solving the linear algebraic system for the components of X.

Note that the process of the solution of the eigenvalue problem that we just outlined is also described in Section 3.4, although there the terms *eigenvalue* and *eigenvector* are not used. Eigenvalues λ are equivalent to the exponents k, and eigenvectors X are equivalent to the vectors $\bar{\alpha}$ in the solution formula (3.15) of the linear, constant-coefficient system (3.14) ((3.14a)) in Section 3.4. Of course, (3.14a) in Section 3.4 coincides with (5.12) in Section 5.2. Now, the construction of the general solution of (3.14a) using k and $\bar{\alpha}$ is described in detail in Section 3.4. Thus absolutely the same construction applies to the solution of the three examples below from Section 5.2, only here we are using λ and X notation instead of k and $\bar{\alpha}$ notation, and fully benefit from using the vector form.

A4.1   EXAMPLE 2.3 FROM SECTION 5.2

Consider equation (A4.1), where $A = \begin{pmatrix} -2 & -2 \\ -1 & -3 \end{pmatrix}$. First, we calculate the characteristic matrix

$$A-\lambda I = \begin{pmatrix} -2 & -2 \\ -1 & -3 \end{pmatrix} - \lambda \begin{pmatrix} 1 & 0 \\ 0 & 1 \end{pmatrix} = \begin{pmatrix} -2 & -2 \\ -1 & -3 \end{pmatrix} - \begin{pmatrix} \lambda & 0 \\ 0 & \lambda \end{pmatrix} = \begin{pmatrix} -2-\lambda & -2 \\ -1 & -3-\lambda \end{pmatrix}. \tag{A4.4}$$

Next, the characteristic equation is

$$\det(A-\lambda I)=(-2-\lambda)(-3-\lambda)-2=\lambda^2+5\lambda+4=0. \tag{A4.5}$$

The characteristic equation is quadratic (polynomial of degree 2), since $n=2$. Solutions of equation (A4.5) are $\lambda_1=-1$, $\lambda_2=-4$. Using equation (A4.4), equation (A4.3) is written as

$$\begin{pmatrix} -2-\lambda & -2 \\ -1 & -3-\lambda \end{pmatrix}\begin{pmatrix} x_1 \\ x_2 \end{pmatrix}=\begin{pmatrix} 0 \\ 0 \end{pmatrix}, \tag{A4.6}$$

where x_1 and x_2 are the components of X. Plugging $\lambda = -1$ in equation (A4.6) gives the linear algebraic system for the unknowns x_1 and x_2:

$$-x_1 - 2x_2 = 0,$$

$$-x_1 - 2x_2 = 0.$$

There is only one equation in this system (the second equation must be discarded as it is precisely the first equation). Thus the system is redundant, or underdetermined—there are more unknowns than there are equations. Redundancy occurs often when solving for eigenvectors. It is known from linear algebra that such systems have infinitely many solutions. For our first eigenvector we may choose, say, $x_1 = -2$ and $x_2 = 1$, that is, $X_1 = \begin{pmatrix} -2 \\ 1 \end{pmatrix}$. Or we may choose any other x_1 and x_2 that solve $-x_1 - 2x_2 = 0$. It is considered a good practice to normalize an eigenvector so that it has length one. We will not do this here. It is clear from all of this that eigenvectors are not unique, as the following theorem states:

Theorem

If X is an eigenvector, then lX is also an eigenvector, where l is any nonzero constant. Also, if X and Y are eigenvectors corresponding to the same eigenvalue λ, then $X + Y$ is also an eigenvector corresponding to λ (provided $X \neq -Y$).

Here, the infinite set of eigenvectors that correspond to the eigenvalue $\lambda_1 = -1$ can be written as $X_I^{(1)} = l\begin{pmatrix} -2 \\ 1 \end{pmatrix}$. Notice that two or more eigenvectors from this set are linearly dependent. Next, plugging $\lambda = -4$ in equation (A4.6) gives another linear algebraic system for the unknowns x_1 and x_2:

$$-x_1 + x_2 = 0,$$

$$2x_1 - 2x_2 = 0.$$

Again this system is redundant: the second equation is the first equation times -2. Discarding the second equation and solving the first one (or doing the opposite) gives, say, $X_2 = \begin{pmatrix} 1 \\ 1 \end{pmatrix}$. The second infinite set of eigenvectors is $X_I^{(2)} = l\begin{pmatrix} 1 \\ 1 \end{pmatrix}$. Again two or more eigenvectors from this set are linearly dependent. Now, the eigenvalue problem has been solved. Note that eigenvalues are real and distinct. It is said that they have the *algebraic multiplicity* one, since each eigenvalue occurs once as the solution of the characteristic equation (the eigenvalue is not repeated). They also have the *geometric multiplicity* one, since there is only one linearly independent eigenvector corresponding to each eigenvalue.

Next, using the eigenvalues and eigenvectors that we determined, we can construct the general solution of the differential system in Example 2.3 by forming the linear combination with arbitrary coefficients C_1 and C_2, as is stipulated by equation (3.18) in Section 3.4:

$$X = C_1 X_1 e^{\lambda_1 t} + C_2 X_2 e^{\lambda_2 t} = C_1 \begin{pmatrix} -2 \\ 1 \end{pmatrix} e^{-t} + C_2 \begin{pmatrix} 1 \\ 1 \end{pmatrix} e^{-4t}.$$

A4.2   EXAMPLE 2.6 FROM SECTION 5.2

Consider equation (A4.1), where $A = \begin{pmatrix} -2 & -3 \\ -3 & -2 \end{pmatrix}$. The characteristic matrix is

$$A - \lambda I = \begin{pmatrix} -2 & -3 \\ 3 & -2 \end{pmatrix} - \lambda \begin{pmatrix} 1 & 0 \\ 0 & 1 \end{pmatrix} = \begin{pmatrix} -2 & -3 \\ 3 & -2 \end{pmatrix} - \begin{pmatrix} \lambda & 0 \\ 0 & \lambda \end{pmatrix} = \begin{pmatrix} -2-\lambda & -3 \\ 3 & -2-\lambda \end{pmatrix}.$$

The characteristic equation for A is

$$\det(A - \lambda I) = (-2-\lambda)^2 + 9 = \lambda^2 + 4\lambda + 13 = 0.$$

Its solutions are $\lambda_1 = -2+3i$, $\lambda_2 = -2-3i$. The linear system for the eigenvector that corresponds to the first eigenvalue is

$$\begin{pmatrix} -2+2-3i & -3 \\ 3 & -2+2-3i \end{pmatrix} \begin{pmatrix} x_1 \\ x_2 \end{pmatrix} = \begin{pmatrix} 0 \\ 0 \end{pmatrix}, \quad \text{or}$$

$$-3ix_1 - 3x_2 = 0,$$

$$3x_1 - 3ix_2 = 0.$$

The system is redundant, because the first equation is $-i$ times the second equation. Discarding, say, the first equation and solving the second one for x_1 gives $x_1 = ix_2$. Thus we can take $X_1 = \begin{pmatrix} i \\ 1 \end{pmatrix}$. The system for the eigenvector that corresponds to the second eigenvalue is

$$3ix_1 - 3x_2 = 0,$$

$$3x_1 + 3ix_2 = 0.$$

This gives $x_1 = -ix_2$, and we can take $X_2 = \begin{pmatrix} -i \\ 1 \end{pmatrix}$. Not surprisingly, $X_2 = X_1^*$. Again the algebraic and the geometric multiplicities of both eigenvalues are equal to one. To find the solution of the ODE system, we first take one eigenvalue and its corresponding eigenvector (say, take λ_1 and X_1) and form the vector complex solution (where the real and the imaginary parts must be separated):

$$X = e^{\lambda_1 t} X_1 = e^{(-2+3i)t} \begin{pmatrix} i \\ 1 \end{pmatrix} = e^{-2t}(\cos 3t + i\sin 3t)\begin{pmatrix} i \\ 1 \end{pmatrix} = e^{-2t}\begin{pmatrix} -\sin 3t \\ \cos 3t \end{pmatrix} + ie^{-2t}\begin{pmatrix} \cos 3t \\ \sin 3t \end{pmatrix}.$$

Multiplying the real and imaginary parts by arbitrary constants and summing up gives the general solution of the ODE system.

Next, we give the formal definition of the algebraic and geometric multiplicity of eigenvalues.

Definition

The order M_λ of an eigenvalue λ as a root of the characteristic polynomial is called the algebraic multiplicity of λ. The number m_λ of linearly independent eigenvectors corresponding to λ is called the geometric multiplicity of λ.

Since the characteristic polynomial has degree n, the sum of all the algebraic multiplicities must equal n. In general, $m_\lambda \leq M_\lambda$. In the two examples above $m_\lambda = M_\lambda = 1$, but the case $m_\lambda < M_\lambda$ occurs frequently. Next, we address this case by considering Example 2.9.

A4.3 EXAMPLE 2.9 FROM SECTION 5.2

Consider equation (A4.1), where $A = \begin{pmatrix} 0 & 1 \\ -4 & -4 \end{pmatrix}$. The characteristic matrix is

$$A - \lambda I = \begin{pmatrix} 0 & 1 \\ -4 & -4 \end{pmatrix} - \lambda \begin{pmatrix} 1 & 0 \\ 0 & 1 \end{pmatrix} = \begin{pmatrix} -\lambda & 1 \\ -4 & -4-\lambda \end{pmatrix}.$$

The characteristic equation for A is

$$\det(A - \lambda I) = \lambda(4+\lambda) + 4 = \lambda^2 + 4\lambda + 4 = 0.$$

Its solutions are $\lambda_1 = \lambda_2 = \lambda = -2$. Thus $M_\lambda = 2$. The linear system for the eigenvector that corresponds to λ is

$$\begin{pmatrix} 2 & 1 \\ -4 & -2 \end{pmatrix}\begin{pmatrix} x_1 \\ x_2 \end{pmatrix} = \begin{pmatrix} 0 \\ 0 \end{pmatrix}, \quad \text{or}$$

$$2x_1 + x_2 = 0,$$

$$-4x_1 - 2x_2 = 0.$$

Again the system is redundant, because the second equation is −2 times the first equation. Discarding, say, the second equation and solving the first one for x_1 gives $x_1 = -x_2/2$. We may take $x_2 = -2$, then $x_1 = 1$. Thus we obtained the eigenvector $X_1 = \begin{pmatrix} 1 \\ -2 \end{pmatrix}$. Other eigenvectors of the system are nonzero multiples of X_1, which does not help since they all are linearly dependent. The method now is to substitute $X_2 = X_1 t e^{\lambda t} + U e^{\lambda t}$ with constant $U = \begin{pmatrix} u_1 \\ u_2 \end{pmatrix}$ in (5.12), $X' = AX$. This gives

$$X_2' = X_1 e^{\lambda t} + \lambda X_1 t e^{\lambda t} + \lambda U e^{\lambda t} = AX_2 = AX_1 t e^{\lambda t} + AU e^{\lambda t} = \lambda X_1 t e^{\lambda t} + AU e^{\lambda t},$$

because X_1 is the eigenvector and thus $AX_1 = \lambda X_1$. $\lambda X_1 t e^{\lambda t}$ cancels from the second and from the last parts of the equality, and then division by $e^{\lambda t}$ gives

$$(A - \lambda I)U = X_1.$$

This is the linear algebraic system for u_1 and u_2. Note that the matrix of this system again is the characteristic matrix!

Plugging here $\lambda = -2$ and $X_1 = \begin{pmatrix} 1 \\ -2 \end{pmatrix}$ gives the redundant system

$$2u_1 + u_2 = 1,$$

$$-4u_1 - 2u_2 = -2.$$

As the solution of one of these equations we can take $\begin{pmatrix} u_1 \\ u_2 \end{pmatrix} = \begin{pmatrix} 0 \\ 1 \end{pmatrix}$, or any other solution vector

that is linearly independent of X_1. Finally, substituting X_2, λ, and U in $X_2 = X_1 t e^{\lambda t} + U e^{\lambda t}$ and forming the linear combination of X_1 and X_2 gives the general solution:

$$X = C_1 \begin{pmatrix} 1 \\ -2 \end{pmatrix} e^{-2t} + C_2 \left[\begin{pmatrix} 1 \\ -2 \end{pmatrix} t + \begin{pmatrix} 0 \\ 1 \end{pmatrix} \right] e^{-2t}.$$

This form of the solution is precisely the same form that we obtained in Section 5.2 by comparing to the solution of the critically damped oscillator.

Appendix 5: How to Use the Software Associated with the Book

The purpose of this appendix is to explain how to use the software associated with the book. The programs included are **TrigSeries**, **Waves**, **Heat**, **Laplace**, and **FourierSeries**. The program **TrigSeries** may be used for the trigonometric Fourier series problems found in Chapter 9. The program **Waves** accompanies Chapters 10 and 11 and is designed to solve the following type of problems: vibrations of infinite or semi-infinite strings, vibrations of finite strings, vibrations of flexible rectangular or circular membranes, and transverse vibrations of thin uniform rods. The **Heat** program is designed to solve the following type of problems, examples of which are found in Chapters 12 and 13: heat conduction within an infinite and semi-infinite rod; heat conduction within a finite uniform rod; and heat conduction within a thin uniform rectangular or circular membrane. The **Laplace** program is designed to solve boundary value problems for elliptic equations over rectangular and circular domains discussed in Chapter 14. Such equations describe problems involving electrostatic or gravitational potentials, steady temperature distributions, and other phenomenon. The **FourierSeries** program is associated with Chapter 8 and provides solutions to three kinds of problems: expansion in a generalized Fourier series in terms of classical orthogonal polynomials (Legendre, Chebyshev of the first and second kind, Jacoby, Laguerre, Hermite), expansion in Fourier-Bessel series, and expansion in Fourier series in terms of associated Legendre functions.

In Section A5.1, we discuss the common features of all the programs. The five sections that follow the overview provide examples of the use of each of the programs to solve problems, some of which were taken from the book problem sets without any changes (and in many cases the same or related examples are provided as library examples); others were modified to show how to use different program options. The programs are also sufficiently flexible to be able to solve many other problems not found in the text. The results of computations can be saved in the users' files to be examined again. The library sets can be extended by the reader. The programs use the same analytical formulas and methods that are presented in the main text. Thus, the programs do the same steps the reader is supposed to do in solving the problems analytically. The only numeric calculation the program performs is the evaluation of the coefficients of Fourier series (with Gauss' method and its modifications) and partial sums.

Each program has an integrated help system where detailed explanation of theoretical background can be found. Help is available by clicking the "Help" button from the main menu of the program. The Help is context dependent, meaning that you will be provided with help associated with the window you are currently using.

A5.1 PROGRAM OVERVIEW

In the following we use the **Heat** program to demonstrate various program menus and screens, but the other programs function in basically the same way. After you start any of the application you will see a main window (Figure A5.1 shows the **Heat** Program main window). From this window you may choose the type of problem you wish to solve (for some programs, for example **TrigSeries**, there is only one choice).

Suppose, for example, you chose the second option, "*Heat conduction within a Finite Rod.*" To solve a particular problem you must next enter the values of the parameters. Starting from the "*Data*" menu there are two basic ways to proceed; by choosing "*Library Example*" or "*New Problem.*" If you choose to load a library example you will be asked to select one of the available problems (shown in Figure A5.2 for the **Heat** program).

A description of each of the examples can be read by clicking the "*Text + Hint*" button. Once you select an example you will see the parameters of the problem in a dialog box with parameters already entered according to the selected example (Figure A5.3 for the **Heat** program).

If you choose "*New problem*" in the previous step, parameters of the problem are not defined and the spaces for input functions and parameters on the previous screen (Figure A5.3) will be left blank. To define boundary values use the drop-down dialog and

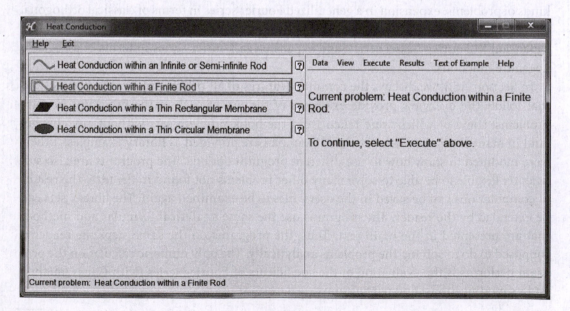

FIGURE A5.1   The **Heat** program main window.

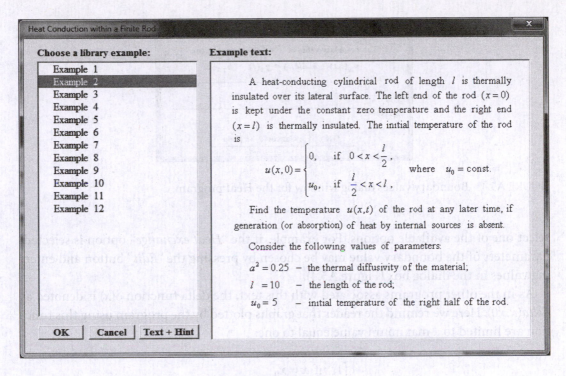

FIGURE A5.2   Library example selection for the **Heat** program.

FIGURE A5.3   Dialog box for problem parameters for the **Heat** program.

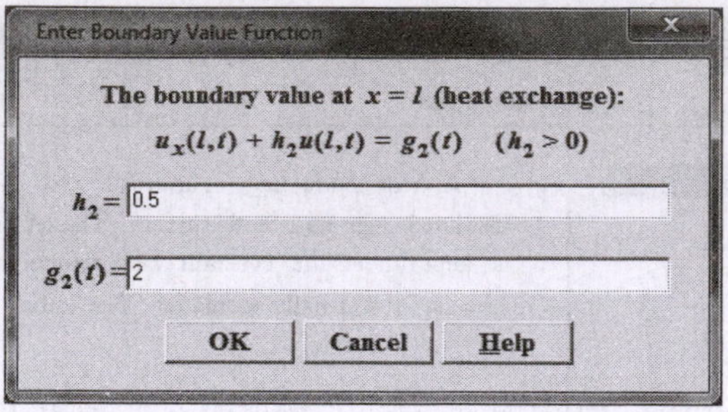

FIGURE A5.4   Boundary values dialog window for the **Heat** program.

select one of the available options. For example, if the "*Heat exchange*" option is selected, parameters of the boundary value may be chosen by pressing the "*Edit*" button and entering values in the dialog box (Figure A5.4).

As in the other programs associated with this text, the delta function $\delta(x)$ is denoted as *delta(x,x0)*. Here we remind the reader that graphs plotted by the program using this function are limited to a maximum value equal to one:

$$\text{delta}(x,x\,0) = \begin{cases} 1, & \text{if } x = x_0 \\ 0, & \text{if } x \neq x_0, \end{cases}$$

$$\text{delta}(x,x\,0) * \text{delta}(y,y\,0) = \begin{cases} 1, & \text{if } x = x_0, \, y = y_0 \\ 0, & \text{if } x \neq x_0 \text{ or } y \neq y_0, \end{cases}$$

but in actual calculations *delta(x,x0)* has the same properties as the "correct" $\delta(x)$.

To solve cases where functions are defined only on intervals, the following function is also helpful:

$$\text{Im } p(x,a,b) = \begin{cases} 1, & \text{if } x \in [a,b], \\ 0, & \text{if } x \notin [a,b]. \end{cases}$$

A5.2   EXAMPLES USING THE PROGRAM TRIGSERIES

The interactive **TrigSeries** program is a convenient instrument for working with trigonometric Fourier series. The function that is to be expanded as a Fourier series can be given either analytically or as a table of values. In the latter case the table can be entered directly from the keyboard or as an ASCII file prepared in advance. The program also includes a formula analyzer with a wide set of functions, allowing one to input initial functions and real parameters analytically (see the help topic "*Mathematical Functions, Operations* *and Constants*" under the "*General Items*" category). The graphical features of the program make it possible to draw graphs of the initial functions, individual members of the series, or certain partial sums of the expansions. A number of examples of how to apply the **TrigSeries** program to solve different problems follow.

Example A5.1

Let the function $f(x) = x$ be defined in the interval $[0,l]$. Expand $f(x)$ in a trigonometric Fourier series using the general (sine and cosine) expansion. Calculate the coefficients of the expansion. Examine the behavior of the individual harmonics of this expansion and of its partial sum $S_n(x)$ using the option "*Choose terms….*" Draw the bar chart of squared amplitudes A_n^2 and the graph of the periodic extensions in the interval $[-2l,3l]$.

Return to the dialog box for entering parameters and solve this problem using even (and then odd) terms only method of expansion. Compare the results of these expansions.

Solution. The program **TrigSeries** has a set of predefined examples of expansions into trigonometric Fourier series and the given problem is from this set. To load the problem select "*Data*" from the main menu, then click "*Library example.*" From the given list of problems select "*Example 1.*" The problem text will appear in the right part of the screen (Figure A5.2). To proceed with the selected problem, click the "OK" button; in the dialog window parameters for the current problem will be displayed (Figure A5.3). Since all parameters for the problem have been entered, proceed by clicking "OK."

At this point, the program is ready to solve the problem with the selected parameters. To see the graph of the given function select the command "*View*" from the main menu (Figure A5.5), and then click "*Return*" to get back to the problem solution screen.

To run the solver choose the "*Execute*" command from the main menu. Results can be seen by selecting "*Results*" → "*Graph of the Partial Sum*" (Figure A5.6). The graph range for displaying the partial sum Sn(x)and the limits for the graph can be adjusted with the help of "*Set Attributes*" command.

To examine the behavior of the individual harmonics of the expansion and of their partial sums use the option "*Choose terms…*" (Figure A5.7). To select (or unselect) the individual term click on the corresponding choice button. The terms whose numbers are selected (red) are included in the partial sum; the ones not selected (white) are excluded. The choice buttons related to terms with zero Fourier coefficients are inactive. Click "OK" to save your selection and return to the graph.

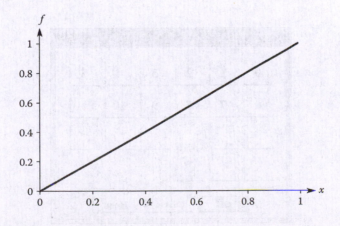

FIGURE A5.5   The given function $f(x)$.

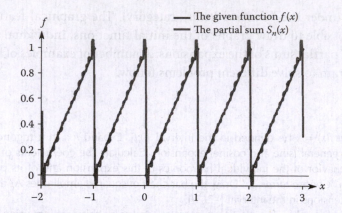

FIGURE A5.6   Graph of the partial sum at $[-2, 3]$ (general expansion), $\delta^2 = 0.0025$.

More results can be displayed with the *"Results"* → *"Bar Chart of Squared Amplitudes"* option. Figure A5.8 depicts the bar chart of squared amplitudes, A_k^2. The bar chart gives a graphic picture of the contribution of individual terms to the Fourier expansion of $f(x)$. A rectangle of a height equal to the squared amplitude, A_k^2 (measured along the vertical axis), for the harmonic with frequency $\omega_k = 2\pi k/T$ at each k value appears in the graph.

To investigate other properties of the solution try these menus:

> *Results → Fourier Coefficients*,
> *Results → Graphs of Orthogonal Functions*,
> *Results → Tabulate Expansion*.

More information on every option can be found by selecting the "Help" menu on the corresponding screen.

To solve this problem using even terms only in the expansion select *"Data"* from the main menu and then click *"Change current problem."* Change the method of expansion on the *"Enter Parameters and Functions of the Problem"* dialog by selecting the option *"Even (cosines only)."* Proceed by clicking the "OK" button and then running the problem solver by selecting *"Execute"* from the main menu. Results of the solution can be displayed by selecting menu *"Results"* → *"Graph of the Partial Sum"* (Figure A5.9).

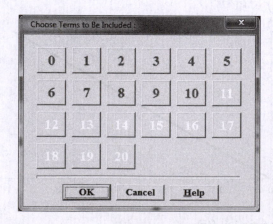

FIGURE A5.7   Dialog window "Choose terms to be included."

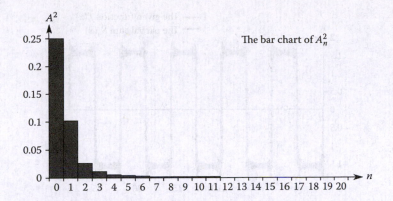

FIGURE A5.8 The bar chart of squared amplitudes A_n^2.

Example A5.2

Let the function

$$f(x) = \begin{cases} -\pi/2, & \text{if } -\pi \le x < 0, \\ \pi/2, & \text{if } \quad 0 \le x \le \pi, \end{cases}$$

be defined in the interval $[-\pi,\pi]$. Expand this function in a trigonometric Fourier series using the general method of expansion ($N = 10, 20, 30$). Note that the Gibbs's phenomenon may be observed in the neighborhood of the points $x = k\pi$ ($k = 0,\pm1,\pm2,\dots$).

Draw the graph of the partial sum of the expansion in the interval $[0,\pi]$, and try to evaluate the coordinates of its extremes with the help of the cursor (use the option "*Graph of the Partial Sum*").

Solution. The given problem is also from the library set. To load it select "*Data*" → "*Library example*" → "*Example 5.*" To proceed with the selected problem, click the "OK" button; on the dialog window, parameters for the current problem will be displayed. The given function, $f(x)$, can be entered using the impulse function, Im $p(x,a,b) = 1$, if $a \le x < b$ and 0 otherwise. Thus,

$$f(x) = (-pi/2)*Imp(x,-pi,0) + (pi/2)*Imp(x,0,pi).$$

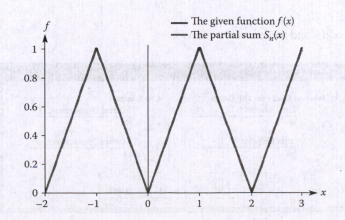

FIGURE A5.9 Graph of the partial sum at $[-2,3]$ (even method), $\delta^2 = 1.7 \cdot 10^{-6}$.

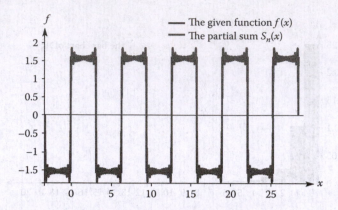

FIGURE A5.10   Graph of the partial sum $S_{30}(x)$ at $[-\pi, 9\pi]$, $\delta^2 = 0.209$.

Since all parameters for the problem have been entered, we can proceed by clicking "OK." To see the given function select "*View*" from the main menu. To run the solver, click the "*Execution*" option.

Figure A5.10 depicts the partial sum $S_N(x)$ ($N = 30$) on the interval $[-\pi, 9\pi]$ (use the command "*Graph of the Partial Sum*").

The expansion gives (an odd) continuation of the given function $f(x)$ on the interval $(-\pi/2, \pi/2)$ to the entire x-axis. Because of the periodicity we can restrict the analysis to the interval $(0, \pi)$. To adjust the parameters of the graph use the "*Set Attributes*" menu item (Figure A5.11).

Figures A5.12, A5.13, and A5.14 show the plots of the partial sum $S_N(x)$ on the interval $[0, \pi]$ for $N = 10$, 20, and 30, respectively. An interesting phenomenon can be observed near points $x = 0$ and $x = \pi$: the graphs of the partial sums near these points oscillate about the straight line $y = \pi/2$. A significant result that can be seen in these graphs is that the amplitudes of these oscillations do not diminish to zero as N increases. To the contrary, the height of the first bump (closest to $x = 0$) approaches the value of $\delta = 0.281$ above the $y = \pi/2$ line. The situation is similar as x approaches the value π from the left. Such a defect of the convergence was first found by J. Gibbs and is known as Gibbs phenomenon.

The height of the first peak of $S_N(x)$ for $N = 10$, 20, and 30 may be evaluated with the help of the screen cursor. We can also evaluate the x-coordinates of the extremes of $S_N(x)$ and verify that the points of maximums and minimums are

$$x_k = k\frac{\pi}{N}$$

(maximums for odd k and minimums for even k).

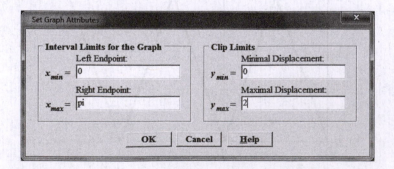

FIGURE A5.11   Graph Attributes dialog.

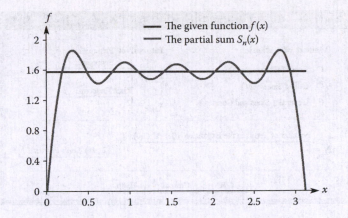

FIGURE A5.12   Graph of the partial sum $S_{10}(x)$, $\delta^2 = 0.626$.

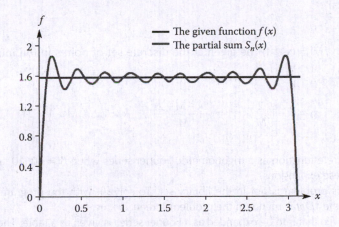

FIGURE A5.13   Graph of the partial sum $S_{20}(x)$, $\delta^2 = 0.314$.

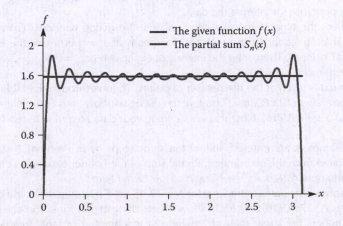

FIGURE A5.14   Graph of the partial sum $S_{30}(x)$, $\delta^2 = 0.209$.

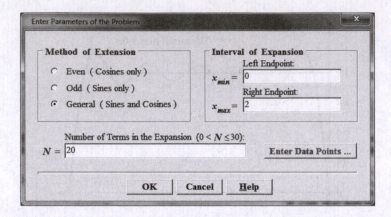

FIGURE A5.15   Parameters of the problem dialog box for Example 3.

Example A5.3

Let the function $f(x) = (x - 1)^2$ be given at the discrete set of points in the interval [0,2]. For example,

$$x_k = (k-1)\frac{2}{M}, \quad y_k = \frac{(2k-M-2)^2}{M^2}, \quad k = 1,2,...,M + 1 = 21.$$

Expand the discrete function as a trigonometric Fourier series when $N = 20$, 10, and 5. Compare the results of these expansions.

Solution. This problem is not in the library set. To solve it with the program select a *"New problem"* from the *"Data"* menu on the problem screen.

The function $f(x)$ that is to be expanded as a Fourier series is given as a table. The table of values can be input directly from the keyboard (using the command *"Input via Keyboard"*) or from an ASCII [.dat] file prepared in advance (using the command *"Read from file"*). The length of the data file cannot exceed 1000 lines (a program limitation). Each line of the file must contain values of x_k and $y_k = f(x_k)$, separated by one or more spaces. The program allows editing of the data and provides some operations for altering the data.

After you select the type of function, you will see the dialog window *"Enter parameters of the problem (Table Function)"* (Figure A5.15). To specify the properties of the expansion click *"General (Sines&Cosines)"*. Then enter the interval of expansion ($x_{min} = 0$, $x_{max} = 2$) and the number of terms in the expansion ($N = 20$).

The function $f(x)$ is given at the discrete set of points. To look through the table function (or to enter it), click the *"Enter Data Points…"* button; the dialog window with coordinates of data points will be displayed (Figure A5.16). Note that after editing your data are put in *ascending order* of the variable x.

When all parameters are entered, follow the same steps as in Problem 1 starting with the *"Execute"* command from the main menu. Partial sum of the Fourier expansion can be displayed by selecting menu option *"Results"* → *"Graph of the Partial Sum."*

Draw the graphs of the partial sum $S_N(x)$ for $N = 20$, 15, and 10 (using the menu option *"Choose terms…"*). Figures A5.17, A5.18, and A5.19 show the resulting expansions. You can see that adding terms whose numbers are larger than $M/2$ (where M is a number of table function points) only makes the precision of approximation worse.

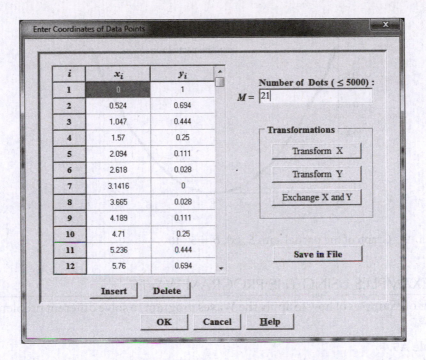

FIGURE A5.16   Dialog window "Enter coordinates of Data Points."

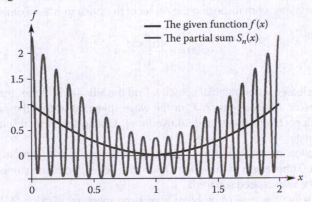

FIGURE A5.17   Graph of the partial sum $S_{20}(x)$, $\delta^2 = 1.08$.

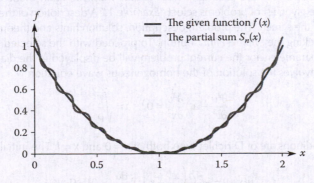

FIGURE A5.18   Graph of the partial sum $S_{15}(x)$, $\delta^2 = 0.0011$.

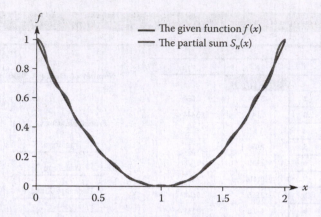

FIGURE A5.19   Graph of the partial sum $S_{10}(x)$, $\delta^2 = 5 \cdot 10^{-5}$.

A5.3   EXAMPLES USING THE PROGRAM WAVES

A number of examples of how to apply the **Waves** program to solve different problems follow.

Example A5.4

An elastic uniform string is fixed at the ends $x = 0$ and $x = l$. The string initially has the form of a quadratic parabola, symmetric with respect to the center of the string (h is a maximal initial deflection):

$$\varphi(x) = -\frac{4h}{l}\left(\frac{x^2}{l} - x\right),$$

At time $t = 0$ it is released without initial velocity. Find the vibrations of the string if the resistance of a medium is absent. Choose tension T (or the wave speed, a) so that the period of vibrations decrease by (1) twice, (2) four times, and (3) five times. Obtain the same results by changing the length, l, of the string.

To begin, the following values of the parameters are assigned: $\rho = 1$ (the mass per unit length of the string), $l = 100$ (the length of the string), $h = 5$ (the deflection of the string in $x = l/2$ at $t = 0$), and $a^2 = T/\rho = 1$ (the wave speed squared).

Solution. First, select the type of problem from main menu by clicking "*Vibrations of a Finite String.*" The program has a set of predefined examples on the selected subject and the given problem is from this set. To load the problem select "*Data*" from the main menu and click "*Library example.*" From the given list of problems select "*Example* 1." A description of the problem is found on the right part of the screen. The problem description, solution hints and theoretical solution can be accessed by clicking the "*Text + Hint*" button. To proceed with the selected example click the "OK" button and parameters for the current problem will be displayed in the dialog window.

This problem involves the solution of the homogeneous wave equation

$$\frac{\partial^2 u}{\partial t^2} - a^2\frac{\partial^2 u}{\partial x^2} = 0, \quad a = \sqrt{\frac{T}{\rho}}.$$

The boundary conditions are of Dirichlet type both at $x = 0$ and $x = l$. The initial conditions are

$$u(x,0) = -\frac{4h}{l}\left(\frac{x^2}{l} - x\right), \quad \frac{\partial u}{\partial t}(x,0) = 0.$$

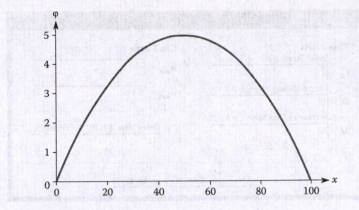

FIGURE A5.20　Initial displacement $\varphi(x)$ for Example A5.4.

Since all parameters for the problem have been entered we may proceed by clicking "OK." At this point the program is ready to solve the problem with the selected parameters. To see the initial deviation of the string select *"View"* → *"First Initial condition"* from the main menu (Figure A5.20); click *"Return"* to go back to the problem solution screen.

To run the solver choose the *"Execute"* command from the main menu. Results can be seen by selecting *"Results"* → *"Evolution of String Profile"* (Figure A5.21).

Graph attributes and parameters of the numerical solver can be adjusted in the *"Set Attributes"* menu. The parameters, initial time, length of time step, number of time steps, graph range, and initial delay may be adjusted (see Figure A5.22).

The following menu options may be used to investigate other properties of the solution:

Results → *Free Vibrations*;
Results → *Graphs of EigenFunctions*;
Results → *Time Traces of String Points*;
Results → *Bar Charts of |V(t)|*;
Results → *Energy of the system* → *Graphs of the Energies*;
Results → *Energy of the system* → *Bar Charts of the Energies*.

More information on each option can be found by selecting the *"Help"* menu on the corresponding screen.

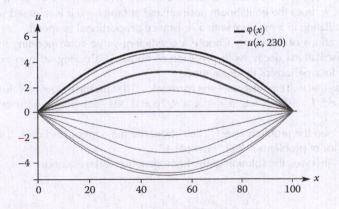

FIGURE A5.21　Evolution of string profile for Example A5.4.

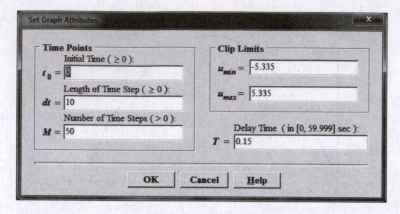

FIGURE A5.22   Set graph attributes dialog box.

As derived in Chapter 3, eigenvalues and eigenfunctions of the given boundary value problem have the form

$$\lambda_n = \left[\frac{n\pi}{l}\right]^2, \quad X_n(x) = \sin\frac{n\pi x}{l}, \quad \|X_n\|^2 = \frac{l}{2}, \quad n = 1,2,3\ldots$$

The frequency of the fundamental (lowest) tone is $\omega_1 = a\sqrt{\lambda_1} = (\pi/l)\sqrt{T/\rho}$. The period of vibrations is defined by the fundamental tone:

$$\tau_1 = \frac{2\pi}{\omega_1} = 2l\sqrt{\frac{\rho}{T}} = 200.$$

To decrease the period of vibrations, τ_1, by two choose the value of wave speed $a = \sqrt{T/\rho} = 2$ or halve the length of the string. To accomplish this, select menu option *"Data"* → *"Change Current Problem,"* and in the parameters of the problem dialog box (Figure A5.3) enter the value $a^2 = T/\rho = 4$ in the edit window *"Wave speed squared a^2"* or the value $l = 50$ in the edit window *"Length of the string."* Now run the solver again.

Example A5.5

An elastic uniform string is fixed at the ends $x = 0$ and $x = l$. At the point $x = x_0$ the string is moved a small distance, h, from the equilibrium position and at time $t = 0$ it is released with zero speed. The string is oscillating in a medium with a resistance proportional to speed.

Find the vibrations of the string. Choose a coefficient value corresponding to medium resistance so that oscillations decay (with precision of about 5%) during (a) two periods; (b) three periods; and (c) four periods of the main mode.

Find a point x_0 that will eliminate all even harmonics. To start, the following values of the parameters are assigned: $l = 100$, $h = 6$, $\rho = 1$, $\kappa = R/2\rho = 0.001$ (damping coefficient), $a^2 = T/\rho = 1$ and $x_0 = 35$.

Solution. To load the problem select *"Data"* from the main menu and click *"Library example."* From the given list of problems select *"Example 4."*

This problem involves the solution of the homogeneous wave equation

$$\frac{\partial^2 u}{\partial t^2} + 2\kappa\frac{\partial u}{\partial t} - a^2\frac{\partial^2 u}{\partial x^2} = 0, \quad a = \sqrt{\frac{T}{\rho}}.$$

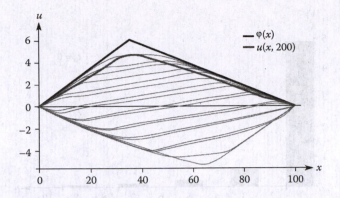

FIGURE A5.23 Evolution of string profile for Example A5.5.

The initial and boundary conditions are

$$u(x,0) = \begin{cases} \dfrac{h}{x_0}\,x, & \text{if} \quad 0 \le x \le x_0, \\[2mm] \dfrac{h}{l-x_0}(l-x), & \text{if} \quad x_0 < x \le l, \end{cases} \qquad \dfrac{\partial u}{\partial t}(x,0) = 0,$$

$$u(0,t) = u(l,t) = 0.$$

The initial deflection of the string can be entered using the *Impulse function* defined by Im $p(x,a,b) = 1$ if $a \le x < b$ and 0 otherwise. For the given parameters the following should be entered in the dialog box:

$$\varphi(x) = \text{Imp}(x,0,35)*6*x/35 + \text{Imp}(x,35,100)*6*(100-x)/(100-35).$$

Since all parameters for the problem have been entered, you may run the solver by choosing the "*Execute*" command from the main menu and then selecting "*Results*" → "*Evolution of String Profile*." The animation sequence in Figure A5.23 shows the profile of the string at times $t_k = k\Delta t$.

To change the decay time of oscillations due to friction use the menu option "*Results*" → "*Time Traces of String Points*." Choosing different values of damping coefficient κ, trace the time evolution of the string until the free oscillations have decayed (with some reasonable precision).

For example, for a damping coefficient $\kappa = 0.004$ the amplitude of oscillations fades to approximately zero during five periods of the main mode (Figure A5.24).

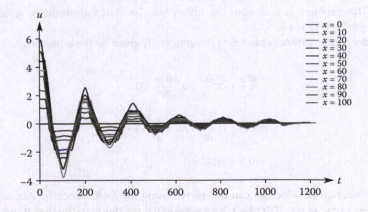

FIGURE A5.24 Time traces of string points for Example A5.5.

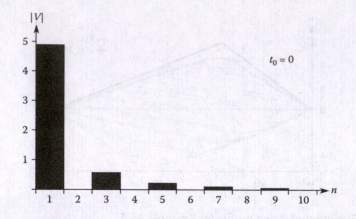

FIGURE A5.25 Bar charts of $|V_n(x)|$ for Example A5.5.

In the expansion of the solution $u(x, t)$ the terms for which $\sin(n\pi x_0/l) = 0$ vanish, that is, the solution does not contain overtones for which the point $x = x_0$ is a node. If x_0 is at the middle of the string, the solution does not contain harmonics with even numbers. To eliminate all even harmonics select the menu *"Data"* → *"Change Current Problem"* and in the parameters of the problem dialog box enter the new initial function

$$\varphi(x) = \mathrm{Imp}(x,0,50)*6*x/50 + \mathrm{Imp}(x,50,100)*6*(100-x)/(100-50).$$

Run the problem solver and then select menu *"Results"* → *"Free Vibrations"* or *"Results"* → *"Bar Charts of |V(t)|"* to ensure that the even harmonic have been eliminated (Figure A5.25).

Example A5.6

An elastic uniform string is fixed at the ends $x = 0$ and $x = l$. At time $t = 0$ it is excited by a sharp blow from a hammer that transmits an impulse l to the string at point x_0. The initial displacement is zero and the string is oscillating in a medium with a resistance proportional to speed. No external forces are present.

Find the vibrations of the string. Find the point where the hammer should strike to minimize the energies of the seventh and eighth harmonics. To start, the following values of the parameters are assigned: $\rho = 1$, $a^2 = 1$, $\kappa = 0.001$, $l = 100$, $I = 10$, and $x_0 = 30$.

Solution. This problem is also from the library set. To load the problem, select *"Data"* → *"Library example"* → *"Example 5."*

The boundary value problem modeling this process is given by the equations

$$\frac{\partial^2 u}{\partial t^2} + 2\kappa \frac{\partial u}{\partial t} - a^2 \frac{\partial^2 u}{\partial x^2} = 0,$$

$$u(x,0) = 0 \quad \frac{\partial u}{\partial t}(x,0) = \frac{I}{\rho}\delta(x - x_0),$$

$$u(0,t) = u(l,t) = 0.$$

The initial distribution of velocities can be entered using the *Delta function*. Accounting for the given parameters type $\psi(x) = 10*\mathrm{delta}(x,30)$ for the initial condition in the dialog box.

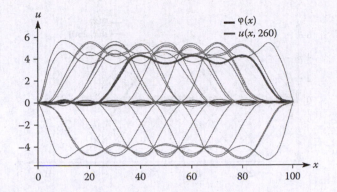

FIGURE A5.26   Evolution of string profile with $x_0 = 30$ for Example A5.6.

To run the solver click *"Execute"* from the main menu; the results can be seen by selecting *"Results"* → *"Evolution of String Profile"* (Figure A5.26).

It should be obvious that if a hammer blow occurs at the node of *n*th harmonic, the energy of this harmonic will be zero. Therefore to decrease the energies of the seventh and eighth overtones the string should be struck between the nodes of the seventh and eighth harmonics, not far from the fixed end. Use the menu option *"Graphs of EigenFunctions"* to find an adequate point for the hammer blow.

Figure A5.27 shows the first eight eigenfunctions $X_n(x)$ for the given problem. Moving the screen pointer to some point between the first nodes of the seventh and eighth harmonics node location we can read the value of coordinate *x* in the Status Bar and choose $x \approx 13.3$ as the location for the hammer blow.

Select menu option *"Data"* → *"Change Current Problem"* and in the parameters of the problem dialog enter the new coordinate x_0 in the initial function: $\psi(x,0) = 10*delta\ (x,13.3)$. The graph shown in Figure A5.28 depicts the animated solution $u(x,t)$ and the bold dark line represents $u(x,t = 250)$ (in software figures are in color and this line is red).

Using menu options *"Bar Charts of |V(t)|"* and *"Bar Charts of the Energies"* we can ensure that the energy in the seventh and eighth harmonics have been substantially decreased compared to other modes (Figure A5.29).

By studying the energy in each harmonic and choosing different values of the impulse, *I* , it can be seen that the energy of the string is proportional to I^2.

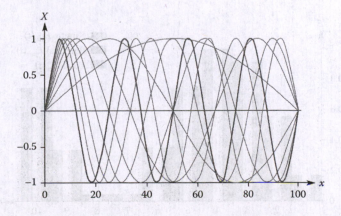

FIGURE A5.27   Eigenfunctions for Example A5.6. The bold line (red line in software figure) represents $X_8(x)$.

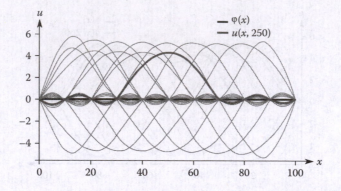

FIGURE A5.28   Evolution of string profile with $x_0 = 13.3$ for Example A5.6.

A5.3.1 Two-Dimensional Problems

All the above examples were one-dimensional problems. The **Waves** program can also solve two-dimensional problems, such as vibrations of thin rectangular or circular membranes. Several examples showing the capacity of the program to deal with these types of problems will be considered below. The theory related to this topic can be found both in the book and in the help system of the program.

Example A5.7

A flexible rectangular membrane is clamped at the edges $x = 0$, $y = 0$, the edge $y = l_y$ is attached elastically, and the edge $x = l_x$ is free. Initially the membrane is at rest in a horizontal plane. Starting at time $t = 0$ a uniformly distributed transversal force with density

$$f(x,y,t) = Ae^{-0.5t}x\sin\frac{4\pi y}{l_y}$$

acts on the membrane.

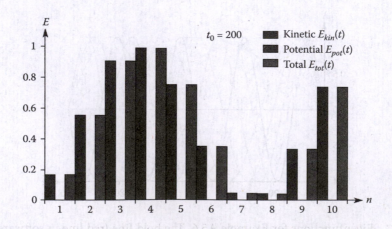

FIGURE A5.29   Bar charts of energies with $x_0 = 13.3$ for Example A5.6.

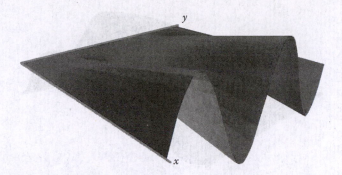

FIGURE A5.30   External force $f(x, y, t)$ at $t = 1.5$ for Example A5.7.

Find the transverse vibrations of the membrane if the resistance of the surrounding medium is proportional to the speed of the membrane. Choose a coefficient of medium resistance so that oscillations decay to zero (with precision of about 5%) during time (a) $t = 20$; (b) $t = 30$; and (c) $t = 40$. The assigned parameter values are $A = 0.5$, $a^2 = 1$, $\kappa = 0.01$, $l_x = 4$, and $l_y = 6$.

Solution. First, select the type of problem from the main menu by clicking "*Vibrations of a Rectangular Membrane.*" The given problem is not from the library set. To solve it, instead of selecting a library example as we did before, select a "*New problem*" from the "*Data*" menu on the problem screen.

The problem involves solving the wave equation,

$$\frac{\partial^2 u}{\partial t^2} + 2\kappa \frac{\partial u}{\partial t} - a^2 \left[\frac{\partial^2 u}{\partial x^2} + \frac{\partial^2 u}{\partial y^2} \right] = Ae^{-0.5t} x \sin \frac{4\pi y}{l_y},$$

under the conditions

$$u(x,y,0) = 0, \quad \frac{\partial u}{\partial t}(x,y,0) = 0,$$

$$u(0,y,t) = 0, \quad \frac{\partial u}{\partial x}(l_x,y,t) = 0, \quad u(x,0,t) = 0, \quad \frac{\partial u}{\partial y}(x,l_y,t) + h_4 u(x,l_y,t) = 0.$$

To specify the boundary condition at the edge $y = l_y$, choose option "3 (elastic fixing)" by selecting this item from the dialog box. Enter the parameters by clicking the "*Edit*" button and typing $h_4 = 0.1$ and $g(x,t) = 0$.

To see the animated surface plot of the external force select "*View*" → "*External force f(x,y,t))*" from the main menu (Figure A5.30) and click "*Return*" to go back to the problem solution screen.

To solve the problem with the help of the program, follow the same steps as in Example A5.6. Results of the solution can be displayed by selecting menu option "*Surface Graphs of Membrane.*" The three-dimensional surface shown in Figure A5.31 depicts the animated solution, $u(x,y,t)$, at $t = 10.5$. Note that the edges of the surface adhere to the given boundary conditions.

The following menus may be used to investigate other properties of the solution:

Results → Free Vibrations;
Results → Surface Graphs of EigenFunctions;
Results → Evolution of Membrane Profile at y = const;
Results → Evolution of Membrane Profile at x = const.

More information on each option can be found by selecting the "*Help*" menu on the corresponding screen.

x

FIGURE A5.31　Surface graph of membrane at $t = 10.5$ for Example A5.7.

Eigenvalues of the given boundary value problem have the form

$$\lambda_{nm} = \lambda_{xn} + \lambda_{ym} \quad \text{where} \quad \lambda_{xn} = \left[\frac{\pi(2n+1)}{2l_x}\right]^2, \quad \lambda_{ym} = \left[\frac{\mu_{ym}}{l_y}\right]^2, \quad n, m = 0, 1, 2, \ldots.$$

and μ_{ym} is the mth root of the equation $\tan \mu_y = -\mu_y /(h_4 l_y)$. Figure A5.32 shows graphs of the functions $\tan \mu_y$ and $-\dfrac{\mu_y}{h_4 l_y}$ used for evaluating eigenvalues. Eigenfunctions of the given boundary value problem are

$$V_{nm}(x, y) = X_n(x)Y_m(y) = \sin\frac{(2n+1)\pi x}{2l_x} \sin\sqrt{\lambda_{ym}}\, y.$$

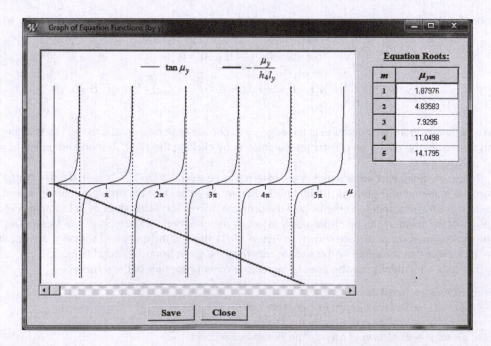

FIGURE A5.32　Graphs of functions for evaluating eigenvalues and the table of roots μ_{ym} for Example A5.7.

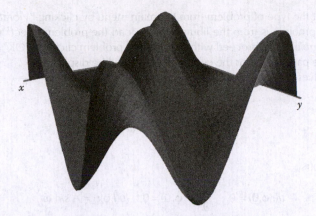

FIGURE A5.33   Eigenfunction $V_{33}(x, y)$ for Example A5.7.

The three-dimensional picture shown in Figure A5.33 depicts the one of eigenfunction for the problem ($\lambda 33 = 5.6019$, $V_{33}(x, y) = \sin\dfrac{5\pi x}{2l_x}\sin\sqrt{\lambda_{y3}}\, y$, $\|V_{33}\|^2 = 6.0569$).

To change the decay time of oscillations due to friction, use the menu option "*Surface Graphs of Membrane,*" "*Evolution of Membrane Profile at y = const,*" or "*Evolution of Membrane Profile at y = const.*" Choosing different values of damping coefficient, κ, trace the time evolution of the membrane until the free oscillations decay (with some reasonable precision). For example, for damping coefficient $\kappa = 0.175$ the amplitude of oscillations fades away during time $t = 30$ (Figure A5.34).

Example A5.8

Find the transversal vibrations of a flexible circular membrane caused by the motion of its periphery under the constraint

$$g(\varphi, t) = A \sin \omega t.$$

The initial displacement and initial velocity are zero. Assume the surrounding medium offers no resistance. The assigned parameter values are $A = 0.2$, $\omega = 3$, $a^2 = 1$, and $l = 2$.

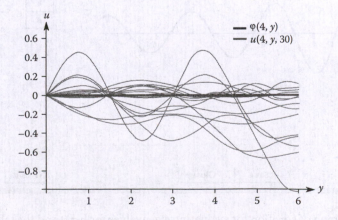

FIGURE A5.34   Evolution of membrane profile at $x = l_x$, $\kappa = 0.175$ for Example A5.7.

Solution. Select the type of problem from the main menu by clicking *"Vibrations of a Circular Membrane."* This problem is from the library set. To load the problem select *"Data"* → *"Library example"* → *"Example 6."* To proceed with the selected problem click the "OK" button and in the dialog window the parameters for the current problem will be displayed.

The problem involves the solution of the equation

$$\frac{\partial^2 u}{\partial t^2} - a^2 \left[\frac{\partial^2 u}{\partial r^2} + \frac{1}{r} \frac{\partial u}{\partial r} \right] = 0$$

under the conditions

$$u(r,\varphi,0) = 0, \quad \frac{\partial u}{\partial t}(r,\varphi,0) = 0, \quad u(l,\varphi,t) = A \sin \omega t.$$

Eigenvalues of the given boundary value problem have the form

$$\lambda_{nm} = \left[\frac{\mu_m^{(n)}}{l} \right]^2, \quad n,m = 0,1,2,...,$$

where $\mu_m^{(n)}$ are positive roots of the equation $J_n(\mu) = 0$. Figure A5.35 shows the graph of function $J_n(\mu)$ for evaluating eigenvalues for the selection $n = 0$.

Eigenfunctions of the given boundary value problem are

$$V_{nm}^{(1)} = J_n \left(\frac{\mu_m^{(n)}}{l} r \right) \cos n\varphi, \quad V_{nm}^{(2)} = J_n \left(\frac{\mu_m^{(n)}}{l} r \right) \sin n\varphi.$$

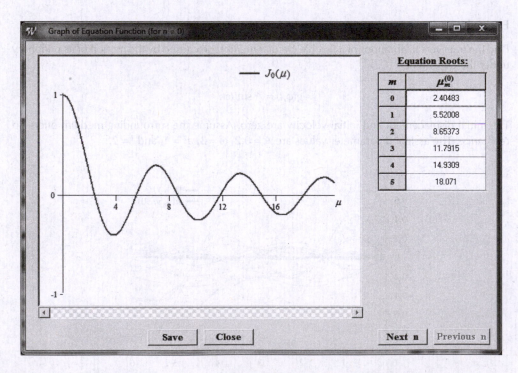

FIGURE A5.35   Graph of equation for evaluating eigenvalues for $n = 0$ and the table of roots $\mu_m^{(0)}$ for Example A5.8.

FIGURE A5.36   Eigenfunction $V_{11}^{(1)}$ for Example A5.8.

The three-dimensional picture shown in Figure A5.36 depicts one of the eigenfunctions for this problem ($\lambda_{11} = 12.3046$, $V_{11}^{(1)} = J_1(\mu_1^{(1)} r/2)\cos\varphi$, $\left\|V_{11}^{(1)}\right\|^2 = 0.565923$).

To solve the problem with the help of the program, follow the same steps as in Problem A5.7. Results of the solution can be displayed by selecting the menu option *"Surface Graphs of Membrane."* The three-dimensional surface shown in Figure A5.37 depicts the animated solution, $u(r,\varphi,t)$, at $t = 13$.

The solution to the given problem can be expressed as the sum of two functions

$$u(r,\varphi,t) = w(r,\varphi,t) + v(r,\varphi,t),$$

where $w(r,\varphi,t) = A\sin\omega t$ is an auxiliary function satisfying the boundary value function and $v(r,\varphi,t)$ is the solution of the boundary value problem with zero boundary conditions where

$$f^*(r,\varphi,t) = A\omega^2\sin\omega t,$$

$$\phi^*(r,\varphi) = 0, \quad \psi^*(r,\varphi) = -A\omega.$$

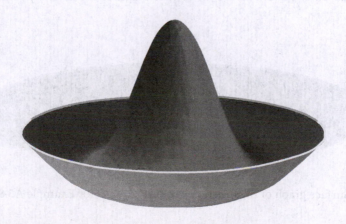

FIGURE A5.37   Surface graph of membrane at $t = 13$ for Example A5.8.

W	0,0	0,1	0,2	0,3	0,4	0,5	☑
	1,0	1,1	1,2	1,3	1,4	1,5	☑
	2,0	2,1	2,2	2,3	2,4	2,5	☑
	3,0	3,1	3,2	3,3	3,4	3,5	☑
	4,0	4,1	4,2	4,3	4,4	4,5	☑
	5,0	5,1	5,2	5,3	5,4	5,5	☑
	☑	☑	☑	☑	☑	☑	

Choose Terms to Be Included :

OK   Cancel   Help

FIGURE A5.38   Selection of terms in the Fourier expansion of the solution.

The program **Waves** allows you to study the behavior of the auxiliary function $w(r,\varphi,t)$ and of any single term of the partial sum of the solution,

$$u_{nm}(r,\varphi,t) = w(r,\varphi,t) + \sum_{i=0}^{n}\sum_{j=0}^{m}\left\{ \left[a_{ij}y_{ij}^{(1)}(t) + b_{ij}y_{ij}^{(2)}(t)\right]V_{ij}^{(1)}(r,\varphi)\right.$$

$$\left. + \left[c_{ij}y_{ij}^{(1)}(t) + d_{ij}y_{ij}^{(2)}(t)\right]V_{ij}^{(2)}(r,\varphi)\right\}.$$

Select the *"Choose Terms..."* option from the graph menu and the *"Choose Terms to Be Included"* dialog box will be displayed (Figure A5.38). To select (or unselect) the individual term click on the corresponding button. The terms whose numbers are selected (red) are included in the partial sum; the ones not selected (white) are excluded. The buttons associated with terms with zero Fourier coefficients are inactive.

The button "**W**" corresponds to the auxiliary function $w(r,\varphi,t)$. Select the button "**W**" and unselect all other buttons to see the surface graph of the auxiliary function $w(r,\varphi,t) = A\ \sin\omega t$.

FIGURE A5.39   Surface graph of the function $w(r, \varphi, t)$ at $t = 13$ for Example A5.8.

A5.4   EXAMPLES USING THE PROGRAM HEAT

Examples of how to use the **Heat** program to solve different problems follow.

Example A5.9

A heat-conducting cylindrical rod of length l is thermally insulated over its lateral surface. The left end of the rod ($x = 0$) is kept at a constant temperature of zero and the right end ($x = l$) is thermally insulated. The initial temperature of the rod is

$$\varphi(x) = \begin{cases} 0, & \text{if} \quad 0 < x < \dfrac{l}{2}, \\ u_0, & \text{if} \quad \dfrac{l}{2} < x < l, \end{cases} \qquad \text{where} \quad u_0 = \text{const.}$$

Find the distribution of temperature, $u(x,t)$, in the rod at later times if generation (or absorption) of heat by internal sources is absent. To start, the following values of the parameters may be assigned: $l = 10$, $u_0 = 5$, $a^2 = 0.25$.

Solution. First, select the type of problem from the main menu by clicking *"Heat Conduction within a Finite Uniform Rod."* The program has a set of predefined examples on the selected subject and the given problem is from this set. To load the problem select *"Data"* from the main menu and click *"Library example."* From the given list of problems select *"Example 2."* The problem description may be read on the right part of the screen. The problem description, solution hints, and theoretical solution can be accessed by clicking the *"Text + Hint"* button. To proceed with the selected problem, click the "OK" button and parameters for the current example will be displayed in the dialog box.

Since all parameters for the problem have been entered we can proceed by clicking "OK." At this point the program is ready to solve the problem with selected parameters. To see the initial temperature distribution select *"View"* → *"Initial condition u(x,0)"* from the main menu, and then click *"Return"* to get back to the problem solution screen.

To run the solver, choose the *"Execute"* command from the main menu and then select *"Results"* → *"Evolution of Rod Temperature."* The animation sequence in Figure A5.41 shows the profile of the rod temperature at times $t_k = k\Delta t$. As can be seen from Figure A5.41, the temperature in the rod is approaching zero along its entire length as time goes to infinity.

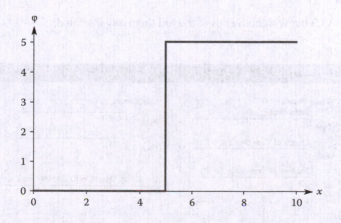

FIGURE A5.40   Initial condition $u(x, 0) = \varphi(x)$ for Example A5.9.

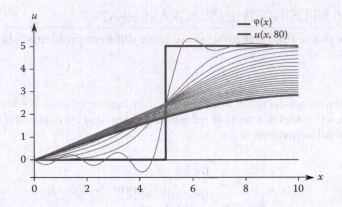

FIGURE A5.41   The animated solution, $u(x, t)$, for Example A5.9.

Graph attributes and parameters of the numerical solver can be adjusted in the *"Set Attributes"* menu. Available parameters are shown in Figure A5.42.

Other properties of the solution may be investigated by accessing the following menus:

Results → *Free Heat Exchange*;
Results → *Graphs of EigenFunctions*;
Results → *Time Traces of Temperature at Rod Points*;
Results → *Bar Charts of* $|V(t)|$.

More information on each option can be found by selecting the *"Help"* menu on the corresponding screen.

Next we consider several problems connected to Example A5.9. They can be solved by changing the initial conditions, boundary conditions, and constants of the problem in the *"Enter Parameters and Functions of the Problem"* dialog box. Available sets of analytic functions and mathematical operators can be found in the program help menu under the topic *"Mathematical Functions, Operations and Constants"* and *"General Items."*

Example A5.10

Repeat example A5.9 but with both ends of the rod thermally insulated.

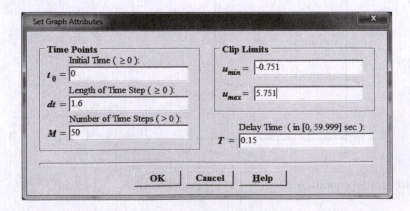

FIGURE A5.42   Set graph attributes dialog window.

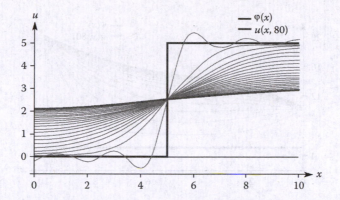

FIGURE A5.43   The animated solution $u(x, t)$ for Example A5.10.

Solution. This problem is not in the library set. To solve it with the help of the program we will need to modify parameters of the existing library problem. First follow the same steps as in problem 1, as far as the *"Enter Parameters and Functions of the Problem"* screen. Here we need to change the boundary condition at the left end of the rod, ($x = 0$). To make boundary conditions appropriate to the given problem, select, for the left end of the rod, "2 (*Given heat flow*)" from the dialog box. To assign a value to a heat flow click *"Edit."* For this case the value should be zero.

At this point we may proceed by clicking the "OK" button and run the problem solver by selecting *"Execute"* from the main menu. Results of the solution can be displayed by selecting the menu option *"Evolution of Rod Temperature."* In this case the temperature tends to a uniform distribution after a significant period of time (Figure A5.43).

Example A5.11

The initial temperature of a slender wire of length l lying along the x-axis is $u(x,0) = u_0$, where u_0 = const and the ends of the wire are kept under the constant temperatures $u(0,t) = u_1 = $ const, $u(l,t)$ = $u_2 = $ const. The wire is experiencing a heat loss through the lateral surface proportional to the difference between the wire temperature and the surrounding temperature. The temperature of the surrounding medium is $u_{md} = $ const. and we introduce the coefficient γ as a heat loss coefficient.

Find the distribution of temperature $u(x,t)$ in the wire for $t>0$ if it is free of internal sources of heat. The assigned values of the parameters are $a^2 = 0.25$ (thermal diffusivity of the material), $l = 10$, $\gamma = h/c\rho = 0.05$, $u_{md} = 15$ (the temperature of the medium), and $u_0 = 5$ (initial temperature of the wire).

Solution. This problem is also given in the library set. To load it, select *"Data"* → *"Library example"* → *"Example 6."*

The problem is an example of the equation

$$\frac{\partial u}{\partial t} = a^2 \frac{\partial^2 u}{\partial x^2} - \gamma \cdot (u - u_{md})$$

under the conditions

$$u(x,0) = u_0, \quad u(0,t) = u_1, \quad u(l,t) = u_2.$$

Since all parameters for the problem have been entered you may run the solver by choosing the *"Execute"* command from the main menu and then selecting *"Results"* → *"Evolution of Rod Temperature."* The animation sequence in Figure A5.44 shows the solution $u(x,t_k)$ at times $t_k = k\Delta t$.

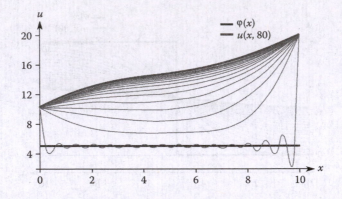

FIGURE A5.44 The animated solution, $u(x, t)$, for Example A5.11.

The menu option *"Time Traces of Temperature at Rod Points"* allows one to investigate the significance of the changes in magnitudes of the diffusivity, a^2, and of the coefficient of lateral heat exchange with a medium, γ, by modifying them and noting their effect on the solution to the problem (Figure A5.45).

More results can be displayed with the *"Bar charts of $|V_n(t)|$"* command. This command produces a sequence of bars of the values

$$V_n(t) = T_n(t) + C_n e^{-(a^2 \lambda_n + \gamma)t}$$

at successive time points. Each bar chart gives a graphic picture of the contribution of individual terms to the solution $u(x,t)$ at a certain time. Bar charts display data by drawing a rectangle of the height equal to $|V_n(t)|$ (measured along the vertical axis) at each n value along the graph. Therefore the height of the nth rectangle corresponds to the absolute value of the amplitude of the nth harmonic at the respective time t.

The sequence of bars (Figure A5.46) shows that the higher-order terms decay more rapidly.

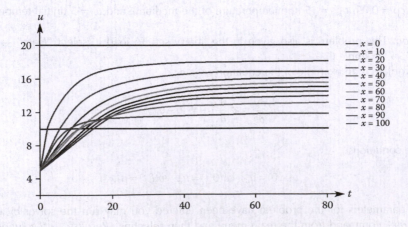

FIGURE A5.45 Time traces of temperature at rod points for Example A5.11.

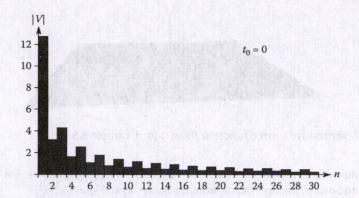

FIGURE A5.46   Bar charts of $|V_n(t)|$ at $t = 0$ for Problem 3.

A5.4.1 Two-Dimensional Problems

All the above examples are one-dimensional problems since we neglected the width of the rod. The **Heat** program can also solve two-dimensional problems, such as heat conduction in thin rectangular or circular membranes. Several examples showing the capacity of the program to deal with these types of problems will be considered below. The theory related to this topic can be found both in the book and in the help system of the program.

Example A5.12

A heat-conducting, thin, uniform, rectangular membrane ($0 \leq x \leq l_x$, $0 \leq y \leq l_y$) is thermally insulated over its lateral faces. One part of the boundary (at $x = 0$ and $y = 0$) is thermally insulated, and the other part (at $x = l_x$ and $y = l_y$) is held at constant zero temperature. The initial temperature distribution within the membrane is $u(x,y,0) = u_0 = $ const. A constant internal source of heat acts at the point (x_0, y_0) of the membrane. The magnitude of this source is $Q = $ const.

Find the temperature distribution $u(x,y,t)$ in the membrane at $t > 0$. Find the time, τ, when the temperature distribution in the membrane arrives at the steady-state mode. Study the influence of the thermal diffusivity coefficient, a^2, on this time. To start, the following values of the parameters are assigned: $a^2 = 0.25$ (thermal diffusivity of the material), $l_x = 4$, $l_y = 6$, $u_0 = 10$, $Q = 50$, $x_0 = 2$, and $y_0 = 3$.

Solution. First, select the type of problem from the main menu by clicking "*Heat Conduction within a Thin Rectangular Membrane*." The present problem is not in the library set. To solve it, instead of selecting a library example as we did before, select a "*New problem*" from the "*Data*" menu on the problem screen.

The problem involves the solution of the equation

$$\frac{\partial u}{\partial t} = a^2 \left[\frac{\partial^2 u}{\partial x^2} + \frac{\partial^2 u}{\partial y^2} \right] + Q \cdot \delta(x - x_0) \cdot \delta(y - y_0)$$

under the conditions

$$u(x,y,0) = u_0,$$

$$\frac{\partial u}{\partial x}(0,y,t) = 0, \quad u(l_x,y,t) = 0, \quad \frac{\partial u}{\partial y}(x,0,t) = 0, \quad u(x,l_y,t) = 0.$$

FIGURE A5.47   Temperature source function $f(x, y, t)$ for Example A5.12.

When all parameters are entered, click the "OK" button. To see the source function select "*View*" → "*Temperature Source Function f(x,y,t)*" from the main menu (Figure A5.47).

To run the solver click "*Execute*" from the main menu. Results can be seen by selecting "*Results*" → "*Evolution of the Membrane Temperature.*" Figure A5.48 shows a snapshot at time $t = 60$ from an animation sequence.

It is convenient to use options "*Evolution of Temperature Profile at x = const*" and "*Evolution of Temperature Profile at y = const*" to find the time, τ, when the temperature distribution in the membrane arrives at the steady-state mode and to study the influence of the thermal diffusivity coefficient, a^2, on this time. Figures A5.49 and A5.50 show the animation sequences for two different values of thermal diffusivity: $a^2 = 0.25$ and $a^2 = 1$. One can see that if a^2 is quadrupled, then the time τ (and the maximal value of temperature) decreases by a factor 4. By changing the dimensions of the membrane the same result can be obtained and studied.

Example A5.13

A heat-conducting thin uniform rectangular membrane ($0 \le x \le l_x, 0 \le y \le l_y$) is thermally insulated over its lateral faces. One part of the boundary ($x = 0, y = 0$) is kept under at constant temperature $u = u_1$, and the other end ($x = l_x, \ y = l_y$) under the constant temperature $u = u_2$. The initial temperature distribution within the membrane is $u(x, y, 0) = u_0 = $ const.

Find the temperature $u(x,y,t)$ of the membrane at any later time if generation (or absorption) of heat by internal sources is absent. The assigned parameter values are $a^2 = 0.25$ (thermal diffusivity of the material), $l_x = 4, l_y = 6, u_0 = 10, u_1 = 20$, and $u_2 = 50$.

Solution. This problem is from the library set. To load it select "*Data*" from the main menu, and then click "*Library example.*" From the given list of problems select "*Example 7.*"

The problem depends upon the solution to the equation

$$\frac{\partial u}{\partial t} = a^2 \left[\frac{\partial^2 u}{\partial x^2} + \frac{\partial^2 u}{\partial y^2} \right]$$

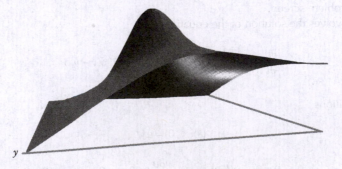

FIGURE A5.48   Surface plot of the solution $u(x, y, t)$ at $t = 60$ for Example A5.12.

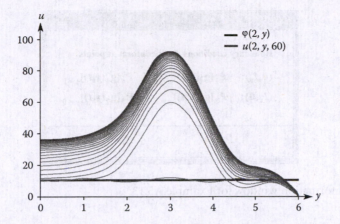

FIGURE A5.49   Animated sequence of the solution profile $u(2, y, t)$; $a^2 = 0.25$.

under the conditions

$$u(x, y, 0) = u_0, \quad u(0, y, t) = u(x, 0, t) = u_1, \quad u(l_x, y, t) = u(x, l_y, t) = u_2.$$

In this problem the boundary value functions don't satisfy the conforming conditions at points $(0, l_y)$ and $(l_x, 0)$:

$$P_1[g_4]_{x=0} = u_2 \neq P_4[g_1]_{y=l_y} = u_1, \quad P_2[g_3]_{x=l_x} = u_1 \neq P_3[g_2]_{y=0} = u_2.$$

As a result, when you run the solver by clicking "*Execute*" from the main menu, the message in Figure A5.51 will be displayed.

Although the boundary conditions do not match, we may still solve the problem with the help of the program by searching for a generalized solution. Start by following the same steps as in Example A5.9. Results of the solution can be displayed by selecting menu option "*Evolution of*

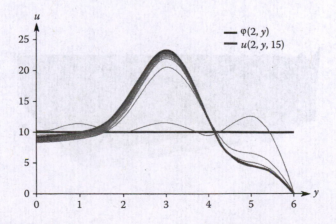

FIGURE A5.50   Animated sequence of the solution profile $u(2, y, t)$; $a^2 = 1$.

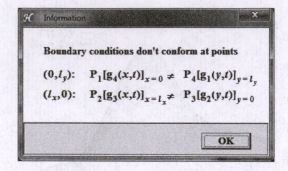

FIGURE A5.51　Information window for Example A5.13.

Membrane Temperature." The three-dimensional surface shown in Figure A5.52 depicts the frame of the animated solution $u(x,y,t)$ at $t = 0.1$. Note that the edges of the surface adhere to the given boundary conditions.

　　The solution to the given problem can be expressed as the sum of three functions

$$u(x, y) = w_1(x, y) + w_2(x, y) + v(x, y),$$

where $w_1(x, y)$ is an auxiliary function satisfying the boundary value functions

$$P_1[w_1]_{x=0} = g_1(y,t) = u_1, \quad P_2[w_1]_{x=l_x} = g_2(y,t) = u_2,$$

$$P_3[w_1]_{y=0} = u_1 + (u_2 - u_1)\frac{x}{l_x}, \quad P_4[w_1]_{y=l_y} = u_1 + (u_2 - u_1)\frac{x}{l_x},$$

$w_2(x, y)$ is a particular solution of Laplace problem with the following boundary conditions

$$P_1[w_2]_{x=0} = 0, \quad P_2[w_2]_{x=l_x} = 0,$$

$$P_3[w_2]_{y=0} = \frac{x}{l_x}(u_1 - u_2), \quad P_4[w_2]_{y=l_y} = (u_2 - u_1)\cdot\left(1 - \frac{x}{l_x}\right).$$

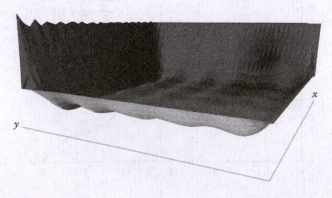

FIGURE A5.52　Distribution of temperature $u(x, y, t)$ at $t = 0.1$ for Example A5.13.

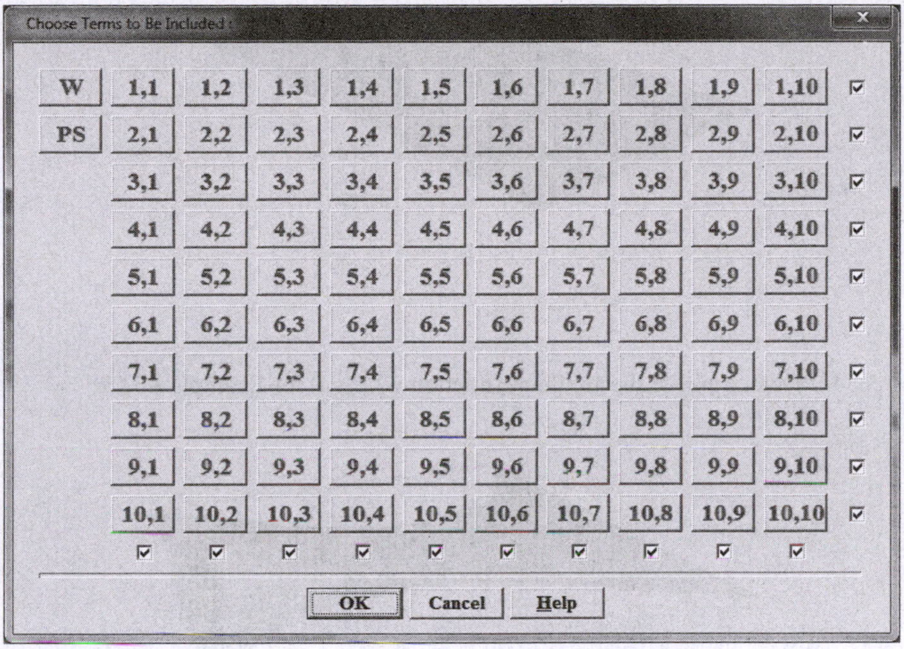

FIGURE A5.53   Selection of terms in the expansion of the solution.

and $v(x,y)$ is a solution of a diffusion equation with zero boundary conditions:

$$v(x,y,t) = \sum_{n=1}^{\infty}\sum_{m=1}^{\infty} C_{nm}e^{-\lambda_{nm}a^2t} \cdot V_{nm}(x,y).$$

The program **Heat** allows you to study the behavior of the functions $w_1(x,y,t)$, $w_2(x,y,t)$ and of any single term of the partial sum of the solution

$$u_{NM}(x,y,t) = w_1(x,y,t) + w_2(x,y,t) + \sum_{n=1}^{N}\sum_{m=1}^{M} C_{nm}e^{-\lambda_{nm}a^2t} \cdot V_{nm}(x,y).$$

Select the "*Choose Terms...*" option from the graph menu and the "*Choose Terms to Be Included*" dialog box will be displayed (Figure A5.53). To select (or unselect) the individual terms click on the corresponding button. The terms whose numbers are selected (red) are included in the partial sum; the ones not selected (white) are excluded. The buttons associated with terms that have zero Fourier coefficients are inactive.

The button "W" corresponds to the auxiliary function $w_1(x,y,t)$ (Figure A5.54). Select the button "W" and unselect all other buttons to see the surface graph of the auxiliary function

$$w_1(x,y,t) = u_1 + (u_2 - u_1)\frac{x}{l_x}.$$

The button "PS" corresponds to the particular solution $w_2(x,y,t)$ (Figure A5.55). This button is active because the boundary value functions don't satisfy the conforming conditions. Select the button "PS" and unselect all other buttons to see the surface graph of the particular solution

$$w_2(x,y,t) = \frac{2}{\pi}(u_2 - u_1)\sum_{n=1}^{\infty}\frac{1}{n \cdot \sinh\sqrt{\lambda_n}l_y}\left\{\sinh\sqrt{\lambda_n}\,y + (-1)^n \sinh\sqrt{\lambda_n}(l_y - y)\right\}\sin\frac{n\pi x}{l_x}.$$

Unselect the buttons "W" and "PS" and the partial sum $v_{NM}(x,y)$ will be displayed (Figure A5.56).

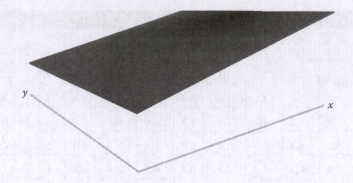

FIGURE A5.54 Surface plot of the auxiliary function $w_1(x, y, t)$ for Example A5.13.

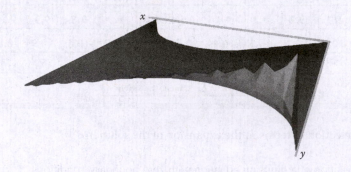

FIGURE A5.55 Surface plot of the particular solution $w_2(x, y, t)$ for Example A5.13.

FIGURE A5.56 Surface plot of the partial sum $v_{NM}(x, y, t)$ for Example A5.13.

Example A5.14

Consider a very long (infinite) cylinder of radius l. At time $t = 0$ let a magnetic field, parallel to cylinder axis, be instantly established outside of the cylinder. The external magnetic field strength is

$$H = H_0 \sin \omega t, \quad H_0 = \text{const}, \quad 0 < t < \infty.$$

Find the magnetic field strength within the cylinder if the initial field is zero. To start, take the following values of the parameters: $a^2 = 0.25$, $l = 4$, $H_0 = 5$, $\omega = 2$.

Solution. This problem is mathematically identical to a heat conduction problem found in the program library of problems. Select the type of the problem from the main menu by clicking "*Heat Conduction within a Thin Circular Membrane.*" To load the appropriate problem select "*Data*" → "*Library example*" → "*Example 6.*"

The problem depends on the solution of the equation

$$\frac{\partial u}{\partial t} = a^2 \left[\frac{\partial^2 u}{\partial r^2} + \frac{1}{r} \frac{\partial u}{\partial r} \right]$$

under the conditions

$$u(r, \varphi, 0) = 0, \quad u(l, \varphi, t) = H_0 \sin \omega t.$$

Eigenvalues of the given boundary value problem have the form

$$\lambda_{nm} = \left[\frac{\mu_m^{(n)}}{l} \right]^2, \quad n, m = 0, 1, 2, \ldots,$$

where $\mu_m^{(n)}$ are positive roots of the equation $J_n(\mu) = 0$. Figure A5.57 shows the graph of the function $J_n(\mu)$ and associated eigenvalues for $n = 2$.

Eigenfunctions of the given boundary value problem are

$$V_{nm}^{(1)} = J_n \left(\frac{\mu_m^{(n)}}{l} r \right) \cos n\varphi, \quad V_{nm}^{(2)} = J_n \left(\frac{\mu_m^{(n)}}{l} r \right) \sin n\varphi.$$

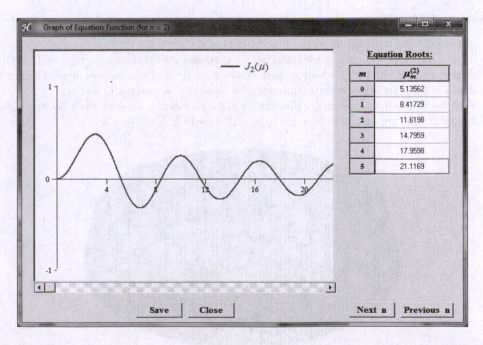

FIGURE A5.57   Graph of equation for evaluating eigenvalues for $n = 2$ and the table of roots $\mu_m^{(2)}$ for Example A5.14.

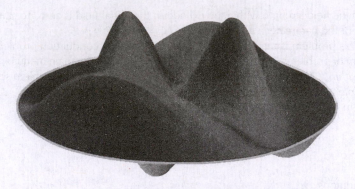

FIGURE A5.58 Eigenfunction $V_{21}^{(1)}(r, \varphi)$ for Example A5.14.

The three-dimensional picture shown in Figure A5.58 depicts the one of eigenfunction for the given problem ($\lambda_{21} = 4.42817 \; V_{21}^{(1)} = J_2(\mu_1^{(2)}r/4)\cos 2\varphi \; \left\| V_{21}^{(1)} \right\|^2 = 1.85061$).

To solve the problem with the help of the program, follow the same steps as in Example A5.9. Results of the solution can be displayed by selecting menu option *"Evolution of Membrane Temperature."* The three-dimensional surface shown in Figure A5.59 depicts the animated solution $u(r,\varphi,t)$ at time $t = 2$.

By changing the values of thermal diffusivity, a^2, radius, l, and parameters H_0 and ω of the boundary value function, different pictures of magnetic field distribution can be obtained and studied.

A5.5 EXAMPLES USING THE PROGRAM LAPLACE

A number of examples of how to apply the **Laplace** program to solve different problems follow.

Example A5.15

A heat-conducting thin uniform rectangular membrane ($0 \le x \le l_x, 0 \le y \le l_y$) is thermally insulated over its lateral faces. One part of the boundary (at $x = 0$ and $y = 0$) is thermally insulated, and the other part (at $x = l_x$ and $y = l_y$) is held at fixed temperature $u = u_0$. One constant internal source of heat acts at the point (x_0, y_0) of the membrane. The value of this source is $Q = $ const.

Find the steady-state temperature distribution in the membrane. To start, the following values of the parameters are assigned: $l_x = 4$, $l_y = 6$, $u_0 = 10$, $Q = 100$, $x_0 = 2$, $y_0 = 3$.

FIGURE A5.59 Distribution of magnetic field, $u(r, \varphi, t)$, at $t = 2$ for Example A5.14.

FIGURE A5.60   Source function $f(x, y)$ for Example A5.15.

Solution. First, select the type of problem from the main menu by clicking *"Elliptic Equations over Rectangular Domains."* The program has a set of predefined examples on the selected subject and the given problem is from this set. To load the problem select *"Data"* from the main menu and click *"Library example."* From the given list of problems select *"Example 6."* A description of the problem can be read in the right part of the screen. The problem description, solution hints, and theoretical solution can be accessed by clicking the *"Text + Hint"* button. To proceed with the selected example, click the "OK" button and in the dialog window parameters for the current problem will be displayed.

The given problem involves the solution of the Poisson equation

$$\frac{\partial^2 u}{\partial x^2} + \frac{\partial^2 u}{\partial y^2} = -Q \cdot \delta(x - x_0)\delta(y - y_0)$$

under the conditions

$$\frac{\partial u}{\partial x}(0, y) = 0, \quad u(l_x, y) = u_0, \quad \frac{\partial u}{\partial y}(x, 0) = 0, \quad u(x, l_y) = u_0.$$

Since all parameters for the problem have been entered we can proceed by clicking "OK." At this point the program is ready to solve the problem with selected parameters. To see the source function select *"View"* → *"Source function f(x,y)"* (Figure A5.60) and click *"Return"* to return to the problem solution screen.

To run the solver choose the *"Execute"* command from the main menu; results can be seen by selecting *"Results"* → *"Surface Plot of the Solution"* (Figure A5.61).

The following menus allow the user to investigate other properties of the solution:

> *Results → Fourier Coefficients;*
> *Results → Surface Plots of EigenFunctions;*
> *Results → Contour Map of the Solution;*
> *Results → Solution Profile at x = const;*
> *Results → Solution Profile at x = const.*

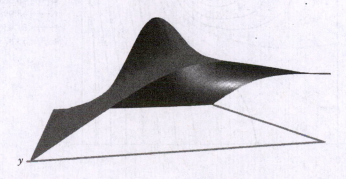

FIGURE A5.61   Surface plot of the solution $u(x, y)$ for Example A5.15.

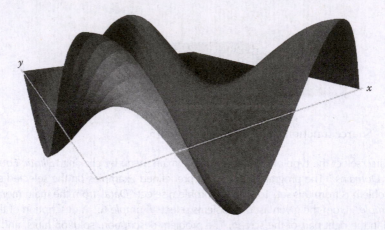

FIGURE A5.62   Eigenfunction $V_{32}(x,y)$; $\|V_{32}\|^2 = 6$, $\lambda_{32} = 4.47216$ for Example A5.15.

More information for each option can be found by selecting the *"Help"* menu on the corresponding screen.

The three-dimensional surface shown in Figure A5.62 depicts one of the eigenfunctions for this problem. The contour map of the solution $u(x,y)$ is shown in Figure A5.63.

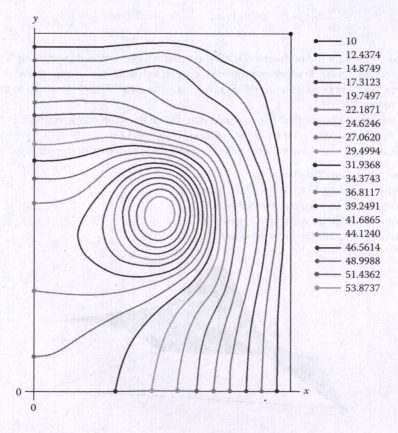

	10
	12.4374
	14.8749
	17.3123
	19.7497
	22.1871
	24.6246
	27.0620
	29.4994
	31.9368
	34.3743
	36.8117
	39.2491
	41.6865
	44.1240
	46.5614
	48.9988
	51.4362
	53.8737

FIGURE A5.63   Contour map of the solution $u(x,y)$ for Example A5.15.

Example A5.16

Consider a heat-conducting uniform rectangular plate ($0 \le x \le l_x$, $0 \le y \le l_y$), thermally insulated over its lateral faces. Let heat be generated throughout the plate; the intensity of internal sources (per unit mass of the plate) is $Q(x,y) = Axy$. The side $y = 0$ of the plate is thermally insulated, the side $x = 0$ is held at fixed temperature $u = u_1$, the side $y = l_y$ has a constant temperature of zero, and the side $x = l_x$ is subjected to convective heat transfer with a medium according to Newton's law of cooling. The temperature of the medium is $u_{md} = $ const. and the coefficient of thermal conductivity is h.

Find the steady-state temperature distribution in the plate. To start, the following values of the parameters are assigned: $A = 5$, $u_1 = 10$, $u_{md} = 40$, $h = 0.5$, $l_x = 4$, $l_y = 6$.

Solution. This problem is also from the library set. To load the problem select "*Data*" from the main menu, and then click "*Library example.*" From the given list of examples select "*Example* 12."

The problem involves the solution of the Poisson equation

$$\frac{\partial^2 u}{\partial x^2} + \frac{\partial^2 u}{\partial y^2} = -Axy$$

under the conditions

$$u\big|_{x=0} = u_1, \quad \frac{\partial u}{\partial x} + h(u - u_{md})\bigg|_{x=l_x} = 0, \quad \frac{\partial u}{\partial y}\bigg|_{y=0} = 0 \quad u\big|_{y=l_y} = 0.$$

Figure A5.64 shows the source function $f(x,y)$ for the problem.

The boundary value functions of the given problem do not satisfy the conforming conditions at points $(0, l_y)$ and (l_x, l_y):

$$P_1[g_4]_{x=0} = 0 \ne P_4[g_1]_{y=l_y} = u_1,$$

$$P_2[g_4]_{x=l_x} = hu_{md} \ne P_4[g_2]_{y=l_y} = 0.$$

As a result, when you run the solver by selecting the "*Execute*" command from the main menu the message in Figure A5.65 will be displayed.

Although the boundary conditions do not match, we may still solve the problem with the help of the program by searching for a generalized solution, following the same steps as in Example A5.15. Results of the solution can be displayed by selecting menu "*Results*" → "*Surface*

FIGURE A5.64   Source function $f(x, y)$ for Example A5.16.

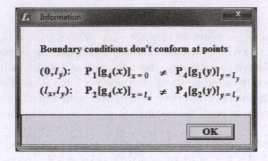

FIGURE A5.65   Information window for Example A5.16.

Plot of the Solution." The three-dimensional surface shown in Figure A5.66 depicts the steady-state temperature distribution $u(x,y)$ in the rectangular plate. Note that the edges of the surface do adhere to the given boundary conditions.

The solution to the given problem can be expressed as the sum of three functions

$$u(x,y) = w_1(x,y) + w_2(x,y) + v(x,y),$$

where $w_1(x,y)$ is an auxiliary function satisfying the conforming boundary value functions

$$P_1[w_1]_{x=0} = u_1, \quad P_2[w_1]_{x=l_x} = hu_{md},$$

$$P_3[w_1]_{y=0} = 0, \quad P_4[w_1]_{y=l_y} = u_1 - (u_1 - hu_{md})\frac{x}{l_x},$$

$w_2(x,y)$ is a particular solution of Laplace problem with the following boundary conditions

$$P_1[w_2]_{x=0} = 0, \quad P_2[w_2]_{x=l_x} = 0,$$

$$P_3[w_2]_{y=0} = 0, \quad P_4[w_2]_{y=l_y} = -u_1 + (u_1 - hu_{md})\frac{x}{l_x}.$$

FIGURE A5.66   Stationary distribution of temperature $u(x,y)$ for Example A5.16.

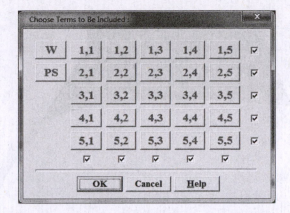

FIGURE A5.67 Selection of terms in the expansion of the solution.

and $v(x,y)$ is a solution of a Poisson equation with zero boundary conditions:

$$v(x,y) = \sum_{n=0}^{\infty}\sum_{m=1}^{\infty} C_{nm}V_{nm}(x,y).$$

The program **Laplace** allows you to study the behavior of the functions $w_1(x, y)$, $w_2(x, y)$ and of any single term of the partial sum of the solution

$$u_{nm}(x,y) = w_1(x,y) + w_2(x,y) + \sum_{i=0}^{n}\sum_{j=0}^{m} C_{ij} \cdot V_{ij}(x,y).$$

Select the *"Choose Terms..."* option from the graph menu and the *Choose Terms to Be Included* dialog box will be displayed (Figure A5.67).

The button "W" corresponds to the auxiliary function $w_1(x, y)$. Select the button "W" and unselect all other buttons to see the surface graph of the auxiliary function (Figure A5.68)

$$w_1(x,y) = u_1 - (u_1 - hu_{md})\frac{x}{l_x}.$$

FIGURE A5.68 Surface plot of the auxiliary function $w_1(x, y)$ for Example A5.16.

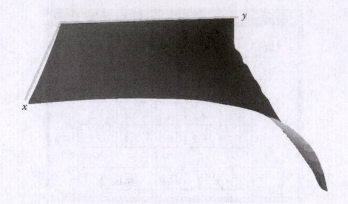

FIGURE A5.69   Surface plot of the particular solution $w_2(x, y)$ for Example A5.16.

The button "PS" corresponds to the particular solution $w_2(x, y)$. This button is active because the boundary value functions do not satisfy the conforming conditions. Select the button "PS" and unselect all other buttons to see the surface graph of the particular solution (Figure A5.69)

$$w_2(x, y) = \sum_{k=1}^{N} A_k \frac{\mathrm{ch}\sqrt{\lambda_k}\, y}{\mathrm{ch}\sqrt{\lambda_k}\, l_y} \sin\sqrt{\lambda_k}\, x.$$

Unselect the buttons "W" and "PS" and the partial sum

$$v_{nm}(x, y) = \sum_{i=0}^{n} \sum_{j=0}^{m} C_{ij} V_{ij}(x, y)$$

will be displayed (Figure A5.70).
All these figures can be rotated to view the shape of the surfaces from different directions.

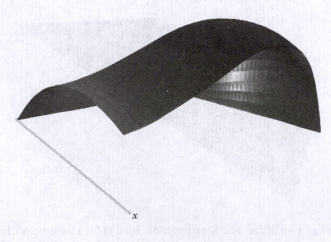

FIGURE A5.70   Surface plot of the partial sum $v_{nm}(x, y)$ for Example A5.16.

Example A5.17

A heat-conducting thin uniform circular membrane of radius l is thermally insulated over its lateral faces. The membrane is subjected to convective heat transfer according to Newton's law at its boundary. The temperature of the medium is u_{md} = const. and the coefficient of thermal conductivity is h.

Let heat be generated throughout the membrane where the intensity of internal sources (per unit mass of the membrane) is

$$f(r,\varphi) = 5r^2 \cos 5\varphi.$$

Find the stationary distribution of temperature within the membrane. To start, take the following values of the parameters: $l = 2$, $u_{md} = 15$, $h = 3$.

Solution. First, select the type of problem from the main menu by clicking *"Elliptic Equations over Circular Domains."* This problem is not from the library set. To solve it, instead of selecting a library example as we did before, select a *"New problem"* from the *"Data"* menu on the problem screen.

The problem involves a solution to the Poisson equation

$$\frac{\partial^2 u}{\partial r^2} + \frac{1}{r}\frac{\partial u}{\partial r} + \frac{1}{r^2}\frac{\partial^2 u}{\partial \varphi^2} = -5r^2 \cos 5\varphi$$

under the condition

$$\frac{\partial u}{\partial r}(l,\varphi) + h[u(l,\varphi) - u_{md}] = 0 \quad \text{or} \quad \frac{\partial u}{\partial r}(l,\varphi) + hu(l,\varphi) = hu_{md}.$$

To specify heat conductivity according to Newton's law at the circular periphery of the membrane, choose option "3 (*Mixed condition*)" for boundary values by selecting this item from the dialog box. Enter the parameters of heat exchange by clicking the *"Edit"* button. Since $u_{md} = 15$ the value of $g(\varphi) = h \cdot u_{md} = 45$ (Figure A5.71).

The intensity of internal sources is determined by function $f(r,\varphi) = 5r^2 \cos 5\varphi$.

When all parameters are entered, click the "OK" button. Figure A5.72 shows the source function for the problem.

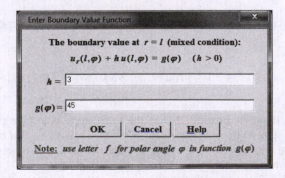

FIGURE A5.71 Parameters of heat exchange on the boundary $r = l$.

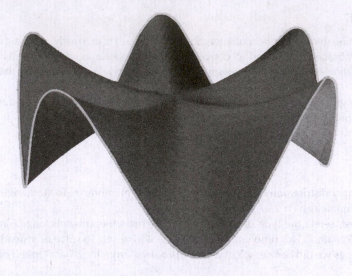

FIGURE A5.72   Source function $f(r, \varphi)$ for Example A5.17.

Eigenvalues of the given boundary value problem have the form

$$\lambda_{nm} = \left[\frac{\mu_m^{(n)}}{l}\right]^2, \quad n,m = 0,1,2,\ldots,$$

where $\mu_m^{(n)}$ are positive roots of the equation

$$\mu J_n'(\mu) + h l J_n(\mu) = 0.$$

Figure A5.73 shows the graph of function $\mu J_n'(\mu) + h l J_n(\mu)$ used for evaluating eigenvalues for the choice $n = 3$.

Eigenfunctions of the given boundary value problem are

$$V_{nm}^{(1)} = J_n\left(\frac{\mu_m^{(n)}}{l}r\right)\cos n\varphi, \quad V_{nm}^{(2)} = J_n\left(\frac{\mu_m^{(n)}}{l}r\right)\sin n\varphi.$$

The three-dimensional picture shown in Figure A5.74 depicts the eigenfunction $V_{32}^{(1)}$ for the current problem:

$$\lambda_{32} = 35.1394, \quad V_{32}^{(1)} = J_3(\mu_2^{(3)}r/2)\cos 3\varphi, \quad \left\|V_{32}^{(1)}\right\|^2 = 0.338524.$$

To solve the problem using the program, follow the same steps as in Example A5.14. Results of the solution can be displayed by selecting menu *"Results"* → *"Surface Plot of the Solution."* The stationary temperature distribution $u(r,\varphi)$ in the circular membrane is represented in the three-dimensional plot shown in Figure A5.75.

More results can be displayed with the *"Results"* → *"Contour Map of the Solution"* option (Figure A5.76).

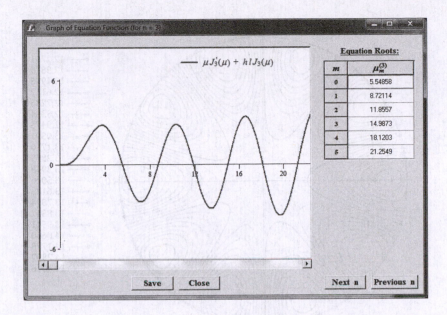

FIGURE A5.73 Graph of equation for evaluating eigenvalues for $n = 3$ and the table of roots $\mu^{(3)}_m$ for Example A5.17.

FIGURE A5.74 Eigenfunction $V^{(1)}_{32}(r, \varphi)$ for Example A5.17.

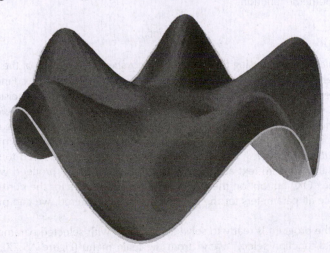

FIGURE A5.75 Stationary distribution of temperature $u(r, \varphi)$ for Example A5.17.

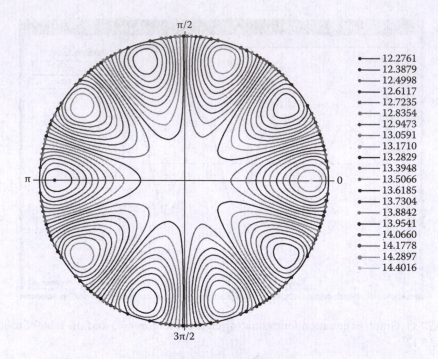

Legend values:
- 12.2761
- 12.3879
- 12.4998
- 12.6117
- 12.7235
- 12.8354
- 12.9473
- 13.0591
- 13.1710
- 13.2829
- 13.3948
- 13.5066
- 13.6185
- 13.7304
- 13.8842
- 13.9541
- 14.0660
- 14.1778
- 14.2897
- 14.4016

FIGURE A5.76   Contour map of the solution $u(r, \varphi)$ for Example A5.17.

A5.6   EXAMPLES USING THE PROGRAM FOURIERSERIES

A number of examples of how to apply the **FourierSeries** program to solve different problems follow. Some of the problems were taken from the book without any changes; others were modified to show how to use different program options.

Example A5.18 (Expansion into Classical Orthogonal Polynomials)

Expand the trigonometric function

$$f(x) = sin(x)$$

defined on the interval $[0,\pi/4]$ using the system of Chebyshev polynomials of the first kind.

Find the partial sum of the expansion obtained in powers of the variable x. Find the number of terms in the power series sufficient to approximate the given function to a precision of $\varepsilon = 10^{-5}$.

Solution. First, select a type of the problem from the main menu: click "*Expansion into Classical Orthogonal Polynomials.*" The program has a set of predefined examples on the selected subject and the given problem is from this set. To load the problem select "*Data*" from the main menu and then click "*Library example.*" From the given list of examples you need to select "*Example 1.*" You can read the problem text at the right part of the screen. To proceed with the selected example, click the "OK" button; on the dialog window, parameters for the current problem will be displayed. Since all parameters for the problem have been entered, we can proceed by clicking "OK."

At this point, the program is ready to solve the problem with selected parameters. To see the graph of the initial function select "*View*" from the main menu (Figure A5.77), and then click "Return" to go back to the problem solution screen.

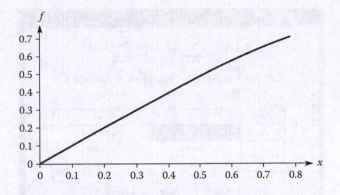

FIGURE A5.77   The given function $f(x)$ for Example A5.18.

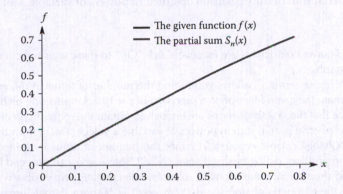

FIGURE A5.78   Graph of the partial sum $S_{10}(x)$, $\delta^2 = 1.04 \cdot 10^{-17}$ for Example A5.18.

To run the solver choose the *"Execute"* command from the main menu. Results can be seen by selecting option *"Graph of the Partial Sum"* (Figure A5.78). The graph range for displaying the partial sum $S_n(x)$ and the limits for the graph can be adjusted with the help of the *"Set Attributes"* command.

To examine the behavior of the individual harmonics of the expansion and of their partial sums use the option *"Choose terms..."* (Figure A5.79). To select (or unselect) the individual term click on the corresponding choose button. The terms whose numbers are selected (red) are included in the partial sum; the ones not selected (white) are excluded. The choose buttons respective to the

FIGURE A5.79   Dialog window "Choose terms to be included."

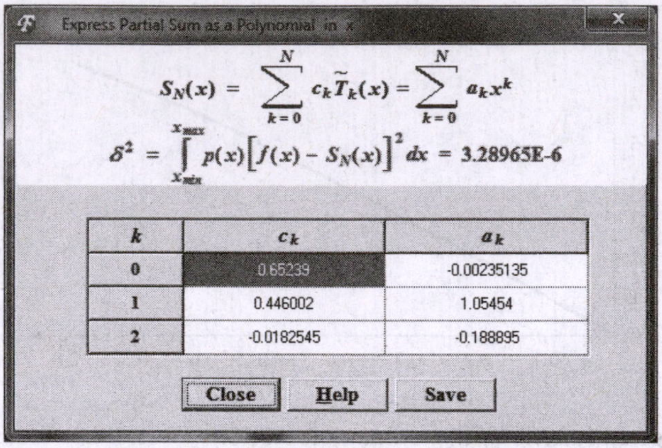

FIGURE A5.80 Partial sum of the expansion obtained in powers of variable x for Example A5.18.

terms with zero Fourier coefficients are inactive. Click "OK" to mark your selection and get back to drawing the graph.

The option "*Choose terms...*" allows you to find the number of terms in the expansion sufficient to approximate the given function to a precision of $\varepsilon = 10^{-5}$. Unselecting higher terms of the expansion we see that the first three terms are enough to obtain the requisite precision.

To find the respective partial sum in powers of variable x select "*Data*" from the main menu, and then click "*Change current problem.*" Change the number of terms in the expansion in the parameters of the problem dialog box and enter "$N = 2$." Now you can proceed by clicking the "OK" button and then running the problem solver by selecting "*Execute*" option from the main menu. To obtain the required polynomial select "*Results*" → "*Express Partial Sum as a Polynomial in x*" (Figure A5.80).

The polynomial approximating the function $f(x) = sin(x)$ to a precision of $\varepsilon = 10^{-5}$ is

$$f(x) = -0.188895x^2 + 1.05454x - 0.00235135.$$

This expansion is valid only in the interval $[0, \pi/4]$. However, you can use the polynomial you obtain for calculating the trigonometric functions of an arbitrary argument if you take advantage of the trigonometric identities connecting the trigonometric functions with $sin(-x)$, $sin(x + \pi/2)$, $sin(x + \pi)$, and so on.

The next example shows that the generalized Fourier series of orthogonal polynomials have some advantage over the Taylor series expansion. To expand the function $f(x)$ defined in the interval $[a,b]$ in a Taylor series, it is necessary that the function have derivatives of all orders in the interval (a,b). Moreover, the Fourier series converges faster for many functions than does the Taylor series.

Example A5.19 (Expansion into Classical Orthogonal Polynomials)

Let the function $f(x) = \dfrac{x(2\pi - x)}{\pi^2}$ be given at the discrete set of points in the interval $[0, 2\pi]$. For example,

$$x_k = (k-1)\frac{2\pi}{M}, \quad y_k = \frac{4(k-1)(M-k+1)}{M^2}, \quad k = 1, 2, \ldots, M+1 = 13.$$

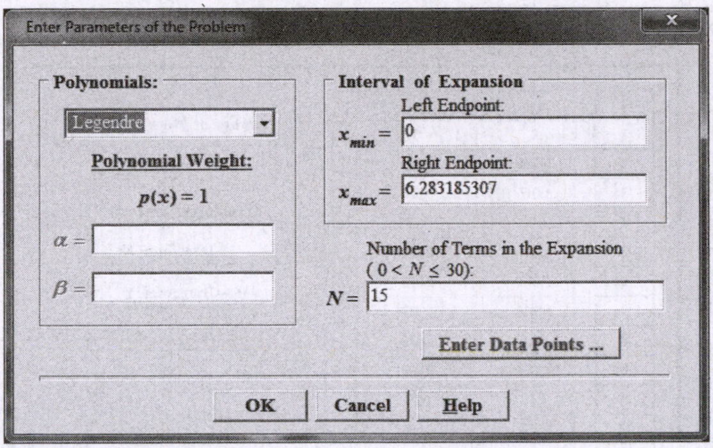

FIGURE A5.81 Parameters of the problem dialog for Example A5.19.

Approximate this discrete function by polynomials of the 3rd, 10th, and 15th order. Find such polynomials with the help of the Fourier expansion by the system of Legendre polynomials, and then find the corresponding partial sums in powers of the variable x. Compare the results of these expansions.

Solution. This problem is not in the library set. To solve it with the program select a *"New problem"* from the *"Data"* menu on the problem screen. The function f(x) that is to be expanded as a Fourier series is given as a table. You can input your table function directly from the keyboard (command *"Input via Keyboard"*) or from an ASCII [.dat] file prepared in advance (using the command *"Read from file"*). The length of your data file must not exceed 1000 lines (a program limitation). Each line of the file must contain values of x_k and $y_k = f(x_k)$, separated by one or more spaces. The program allows editing of the data and provides some operations for altering the data.

After you select the type of function, you will see the dialog window *"Enter parameters of the problem (Table Function)"* (Figure A5.81). To specify the type of polynomial, choose option *"Legendre"* by selecting this item from the combo box. Then enter the interval of expansion ($x_{min} = 0$, $x_{min} = 2\pi$) and the number of terms in the expansion ($N = 15$).

The function f(x) is given at the discrete set of points. To look through the table of points (or to enter it), click the *"Enter Data Points…"* button and the dialog window with coordinates of data points will be displayed (Figure A5.82). Note that after editing, the data is put in *ascending order* of the variable x.

When all parameters are entered follow the same steps as in Example A5.18 and select the *"Execute"* command from the main menu. A partial sum of the Fourier expansion can be displayed by selecting menu option *"Results"* → *"Graph of the Partial Sum."*

Draw the graphs of the partial sum $S_N(x)$ for N = 15, 10, and 3 (use the menu option *"Choose terms…"*). Figures A5.83 and A5.84 show two of the calculated expansions. You can see that adding terms only makes the precision of approximation worse in this example.

Example A5.20 (Expansion into Fourier-Bessel Series).

Expand the function

$$f(x) = \begin{cases} x^2, & \text{if } 0 \le x < 1 \\ x, & \text{if } 1 \le x \le 2 \end{cases}$$

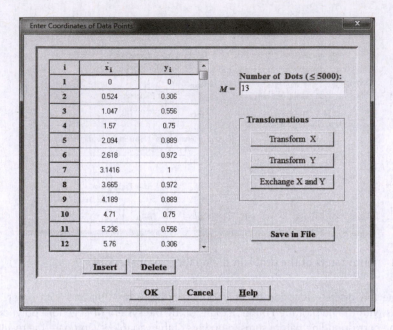

FIGURE A5.82   Dialog window "Enter coordinates of Data Points."

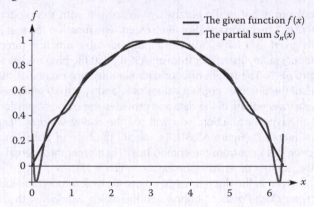

FIGURE A5.83   Graph of the partial sum $S_{15}(x)$, $\delta^2 = 0.039$ for Example A5.19.

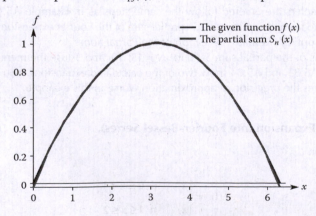

FIGURE A5.84   Graph of the partial sum $S_3(x)$, $\delta^2 = 0.00085$ for Example A5.19.

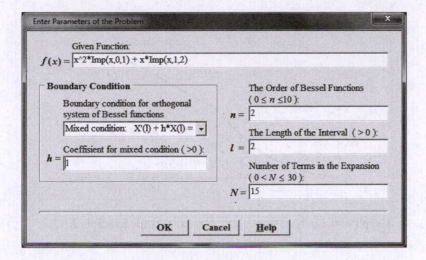

FIGURE A5.85 Parameters of the problem dialog box for Example A5.20.

defined at the interval [0,2] using the system of Bessel functions of the first kind of order $n = 2$

$$X_k(x) = J_2\left(\frac{\mu_k^{(2)}}{l} x\right),$$

where $\mu_k^{(2)}$ are positive roots of the equation $\mu J_2'(\mu) + h l J_2(\mu) = 0$.

Search for this problem when $h = 0.1, 1, 10$.

Solution. Select the type of the problem from the main menu by clicking *"Expansion into Fourier-Bessel Series."* The given problem is from the library set. To load it select *"Data"* → *"Library problem"* → *"Problem 6."* To proceed with the selected problem click the "OK" button; on the dialog window parameters for the current problem will be displayed (Figure A5.85).

In the expression for the given function

$$f(x) = x^2 * \text{Imp}(x,0,1) + x * \text{Imp}(x,1,2),$$

Imp(x,a,b) is an impulse function where $\text{Imp}(x,a,b) = 1$ if $a \le x < b$ and 0 otherwise.

Since all parameters for the problem have been entered we can proceed by clicking "OK." To see the given function select *"View"* from the main menu, and then click *"Return"* to go back to the problem solution screen. To look over the values of coefficients C_k in the expansion select *"Results"* → *"Fourier Coefficients."*

To run the solver click the *"Execution"* option; results can be seen by selecting *"Results"* → *"Graph of the Partial Sum"* (Figure A5.86).

Figure A5.87 depicts the bar chart of Fourier coefficients, C_k (option *"Bar Chart of Fourier Coefficients"*). The bar chart gives a graphic picture of the contribution of individual terms to the Fourier-Bessel expansion of $f(x)$. The bar chart displays data by drawing a rectangle of the height equal to the kth Fourier coefficient (measured along the vertical axis) at each k value along the graph. The height of the kth rectangle corresponds to the value of the Fourier coefficient C_k of the general expansion of $f(x)$.

Menu option *"Results"* → *"Graphs of Bessel Functions"* allows the user to look through the plots of Bessel functions of the first kind of the given order $m = 2$ (functions of the orthonormal system)

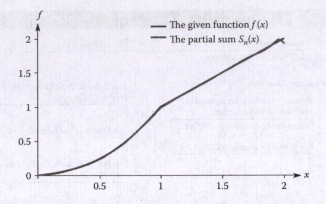

FIGURE A5.86   Graph of the partial sum $X_{15}(x)$, $\delta^2 = 0.00021$ for Example A5.20.

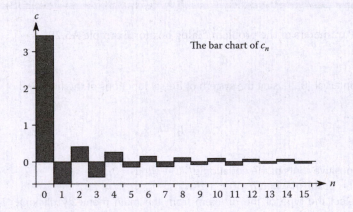

FIGURE A5.87   Bar chart of Fourier coefficients, c_k for Example A5.20.

for the successive values of $k = 0,1,...N$. One of the Bessel functions for the problem is represented in Figure A5.88 is

$$X_{10}(x) = J_2(\mu_{10}^{(2)}x/l),$$

where $\mu_{10}^{(2)}$ is the 10th root of the equation $\mu J_2'(\mu) + hlJ_2(\mu) = 0$.

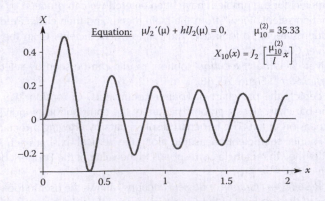

FIGURE A5.88   Graph of Bessel function $X_{10}(x)$ for Example A5.20.

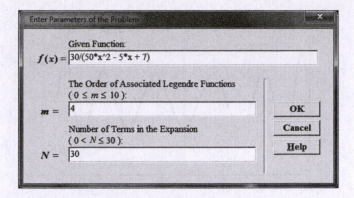

FIGURE A5.89   Parameters of the problem dialog box for Example A5.21.

Example A5.21 (Expansion into Associated Legendre Functions)

Expand the function

$$f(x) = \frac{30}{50x^2 - 5x + 7}$$

defined in the interval $(-1,1)$ using the system of associated Legendre functions of order $m = 4$, which in normalized form are

$$\hat{X}_k(x) = \hat{P}^4_{4+k}(x) = \sqrt{\frac{2k+9}{2} \cdot \frac{k!}{(8+k)!}} \; P^4_{4+k}(x).$$

Solution. This problem is not in the library set. To solve it with the program select a *"New problem"* → *"Formula for f(x)"* from the *"Data"* menu on the problem screen, and then enter the given function and parameters (Figure A5.89).

To run the solver select the *"Execute"* command. Figure A5.90 depicts the partial sum of the expansion.

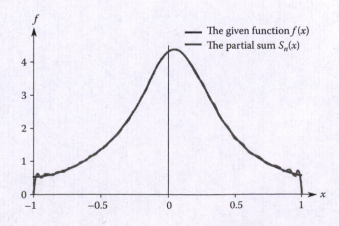

FIGURE A5.90   Graph of the partial sum $S_{30}(x)$, $\delta^2 = 0.0036$ for Example A5.21.

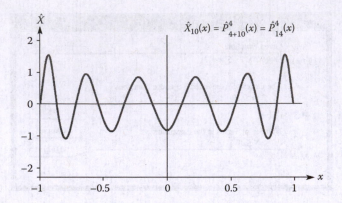

FIGURE A5.91   Graph of associated Legendre function, $X_{10}(x)$, for Example A5.21.

Use menu option *"Results"* → *"Graphs of Associated Legendre Functions"* to look through the plots of associated Legendre functions of the first kind of the given order $m = 4$ (functions of the orthonormal system) for the successive values of $k = 0,1,...,N$. One of the associated Legendre functions $\hat{X}_{10}(x) = \hat{P}_{14}^4(x)$ for the problem is represented in Figure A5.91.

Bibliography

The textbook is self-contained, but in the books listed below one can find proofs of many theorems, deeper mathematical analysis of the issues linked to existence and uniqueness of solutions (on advanced graduate level), and other topics from PDEs and ODEs. As a starting point for readers who are searching for more material on PDEs and special functions, we recommend the first reference—this text is written in the same style and uses the same approaches and notations as the current one.

[1] Henner, V., Belozerova, T., and Forinash, K. *Mathematical Methods in Physics Partial Differential Equations, Fourier Series, and Special Functions*. Wellesley, MA: A K Peters, Ltd., 2009.

ODES

[2] Boyce, W. E. and DiPrima, R. C. *Elementary Differential Equations and Boundary Value Problems*. Hoboken, NJ: Wiley, 2008.

[3] Brauer, F. and Nohel, J. *The Qualitative Theory of Ordinary Differential Equations*. New York: Benjamin, 1969; New York: Dover, 1989.

[4] Coddington, E. A. *An Introduction to Ordinary Differential Equations*. Englewood Cliffs, NJ: Prentice Hall, 1961; New York: Dover, 1989.

[5] Coddington, E. A. and Levinson, L., *Theory of Ordinary Differential Equations*. Malabar, FL: Krieger Publication Company, 1984.

[6] Drazin, P. G. *Nonlinear Systems*. Cambridge, UK: Cambridge University Press, 1992.

[7] Dyke, P. P. G. *An Introduction to Laplace Transforms and Fourier Series*. London: Springer, 2002.

[8] Jordan, D. W. and Smith, P. *Nonlinear Ordinary Differential Equations: An Introduction for Scientists and Engineers*. Oxford, UK: Oxford University Press, 2007.

[9] Kuhfittig, P. K. F. *Introduction to the Laplace Transform*. New York: Plenum, 1978.

[10] Strogatz, S. H. *Nonlinear Dynamics and Chaos: With Applications to Physics, Biology, Chemistry, and Engineering*. Reading, MA: Addison-Wesley, 1994; Boulder, CO: Westview Press, 2001.

PDES

[11] Churchill, R. V. and Brown, J. W. *Fourier Series and Boundary Value Problems*. New York: McGraw Hill, 1996.

[12] Courant, R. and Hilbert, D. *Methods of Mathematical Physics*, Volume 1. Hoboken, NJ: Wiley-Interscience, 1989.

[13] Evans, L. C. *Partial Differential Equations*. Providence, RI: American Mathematical Society, 2010.

[14] Haberman, R. *Applied Partial Differential Equations: With Fourier Series and Boundary Value Problems*. Englewood Cliffs, NJ: Prentice Hall, 2003.

[15] Kevorkian, J. *Partial Differential Equations: Analytical Solution Techniques*. New York: Springer, 2010.

[16] Strauss, W. A. *Partial Differential Equations*. Hoboken, NJ: Wiley, 1992.

[17] Zachmanoglou, E. C. and Thoe, D. W. *Introduction to Partial Differential Equations with Applications*. New York: Dover, 1989.

PDES AND ODES HANDBOOK

[18] Zwillinger, D. *Handbook of Differential Equations*. San Diego: Academic Press, 1998.

INDEX

A

Autonomous
  Equations, 135–136, 139–140, 144
  Systems, vii, 153, 156, 162

B

Bernoulli equation, 24–25
Bessel, 203, 205, 207, 209, 211, 213, 215, 217, 219,
  221, 223, 225, 227, 229, 231, 233, 235, 237,
  239, 241, 386
  Equation, vii, 204, 208, 213–214, 219, 224, 376,
  540
  Functions, viii, 122–123, 131–132, 201, 208, 210,
  213–214, 216, 217–218, 225–226, 473, 477,
  532, 538, 540, 619–620
  Spherical, 209, 219–220
Boundary conditions
  Dirichlet, 119, 215, 292, 307, 332–334, 347, 373,
  378, 395, 497, 526
  Homogeneous, ix, 118, 213, 304–306, 311, 316,
  319, 323–324, 332, 349, 356, 359–360, 362,
  367–368, 374, 376, 378, 388, 411–412, 419,
  423, 449, 459, 473, 483, 499–500, 502,
  505, 526
  Neumann, 365, 408–409, 427, 446, 471, 504, 510,
  513, 536, 550–552, 554–555, 558–561
  Nonhomogeneous, ix–x, 323, 326, 347, 366, 371,
  393, 396, 422, 424, 458, 464, 471, 483,
  504–505, 533, 553
  Mixed, 115, 119, 214, 216, 292, 306–307, 315, 347,
  373, 378, 387, 395, 437, 522
  Periodic, 121, 133, 328, 376
Boundary value problems, v–vi, viii–ix, 1, 4,
  113–119, 121, 123, 125, 127, 129, 131, 133,
  201, 209, 213, 219, 226, 283, 294, 303, 405,
  450–451, 473, 498, 501, 522, 524, 534, 537,
  623

C

CAS (computer algebra systems), 153, 162
Cauchy's
  Problem, 4–6, 30, 32, 36–37, 44, 78, 293, 298,
  318, 361, 391
  Theorem, 80, 86
Completeness, 216, 306, 420
Convergence, 123, 165–166, 197, 216, 232, 248–250,
  419, 522, 576
  Rate of, 249–251, 257
  Uniform, 248, 310, 389, 413
Cooling, law of, 416, 440, 496, 607
Coordinates, polar, 208, 373, 388, 470, 474,
  518–519, 532
Cosine series, 254, 277

D

D'Alembert's
  Equation, 283
  Method, viii, 298
Delay Theorem, 167
Diffusion, 283–284, 405–406, 408, 410, 439, 445
Direction field, 7–10, 30, 32, 149, 151–152
Dirichlet, 306, 311, 332–334, 525, 532
  Condition, 408–409, 446, 471, 504, 510, 513,
  515, 536, 550–552, 554–555, 558–560
  Problem, 311, 496, 498, 529
  Theorem, 250
  Dirichlet, 537, 541
  Exterior, 496, 520, 526, 528, 531
  Interior, 496, 497, 520, 521, 523–524,
  526–529, 531, 533

E

Eigenfunction expansion, 130, 219

Eigenfunctions, xi, 119–123, 125–126, 128–131, 133, 183, 185, 187–188, 214, 305–311, 314–315, 317, 319, 322, 326, 329, 336, 351–352, 354–358, 362, 365, 376–378, 382, 384–387, 392, 397–398, 412–413, 415, 417, 421, 425, 427–428, 431, 437, 439, 450–452, 457, 462–463, 465–466, 469, 473–474, 476, 480, 485–486, 503, 508–509, 519, 543, 549–552, 511, 588–591, 603–604
Eigenvalue problem, 226, 375, 563–565
Eigenvalues, xi, 119, 121–122, 125–132, 133, 145, 147, 151–152, 158–159, 162, 184–185, 204, 214–216, 226, 305–306, 308, 311, 314–315, 319, 322, 326, 329, 351–352, 355–356, 358, 362, 365, 376–378, 384, 392, 412, 414–415, 417, 421, 425, 427–428, 431, 434, 437, 450–453, 462–463, 469, 473–474, 503–504, 508–509, 519, 535–536, 540–541, 549–552, 557–558, 563–567
  Complex, vii, 563
  Discrete, 122, 126
  Distinct, vii, 158
Eigenvectors, 153, 352, 563–568
Electrostatic, 284, 495–496, 527
Equation
  Characteristic, vi, 59–61, 63–69, 89, 93, 95–96, 98, 100, 102, 104, 106, 114–116, 145–147, 149, 433, 564–567
  Diffusion, ix, 407, 494, 601
  Eigenvalue of, 308, 314, 387, 417
  Elliptic, x, 493, 495, 497, 499, 501, 503, 505, 507, 509, 511, 513, 515, 517, 519, 521, 523, 525, 527, 529, 531, 533, 535, 537, 539, 541, 543, 545, 547, 569, 605, 611
  Exact, v, 27
  Homogeneous, v–vi, ix, 21–23, 47, 50–51, 53–54, 57, 63, 67–69, 108, 114, 116, 119, 183, 185, 187–188, 190, 316, 345, 349, 359, 374, 389, 411, 419, 433, 449, 472–473
  Hyperbolic
    First-order, v–vi, 4–5, 11, 21, 27, 30, 47, 78, 135, 140–141, 153, 156, 411
    Second-order, 44, 47, 78, 84, 144, 153, 411
  Nonhomogeneous, ix, 47, 53, 114, 116, 183, 188, 272, 316–317, 319, 323, 360, 362, 372, 388–390, 392, 419, 433, 449, 454, 472, 477, 479, 493, 526
  Parabolic, 405, 407, 410, 433, 445, 532
Equilibrium, 142, 147, 159, 175, 287, 294, 311, 334, 339, 343–344, 374, 381, 400, 406, 414–415, 464, 481–482, 494
  Point, 135, 139–142, 144, 146, 148, 152–153, 157, 159, 162–163

Solutions, 135–139, 140, 156, 160
Euler's
  Equation, 70, 520
  Formula, 6, 262
  Method, 6
Expansion
  Eigenfunction, 232, 239, 418
  Method of, 277–281, 573–575
Extension, periodic, 251, 573

F

Factor, integrating, 23–24, 28
Formulas, recurrence, 211, 227–228, 231, 239, 538
Fourier-Bessel series, 132, 213, 215–218, 569, 617, 619
Fourier
  Coefficients, 124, 247–248, 251–252, 354, 529, 574, 605, 619–620
  Expansion, 131, 189, 215, 247, 256, 262, 269, 281, 380, 448, 574, 578, 592, 617
  Method, ix–x, 118, 298, 303–304, 316, 322–323, 348–349, 358, 366–367, 373, 388, 393–394, 411, 422, 448–449, 453, 458, 472, 477, 483, 519
  Transform, 273–276
Fourier-Legendre series, viii, 131, 231, 233–324, 236, 239–241
Fourier series, v, viii, 1, 188–189, 243, 245, 247–255, 257, 259, 261, 263–270, 273, 275, 277, 279, 281, 310, 317, 348, 353–354, 360, 380, 390, 413, 420, 454, 542, 569, 572, 578, 616–617, 623
  Convergence of, 249
  Generalized, viii, 122, 215, 267, 281, 569, 616
  Trigonometric, 123, 216, 267–269, 277–281, 572–573, 575, 578
Fredholm equations, vii, 182, 188–189, 193
Functions
  Basis, 118, 201, 216, 352–353, 360, 390, 542
  Convolution of, 168, 197
  Cylindrical, 208–209
  Delta, 169, 275, 421, 572, 584
  Harmonic, 497, 514, 531
  Orthogonal system of, 310, 354, 380
  Weight, 119, 121–122, 131, 216, 239

G

Gamma function, viii, 207–208, 221, 223
General solution, 12, 14, 18, 24–25, 29, 45, 47, 52, 61, 64, 69, 81, 113, 147, 151–152
Gibbs's phenomenon, 278, 575–576

H

Harmonics, 244, 270, 273, 277, 311, 321, 332–333, 335, 398, 402, 493, 573–574, 584–585, 615
Heat
    Conduction, ix, 125, 130–131, 226, 405, 422, 445, 448, 454, 458, 470–471, 475, 483, 486, 569–570, 593, 597, 603
    Equation, 284, 406–407, 410, 470, 494
        Homogeneous, 421, 494
    Sources, 413, 439, 441, 445, 448, 486, 489–490, 494, 543
    Transfer, 440, 442–443, 488
        Convective, 440–444, 487–488, 490–491, 544–545, 547, 607, 611
Helmholtz equations, 284, 456
Hyperbolic equations, 285, 295, 433
    One-dimensional, 289, 291, 293, 295, 297, 299, 301, 303, 305, 307, 309, 311, 313, 315, 317, 319, 321, 323, 325, 327, 329, 331, 333, 335, 337, 339, 341
    Two-dimensional, ix, 343, 345, 347, 349, 351, 353, 355, 357, 359, 361, 363, 365, 367, 369, 371, 373, 375, 377, 379, 381, 383, 385, 387, 389, 391, 393, 395, 397, 399, 401, 403, 445

I

Inhomogeneous equations, v–vi, 21–24, 52, 54–56, 57–58, 62–63, 65, 69, 88
    Linear, 24, 28, 56
    Particular solution of, 62, 69, 108
Integral curves, 3, 6–8, 10, 13, 15, 17, 30, 32, 44, 48, 80, 114, 137
Internal sources, 443, 449, 453, 472, 477, 487–488, 491, 593, 595, 598
Iterative method, vii, 191–193
Isoclines, 7–10

K

Kernel, vii, 182, 187–188, 191, 196

L

Laplace, 165–167, 197, 284, 493, 495–496, 532, 569, 623
    Equation, xi, 284, 406, 462, 464, 468, 493–501, 515, 518–522, 525–527, 529, 531, 557
    Problem, 508, 517–518, 520, 557, 600, 608
    Transform, vii, 12, 165–170, 171, 173, 174–176, 196–199, 206, 623
Laplacian, 494, 503, 519, 532, 535

Legendre

Equation, vii, 203, 205, 207, 209, 211, 213, 215, 217, 219, 221, 223, 225, 226, 227, 229, 231, 233, 235, 237, 239, 241
    Functions, 201, 241
        Associated, 237–240, 569, 622
    Polynomials, viii, 229–231, 235–236, 240, 617
Linear
    Combination, 50–52, 54, 69, 87–88, 91, 93–96, 121, 158, 204, 209, 292, 379, 566, 568
    Equations, first-order, 20, 86, 108, 412
    Equations, second-order, 49, 54, 58, 68, 113, 118, 153, 170, 317, 379

M

Maxwell's equations, 283–284
Mixed condition, 409, 446, 471, 504, 536, 550–552, 554–556, 559–561, 611
Motion, equation of, 143, 344, 348, 373, 379

N

Neumann
    Functions, 209, 213, 217, 219
    Problem, 522, 525
    Series, 193–194
Newton's law, 409–410, 440, 442–443, 488, 492, 496, 607, 611
Nodes, 152, 158, 311–312, 356, 584–585
    Degenerate, 138–140, 152–153, 155
    Improper, 152, 160, 163
Nonlinear systems, 159, 161, 163, 623
Numerical solutions, 12–13, 15, 17, 203

O

ODEs, v, vii, 1, 3–4, 6, 12, 33, 37, 39, 72, 77, 86, 118, 163, 165, 167, 169–171, 173, 175, 177, 201, 203, 318, 420, 433, 438, 455, 479, 481–482, 497, 623
    First-order, 3, 5, 37, 160, 179
    Second-order, vi, 115, 117, 119, 121, 123, 125, 127, 129, 131, 133
    Series solutions of, 203, 205, 207, 209, 211, 213, 215, 217, 219, 221, 223, 225, 227, 229, 231, 233, 235, 237, 239, 241
    Solutions, stability of, 137, 139, 141, 143, 145, 147, 149, 151, 153, 155, 157, 159, 161, 163
Orbital stability, vii, 156–158
Order
    Linear equation of, 54, 68
    Reduction of, 46–48

Orthogonal functions, 118, 216, 267–268, 273, 352, 574
    Complete set of, 266, 355
Orthogonal polynomials, 122–123, 616
    Classical, 569, 614, 616
Orthogonality, 123, 125, 232, 239, 267–268, 310, 332, 354, 413, 452, 478, 535, 543

P

Parabolic equations
    One-dimensional, 407, 409, 411, 413, 415, 417, 419, 421, 423, 425, 427, 429, 431, 433, 435, 437, 439, 441, 443
    Two-dimensional, 447, 449, 451, 453, 455, 457, 459, 461, 463, 465, 467, 469, 471, 473, 475, 477, 479, 481, 483, 485, 487, 489, 491
Partial sum, 131, 216–218, 232–233, 236, 240–241, 249, 257, 259–260, 277–279, 569, 573–580, 578, 592, 601, 609–610, 614–615, 617–618, 620–621
PDEs, 3, 29, 118, 170, 179, 283, 291, 303, 493, 623–624
Phase
    Line, 135–140
    Orbit, 141, 157
    Plane, 83, 141–142, 157
    Portrait, 141, 143–146, 148–153, 162
    Trajectories, 78, 113, 141–143, 159
Picard's Theorem, 5–6, 12–13, 13–15, 44
Points, critical, 135, 141–142, 160–163
Poisson
    Equation, 284, 493–496, 502–504, 513, 526–527, 532–533, 537, 605, 607, 609, 611
    Problem, 511, 514
Polynomials, characteristic, 159, 564, 567
Power series, 205, 210, 227, 235, 614
Principle of superposition, 54, 311
Program
    FourierSeries, 217, 219, 225, 233, 241, 569, 614
    Heat, 125, 130–131, 414, 416, 418, 422, 426, 428–430, 432, 435, 438–439, 453, 456, 458, 466, 470, 475, 477, 481–482, 485–486, 601
    Laplace, 502, 512, 515, 518, 525, 533, 539, 543, 569, 604, 609
    TrigSeries, 257, 261–262, 277–278, 569, 573
    Waves, 125, 130–131, 302–303, 308, 311–315, 320–321, 322, 327–329, 331, 334–338, 357, 363, 383, 385–386, 388, 393, 397–403, 569, 580, 586, 592

R

Radioactive decay, 35
Recurrence formula, 206, 211, 220, 227–229, 231, 235, 239, 538
Regularity, condition of, 116, 119
Resonance, 321, 327–328, 340, 363–364
Riccati equation, 25
Rodrigues' formula, 230

S

Saddle, 146–147, 153–154, 157–158, 162–163
Second-order differential equations, 45, 47, 49, 51, 53, 55, 57, 59, 61, 63, 65, 67, 69, 71, 73, 75
Sensitivity, analysis of, 159, 162
Separable equations, v, 11, 17
Separation of variables, 11, 303, 348, 352, 411, 449, 478, 498, 519, 539–540
Series expansion, 203, 211, 241, 336, 398, 439, 486, 543
Set
    Complete, 215–216, 232, 239, 310, 352–353
    Fundamental, 87–88, 90, 93, 95, 500
Shift Theorem, 168
Sine series, 254, 277–279
Sink, spiral, 148–149, 154–155, 157–159, 162–163
Small perturbations, vii, 157, 159, 162
Source, 138–140, 146–147, 154, 157–158, 162–163, 297, 341, 442–443, 487, 490, 494, 544, 547, 597, 604
    Function, 607, 611
    Spiral, 148–149, 154–155, 157–159
Spaces, 80, 268, 407, 570, 578, 617
    Empty, 284
    Linear Euclidian, 123
Spectrum, 183, 274, 563–564
Stability, vi, 7, 138, 156–161, 290
    Asymptotic, 156, 159
    Diagram, vii, 157, 159
    Lyapunov, vii, 156–158, 160
Sturm-Liouville
    Operator, 119, 125, 129
    Problems, vi, xi, 118–119, 121–122, 124–127, 130–133, 214, 224, 226, 231, 239, 304–305, 307, 351, 376–377, 412, 462, 503, 508, 549, 557
        Regular, 121–123
        Singular, 121–123
    Theory, vi, 113, 115, 117, 119, 121, 123, 125, 127, 129, 131, 133, 215, 307

System
    Fundamental, 52, 55, 97, 463, 499, 501, 509, 558–561
    Inhomogeneous, 108, 111
    Linearized, 161–162
    Orthonormal, 619, 622

T

Temperature distribution, 439, 446, 466–467, 471, 486, 489–492, 496, 543–544, 569, 593, 595, 597–598, 600
    Steady-state, 543–544, 546–547, 604, 607–608
Trace-determinant plane, vii, 159
Transform, 12, 166, 169, 170, 174, 197–198, 206

U

Undetermined coefficients, method of, vi, 62–63, 65, 69, 96, 527

Uniqueness, 5, 15, 45, 50, 80
    Theorem, 45, 50, 80

V

Variation of parameters, method of, 21, 24, 28, 53, 55, 62, 65, 69
Vector field, 141–144, 146, 148–153
Volterra equations, 194–196

W

Wave equation, viii, 284, 290, 293, 297–298, 326, 333, 337, 587
    Homogeneous, 289, 293, 311, 317, 580, 582
Waves, viii, 118, 283–284, 289, 298–302, 311, 335–336, 338, 340, 398, 402, 569, 580, 582
    Standing, 311, 354, 380
Wronskian, 51, 55